FORMAL DESCRIPTION TECHNIQUES AND PROTOCOL SPECIFICATION, TESTING AND VERIFICATION

IFIP – The International Federation for Information Processing

IFIP was founded in 1960 under the auspices of UNESCO, following the First World Computer Congress held in Paris the previous year. An umbrella organization for societies working in information processing, IFIP's aim is two-fold: to support information processing within its member countries and to encourage technology transfer to developing nations. As its mission statement clearly states,

> IFIP's mission is to be the leading, truly international, apolitical organization which encourages and assists in the development, exploitation and application of information technology for the benefit of all people.

IFIP is a non-profitmaking organization, run almost solely by 2500 volunteers. It operates through a number of technical committees, which organize events and publications. IFIP's events range from an international congress to local seminars, but the most important are:

- the IFIP World Computer Congress, held every second year;
- open conferences;
- working conferences.

The flagship event is the IFIP World Computer Congress, at which both invited and contributed papers are presented. Contributed papers are rigorously refereed and the rejection rate is high.

As with the Congress, participation in the open conferences is open to all and papers may be invited or submitted. Again, submitted papers are stringently refereed.

The working conferences are structured differently. They are usually run by a working group and attendance is small and by invitation only. Their purpose is to create an atmosphere conducive to innovation and development. Refereeing is less rigorous and papers are subjected to extensive group discussion.

Publications arising from IFIP events vary. The papers presented at the IFIP World Computer Congress and at open conferences are published as conference proceedings, while the results of the working conferences are often published as collections of selected and edited papers.

Any national society whose primary activity is in information may apply to become a full member of IFIP, although full membership is restricted to one society per country. Full members are entitled to vote at the annual General Assembly, National societies preferring a less committed involvement may apply for associate or corresponding membership. Associate members enjoy the same benefits as full members, but without voting rights. Corresponding members are not represented in IFIP bodies. Affiliated membership is open to non-national societies, and individual and honorary membership schemes are also offered.

FORMAL DESCRIPTION TECHNIQUES AND PROTOCOL SPECIFICATION, TESTING AND VERIFICATION

FORTE XI/PSTV XVIII'98

IFIP TC6 WG6.1 Joint International Conference on Formal Description Techniques for Distributed Systems and Communication Protocols (FORTE XI) and Protocol Specification, Testing and Verification (PSTV XVIII)

3-6 November 1998, Paris, France

edited by

Stan Budkowski
*Institut national des télé*communications (INT)
Evry
France
Ana Cavalli
*Institut national des télé*communications (INT)
Evry
France
Elie Najm
*Ecole Nationale Supérieure des Télé*communications (ENST)
Paris
France

SPRINGER SCIENCE+BUSINESS MEDIA, LLC

Library of Congress Cataloging-in-Publication Data

A C.I.P. Catalogue record for this book is available
from the Library of Congress.

DOI 10.1007/978-0-387-35394-4

Printed on acid-free paper.
www.springer.com/mycopy

CONTENTS

PART SEVEN Case Studies

Invited Talk III

PART EIGHT Test 2

PART NINE Hardware/Software Development

PREFACE

The Joint International Conference on FORMAL DESCRIPTION TECHNIQUES for Distributed Systems and Communication Protocols, and PROTOCOL SPECIFICATION, TESTING AND VERIFICATION (FORTE/PSTV'98), is to be held this year at Ecole Nationale Supérieure des Télécommunications, Paris, France, from 3 to 6 November, 1998. FORTE/PSTV'98 is the eleventh/eighteenth, respectively, of a series of annual meetings sponsored by the IFIP Working Group TC6/WG6.1 dedicated to "Architecture and Protocols for Computer Networks".

The conference provides a forum for researchers and practitioners from universities and industry to meet and to advance technologies in the areas of specification, testing, and verification of distributed systems and communication protocols.

A total of 83 papers have been submitted to FORTE/PSTV'98 and all of them were reviewed by the members of the Program Committee and additional reviewers. The completed reviewers list is included in this Proceedings. Based on these reviews, the Program Committee selected 26 for presentation at the conference. Three specially invited papers complete the Conference Program, with ten sessions :

- FDTs Extensions (PART ONE),
- Verification 1 (PART TWO),
- Test 1 (PART THREE),
- Methodology 1 (PART FOUR),
- Methodology 2 (PART FIVE),
- Verification 2 (PART SIX),
- Case Studies (PART SEVEN),
- Test 2 (PART EIGHT),
- Hardware/Software Development (PART NINE),
- Real-Time & Performance (PART TEN).

FORTE/PSTV'98 has received financial support from the European Commission to help researchers and students from Central and Eastern European countries to participate in the conference. We also gratefully acknowledge the financial support from CNET- France Télécom, EDF and Verilog.

FORTE/PSTV'98 could not take place without the effort of all the PC members, reviewers and people participating in the local organization. The editors wish to thank all of them.

Stanislaw Budkowski
Ana Cavalli
Elie Najm

Comittee Members and Invited Speakers

General Chair
Elie Najm - ENST

Program Committee Co-Chairs
Stanislaw Budkowski - INT
Ana Cavalli - INT

Program Committee

P. Amer (Univ. of Delaware, USA), J.W Atwood (Concordia Univ., Canada), G. v. Bochmann (Univ. of Ottawa, Canada), T. Bolognesi (IEI, Italy), H.Bowman (Univ. of Kent at Canterbury, UK), E. Brinksma (Univ. of Twente,Netherlands), R. Castanet (Univ. of Bordeaux, France), O. Catrina (Univ.of Bucharest, Romania), S. T. Chanson (Univ. of Sci. and Tech., Hong Kong),J.-P. Courtiat (LAAS-CNRS, France), P. Dembinski (IPIPAN, Poland), M. Diaz (LAAS, France), R. Dssouli (Univ. of Montreal, Canada), S. Fischer (Univ.of Mannheim, Germany), R. Gorrieri (Univ. of Bologna, Italy), R. Gotzhein (Univ. of Kaiserslautern, Germany), R. Groz (CNET, France), T. Higashino (Osaka Univ. Japan), D. Hogrefe (Univ. of Luebeck, Germany), S. P. Iyer (North Carolina State Univ.), M. C. Kim (IC Univ., Korea), P. Kritzinger (Univ. of Cape Town, South Africa), R. Lai (La Trobe Univ. Australia), G. Leduc (Univ. of Liege, Belgium), D. Lee (Bell Lab., USA), S. Leue (Univ.of Waterloo, Canada), L. Logrippo (Univ. of Ottawa, Canada), J. de Meer(GDM-FOKUS, Grmany), T. Mizuno (Shizuoka Univ., Japan), A. Petrenko (CRIM,Canada), J. Quemada (ETSI Telecom., Spain), O. Rafiq (Univ. of Pau, France), H. Rudin (IBM, Switzerland), A. Shaff (LORIA, France), D. Sidhu (Univ.of Maryland-BC, USA), N. Shiratori (Tohuku Univ., Japan), J.B. Stefani (CNET,France), K. Suzuki (KDD, Japan), K. Tarnay (Univ. of Budapest, Hungary), R. Tenney (Univ. of Massachusetts, Boston, USA), A. Togashi (Shizuoka Univ.,Japan), K. Turner (Univ. of Stirling, UK), S. T. Vuong (Univ. of BritishColumbia, Canada), N. Yevtuchenko (Tomsk Univ., Russia), J. Wu (TsinghuaUniv., China)

Invited Speakers
P. J. Kuehn (University of Stuttgart, Germany), J. Meseguer (SRI International, USA)

LIST OF REFEREES

E. Abdeslam
F. Accordino
P. Amer
D. Amyot
W. Atwood
G. Banerjee
H. Ben-Abdallah
M. Bernardo
C. Besse
P. Bieber
G.v. Bochmann
E. Boiten
D. Bolignano
T. Bolognesi
E. Borcoci
M. Born
C. Bourhfir
H. Bowman
A. Bracciali
M. Bravetti
J. Bredereke
E. Brinksma
P. Brisset
J.M. Bruel
J. Bryans
N. Busi
L. Cacciari
M. Caneve
R. Castanet
O. Catrina
S. Chanson
O. Charles
I. Chrismeht
P. Ciancarini
S. Cimato
R. G. Clark
J.-P. Courtiat
G. Csopaki
P. D'Argenio
P. Dembinski
J. Derrick
S. Dibuz
I. Dinsenmeyer
J. Drissi
R. Dssouli
F. Dubois
H. Erdogmus
G. Faconti
M. Fecko
O. Festor
S. Fischer
R. Focardi
D. Frutos-Escrig
B. Geppert
B. Ghribi
S. Gnesi
R. Gorrieri
R. Gotzhein
J. Grabowski
P. Gradit
R. Groz
T. Hasegawa
L. Heerink
S. Heymer
D. Hogrefe
A. Hoffmann
G. Huecas
P. Hunel
A. Idoue
P. Iyer
V. Jones
G. Juanole
H. Kabutz
M. Kamel
S. Kang
G. Karjoth
Y. Katsnelson
F. Khendek
M. Kim
S.U. Kim
B. Koch
B. Kodeswaran
P. Kritzinger
R. Kubiak
L. Lacayrelle
R. Lai
R. Langerak
D. Latella
S. Lazar
G. Leduc
D. Lee
S. Leue
L. F. Llana Diaz
A. Loureiro
L. Logrippo
S. Maharaj
N. Marti-Oliet
M. Massink
A. Mazurkiewicz
F. Mazzanti
J. de Meer
T. Mizuno
E. Montes de Oca
A. Naik
E. Najm
A. Nakata
C. M. Nieto
M. Nunez
N. Okazaki
R. Park Mi
R. Pasini
W. Penczek
C. Peper
A. Petrenko
R. Pugliese
J. Quemada
O. Rafiq
M. Rezai
T. Robles
F. Roessler
H. Rudin
T. C. Ruys
P. de Saqui-Sanes
A. Schaff
M. Schmitt
A. Selcuk
D. Sidhu
C. E. Shankland
J. Shin
J. Sincennes
R. Sinnott
X. Song
K. Tarnay
J. Templemore-Finlayson
R. L. Tenney
J. Theess
K. Toshihiko
J. Tretmans
K. J. Turner
A. Ulrich
U. Uyar
R. Valette
F. Vernadat
D. Vincent
R. de Vries
S. Vuong
J. Winkowski
J. Wu
J. Wytrebowicz
K. Yasumoto
N. Yevtushenko
G. Zavattaro
J. Zhu
W. Zuberek

PART ONE

FDTs Extensions

AN ALGEBRAIC SEMANTICS FOR MESSAGE SEQUENCE CHART DOCUMENTS

Thomas Gehrke, Michaela Huhn*,
Arend Rensink, Heike Wehrheim°

Universität Hildesheim, Institut für Informatik (Prof. U. Goltz),
Postfach 101363, D–31113 Hildesheim, Germany
*Universität Karlsruhe, Institut für Rechnerentwurf und Fehlertoleranz (Prof. D. Schmid),
Postfach 6980, D–76128 Karlsruhe, Germany
°Universität Oldenburg, Fachbereich Informatik, Abt. Semantik (Prof. E-R. Olderog),
Postfach 2503, D–26111 Oldenburg,Germany

gehrke|rensink@informatik.uni-hildesheim.de,
Michaela.Huhn@informatik.uni-karlsruhe.de,
Heike.Wehrheim@informatik.uni-oldenburg.de

Abstract: Message Sequence Charts (MSCs) are a graphical and textual language for the specification of message passing systems, in particular telecommunication systems. MSCs are standardised by the Internal Telecommunication Union in standard Z.120. Included in the standard is a formal semantics for MSCs by means of a process algebra. This semantics covers the complete language of single MSCs but lacks an interpretation for *conditions* which are used as continuation points of MSCs within an MSC document (a collection of MSCs). In this paper, we give a process algebraic semantics for basic MSCs including conditions, enabling the formal interpretation of entire MSC documents.

1 INTRODUCTION

Message Sequence Charts (MSCs) are a widely used formalism for the specification of the communication behaviour of reactive systems. It allows for the graphical and textual representation of the communication structure of systems. MSCs focus on the temporal ordering of interaction among system components by specifying the executable traces of a system. They are for instance used as a graphical representation of executable traces of SDL [17] specifications but also as a specification language in their own right. The language has been standardised in the standard ITU-T Z.120 of the International Telecommunication

Union [7]. Communication in MSCs is asynchronous; the synchronous variant are Interworkings [15]. MSC-like diagrams have nowadays also been incorporated into various object-oriented specification technique, like for instance the Unified Modeling Language (UML) [18].

In order to give a precise meaning to MSCs and to allow verification, a formal semantics is needed. In the standard, the semantics of MSCs is given via a translation, originally developed in [13], transforming the textual representation of MSCs into a process algebra, based on [1], for which an axiomatic semantics exists. Other semantics based on Petri-nets or Büchi automata can be found in [11, 10, 6]. However, the standard process algebra semantics does not capture a specific feature of MSCs called *conditions*. Conditions are a rudimentary form of MSC composition: Within an MSC document (a collection of MSC diagrams), a condition describes possible continuation points of system behaviour. Every MSC describes a part of the interaction behaviour of the system, and an MSC ending with a specific condition can be "glued together" with every other MSC starting with this condition. This gives rise to a form of sequential composition, followed by a choice of follow-up MSCs, which is indispensable for the specification of infinite behaviour by means of a set of finite MSCs. The only semantics incorporating this interpretation of conditions is the automata semantics of [11]. The latter, however, presumes finite-stateness of the system under consideration, which in our opinion is not *a priori* fixed, given that MSCs are based on asynchronous communication and, even more important, conditions easily allow the specification of non-regular behaviour.

In this paper, we therefore present an alternative proposal for a process algebra semantics for MSCs, which is capable of handling the composition of MSCs via conditions. Conditions are translated into process names, which are interpreted according to the semantics of the MSCs starting with the condition. The composition operator used for glueing MSCs together is a form of *weak sequential composition* based on [19], which is essentially sequential composition on the level of *instances* (the MSC equivalent of a sequential process). This operator captures precisely the right interplay between sequential composition and the choice of the follow-up MSC.

Apart from communication behaviour, other aspects specifiable in MSCs are:

Local actions: actions going on within a single instance that do not influence and cannot be influenced by the environment. Our treatment coincides with the standard.

Timer actions: the setting of timers and the resulting timeouts. Since we have no timing aspects in our model, and timer actions are local to instances, their formalisation would coincide with that of local actions (see above). For that reason, we ignore timer actions in this paper.

General ordering: explicit orderings of actions of different instances, by unspecified means. We have not attempted to model such arbitrary ordering; in fact, we know of no formal semantics to date. A straightforward formalisation would be to use a special kind of communication for this purpose and hide it (in a process algebra sense) afterwards.

Coregions: segments of a given instance where the ordering of the actions is not fixed but may be arbitrarily interleaved. We model this by non-synchronising parallel composition, coincident with the standard treatment.

Instance creation: the generation of a new instance and its eventual termination. We ignore instance creation in this paper; in the conclusions we briefly discuss how it might be integrated, using a technique inspired by the standard semantics.

Instance decomposition: the replacement, within a given MSC, of a single instance by an entire sub-MSC, with corresponding redirection of messages sent to or received from the refined instance. Because of the consistency requirements involved, as well as the issue of redirection and various other technical questions (not all of which are answered or even addressed in the official standard), decomposition is a very complex matter. We intend to investigate instance decomposition in the future but omit it for now.

More involved structuring mechanisms mentioned in [7], also not modelled here, include *inline expressions*, *MSC references* and *High-level MSCs*. A semantics for the latter has recently been proposed in [14]. High-level MSCs involve the explicit composition of basic MSCs in a flow chart style, as opposed to their implicit composition through conditions. The formalisation of sequential composition in High-level MSCs in [14] is also based on [19], just as our approach, with the difference that we explicitly recognise the localities of the MSC instances, whereas they model them indirectly through *dependencies* – which is closer to the formalisation in [19] and more powerful than our locality-based approach, but for the purpose of formalising MSCs poses unwarranted complications. The combination of the standard basic MSC semantics in [7] and the High-level MSC semantics in [14] gives rise to a framework that is significantly more complicated than the one we present here.

In Section 2, we start with a description of MSC documents. In Section 3, we present our process algebra with its structural operational semantics. Section 4 is concerned with a translation of MSC documents into our process algebra. Section 5 discusses the correspondence of our semantics with the standard one for single basic MSCs according to [13] and shows some algebraic properties of our translation. Finally, Section 6 contains conclusions and discusses the relation of our work with [14] in somewhat more detail.

2 MESSAGE SEQUENCE CHART DOCUMENTS

We start with a brief description of the functionality of MSC documents, as far as it is relevant for our approach. According to the ITU-T standard Z.120 [7], a Message Sequence Chart document consists of a collection of Message Sequence Charts and Message Sequence Chart diagrams (i.e., high-level MSCs). In this paper, we merely consider MSCs. They specify communication/interaction scenarios among a set of *instances* exchanging *messages*. Instances can be seen as processes residing on different locations. In the graphical charts, the temporal behaviour of one instance is written along a vertical axis. The execution of different instances is assumed to be asynchronous, thus an axis denotes the *local* time of the associated instance running from top to bottom. The behaviour of the external *environment* of a system is modelled as a specific instance in MSCs whose time axis is the frame of the MSC. See for instance the MSC `Alt2` in Figure 1, in which an instance communicates with the environment.

Figure 1: Example: A simple MSC document

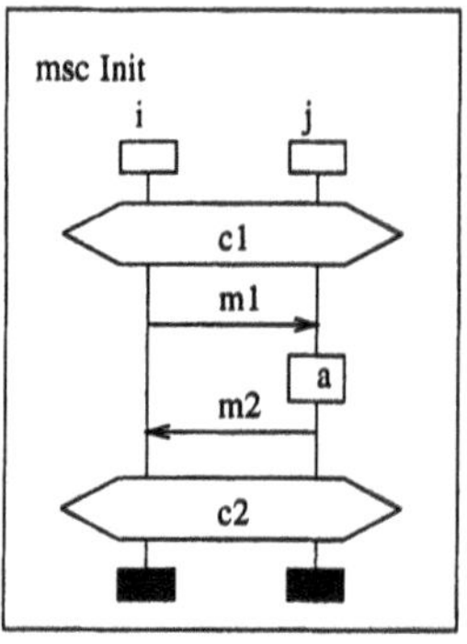

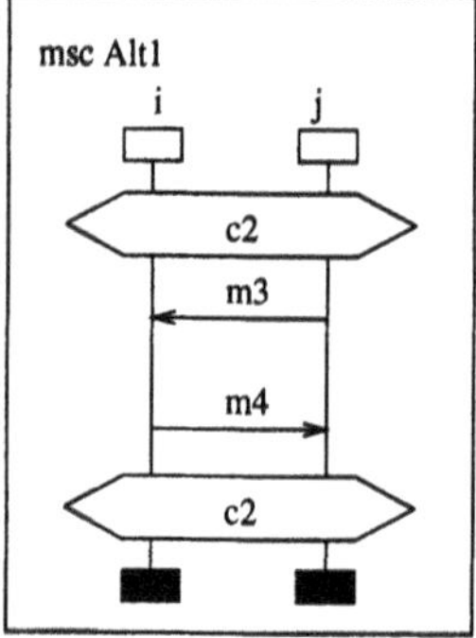

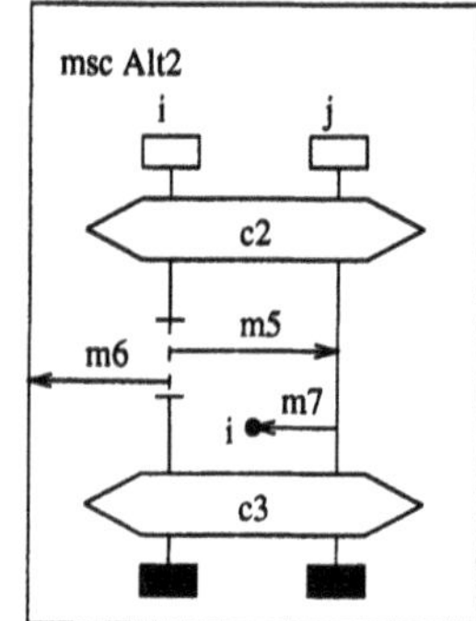

The actions an instance may execute are depicted as follows: Communications are denoted by horizontal or diagonal arrows linking the sender of a message to the receiver. The message to be sent is written as a name upon the arrow. For instance, in the MSC `Init` in Figure 1, instances i and j exchange messages $m1$ and $m2$. In case a message is lost, the head of the arrow does not end at the receiver instance, but in a bullet with the name of the intended receiver associated. Vice versa, a message may be spontaneously generated. *Internal* (or *local*) actions are drawn as rectangles containing the name of the action inside; see for instance the action a in the MSC `Init`. In the ITU-T standard Z.120, internal actions are simply called "actions" and describe some activity local to an instance. A further class of actions are *timer* actions; however, as mentioned in the introduction, we follow [13] and (essentially) [7] in regarding these as special cases of internal actions.

Instances are assumed to run sequentially. Thus, the order of action occurrences along the time axis specifies a *total* order of execution for an instance. An exception to this rule are so-called *coregions*, which specify unordered events of an instance. In a coregion, the time axis is depicted as a dashed line. See for instance the sending of $m5$ and $m6$ in `Alt2`.

An important concept for the composition of MSCs are *conditions*. Graphically, conditions are represented as horizontally elongated hexagons, with the name of the condition inside. Conditions can be *shared* between instances; a condition that is shared by all instances of an MSC is called *global*. A global condition is *initial* if it is the first item of all instances, and *final* if it is the last. According to the ITU standard [7], conditions can be used either for informal annotations or for the composition of different MSCs of a document. The latter role is reserved for (global) initial and final conditions: a chart ending with a specific global condition can be continued by any other chart starting with the same global condition. Thus, conditions determine possible continuations of MSCs. Since the standard semantics does not take MSC composition into account, it treats all kinds of conditions as empty steps. In this paper, we deal with global conditions only; local conditions would still be translated to empty steps.

It is important to note that conditions are *not* intended as a synchronisation device: although they are sometimes referred to as "global states" in the standard, there is *no* requirement that all instances partaking in a condition are simultaneously at that point of their time axis.

Figure 1 shows an MSC document which consists of three MSCs connected by conditions. The MSCs describe the interaction between two instances i and j. Assume that execution is started at the initial MSC `Init`.[1] The final condition $c2$ of `Init` is also the initial condition of the MSCs `Alt1` and `Alt2`; therefore, after performing `Init` either `Alt1` or `Alt2` can be executed. In `Alt1`, the condition $c2$ is also its final condition, which means that this MSC describes a loop.

Events which are not causally related may be executed independently; for example, after composing `Init` and `Alt1` through the condition $c2$, the output event of message $m3$ and the input event of message $m2$ are independent, which means that they may occur in any order. In particular, since the condition $c2$ is not a synchronisation point, there is nothing to prevent the sending of $m4$ (by instance j) to occur before the reception of $m2$ (by instance i). The same holds for the input event of $m4$ and the output events of $m5$ and $m6$.

In the case where there are several "follow-up" MSCs with the same initial condition, there is an interesting, subtle issue involved in their composition: namely, the choice of the actual follow-up MSC during a concrete run of the system. For instance, in Figure 1, after `Init` has finished $m3$ can be sent; this automatically involves continuing with `Alt1` and rejecting `Alt2`. Another possibility is that either $m5$ or $m6$ is sent, deciding the choice in favour of `Alt2`. However, *sending* $m3$ *and* $m5/m6$ *are independent actions*, since they occur in different instances. Yet the case where instances i and j simultaneously decide to send $m3$ resp. $m5/m6$ is not a valid system run. The choice between the two follow-up charts is apparently taken on a global level. This is an effect that any formal semantics has to take into account; for instance, [12] follows precisely this solution.[2]

Finally, we want to comment on an issue raised by [12]. In our opinion, and consequently in our semantics, MSC documents do not *a priori* specify finite-state systems, even though, of course, finite-state behaviour is a desirable property. For a counterexample, consider an MSC over two instances with

[1]Note that the standard does not prescribe an unambiguous starting point; rather, single MSCs are thought to represent possible fragments of behaviour.

[2]Another point of view is that such a global choice is a specification error and should be ruled out in advance. However, we are of the opinion that formulating a semantics comes properly *before* deciding whether the specified behaviour is implementable.

identical initial and final condition (i.e., a loop) containing just one message sent from one instance to the other. The execution traces of this MSC are all traces σ over $\{in(i,m)@j, out(j,m)@i\}$ such that in each prefix of σ the number of $in(i,m)@j$'s never exceeds the number of $out(j,m)@i$ and finally in σ the amount of $in(i,m)@j$'s and $out(j,m)@i$'s is equal. Clearly, this is not a finite-state recognisable language. In fact, we conjecture that it is undecidable whether the behaviour specified by an MSC document is finite-state. As in the issue of global choice, we feel that the semantics should be well-defined regardless of whether or not the behaviour is finite-state.

Table 2: Textual representation of MSC documents.

```
<document>    ::= mscdocument <docid>; { <msc>; }* endmscdocument
<msc>         ::= msc <mscid>; { <inst def>; }* endmsc
<inst def>    ::= instance <iid>; { <event>; }* endinstance
<event>       ::= <comm event>
                | action <aid>
                | condition <cid> shared all
                | concurrent { <comm event>; }* endconcurrent
<comm event>  ::= in <mid> from [found] <address>
                | out <mid> to [lost] <address>
<address>     ::= <iid> | env
```

The relevant fragment of the textual grammar of MSCs is reproduced in Table 2.[3] The grammar contains the following undefined non-terminals:

- <docid>: MSC document identifiers, ranged over by $\mathcal{D}$;
- <mscid>: message sequence chart identifiers, ranged over by M;
- <iid>: instance identifiers, collected in *Inst*, such that $env \notin Inst$. We denote $Addr = Inst \cup \{env\}$ for the set of *addresses*, ranged over by i, j.
- <mid>: message identifiers, collected in *Mess* and ranged over by m;
- <aid>: internal action identifiers, collected in *Int* and ranged over by τ;
- <cid>: condition identifiers, collected in *Cond* and ranged over by c, d.

It should be clear that not every syntactically correct MSC document is acceptable: the above discussion contains a large number of "consistency requirements". In fact, one additionally needs a static semantics for MSC documents, for instance in the form of a type system, such that a document is well-formed (well-typed) if and only if it satisfies all those criteria. Some of the crucial points are:

- Outgoing and incoming messages must be matched precisely;
- The ordering imposed by messages may not be circular;

[3] In fact, the grammar is an adapted version of that in [7], but the syntax it generates is a subset of the standard.

- Global conditions (which are the only ones we model) must occur in all instances;
- All MSCs in a document must have the same set of instances.

3 THE PROCESS ALGEBRA $\mathcal{L}_{MSC}$

We now introduce the process algebra $\mathcal{L}_{MSC}$ into which we will translate MSC documents. The basic building blocks are *events* of one of the following kind:

- $out(i, m)$: normal output of $m \in Mess$ to $i \in Addr$;
- $lost(i, m)$: discarded output of $m \in Mess$ to $i \in Addr$;
- $in(i, m)$: normal reception of $m \in Mess$ from $i \in Addr$;
- $found(i, m)$: spontaneous input of $m \in Mess$ from $i \in Addr$;
- $act(\tau)$: an internal action $\tau \in Int$;
- $\surd$: a termination event.

Events of the first five kinds are collected in *Evt*, ranged over by e; furthermore, $Evt_{\surd} = Evt \cup \{\surd\}$ is ranged over by α. In each case, $\alpha@j$ denotes the event α taking place at instance $j \in Inst$, and $\alpha@Inst$ the collection of all such $\alpha@j$. Additionally, we assume a set of process names *Names* to allow recursive definitions. $\mathcal{L}_{MSC}$ is then given by the following abstract grammar:

$$B ::= \delta \mid \varepsilon \mid e@i \mid B \cdot B \mid B + B \mid B \parallel_A B \mid X \mid \perp ,$$

where $A \subseteq Evt@Inst$ and $X \in Names$. The operators have the following intuition:

- δ is an empty, non-terminated (i.e., deadlocked) process.
- ε is an empty process, terminated at all instances (corresponding to "skip").
- $e@i$ specifies an event at instance i, and is, furthermore, *terminated* at all instances *except for* i.
- $\cdot$ specifies *weak sequential composition* of its operands. It has been introduced in [19] in a more general setting and we adapt it here to handle MSC composition. A similar operator has also been used for Interworking composition (which are the synchronous variant of MSCs) in [15] and as a sequencing operator on High-level MSCs [14]. The effect is that of ordinary sequential composition *within each instance*, whereas different instances are allowed to proceed independently. That is, termination of the first operand at a given instance i (signalled by a $\surd@i$-transition) allows the second operand to perform events $\alpha@i$ but not $\alpha@j$ for $j \neq i$. (It is important to keep in mind that a term may be terminated at one instance but not at another, so that termination of a term at one instance does not imply the inability to do some real event at another; witness $e@i$ above.)

- $+$ specifies a *choice* between its operands, which is resolved by the first non-termination event that occurs. The behaviour with respect to termination is somewhat more complex, based on the principles developed in [19]: $B_1 + B_2$ terminates at i if either of its operands does, but the choice is only resolved thereby if the other operand does *not* terminate at i. We let $\sum_{n \in N} B_n$ denote a choice over all terms B_n where n is out of some finite set N; consequently, $\sum_{n \in \emptyset} B_n$ equals δ.
- $\|_A$ is a TCSP-parallel composition [2] requiring synchronisation on all events in A; that is, the operands may do events $e@i \in A$ together (i.e., both at the same time) or events $e@i \notin A$ on their own. Moreover, we require implicit synchronisation on termination, i.e., events $\surd@i$ may also only be performed by both operands together.
- $X \in Names$ stands for the invocation of the process names X. Processes are defined by a process environment $\theta : Names \to \mathcal{L}_{MSC}$, which is assumed to be given.
- Finally, $\perp$ stands for an empty *message pool.* This is the only really non-standard operator in $\mathcal{L}_{MSC}$; it is defined especially to model MSCs. Message pools will be used for the modelling of the asynchronous communication of MSCs using the synchronous communication of $\mathcal{L}_{MSC}$. Message pools are inspired by the modelling of asynchronous communication in of coordination languages; see, e.g., [3]. A message pool is used to buffer the sent but not yet received messages. Operationally, this is done by generating a parallel $in(i, m)@j$-event whenever an $out(j, m)@i$-event is sent.

As usual, the formal semantics of $\mathcal{L}_{MSC}$ will be derived via SOS rules generating a labelled transition system over $Evt_{\surd}@Inst$. We recall the general definition:

1 Definition. A *labelled transition system* over a set of labels L is a tuple $\langle S, \to, q \rangle$ such that S is a set of *states*, $\to \subseteq S \times L \times S$ is a *transition relation* and $q \in S$ is the *initial state.*

In our case, the labels are taken form the localised alphabet $Evt_{\surd}@Inst$. A transition $B \xrightarrow{\surd@i} B'$ means that the term B may signal termination of instance i, thereby evolving into B'. For instance, a term B not referring to instance i in any of its events (and not containing δ) may always signal termination of i; it in fact has no information on i and assumes it to be terminated. Table 3 gives the structural operational semantics of $\mathcal{L}_{MSC}$. This gives rise to a transition system semantics for each $\mathcal{L}_{MSC}$-term. We briefly discuss the intuitive meaning of some of the operational rules. The first rule for weak sequential composition equals the one for (normal) strong sequential composition: the first process is allowed to proceed. The second rule, on the other hand, describes the fact that the second component may also execute events at instance i if in the first component instance i is terminated, as specified by the second rule. For a term that only refers to events from one instance i, weak sequential composition coincides with strong sequential composition; i.e., the second operand starts execution only if the first one is completely terminated.

Table 3: Operational semantics for $\mathcal{L}_{MSC}$

Empty process	$$\frac{}{\varepsilon \xrightarrow{\surd @i} \varepsilon}$$
Events	$$\frac{}{e@i \xrightarrow{e@i} \varepsilon} \qquad \frac{j \neq i}{e@j \xrightarrow{\surd @i} e@j}$$
Weak sequential composition	$$\frac{B_1 \xrightarrow{e@i} B_1'}{B_1 \cdot B_2 \xrightarrow{e@i} B_1' \cdot B_2} \qquad \frac{B_1 \xrightarrow{\surd @i} B_1' \quad B_2 \xrightarrow{\alpha @i} B_2'}{B_1 \cdot B_2 \xrightarrow{\alpha @i} B_1' \cdot B_2'}$$
Choice	$$\frac{B_1 \xrightarrow{e@i} B_1'}{B_1 + B_2 \xrightarrow{a@i} B_1'} \qquad \frac{B_2 \xrightarrow{e@i} B_2'}{B_1 + B_2 \xrightarrow{a@i} B_2'}$$ $$\frac{B_1 \xrightarrow{\surd @i} B_1' \quad B_2 \xrightarrow{\surd @i} B_2'}{B_1 + B_2 \xrightarrow{\surd @i} B_1' + B_2'}$$ $$\frac{B_1 \xrightarrow{\surd @i} B_1' \quad B_2 \not\xrightarrow{\surd @i}}{B_1 + B_2 \xrightarrow{\surd @i} B_1'} \qquad \frac{B_2 \xrightarrow{\surd @i} B_2' \quad B_1 \not\xrightarrow{\surd @i}}{B_1 + B_2 \xrightarrow{\surd @i} B_2'}$$
Parallel composition	$$\frac{B_1 \xrightarrow{e@i} B_1' \quad e@i \notin A}{B_1 \parallel_A B_2 \xrightarrow{e@i} B_1' \parallel_A B_2} \qquad \frac{B_2 \xrightarrow{e@i} B_2' \quad e@i \notin A}{B_1 \parallel_A B_2 \xrightarrow{e@i} B_1 \parallel_A B_2'}$$ $$\frac{B_1 \xrightarrow{\alpha @i} B_1' \quad B_2 \xrightarrow{\alpha @i} B_2' \quad \alpha @i \in A \cup \surd @Inst}{B_1 \parallel_A B_2 \xrightarrow{\alpha @i} B_1' \parallel_A B_2'}$$
Process names	$$\frac{\theta(X) \xrightarrow{\alpha @i} B'}{X \xrightarrow{\alpha @i} B'}$$
Empty pool	$$\frac{j \in Inst}{\perp \xrightarrow{out(j,m)@i} in(i,m)@j \parallel_\emptyset \perp} \qquad \frac{}{\perp \xrightarrow{\surd @i} \perp}$$

Choices can be resolved both by normal events and by termination signals within one operand, i.e. a term containing choices is terminated if one of its components is. This is in accordance with the usual interplay of termination and choice; see, for instance, [1]. However, if both operands of a choice may terminate for an instance i, the choice is not yet resolved and both components execute their termination transition. In this way, we avoid that a choice can be resolved by the termination of an instance not participating in an execution. For instance, using the rules in Table 3, we can derive $(e@1+e'@2)\cdot e''@3 \xrightarrow{e''@3} (e@1+e'@2)\cdot\varepsilon$ rather than $(e@1+e'@2)\cdot e''@3 \xrightarrow{e''@3} e@1\cdot\varepsilon$ or $(e@1+e'@2)\cdot e''@3 \xrightarrow{e''@3} e'@2\cdot\varepsilon$.

Process names behave according to their instantiation by θ.

The semantics of $\perp$ shows a process algebraic modelling of a message pool: If the pool receives a message from a sender, i.e., the pool performs $out(j,m)@i$, it stores it for the receiver by evolving into the term $in(i,m)@j \parallel_\emptyset \bot$. Delivering the message to the receiver will be done by synchronising on $in(i,m)@j$; this does not cause any interference with the rest of the message pool. The details of the semantics of $\bot$ will become clearer in the next section. There we describe how MSC documents can be translated into $\mathcal{L}_{MSC}$.

4 TRANSLATION OF MSC DOCUMENTS INTO $\mathcal{L}_{MSC}$

We assume that the document in question has instances *Inst* (i.e., all MSCs in the document contain definitions of the instances *Inst*), messages *Mess* and conditions *Cond*; we let $Cond \subseteq Names$, i.e., conditions serve as names for recursive processes. For the corresponding process definition θ see below. We let $init(M)$ equal the initial condition of an MSC M, if it exists, and ε otherwise; likewise, $fin(M)$ equals the final condition of M, if it exists, and ε otherwise. The translation is defined in Table 4. MSCs, instances and events are mapped

Table 4: Translation functions.

$[\cdot]_{\texttt{doc}} : \texttt{<doc>} \to 2^{\mathcal{L}_{MSC}}$ $[\texttt{document } \mathcal{D}\texttt{; <m>}_1\texttt{; }\cdots\texttt{; <m>}_n\texttt{; endmscdocument}]_{\texttt{doc}} = \{[\texttt{<m>}_k]_{\texttt{msc}} \mid 1 \le k \le n\}$
$[\cdot]_{\texttt{msc}} : \texttt{<msc>} \to \mathcal{L}_{MSC}$ $[\texttt{msc } M\texttt{; <i>}_1\texttt{; }\cdots\texttt{; <i>}_n\texttt{; endmsc}]_{\texttt{msc}} = (\ \bot \parallel_{pool}$ $([\texttt{<i>}_1]_{\texttt{inst}} \parallel_\emptyset \cdots \parallel_\emptyset [\texttt{<i>}_n]_{\texttt{inst}})$ $)\cdot fin(M)$ where $pool = \{in(i,m)@j, out(i,m)@j \mid i,j \in Inst, m \in Mess\}$
$[\cdot]_{\texttt{inst}} : \texttt{<inst def>} \to \mathcal{L}_{MSC}$ $[\texttt{instance } i\texttt{; <e>}_1\texttt{; }\cdots\texttt{; <e>}_n\texttt{; endinstance}]_{\texttt{inst}} = [\texttt{<e>}_1]^i_{\texttt{event}} \cdot \ldots \cdot [\texttt{<e>}_n]^i_{\texttt{event}}$
$[\cdot]^i_{\texttt{event}} : \texttt{<event>} \to \mathcal{L}_{MSC}$ $[\texttt{<ce>}]^i_{\texttt{event}} = [\texttt{<ce>}]^i_{\texttt{comm}}$ $[\texttt{action } \tau]^i_{\texttt{event}} = act(\tau)@i$ $[\texttt{condition } c \texttt{ shared all}]^i_{\texttt{event}} = \varepsilon$ $[\texttt{concurrent <ce>}_1\texttt{; }\cdots\texttt{;<ce>}_n\texttt{; endconcurrent}]^i_{\texttt{event}} = [\texttt{<ce>}_1]^i_{\texttt{comm}} \parallel_\emptyset \cdots \parallel_\emptyset [\texttt{<ce>}_n]^i_{\texttt{comm}}$
$[\cdot]_{\texttt{comm}} : \texttt{<comm event>} \to \mathcal{L}_{MSC}$ $[\texttt{in } m \texttt{ from } j]^i_{\texttt{comm}} = in(j,m)@i$ $[\texttt{in } m \texttt{ from found } j]^i_{\texttt{comm}} = found(j,m)@i$ $[\texttt{out } m \texttt{ to } j]^i_{\texttt{comm}} = out(j,m)@i$ $[\texttt{out } m \texttt{ to lost } j]^i_{\texttt{comm}} = lost(j,m)@i$

to $\mathcal{L}_{MSC}$-terms and MSC documents to *sets* of $\mathcal{L}_{MSC}$-terms. The following abbreviations occur in the table:

- $B_1 \cdot \ldots \cdot B_n$, denoting the right-associative weak sequential composition of a series of terms B_k for $1 \le k \le n$. The combined term equals B_1 if $n = 1$ and ε if $n = 0$.

- $B_1 \parallel_\emptyset \cdots \parallel_\emptyset B_n$, denoting the right-associative parallel composition of the terms B_k, $1 \le k \le n$. The combined term equals B_1 if $n = 1$ and ε if $n = 0$.

B^M denotes the translation of an MSC M of a given document $\mathcal{D}$ according to Table 4, and B_i^M denotes the sub-term of instance i of MSC M. The terms resulting from the translation of $\mathcal{D}$ are to be interpreted in a process environment $\theta_{\mathcal{D}}$ defined as follows:

$$\text{for all } c \in \mathit{Cond} \qquad \theta_{\mathcal{D}}: c \mapsto \textstyle\sum_{init(M)=c} B^M \ .$$

In words, the behaviour of a condition equals the sum of the MSCs of which it is the initial condition. Since the translation of an MSC ends in the invocation of its final condition (if any), the "glueing together" of MSCs works as planned: after an MSC is terminated, a choice of continuations exists, as determined by the condition names. Moreover, since we are using weak composition, termination is local to an instance.

2 Example. For the MSC document in Figure 1 we get the following instance and MSC terms:

$$\begin{array}{lcl}
B_i^{\mathtt{Init}} & = & out(j,m1)@i \cdot in(j,m2)@i \\
B_j^{\mathtt{Init}} & = & in(i,m1)@j \cdot act(a)@j \cdot out(i,m2)@j \\
B_i^{\mathtt{Alt1}} & = & in(j,m3)@i \cdot out(j,m4)@i \\
B_j^{\mathtt{Alt1}} & = & out(i,m3)@j \cdot in(i,m4)@j \\
B_i^{\mathtt{Alt2}} & = & out(j,m5)@i \parallel_{\emptyset} out(env,m6)@i \\
B_j^{\mathtt{Alt2}} & = & in(i,m5)@j \cdot lost(i,m7)@j \\
B^{\mathtt{Alt2}} & = & (\perp \parallel_{pool} (out(j,m5)@i \parallel_{\emptyset} out(env,m6)@i \parallel_{\emptyset} \\
 & & \qquad in(i,m5)@j \cdot lost(i,m7)@j)) \cdot c3
\end{array}$$

Moreover, we get the following process environment:

$$\begin{array}{rcl}
\theta_{\mathcal{D}}: c1 & \mapsto & (\perp \parallel_{pool} (out(j,m1)@i \cdot in(j,m2)@i \parallel_{\emptyset} \\
 & & \qquad in(i,m1)@j \cdot act(\tau)@j \cdot out(i,m2)@j)) \cdot c2 \\
c2 & \mapsto & (\perp \parallel_{pool} (in(j,m3)@i \cdot out(j,m4)@i \parallel_{\emptyset} \\
 & & \qquad out(i,m3)@j \cdot in(i,m4)@j)) \cdot c2 \\
 & & + (\perp \parallel_{pool} ((out(j,m5)@i \parallel_{\emptyset} out(env,m6)@i) \parallel_{\emptyset} \\
 & & \qquad in(i,m5)@j \cdot lost(i,m7)@j)) \cdot c3 \\
c3 & \mapsto & \delta
\end{array}$$

It can be seen from the translation that the instances may proceed independently in parallel (except that they have to synchronise on termination) but have to synchronise with the message pool on all communication (and termination) events. Thus, the role of $\perp$ is as described in the previous section: It takes the message from a sender by performing a synchronised $out(i,m)@j$-event and stores the corresponding receive event $in(j,m)@i$ until the receiving instance, j, wants to synchronise on that event.

In Fig. 5 the transition system for our example is given, as derived from the structural operational semantics. The states corresponding to the three MSCs are marked, as are the states corresponding to the conditions. An interesting parts of the behaviour is the region surrounding the state marked $c2$.

- The global choice between the outgoing $out(i,m3)@j$- and $out(j,m5)@i$-transitions is as described in Section 2.
- Starting from `Init`, this choice can be taken prematurely in favour of $out(i,m3)@j$, which can already be done before $in(j,m2)@i$ occurs. This is the effect of weak sequential composition.

Other noteworthy aspects are that the event $in(i,m4)@j$ can be executed concurrently to the events $out(j,m5)@i$ and $out(env,m6)@i$, and that the combined behaviour may loop around, i.e., is infinite.

Figure 5: Example: Transition system for the document in Fig. 1.

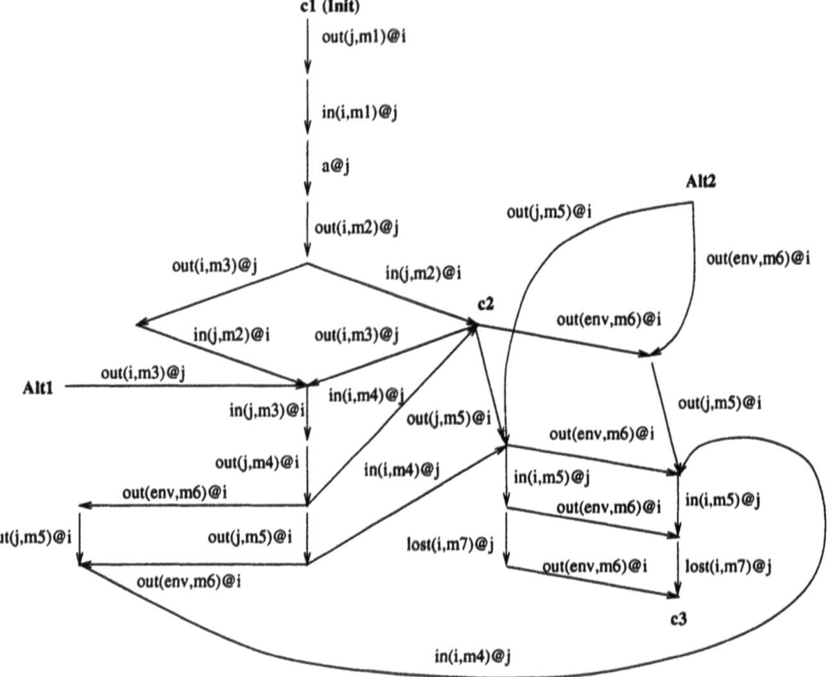

5 RESULTS

In this section we discuss two issues concerning our semantics: its consistency with the standard semantics of [13], and an algebraic property concerning message pools.

The standard semantics The standard MSC semantics in [7], based on the work of Mauw and Reniers in [13], is also process algebraic. We show consistency of our semantics with the standard by comparing the resulting transition systems up to strong bisimulation equivalence [16]. Since the standard semantics does not capture continuations via conditions, a comparison can be made only for single MSCs and not for documents.

For the purpose of comparison, we recall the relevant part of [13]. We refrain from giving the full semantics but instead focus on the major differences. The

Table 6: Operational semantics for additional operators.

Termination	$\dfrac{}{\varepsilon\downarrow}$	$\dfrac{B_1\downarrow \quad B_2\downarrow}{(B_1;B_2)\downarrow}$	$\dfrac{B_1\downarrow \quad B_2\downarrow}{(B_1 \parallel B_2)\downarrow}$	$\dfrac{B\downarrow}{\lambda_M(B)\downarrow}$
Sequential composition	$\dfrac{B_1 \xrightarrow{e@i} B_1'}{B_1;B_2 \xrightarrow{e@i} B_1';B_2}$	$\dfrac{B_1\downarrow \quad B_2 \xrightarrow{e@i} B_2'}{B_1;B_2 \xrightarrow{e@i} B_2'}$		
State operator	$\dfrac{\alpha@i \notin pool \quad B \xrightarrow{e@i} B'}{\lambda_M(B) \xrightarrow{e@i} \lambda_M(B')}$	$\dfrac{B \xrightarrow{out(j,m)@i} B'}{\lambda_M(B) \xrightarrow{out(j,m)@i} \lambda_{M+out(j,m)@i}(B')}$	$\dfrac{out(j,m)@i \in M \quad B \xrightarrow{in(i,m)@j} B'}{\lambda_M(B) \xrightarrow{in(i,m)@j} \lambda_{M-out(j,m)@i}(B')}$	

semantics of events is essentially the same up to some renaming. Coregions are not handled in [13], but in [7] they are translated into a free-merge of (the semantics of) all events in the coregion. The free-merge of ACP (denoted $\parallel$) coincides with TCSP parallel composition with an empty synchronisation set ($\parallel_\emptyset$ above) —except for termination, on which more below. From now on, we ignore the differences in the translation of events.

The major differences start on the level of instances. In [13], instances are interpreted as the *strong* sequential composition of their events, in contrast to weak sequential in our semantics. To define the operational semantics of strong sequential composition, instead of termination transitions local to instances ($\xrightarrow{\surd@i}$ above), Mauw and Reniers use a global termination predicate: $B\downarrow$ denotes the successful termination of process B. The operational semantics are given in Table 6. To avoid confusion, we use ; to denote strong sequential composition instead of $\cdot$ as in [13].

The second difference concerns the composition of instances into MSCs. For modelling asynchronous communication, Mauw and Reniers use a *state operator* λ_M [1]. This operator plays the role of the *message pool* in our semantics. M is a multiset containing *out*-events: If $\lambda_M(B)$ performs an *out*-event, this event is added to the set M. An *in*-event of B can only be performed if the corresponding *out*-event is an element of M; this element is then removed. (We denote addition and subtraction of multisets by $+$ and $-$, respectively). Therefore, the state operator ensures that the sending of a message occurs before its receipt.

The translation function of [13] is given by

$$S[\![\texttt{msc}\ M;\ \texttt{<i>}_1;\ \cdots;\ \texttt{<i>}_n;\ \texttt{endmsc}]\!]_{\texttt{msc}} = \lambda_\emptyset(S[\![\texttt{<i>}_1]\!]_{\texttt{inst}} \parallel \cdots \parallel S[\![\texttt{<i>}_n]\!]_{\texttt{inst}})$$

$$S[\![\texttt{instance}\ i;\ \texttt{<e>}_1;\ \cdots;\ \texttt{<e>}_n;\ \texttt{endinstance}]\!]_{\texttt{inst}} = [\![\texttt{<e>}_1]\!]^i_{\texttt{event}};\cdots;[\![\texttt{<e>}_n]\!]^i_{\texttt{event}}$$

The function $S[\![\cdot]\!]_{\texttt{inst}}$ translates a single instance i into a process term which consists of the strong sequential composition of the events performed by i. As before, the events are translated by the function $[\![\cdot]\!]^i_{\texttt{event}}$ given in Table 4. The function $S[\![\cdot]\!]_{\texttt{msc}}$ translates a MSC into a parallel composition (free merge) of the terms resulting from the translation of the instances. Moreover, the term is enclosed in a state operator $\lambda_\emptyset$ to achieve the correct causal order of corresponding *out*- and *in*-events.

As mentioned above, we compare our semantics with the standard one up to strong bisimulation, where, however, we have to take the different notions of termination into account somehow. The natural thing is to consider the global termination used in the standard semantics (as indicated by $B\downarrow$) to be equivalent to the termination of all instances ($B \xrightarrow{\surd @ i} B$ for all $i \in Inst$). This gives rise to the following definition:

3 Definition. Two transition systems $T_i = \langle S_i, \rightarrow, q_i \rangle$ for $i = 1, 2$ are called *bisimilar*, denoted $T_1 \sim T_2$, if there exists a relation $\mathcal{R} \subseteq S_1 \times S_2$ such that $(q_1, q_2) \in \mathcal{R}$ and whenever $(s_1, s_2) \in \mathcal{R}$ we have

- $s_1 \xrightarrow{\alpha @ i} s_1'$ implies $\exists s_2' \colon s_2 \xrightarrow{\alpha @ i} s_2'$ such that $(s_1', s_2') \in \mathcal{R}$;
- $s_2 \xrightarrow{\alpha @ i} s_2'$ implies $\exists s_1' \colon s_1 \xrightarrow{\alpha @ i} s_1'$ such that $(s_1', s_2') \in \mathcal{R}$.

Note that our operational semantics (Table 3) as well as the standard one (Table 6) fall into the SOS format of [20], and therefore bisimulation is a congruence. Consistency of the semantics holds for all *well-formed* MSC definitions, which we defined above to mean that for every outgoing message, a corresponding incoming one is specified as well. Since we are just comparing single MSCs, we interpret all process names (i.e., MSC conditions) in the environment $\theta_{\mathcal{D}}$ where $\theta_{\mathcal{D}}(c) = \varepsilon$ (no continuation).

1 Theorem. Let <i> be an instance definition and <m> a well-formed MSC definition. Then **[**<i>**]**$_{\texttt{inst}}$ $\sim$ S**[**<i>**]**$_{\texttt{inst}}$ and **[**<m>**]**$_{\texttt{msc}}$ $\sim$ S**[**<m>**]**$_{\texttt{msc}}$.

The proof can be found in the full version of the paper [5].

Distribution of the message pool Our translation of MSC documents into $\mathcal{L}_{MSC}$-terms introduces a message pool per MSC: control is transferred from one MSC to the next (as directed by the conditions) only if the message pool of the first contains no more remaining elements. This does not correspond to the actual assumption about the communication structure underlying MSCs, where there is a single global medium. Now we indicate that our model is equivalent to such a global medium. The crucial algebraic properties necessary to show this are the following:

$$(B_1 \parallel_{pool} \perp) \cdot (B_2 \parallel_{pool} \perp) \quad = \quad (B_1 \cdot B_2) \parallel_{pool} \perp \qquad (1)$$

$$(B_1 \parallel_{pool} \perp) + (B_2 \parallel_{pool} \perp) \quad = \quad (B_1 + B_2) \parallel_{pool} \perp \qquad (2)$$

The first of these states that communication with an empty message pool may be distributed over weak sequential composition. This is valid up to $\sim$ if B_1 and B_2 are well-formed in the sense that they contain as many $out(i, m)@j$-events as $in(j, m)@i$-events. In fact, (1) is reminiscent of the ***communication closed layers*** law of [4], which also plays an important role in the work of Janssen and Zwiers [8]. The second equation states a similar distribution property for the choice operator.

As a consequence of these laws, we can give an alternative translation function featuring a global message pool rather than local ones to each MSC, by replacing the functions $[\![\cdot]\!]_{\texttt{doc}}$ and $[\![\cdot]\!]_{\texttt{msc}}$ of Table 4 by the following:

$$G[\![\texttt{document } \mathcal{D};\ \texttt{<m>}_1;\cdots;\texttt{<m>}_n;\texttt{endmscdocument}]\!]_{\texttt{doc}} = \{\bot \parallel_{pool} G[\![\texttt{<m>}_k]\!]_{\texttt{msc}} \mid 1 \le k \le n\}$$

$$G[\![\texttt{msc } M;\ \texttt{<i>}_1;\ \cdots;\ \texttt{<i>}_n;\ \texttt{endmsc}]\!]_{\texttt{msc}} = ([\![\texttt{<i>}_1]\!]_{\texttt{inst}} \parallel_{\emptyset} \cdots \parallel_{\emptyset} [\![\texttt{<i>}_n]\!]_{\texttt{inst}}) \cdot \mathit{fin}(M)$$

The following theorem states the equivalence of $[\![\cdot]\!]$ and $G[\![\cdot]\!]$

2 Theorem. Let $\texttt{<m>}$ be an MSC definition. Then $\bot \parallel_{pool} G[\![\texttt{<m>}]\!]_{\texttt{msc}} \sim [\![\texttt{<m>}]\!]_{\texttt{msc}}$.

We regard the fact that this equivalence can be shown by relying on the algebraic distribution properties (1) and (2) as evidence of the power of the process algebraic approach.

6 CONCLUSION

We have presented a structural operational semantics for Message Sequence Chart documents based on process algebras. The semantics, which is consistent with the standard semantics, covers all basic MSC features *including* the use of conditions as composition operators. Instance creation can quite easily be mimicked with asynchronous communication. The creating instance sends a special message `create` to the instance to be created which as an initial action has to perform a receive action of the create message (a kind of "await creation"). Instance decomposition can be incorporated in a way similar to the standard semantics, namely by syntactic substitution in the MSC term.

The main advantage of our semantics is the conceptually clear modelling of the issues termination and sequential composition. Sequential composition of MSCs in the context of high-level MSCs, also based on [19], has been given in [14]. In contrast to [14], we use a single concept of termination and just one sequential composition operator; the latter can be used both on the level of instances and for MSCs. Moreover, we have adapted the weak sequential composition operator of [19] to the specific setting of MSCs, allowing to replace the complex notion of "permission" by a concept of *local termination* which is the natural translation of termination into the area of MSCs. Starting from our semantics, it should be easy to develop a semantics for high-level MSCs; sequential composition in high-level MSCs is already present here.

Acknowledgements. We are grateful to Peter Niebert for cooperation in an initial draft version of this paper.

References

[1] J. C. M. Baeten and W. P. Weijland. *Process Algebra.* Cambridge University Press, 1990.

[2] S. D. Brookes, C. A. R. Hoare, and A. W. Roscoe. A theory of communicating sequential processes. *J. ACM*, 31(3):560–599, July 1984.

[3] N. Busi, R. Gorrieri, and G. Zavattaro. A Process Algebraic View of Linda Coordination Primitives. Technical Report UBLCS-97-6, University of Bologna, May 1997. To appear in: Theoretical Computer Science.

[4] T. Elrad and N. Francez. Decomposition of distributed progams into communication closed layers. *Science of Computer Programming*, 2, 1982.

[5] T. Gehrke, M. Huhn, A. Rensink, and H. Wehrheim. An Algebraic Semantics for Message Sequence Chart Documents. Hildesheimer Informatik-Bericht 5/98, Universität Hildesheim, Institut für Informatik, May 1998.

[6] P. Graubmann, E. Rudolph, and J. Grabowski. Towards a Petri Net Based Semantics Definition for Message Sequence Charts. In *Proceedings of the 6th SDL Forum (SDL '93)*, 1993.

[7] International Telecommunication Union. *Message Sequence Chart (MSC)*, z.120 edition, 1996.

[8] W. Janssen and J. Zwiers. From sequential layers to distributed processes. In *Principles of Distributed Computing*, pp. 215–227. ACM, 1992.

[9] B. Jonsson and J. Parrow, eds., *Concur '94: Concurrency Theory*, volume 836 of *Lecture Notes in Computer Science*. Springer-Verlag, 1994.

[10] P. Ladkin and S. Leue. Interpreting message flow graphs. *Formal Aspects of Computing*, 7(5):473–509, 1995.

[11] P. B. Ladkin and S. Leue. What do Message Sequence Charts mean? In R. L. Tenney, P. D. Amer, and M. Ü. Uyar, eds., *Formal Description Techniques VI*, IFIP Transactions C, pp. 301–316. North-Holland, 1994.

[12] P. B. Ladkin and S. Leue. Four issues concerning the semantics of Message Flow Graphs. In D. Hogrefe and S. Leue, eds., *Formal Description Techniques VII*, pp. 355–369. Chapman & Hall, 1995.

[13] S. Mauw and M. Reniers. An Algebraic Semantics of Basic Message Sequence Charts. *The Computer Journal*, 37(4):269–277, 1994.

[14] S. Mauw and M. Reniers. High-level message sequence charts. In A.Cavalli and A.Sarma, eds., *SDL'97: Time for testing - SDL, MSC and Trends.* Elsevier, 1997.

[15] S. Mauw, M. v. Wijk, and T. Winter. A formal semantics of synchronous Interworkings. In O. Færgemand and A. Sarma, eds., *SDL'93: Using Objects*, pp. 167–178, North-Holland, 1993.

[16] R. Milner. *Communication and Concurrency.* Prentice-Hall, 1989.

[17] A. Olsen, O. Færgemand, B. Møller-Pedersen, R. Reed, and J. Smith. *Systems Engineering Using SDL-92.* North Holland, 1994.

[18] Rational Software Corporation. *UML Notation Guide – version 1.1*, Sept. 1997. URL: http://www.rational.com/uml.

[19] A. Rensink and H. Wehrheim. Weak sequential composition in process algebras. In Jonsson and Parrow [9], pp. 226–241.

[20] C. Verhoef. A congruence theorem for structured operational semantics with predicates and negative premises. In Jonsson and Parrow [9], pp. 433–448.

Open Estelle —
An FDT for Open Distributed Systems

J. Thees, R. Gotzhein
University of Kaiserslautern
Postfach 3049, D-67653 Kaiserslautern, Germany
e-mail: {thees, gotzhein}@informatik.uni-kl.de

Abstract

An Estelle specification describes a system of communicating components (module instances). The specified system is *closed*, i.e. it has no ability to interact with some environment. Because of this restriction, *open* systems can only be specified together with and incorporated into an (Estelle) environment.

To overcome this restriction, we introduce a compatible extension of Estelle, called '*Open Estelle*'. It allows one to specify *open systems* (i.e. systems that have the ability to communicate with any environment through a well-defined external interface), as well as their *formal incorporation* into different environments. We define a *formal syntax* and a *formal semantics* for Open Estelle, both based on and extending the syntax and semantics of Estelle. Furthermore, we present a set of tools, including a compiler for the automatic generation of implementations directly from Open Estelle specifications.

Keywords

FDTs, Estelle, extensions of FDTs, semantic model, FDT-based implementation

1. Introduction

Estelle [ISO97, DeBu89] is a formal description technique (FDT), internationally standardized since 1989. An Estelle specification describes a hierarchical system of indeterministic components called 'module instances'. Every module instance has a well-defined external interface, through which it may interact with other module instances of the same system. However, *systems* formally specified in Estelle are *closed*, i.e. they have no abilities for external communication. Because of this restriction, *open systems* can be described *formally* only together with and embedded into a *concrete environment*[2] such that the resulting system is again closed in the described sense. This means

1. This work has been supported by the Deutsche Forschungsgemeinschaft (DFG) under grant Go 503/4-1.

that the environment has to be known in advance, and that it must already be determined when the open system is specified.

In [GoRoTh96], pragmatic approaches to the specification of systems with the ability to interact with pre-existing real-world environments are studied. These approaches are based on the use of primitive functions and procedures or pragmatic compiler modifications. According to the Estelle standard, these specifications do not possess a formal semantics[3]. Also, the *incorporation* of systems specified in such a manner into a formal context (e.g. another Estelle specification) is not supported by these approaches.

The formal description technique *SDL* [ITU94] supports the specification of open systems syntactically, since it is possible to connect channels to the system boundary. Also, open systems are assigned a formal semantics in terms of a set of communicating Meta-IV processes. However, communications are purely internal, because signal exchange with the environment is not expressed. Furthermore, the composition of open system specifications is neither supported syntactically nor semantically.

The ability to *formally* specify open systems in Estelle independently of any environment, and to specify their incorporation into different concrete environments, has the following benefits:

- The *expressiveness* and the *abstraction level* of Estelle is enhanced. For instance, protocol machines can be formally described and analyzed *independently* of concrete user modules or network modules. Their formal semantics takes all possible environment behavior into account, and is not restricted to a single, explicitly specified environment.
- The *decomposition of large systems* into sets of components is supported. This improves the practical development of systems with Estelle, as it allows one to specify and analyze system components separately, and to compose them afterwards. Also, *component reuse* is supported, because components are specified independently of their environment.
- The formal description of an open system can serve as a basis to generate an *implementation* with the ability to communicate with pre-existing environments. Unlike the pragmatic approaches mentioned before, the Estelle specification on which the implementation is based has a formal semantics.

In this paper, we introduce a syntactic and semantic extension of Estelle called '*Open Estelle*'. It makes it possible to specify open systems (i.e. systems that have the ability to interact with any environment) through a well-defined external interface, as well as their incorporation into different environments. In Section 2, we explain the basic concepts of Open Estelle. Section 3 introduces the syntactic language elements of Open

---

2. An environment specified manually or generated with tools like the '*Universal Test Drivers Generator*' (UTDG, [LLM91,INT97]); see Sections 2.2 and 2.4.
3. Primitive functions and procedures (and also the specification containing them) have no meaning, unless a 'rigorous, implementation independent (e.g. mathematical) definition of the relevant block is supplied by the specifier' (Clause 8.2.4.3 of [ISO97]). Such a definition may, however, be difficult to obtain for communication with some environment realized, for instance, via primitive functions.

Estelle. In Section 4, we define a formal semantics for Open Estelle, which is based on the Standard[4] Estelle semantics. Section 5 addresses implementation issues, including tool support for Open Estelle.

2. Basic Concepts of Open Estelle

The main objective of Open Estelle is to provide language support for the formal description of open systems and their incorporation into different environments. This means that an open system not only has a well defined syntactic and semantic interpretation *independently of its environment*, but also *as part of a compound system*. In the following, we introduce some basic concepts of the syntactic and semantic interpretation of open systems, and their incorporation in Open Estelle.

2.1. Representation of Open Systems

A basic design decision of Open Estelle concerns the representation of an open system. Standard Estelle already possesses an abstraction to describe encapsulated systems with well-defined external interfaces and their aggregation into more complex systems: *modules*. Module instances could in principle supply a suitable representation of open systems in Estelle: the module header describes the external interface of the module instance without regard to its internal description, the module body describes the module instance behavior.

Note that a module definition *describes* an open system, while a module instance *is* an open system. At any point in time, a module instance (i.e. an open system) is surrounded by a collection of other module instances. We call this collection of other module instances *environment*, and we say that the open system is *incorporated* into an environment when placing it into a collection of surrounding module instances.

Obviously, modules are a suitable and natural means to describe open systems, and in fact Open Estelle basically describes open systems as Standard Estelle modules. However, in Standard Estelle, a module can only be defined and used inside the specification of a closed system. Furthermore, being a fragment of a specification, it has no formal semantics. Therefore, based on Standard Estelle, the following aspects have to be addressed:

- syntactic extensions to define modules that describe open systems independently of their environment,
- a formal semantics for open systems independent of their environment, and also as part of a compound system,
- a mechanism to allow other specifications (e.g. specifications that define a specific environment) to import the definitions of open systems.

---

4. By *Standard Estelle*, we refer to the Estelle language defined in [ISO97], whereas the proposed extension is called *Open Estelle*.

2.2. Formal Import vs. Textual Inclusion

As noted above, the formal description of an open system must have a *unique syntactic and semantic interpretation* for the open system itself, as well as for its incorporation into all potential environments. While the semantic interpretation of an open system (its *'behavior'*) depends on the possible interactions with all environments, its syntactic interpretation should be independent of the incorporating context. In particular, internal definitions of any potential environment incorporating an open system should not have any impact on the interpretation of the open system or its external interface.

The following examples show that this requirement rules out solutions that are based on *simple textual inclusion* of open system specifications into environment specifications. In these examples, we assume that open systems are defined by an Estelle specification fragment[5]. We then consider the consequences of textually including (preprocessor '#include' statement) these fragments into an incorporating Estelle specification.

Example 1: *Interference between an imported interface and its importing environment.*

Let the Estelle specification fragment OS define an open system that refers to a predefined function, procedure, or type. Furthermore, let Estelle specification S redefine this identifier, e.g like 'integer' shown in the following specification fragment:

```
SPECIFICATION S;
    TYPE integer = (zero, one, two); (* predefined type integer is redefined in this scope *)
    MODULE M FOR {...};
        #include OS                   (* textually include specification text fragment OS *)
    END;
END.
```

The textual inclusion of OS into module S.M causes the definitions given inside OS to be interpreted in the context of module M, i.e. with a redefined type 'integer'. Since OS refers to the identifier 'integer', it will no longer refer to the predefined type, but to the redefinition. This may have a massive impact on the syntactic and semantic interpretation of OS.

Example 2: *Interference between two imported open systems.*

Let the Estelle specification fragments OS1 and OS2 define two open systems that (by coincidence) define the same identifier (e.g. different types named PDU for different 'Protocol Data Units'). Textual inclusion of both open systems into the same context leads to a syntactically incorrect specification, since it contains an inadmissible identifier redefinition.

Obviously, the syntactic interpretation of the definitions given inside the textually included open system descriptions depends on the including context and the combina-

---

5. e.g., a module definition together with all necessary constant, type, channel and module header definitions

tion of included open system descriptions. This makes the syntactic definition of these open systems *non-formal*. To obtain a *formal* definition of an open system, it has to be interpreted strictly independently of any environment that makes use of it; in particular, the name-spaces of different open systems and importing environments must be separated. In Section 3, we will see that Open Estelle avoids these problems, since

a) the syntactic definitions given inside the description of an open system are *interpreted in a unique way*, independently of a concrete importing environment (see Section 3.1 and 3.2);
b) the definitions given inside an open system are not included textually, but they are *imported* in a well-defined way; this formal import only grants access to the (unique) definitions of the open system (see Section 3.3);
c) identifiers imported from an open system description are *qualified* by the name of the open system (e.g. '**OS1::PDU**' and '**OS2::PDU**' in Example 2); this avoids naming conflicts between different interfaces or with identifiers inside the importing environment; furthermore, it improves the readability of the specification, since the origin of each identifier is explicitly stated (see Section 3.3).

2.3. Separation between Declaration and Definition of Open Systems

In Open Estelle, the description of an open system consists of the *external interface* and the *internal description*, which are textually separated. Both are syntactically independent of any possible environment using the open system. Furthermore, Open Estelle environments incorporating an open system only refer to its external interface (see Figure 1).

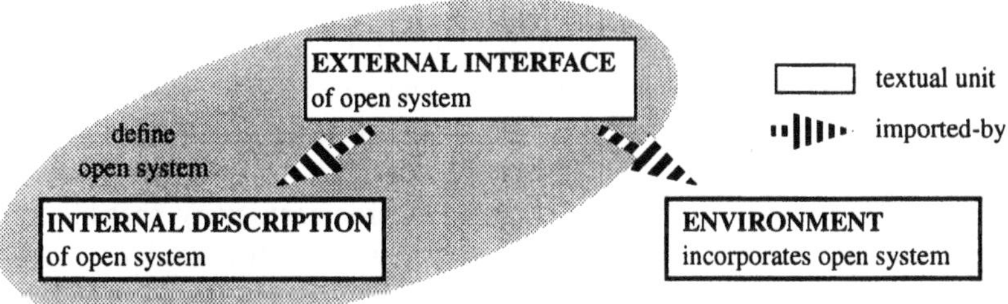

Figure 1. Independent Descriptions for Open System and Environment.

An 'external interface' *declares* open systems (i.e. module body declarations) together with all necessary underlying definitions (i.e. module headers, channels, types and constants; see Section 3.1).[6] An 'internal description' *defines* these open systems by adding an internal description (see Section 3.2). This separation makes it possible to develop and compile open systems and importing environments independently, as soon as their common interface has been defined. Furthermore, it supports the separation of concerns between open systems and their possible environments.

---

6. External interfaces may import other external interfaces and make use of their definitions (Section 3.3).

The textual separation between the external interface and the internal description of a system is similar to the concepts found in Modula-2 [Wir85]. Also, the distinction between a declaration and a definition of a syntactic object is similar to Modula-2, C or C++: a *declaration* defines name and type (i.e. the external interface) of an object, whereas a *definition* adds internal details.[7]

2.4. Semantics of Open Systems

Open Estelle has been designed to formally describe open systems and their incorporation into any environment. Consequently, an open system should have a formal semantics independent of an environment, and also as part of a compound system.

The *incorporation* of an open system into an Estelle environment is already semantically covered by the standard Estelle semantics, since the instantiation of an imported open system (i.e. of an Estelle module) creates the same module structure as the instantiation of a normal (textually included) local child module. This produces an intuitive semantics of the application of open systems. Even more, this semantics makes it possible to fuse the description of an importing environment and its imported open systems into one single specification that is semantically equivalent. If textual interferences (see Section 2.2) are excluded (e.g. by appropriate renaming of identifiers defined in open systems and their importing environment), the import of an open system can be replaced essentially by a semantically equivalent textually included child module definition. In Section 5.1 we will present a tool that implements this fusion.

An important aspect of Open Estelle is the definition of a semantics for an open systems *independent of its environment*. In the special case of an open system *without* any external interaction points and exported variables, its semantics is identical to the semantics of a corresponding (closed) Estelle specification, independent of any concrete environment, since there are no means for interactions with an environment. But if the open system *has* external interaction points or exported variables, a concrete environment can interact by modifying exported variables and/or sending messages through external interaction points. Since the semantics of an open system can not refer to the specific restrictions of a concrete environment, every possible influence through the external interface of the open system has to be considered, i.e. every possible sequence of messages received through external interaction points and every possible modification of exported variables[8]. If the open system is incorporated into a concrete environment, this maximum set of possible interaction is reduced to a subset of interactions with this particular environment. In Section 4, we will describe this concept in detail.

---

7. E.g. in C 'int succ(int n);' is a *function declaration* and 'int succ(int n) {return n+1;}' its *function definition.*

8. This kind of maximum interaction is also the foundation of the *'Universal Test Drivers Generator'* (UTDG, [LLM91,INT97]). UTDG generates modules that can send or accept any interaction through their external interaction points or execute any modification of their external variables. It was developed to simulate unspecified system components during tests of Estelle specifications. But since it does not solve the syntactic problems presented in Section 2.2, it is not directly suitable for the formal specification of open systems.

3. The Syntax of Open Estelle

In this section, we introduce the syntactic extensions of Estelle for the description of the external interface of open systems (Section 3.1), the internal description of open systems (Section 3.2), and the incorporation of open systems into different environments (Section 3.3) through examples. These extensions are based on the syntactic concept of modules. The formal definition of the Open Estelle syntax is given in [ThGo97a].

3.1. Interface Description of Open Systems

The declaration (and thus the external interface) of an open system is given in an INTERFACE-DEFINITION residing in a *separate textual unit*[9]. An INTERFACE-DEFINITION is a container for the *declaration* (not definition) of a set of open systems and for all definitions that are necessary to describe their type and thus their external interface (i.e. module headers, channels, types and constants).

```
INTERFACE binaryService;
    TYPE  tOperand = RECORD x1, x2: REAL; END;
          tResult = REAL;
    CHANNEL binaryServiceChannel(user, provider)
          BY user: request(x: tOperand);
          BY provider: respond(y: tResult);
    MODULE binaryOperatorHeader ACTIVITY;
          IP toUser: binaryServiceChannel(provider) COMMON QUEUE;
          END;
    BODY binaryOperator FOR binaryOperatorHeader; { this declares the open system }
          EXTERNAL;                               { 'binaryService::binaryOperator' }
END. { end of interface 'binaryService' }
```

Figure 2. Example for an Interface Definition.

The INTERFACE-DEFINITION is one of the two new start non-terminals of the Open Estelle syntax. Interfaces are uniquely defined, independently of any importing environment (see Section 3.3). This is important for a formal description of the exported definitions. An interface starts with the new keyword '**INTERFACE**', followed by its name. The *declaration* of an open system inside of an interface is syntactically represented by a MODULE-BODY-DECLARATION, which is a MODULE-BODY-DEFINITION containing the keyword '**EXTERNAL**' (see Figure 2). This declaration refers to a module header that describes the external interface, but it gives no internal description of the module body.

Analogous to Standard Estelle, the appearance of the keyword '**EXTERNAL**' inside a MODULE-BODY-DECLARATION of an INTERFACE-DEFINITION leads to a syntactically correct, but incomplete description. But in contrast to a SPECIFICATION or a BEHAVIOR-DEFINITION, where only a textual modification can remove this incom-

9. Since the Estelle standard does not state a representation for a specification text, we use the term '*textual unit*' for self-contained syntactic objects (e.g. a specification). In a UNIX environment such a textual unit is usually represented as an ASCII-text file.

pleteness, the attachment[10] of an appropriate BEHAVIOR-DEFINITION (see Section 3.2) leads to a complete description of the declared open systems.

The reader should note that the interface definition itself describes no behavior (no transitions) and no state (no variables or control states). Consequently, it is only a container for a set of definitions that may be *imported* by other Estelle components (see Section 3.3). It is even possible for an interface definition to import another interface definition. This allows two (otherwise independent) interface definitions to share common definitions such as types, constants and channels (see Figure 5) and therefore supports a clean separation of concerns between open systems that refer to common definitions.

3.2. Internal Description of Open Systems

The internal description of an open system is given by a MODULE-BODY-DEFINITION inside of a BEHAVIOR-DEFINITION residing in a *separate textual unit.* As suggested by the term BEHAVIOR-DEFINITION, from an abstract point of view the internal description of an open system only determines the *behavior* of the open system, because the syntactic interface is already determined by the interface definition.

The BEHAVIOR-DEFINITION is the second new start non-terminal of the Open Estelle syntax. It begins with the new keyword '**BEHAVIOR**', followed by its name and a reference to its interface (see Figure 3). The definitions of its interface are imported implicitly (see Section 3.3).

```
BEHAVIOR binaryAdder FOR binaryService;
    BODY binaryOperator FOR binaryService::binaryOperatorHeader;
            { this defines the open system 'binaryService::binaryOperator' }
            TRANS
                    WHEN toUser.request(x: tOperand)
                    BEGIN
                            OUTPUT toUser.respond( x.x1 + x.x2 );
                    END; { end of transition-block }
    END; { end of module-body 'binaryOperator;' }
END. { end of interface 'binaryAdder' }
```

Figure 3. Example for a Behavior-Definition (see Figure 2).

A BEHAVIOR-DEFINITION is a container for a set of open system definitions: for every MODULE-BODY-DECLARATION of its interface, it contains exactly one *matching*[11] MODULE-BODY-DEFINITION, and vice versa. The module body defining the open system can make use of all (Open) Estelle constructs for module bodies in its internal description, for example, it can be sub-structured into child modules and it can even import and use other open systems. This supports a very flexible internal structuring of open systems.

---

10. The attachment of a BEHAVIOR-DEFINITION to an INTERFACE-DEFINITION is a matter of the interpreting context, since this operation involves different textual units (e.g. UNIX text-files).
11. They have the same name and refer to the same header-definition.

3.3. Incorporation of Open Systems into Open Estelle Environments

To be able to incorporate[12] an open system into an environment specified in Open Estelle, the interface that declares this open system has to be *imported* into this environment. The import of an interface is described syntactically by an IMPORT-STATEMENT inside either a specification, a module-body, an interface, or a behavior-definition (we will refer to each of them as *importing environment*). An IMPORT-STATEMENT consists of the new keyword '**IMPORT**', followed by a list of identifiers (see Figure 4), which uniquely identify the imported interfaces with the respective names.[13]

```
SPECIFICATION test;
    IMPORT binaryService; { imports definition of interface 'binaryService' }
    MODVAR mv: binaryService::binaryOperatorHeader;
    INITIALIZE
        BEGIN
            { create open system by instantiating module 'binaryService::binaryOperator': }
            INIT mv WITH binaryService::binaryOperator;
            { the open system can now be handled like a regular module-instance }
        END;
    { ... }
END. { end of specification 'test' }
```

Figure 4. Example for the incorporation of an open system (see Figure 2).

The import of an interface defines a *qualified visibility* in the importing environment for all definitions of the interface. This includes (apart from constants, types, channels and module headers) also the open systems declared inside the interface. These definitions can be referred to by means of their *qualified name*, which consists of their unqualified name (given in their definition), preceded by the name of their defining interface and the new symbol '::'. For example the identifier 'binaryService::binaryOperator' refers to the definition of 'binaryOperator' inside the interface 'binaryService'. The reader should note that since the interface identifier uniquely identifies an interface, and since the unqualified name is unique for this interface, every qualified identifier is globally unique.

The open system itself is originated when a new instance of its describing module-definition is created by an appropriate INIT-statement (see Figure 4). It is possible to dynamically create several open systems (i.e. module instances) from the same open system description (i.e. module). In general, all mechanisms that Standard Estelle offers for the handling of instances of locally defined child-modules can also be used with respect to an open system, including communication with the open system or its termination.

---

12. By '*incorporate*', we refer to the embedding of the open system (i.e., the *instance*) into the instance of an environment. This should be distinguished from a *textual inclusion* of the description of the open system into the description of an Open Estelle environment.
13. The unique mapping of an interface-identifier to an interface is a matter of the interpreting context, since this operation involves different textual units (e.g. UNIX text-files).

Importing an interface is not only useful to get access to the open systems defined inside of it. Since the import of an interface makes *all* of its definitions visible to the importing environment, interfaces are also useful for the global definition of constants, types, channels and headers, which can be used in different specifications, behavior definitions, or even interfaces. This aspect is particularly important for the separation of concerns of independent interfaces with common definitions; e.g. in Figure 5 the separated interfaces intC and intS declare open systems with compatible external interaction points. This is possible because both of them import interface intA, which contains the appropriate channel-definition. Thus it supports the specification of a system of interfaces that exactly models the type dependencies of the global system.

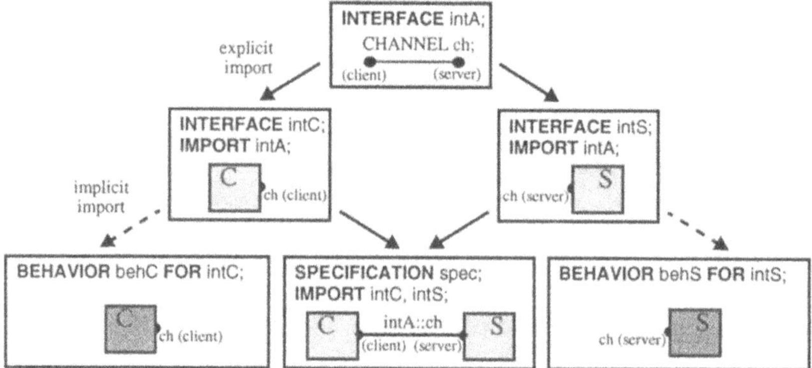

Figure 5. Multiple interfaces with common definitions.

3.4. Module Attribution

The module attribution rules given in Clauses 5.2.1 and 7.3.6.2 of [ISO97] are not directly suitable for the application in Open Estelle, since they only relate to the textual nesting of modules. Therefore, in [ThGo97a] we introduce extended module attribution rules, which imply the same attribution scheme for the dynamic module instance tree as the original rules. It is possible to specify both, open systems that contain one or more subsystems[14] (which can be incorporated into un-attributed environments) and open systems that can be part of other subsystems. In particular, it is possible to specify open systems that can be incorporated into *arbitrarily attributed importing environments*.

4. Semantics of Open Estelle

Estelle qualifies as a formal description technique, meaning that it has both a formal syntax *and* a formal semantics. The proposed language extension should therefore cover both aspects to preserve the formality of Estelle. In order to define the meaning of Open Estelle specifications formally, we have devised a semantics that models the

14. With the term *'subsystem'* we refer to instances of system modules.

interaction between an open system and an environment. More precisely, the semantics captures all potential behavior of an open system in *all possible* environments; this potential behavior is reduced when the open system is incorporated into a specific environment.

In the following, we describe this semantics in some detail. We argue why despite the different nature of Open Estelle it is feasible to define its semantics in terms of computations as in the Standard Estelle semantics. This has the advantage that the language extension is fully compatible with Standard Estelle, and that it will be straightforward to argue why the incorporation discussed above is complete and sound.

4.1. Formal definition of the Open Estelle semantics

According to the Estelle standard [ISO97], the semantics of a Standard Estelle specification SP is formally given by the set of its *computations*. Computations are sequences $<sit_0, sit_1, \ldots>$ of so-called *global situations*, where sit_0 is initial, and for all $j>0$, sit_j is a possible next global situation of sit_{j-1} with respect to the next-state relation. Roughly speaking, a global situation comprises the local states of all module instances (state of input queues, values of local variables, module structure, connection structure) and sets of transitions selected for firing. A next global situation results from either firing a previously selected transition, or selecting a set of transitions according to certain rules. Thus, an Estelle specification defines a *transition system* (S, δ), where S and δ are related to the set of global situations and the next-state relation, respectively. Concurrency is modeled by interleaving.

We will now generalize this semantics to capture the meaning of Open Estelle specifications. In particular, the formal semantics of Open Estelle will reduce to the Estelle semantics in the special case of a closed system, namely systems without external interaction points and without exported variables.

To start with, let us consider the definition of the *next global situations* and *computations* as given in the Estelle standard:

> **Definition 1** (*next global situations, computation*; Clause 5.3.4, [ISO97]):
> Given a global situation, $sit = (gid_{SP}; A_1, \ldots, A_n)$, the *set of next global situations* is described as follows:
> (a) For every $i = 1, \ldots, n$: if $A_i = \varnothing$, then for every $AS(gid_{SP}/S_i) \in AS^*(gid_{SP}/S_i)$, $(gid_{SP}; A_1, \ldots, AS(gid_{SP}/S_i), \ldots, A_n)$ is a next global situation of sit.
> (b) For every $i = 1, \ldots, n$: if $A_i \neq \varnothing$, then for every $t \in A_i$, $(t(gid_{SP}); A_1, \ldots, A_i \setminus \{t\}, \ldots, A_n)$ is a next global situation of sit.
>
> NOTE — There are as many next global situations of the situation sit as there are possible choices of next transition t (and its results) in case (b), and different empty sets A_i in sit, for the case (a). In addition, in case (a), all possible choices of the set AS resulting from non-determinism of each component process in the system rooted at S_i must be taken into account.
> NOTE — [...]

A sequence of global situations of SP, $sit_0, sit_1, \ldots, sit_j, \ldots$ is called a *computation of SP* if and only if sit_0 is initial, and for every $j > 0$, sit_j is one of the next global situations of sit_{j-1} as described by (a) or (b) above.

For each subsystem represented by a system module, either a new set of fireable transitions is selected, or an already selected transition is executed. If a transition t of module instance P is fired, the outputs of t are transmitted to the destination queues as part of the global effects of t, as defined by $transmission_P(gid'_{SP})$, where $gid'_{SP} \in [t]_P(gid_{SP})$ (see Clause 9.5.4, [ISO97]). If, for a given output, no destination queue is defined, then the interaction is discarded. Note that all destination queues are represented in gid'_{SP}, therefore, all global effects of t can be applied directly. Furthermore, exported variables may be modified as part of the local effects of t.

In order to generalize this definition to Open Estelle specifications, we have to incorporate the interaction of the specified system into its environment in some way. More specifically, we have to incorporate receptions from, transmissions to, and modifications of local variables by the environment. Recall that these interactions occur through external interaction points and exported variables. However, as the environment is not determined, we have to consider all possible environments, which amounts to taking all possible receptions, transmissions, and assignments to exported variables into account[15]. We will first deal with receptions from the environment and assignments to exported variables by extending Definition 1. Then, we will address transmissions to the environment.

All interactions with the environment occur through external interaction points and exported variables. As these interaction points and the exported variables are typed, the set of interactions that may be received from the environment by module instance P through an external interaction point ip ($receive_P(ip)$, see Clauses 9.3.1 and 9.4.3, [ISO97]), as well as the set of values that may be assigned to an exported variable e [16], are determined. To take all environments into account, we model all possible receptions and assignments by extending the set of next global situations as defined below. Furthermore, we consider global situations w.r.t. an arbitrary module instance P, not just the specification SP.

Definition 2 (*next global situations, potential computation*; Open Estelle):
Given a global situation, $sit_P = (gid_P; A_1,...,A_n)$, of a module instance P, the *set of next global situations* is described as follows:

(a) For every $i = 1,...,n$: if $A_i = \varnothing$, then for every $AS(gid_P/S_i) \in AS^*(gid_P/S_i)$: $(gid_P; A_1,...,AS(gid_P/S_i),...,A_n)$ is a next global situation of sit_P.
(b) For every $i = 1,...,n$: if $A_i \neq \varnothing$, then for every $t \in A_i$: $(t(gid_P); A_1,...,A_i \setminus \{t\},...,A_n)$ is a next global situation of sit_P.
(c) For every $gid_P' \in env\_mod^+(gid_P)$: $(gid_P'; A_1,...,A_n)$ is a next global situation of sit_P. env_mod is defined as follows[17]:

---

15. Note that the effects of a 'terminate' or 'release' statement referring to the open system have no impact on the set of computations. In this paper, for reasons of simplicity, we do not consider the effects of 'attach' and 'detach' statements of the environment involving (directly or indirectly) external interaction points of open systems. Also, for the same reasons, we do not consider the effects of connecting two external interaction points of the same open system.

16. We assume that the type of e is well-defined, i.e. its declaration does not include type-identifiers associated with the '...' construct.

(c1) For every $ip \in EIP_P$: for every $<m,v_1,\ldots,v_k> \in receive_P(ip)$: $received_P(gid_P, <m,v_1,\ldots,v_k>) \in env\_mod(gid_P)$.
(c2) For every $e \in EV\text{-}id_M$, where $P \in INST(M,B,E)$, and e is of type T: for every $v \in E(T)$: $assign_P(gid_P, e, v) \in env\_mod(gid_P)$.

NOTE — $sit_P = (gid_P; A_1,\ldots,A_n)$ is the global situation of module instance P, where gid_P is defined as usual, and each A_i is a set of transitions of the component instances rooted at (α) P, if P is attributed, or (β) S_i, if P is not attributed, and S_i are system modules. This generalizes the definition of global situations to arbitrary module instances. If P = SP, the definition of sit_P is identical to that of sit in the Estelle standard.

NOTE — There are as many next global situations of the situation sit_P as there are possible choices of (a) different empty sets A_i in sit_P and possible choices of the set AS resulting from non-determinism of each component process in the system rooted at S_i, (b) next transitions t (and their results), (c) inputs from the environment and assignments to exported variables of P by the environment.

NOTE — For closed systems (i.e. module instances that have no external interaction points and no exported variables) Clause (c) of the definition above has no effects. Consequently, the 'next global situations' relation given above reduces to the one of Definition 1.

NOTE — EIP_P is the set of external interaction points of instance P (Clause 9.4.3, [ISO97]). If P = SP, then EIP_P is empty, and no inputs will be received from the environment. $EV\text{-}id_M$ is the set of exported variables of P as declared in its module header M (see Clause 9.4.1, [ISO97]). If P = SP, then no exported variables are defined. If EIP_P and $EV\text{-}id_M$ are both empty, Definition 2 is equivalent to Definition 1 of the Estelle standard.

NOTE — $receive_P(ip)$ is the subset of Interactions (Clause 9.3.1, [ISO97]) that the instance P can receive through interaction point ip (Clause 9.4.3, [ISO97]). $received_P(gid_P)$ is obtained from gid_P by replacing s'.ie(ip') by append(<ip',ip,m,v_1, ..., v_k>, s'.ie(ip')), where downattach(ip) = ip', ip' in $EIP_{P'}$, and s' is the local state of P' (see Clause 9.5.4, [ISO97]). This includes the special case that ip is not attached, i.e. downattach(ip) = ip.

NOTE — $assign_P(gid_P, e, v)$ is a new gid of P where the difference with gid_P is expressed by $s.Loc(alloc_B(e)) := v$, where s is the local state of P (see Clause 9.5.4, [ISO97]).

A sequence of global situations of P, $<sit_0, sit_1, \ldots, sit_j, \ldots>$ is called a *potential computation of P* if and only if sit_0 is initial[18], and for every $j > 0$, sit_j is one of the next global situations of sit_{j-1} as described by (a), (b), or (c) above.

Compared to the Standard Estelle semantics, there are two important differences. The first difference concerns the definition of global situations sit_P for arbitrary module instances P, i.e. not only for the distinguished module instance SP as in Definition 1. In order to model the execution of an open system, we generalize the notion of global situation by combining gid_P with sets of transitions selected for execution.

The second difference concerns the reception of interactions from the environment and the assignments to exported variables of module instance P. This is described by the transitive closure of a function env_mod in (c). Interactions can be received

---

17. The function env_mod: $EIP_P \rightarrow \wp(EIP_P)$ defines a relation $env\_mod_R = \{(x,y) \in EIP_P \times EIP_P \mid y \in env\_mod(x)\}$. Let $env\_mod_R^+$ be the transitive closure of env_mod_R. Then the function **env_mod**$^+$: $EIP_P \rightarrow \wp(EIP_P)$, $x \rightarrow \{y \in EIP_P \mid (x,y) \in env\_mod_R^+\}$ defines a *transitive closure* of function env_mod, i.e. the result of one or more consecutive events as described in (c1) and/or (c2).
18. The initial global situations of an open system take into account all possible actual module parameters and all possible modifications by the environment according to (c) above, which may take place during the execution of a transition executing the init-statement of the open system.

through and only through the external interaction points of module instance P, as expressed in (c1). Note that env_mod ranges over all ip $\in$ EIP_P and all <m,v_1,...,v_k> $\in$ $receive_P$(ip). Thus, all possible environments are taken into account. If an external interaction point of P is attached, then any reception is appended to the destination queue at the end of the attach chain. Assignments to exported variables of module instance P are expressed by (c2). Note that env_mod ranges over all e $\in$ EV-id_M and all v $\in$E(T), where e is of type T. Thus, again, all possible environments are taken into account.

Since the behavior that is described by the definition of the next global situation may in general only occur when the open system is composed with some environment, we use the term *potential computation of P* to denote a sequence of global situations satisfying the aforementioned condition. The reason is that in our definition, the behavior of the environments as captured by (c) does not yet exist. Once the environment is completely determined, the *set of potential computations* is reduced to a *set of computations* in the sense of [ISO97]. This is, for instance, the case if we consider a closed system. Here, the next global situations are completely described by (a) and (b), which gives evidence that the semantics of Open Estelle is indeed compatible with the semantics of Standard Estelle.

Having dealt with receptions from the environment and assignments to exported variables, we will now address transmissions to the environment. In the Standard Estelle semantics, all outputs of a transition t of P are collected in the local state component s.out as the result of its local effects defined by $[t]_P(s)$ (see Clause 9.6.6.5, [ISO97]). The function $transmission_P$ then defines the global effects by appending, in the same order, the elements of the sequence s.out to the destination queues (Clause 9.5.4, [ISO97]). If for some output at interaction point ip, there is no destination queue, i.e. linked(ip,gid_{SP}) is false, then that output is discarded (see Clauses 9.5.3 and 9.5.4, [ISO97]).

The treatment of transmissions in Open Estelle directly follows the Standard Estelle semantics. If the destination queue of an output belongs to the open system, the meaning of a transmission is just the same. However, if the destination queue is outside the open system, i.e. an output is made at some external interaction point of the root module P, or at some interaction point that is attached to it, then the output leaves the open system[19]. This means that the effects of the output on a possible environment are not visible in the open system, and therefore not represented in its state. Transmissions to the environment can therefore be modeled by discarding the corresponding outputs. As this is already handled by the definitions in [ISO97] (see Clauses 9.5.3 an 9.5.4), the definitions of the Standard Estelle semantics can be used[20].

19. As already noted above, we do not consider the effects of connecting two external interaction points of the same open system in this paper.

20. To be precise, the functions $sent_P$, $received_P$ and $transmission_P$ have to be applied to $gid_{P'}$, where P' is the root module instance of the open system, instead of gid_{SP} (see Clause 9.5.4, [ISO97]).

4.2. Completeness and Soundness of the Open Estelle Semantics

As already noted in Section 2.4, the semantics of the *incorporation* of an open system into an Estelle environment is reduced to the Standard Estelle semantics by construction, since the instantiation of an imported open system (i.e. of an Estelle module) creates the same module structure as the instantiation of a normal (textually included) local child module. It remains to show that the semantics of an open system on its own as defined in Section 4.2 is correct and sound.

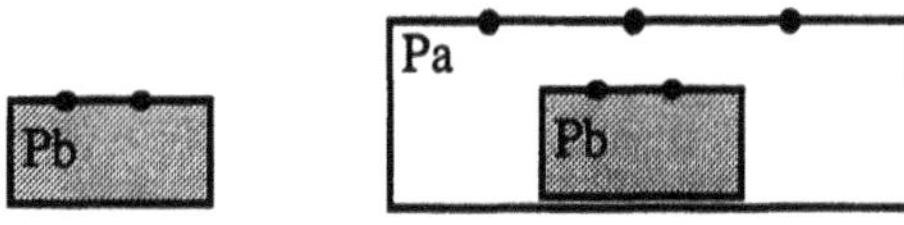

Figure 6. Open systems Pa and Pb.

If we consider the semantics of open system Pb on its own in comparison to its incorporation into an (open or closed) environment Pa (see Figure 6), module Pa partially (if open itself) or completely (if closed) determines the environment of Pb. *Completeness* of the Open Estelle semantics means that the potential computations of Pa, when 'projected onto' the incorporated Pb, are a subset of the potential computations of Pb on its own. In other words, the computations of the open system Pb are reduced when it is incorporated into some environment. *Soundness* means that only potential computations are defined for Pb that are possible in some environment Pa. The proof of these properties is addressed in [ThGo97a].

5. Implementation Issues

Estelle specifications do not only describe systems formally, they can also serve as a basis for an automated implementation. Several tools for the automatic implementation of Estelle specifications are currently available (e.g. EDT [Bud92], PET/DINGO [SiSt93]). Consequently, the automatic creation of implementations directly from specifications is also an important objective of the extension of Estelle, especially since it is just the openness of systems described with Open Estelle that establishes a well-defined relationship with their implementations: open systems formally specified with Open Estelle can be implemented with the ability to interact with real-world systems (such as operating systems, communication networks or applications) through their *well-defined external interfaces* and with a *well-defined semantics*. This opens the possibility for a direct practical incorporation of formally specified open systems into real-world environments.

5.1. Tool Support

To create a platform for practical experiments, we have developed a tool set for the processing of Open Estelle sources. The *front end* of this tool set is a compiler that translates Open Estelle sources (i.e. interfaces, specifications, and behavior-definitions) into a binary intermediate form, which can be processed by the other tools. This front end was developed out of the existing Estelle compiler front end PET ('Portable Estelle Translator', [SiSt93]).

Further, we have developed the *eXperimental Estelle Compiler* (*XEC*, [ThGo97b, ThGo98, The98]), a completely new optimizing code generator for Estelle and Open Estelle, which creates C++-code for open systems and importing environments independently of one another (see Figure 7). It is possible to compile these components separately and to delay the 'fusion' of open systems and their environments to the moment of linking the created machine-object-files into an executable. The compiled open systems can be used also by hand-crafted environments and therefore incorporated into and communicating with real-world environments.

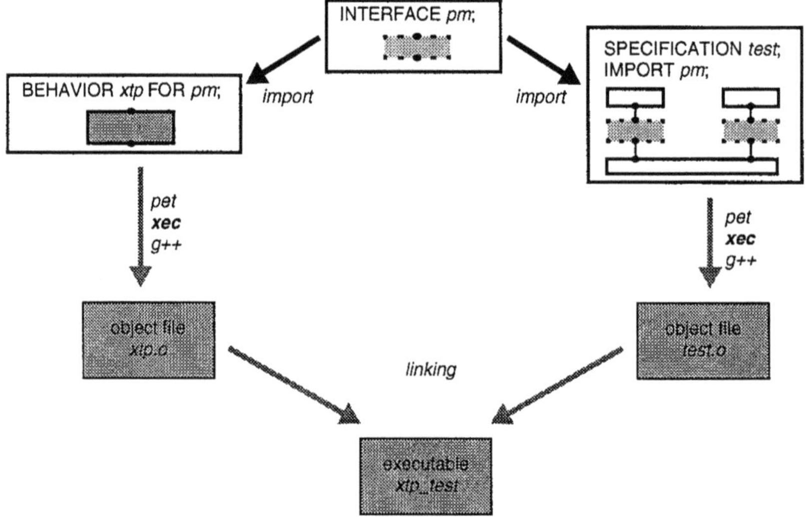

Figure 7. Independent compilation of open systems and a compatible environments.

Another Open Estelle tool supports the *textual embedding* of a set of open systems into an Open Estelle environment that incorporates them (see Section 2). This method finally leads to an equivalent *closed Standard Estelle specification*, which includes the formerly open systems in form of local module bodies (see Figure 8). The tool enables the application of existing formal methods for Standard Estelle to systems consisting of a set of Open Estelle descriptions.

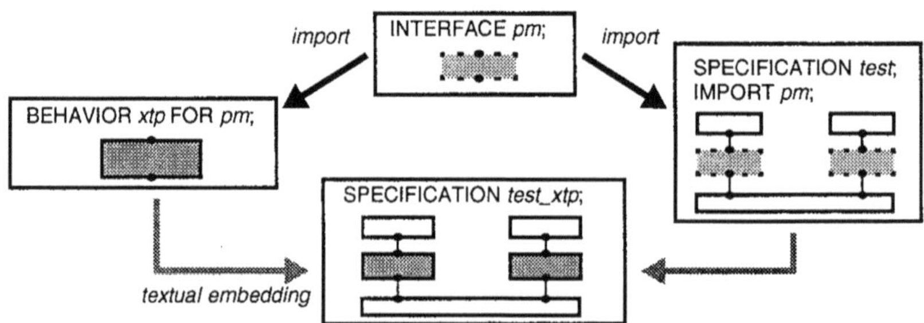

Figure 8. Embedding of an open system into a compatible environment.

5.2. Case Study

As a case study for the practical use of Open Estelle and of the tools presented above, we have split an existing, relatively large Estelle specification (more than 7500 lines of specification text) of the *Xpress Transport Protocol* (XTP, [XTP95]) into a set of open systems specified in Open Estelle. We have incorporated the XTP protocol machines into different Open Estelle and real-world environments, and with different client and network implementations without having to re-compile the generated machine object file (see Figure 7).

In particular, it is not necessary to re-compile the whole system after a modification that only affects parts of the system. In our case study the relatively complex XTP protocol machine was specified as open system, independently of the application context (e.g. specification 'test' in Figure 7) that import it. After a modification of the application context (including all user and network modules), only the context had to be re-compiled. This reduced the turnaround-times significantly: the re-compilation and linking of specification 'test' took less than 10 seconds[21] (1.0 seconds PET/XEC, 5.8 seconds g++, and 3.0 seconds linking). This is an effective **speedup of factor 13** compared to the otherwise necessary complete re-compilation of the original Standard Estelle specification with the textually embedded XTP protocol machine (17.1 seconds PET/XEC, 110.0 seconds g++, and 4.6 seconds linking).

The case study demonstrated some of the benefits of Open Estelle for the specification of large systems: one single compilation of the relatively complex protocol machine can serve all implementations of Estelle or real-world environments that incorporate it. A further structuring of the XTP protocol machine itself into smaller open systems additionally supported the maintenance and development of the protocol machine, because it simplified the individual compilation units, shortened the turnaround times after local modifications of components of the protocol machine and clarified the type dependences between the components. In particular, it led to a clean separation between private definitions of a module and definitions that are exported to the former child modules.

6. Summary and Outlook

In this paper, we have introduced an extension of Estelle, called '*Open Estelle*'. It allows one to specify *open systems* and their *formal incorporation* into different environments. We have defined a *formal syntax* and a *formal semantics* for Open Estelle. The extension is compatible with Estelle both syntactically and semantically, i.e. Estelle is a subset of Open Estelle. In particular, the formal semantics of Open Estelle reduces to the Estelle semantics in the special case of a closed system. Furthermore, we have developed a set of tools supporting Open Estelle, including the new code generator *XEC (eXperimental Estelle Compiler)* that creates efficient implementations of open systems, which can be incorporated into different Open Estelle and hand-crafted

---

21. Platform: Sun SPARCstation 20, Solaris v2.5, PET v2.0, XEC v1.14, g++ v2.8 (without optimizations)

environments. We have demonstrated the use of Open Estelle and of our tools by means of a case study with the Xpress Transport Protocol.

There are some aspects of Open Estelle that need further consideration. An important issue that is directly related to the Estelle semantics is the validation of correctness, based on an *implements-relation*. We expect that the Open Estelle semantics will give rise to an *abstract semantics* in terms of input/output-behavior, which may serve as a basis for a correctness notion. We plan to investigate this issue in more detail.

7. Acknowledgments

The case study discussed in Section 5.2 was based on an Estelle specification of XTP 4.0 that was kindly made available by Stanislaw Budkowski (INT, France) and Octavian Catrina (PU Bucharest, Romania).

8. References

[Bud92] Budkowski, S.: *Estelle Development Toolset*, Computer Networks and ISDN Systems, Vol. 25, No.1, 1992

[DeBu89] Dembinski, P., Budkowski, S.: *Specification Language Estelle*, in: M. Diaz et al. (eds.), 'The Formal Description Technique Estelle', North-Holland, 1989

[GoRoTh96] Gotzhein, R., Rößler, F., Thees, J.: *Towards 'Open Estelle'*, 6. GI/ITG Workshop 'Formal Description Techniques for Distributed Systems', Erlangen-Nürnberg, Germany, May 1996

[ISO97] ISO/TC97/SC21: *Estelle - A Formal Description Technique Based on an Extended State Transition Model*, ISO/TC97/SC21, IS 9074, 1997

[INT97] INT: *Universal Test Drivers Generator — User Reference Manual,* Estelle Development Toolset 4.1, Institut National des Telecommunications, France, 1997

[ITU94] ITU-T: *CCITT Specification and Description Language (SDL)*, Recommendation Z.100 (03/03), 1994

[LLM91] E. Lallet, Ch. A. Lebrun, J.-F. Martin, *Un outils de génération automatique de l'environnement d'exécution de specification Estelle*, Acte du 'Colloque Francophone sur l'Ingénierie des Protocoles (CFIP'91)', Pau, 17-19 Sept. 1991

[SiSt93] Sijelmassi, R., Strausser, B.: *The PET and DINGO tools for deriving distributed implementations from Estelle*, Computer Networks and ISDN Systems 25, 1993

[The98] J. Thees: *The eXperimental Estelle Compiler — User Manual*, Univ. of Kaiserslautern, Germany, 1998, (see also '**http://rn.informatik.uni-kl.de/activities/Estelle/XEC.html**')

[ThGo97a] J. Thees, R. Gotzhein: *A Formal Syntax and a Formal Semantics for Open Estelle*, Internal Report No. 292/97, Dept. of Comp. Sci., Univ. of Kaiserslautern, Germany, 1997

[ThGo97b] J. Thees, R. Gotzhein: *Generation of Efficient Protocol Implementations - an Experimental Code Generator for Estelle and its Application to XTP,* 6th Open Workshop on 'High Speed Networks', Stuttgart, October 8-9, 1997

[ThGo98] J. Thees, R. Gotzhein: *The eXperimental Estelle Compiler - Automatic Generation of Implementations from Formal Specifications*, Second Workshop on 'Formal Methods in Software Practice (FMSP'98)', Clearwater, USA, 1998

[Wir85] Wirth, N.: *Programming in Modula-2*, Springer, Stuttgart, 1985

[XTP95] XTP Forum, *Xpress Transport Protocol Specification*, XTP Rev. 4.0, XTP Forum, Santa Barbara, USA, 1995

A GRAPHICAL REPRESENTATION AND PROTOTYPE EDITOR FOR THE FORMAL DESCRIPTION TECHNIQUE ESTELLE *

Justin Templemore-Finlayson
Jean-luc Raffy, Pieter Kritzinger and Stan Budkowski

Data Network Architectures Laboratory
University of Cape Town
South Africa
Software-Networks Department
Institut National des Télécommunications
France

jtemplem@cs.uct.ac.za

Abstract:

This paper presents a graphical representation for the Formal Description Technique Estelle, which has up until now been restricted to a textual syntax based on the general programming language Pascal. The graphical syntax and semantics of the representation are fully descriptive of a general Estelle system and complies with the ISO Estelle standard [ISO97].

This representation aims to become a standard technique for formally specifying and simultaneously documenting concurrent, communicating systems. Conventions introduced in existing graphical Estelle tools are re-used, as are graphical concepts which have become familiar through widespread use of the similar Specification and Description Language (SDL).

In addition, we present a syntax-directed editor based on the graphical syntax. The editor enables practical application of the graphical technique, by translating a graphical description to the equivalent textual form which can be input into existing Estelle tools. The reverse function makes it possible to view and navigate existing textual Estelle specifications more effectively.

*This project is partially funded by CNet France, the French Government and the South African Foundation for Research and Development (FRD).

1.1 INTRODUCTION

Estelle is a well-defined Formal Description Technique (FDT) used for defining and analyzing concurrent, communicating systems. It shares this function with a number of other techniques, including the Specification and Description Language [ITU89], LOTOS [BB87] and Petri-nets [Pet62].

Estelle was specifically developed by the International Standards Organisation (ISO) for the purpose of describing the protocols and services of the Open Systems Intercommunication Model [ISO94]. It is also an implementation-oriented language based on Pascal [ISO83] which can be used as a basis for reliable protocol and service implementation. Estelle is standardized by the ISO and the latest standard ([ISO97]) was produced last year.

Estelle is however considered to be lagging behind other Formal Description Techniques such as the SDL, due to its lack of a standard graphical representation. SDL which is very similar to Estelle experiences much greater popularity, and one reason for this is considered to be the existence of a standard graphical notation and support tools for SDL.

A graphical technique is usually preferable to a textual one due to its intuitive and pictorial nature. A graphical representation is able to better structure conceptual ideas than pure text, and is closer to the abstract meaning of that which is being described.

In fact, graphical techniques have been in use since long before the development of early digital computers, whose limitations of graphical hardware and software forced the development of textual techniques. Modern graphical interfaces have removed this limitation, and graphical techniques have once again become popular.

A graphical technique can be used in two ways in the context of concurrent, communicating systems:

- to statically *define* the unique structure specifying the functional behaviour of a system, or to
- dynamically *describe* (as opposed to "define") the non-deterministic behaviour of the system.

Techniques which fall into the first class have a formal syntax and semantics, and can be used for formally defining systems. This is not true of methods in the second class, which can only describe behaviour instances of a system, i.e. the results of a static definition.

In this paper we introduce a standard, *definitive* graphical technique for Estelle which is currently being developed. We refer to this notation as the **Estelle/GR** (graphical representation), as opposed to the **Estelle/PR** (phrasal representation as defined in the ISO Estelle standard. The Estelle/GR has a formally-defined syntax and semantics, and is compliant with the ISO Estelle/PR definition.

In addition, the second part of this paper presents a prototype editor based on the Estelle/GR.

The editor makes use of the features of modern graphical interfaces, such as multiple, linked windows and virtual scrolling canvases to provide powerful manipulation of large graphical documents.

The editor combines the utility of the graphical technique with the power of existing Estelle/PR software tools, by providing translation of the graphical document into equivalent Estelle/PR text. The reverse function furthermore allows the graphical technique to be applied to existing Estelle/PR specification documents.

1.1.1 Existing Estelle visualisations

Only two tools exist which attempt to implement graphical techniques for Estelle. Neither of these tools attempts to produce a standardized notation, and neither provides the ability to graphically define a general Estelle specification.

P. Amer *et al* present **GROPE** in [NP93, NP90]. GROPE simulates the execution of a general Estelle specification graphically. It which uses a *descriptive* notation to present the dynamic state of an Estelle system and animates changes in this state. It visualizes the structure of the specification using the common SDL-like nested box notation, and describes each internal module behavior separately in the form of a simple finite state automaton. We re-use some of the abstraction concepts of the GROPE automaton description in the Estelle editor. GROPE is not publicly available.

ASA+ is maintained by Verilog, and is an editor for a non-standard variation of Estelle called LSA+. It uses a partially graphical notation which defines the internal structure of an individual module using a nested box notation, and specifies hierarchical module relationships in a tree form, but specifies internal module behavior using a modified Estelle/PR syntax. The usefulness of ASA+ is limited by the its non-standard extensions and restrictions, and it is not widely used.

1.1.2 Paper outline

Section 1.2 of this paper is a brief introduction to the Estelle model, which provides a starting point for describing graphical representation and structure of the editor. **Sections 1.3, 1.4 and 1.5** introduce the main points Estelle/GR syntax using the Trivial File Transfer Protocol to provide examples. **Section 1.6** discusses the main functions of the prototype editor, and includes some screenshots showing it in use. **Section 1.7** concludes.

Space considerations unfortunately prevent the formal definition of the full Estelle/GR syntax and semantics. Interested readers however are welcome to contact the primary author for further information.

1.2 THE ESTELLE MODEL

An Estelle specification describes a hierarchically structured system of non-deterministic sequential components exchanging messages through bidirectional links between their ports.

Each component is an *instance* of a *generic* module definition. A generic module statically defines a *header* and several associated *body* parts and an instance is dynamically created from a given header-body pair of a generic module definition.

The header part defines the external communication interfaces of the module and specifies the parallel or sequential non-deterministic execution of its child modules; the body part defines its internal behaviour and recursively defines generic children modules.

Estelle uses an extended finite state machine (EFSM) paradigm to describe module behaviour. Each module body defines an EFSM which receives inputs and sends outputs via the communication interfaces of the header. Each transition has a set of conditions which must be satisfied before the transition may fire, and a set of actions which are executed atomically on firing.

An Estelle system has a *dynamic structure* which sets it apart from other FDTs such as SDL, which have a *static structure*. An instance can dynamically create and terminate or release children instances and change their intercommunication structure. As an instance can be instantiated with alternative bodies, it is possible to change the entire sub-hierarchy of instances.

The global behaviour of the system is characterized by the interaction of the executing modules, subject to the concurrency imposed on sibling modules by their parent module, and a priority principle that grants parent instances execution priority over its children.

1.3 THE ESTELLE/GR

The Estelle/GR separates the definition of system structure and behavior, as these two parts are syntactically unrelated. Three notations are used, including two alternative representations for a system structure:

1. A **structure tree diagram** defines a static generic module hierarchy in a tree structure. It can define the dynamic structure of any general Estelle system.

2. An **instance block diagram** defines the initial instance hierarchy and communication structure of an Estelle system which has a statically-determinable initial state. It is used to define Estelle systems limited a static structure, and uses a nested box notation similar to that of ASA+ and SDL.

3. A **transition tree** defines a transition in the internal EFSM of a module body (or instance) of the specification structure. A series of transition trees defines a *forest*, which defines the full EFSM. The transition tree is represented using an SDL-like flow diagram notation.

The editor introduces a further informal **simplified automaton notation** which describes the internal behavior of a module in a simpler automaton form.

A *complete* complete Estelle/GR specification consists of a single **structure tree** or **instance block diagram**, and a **transition tree forest** defining the internal behaviour of each module body or instance occurring in the structure diagram.

In the following sections we define each of these syntaxes using the Trivial File Transfer Protocol to provide illustrative examples. The relative merits of the two structure notations are also discussed. We briefly introduce the TFTP protocol used before continuing.

1.3.1 The Trivial File Transfer Protocol

The Trivial File Transfer Protocol (TFTP) is a simple protocol for reading and writing files across a network. It is defined in [Chi94]. Here we define a version of the TFTP in which the Reader and Writer are implemented as separate 'request handlers' within Initiator and Responder objects, and which are created when the User generates either a read (READ_RQST) or write (WRITE_RQST) request, and destroyed immediately on completion. We only describe and use examples of Writer behavior in this paper.

A User acts as both the Client and a Server. The Client sends requests to the Initiator object, and the Server responds to requests from the Responder object. All network communication takes place between the Initiator and Responder. Our implementation assumes an error-free network link, and connects the Initiator directly to the Responder.

The Client initiates a connection by sending a write request (WRITE_RQST) to the Initiator. If the Initiator is not busy with an existing transfer, it creates a Write Handler, and establishes a link between it and the Responder. The Write handler then forwards the write request to the Responder, which if it is not busy and the Server grants permission, creates a Write Handler to process the request.

The Initiator and Responder Handlers communicate directly to exchange data. Synchronized acknowledgment and timeouts are ensure data integrity. The Initiator handler sends DATA packets, which the Responder handler acknowledges with ACK (acknowledgment) packets. Any errors result in disconnection and the sending of an ERROR packet.

On successful completion of the transfer, or receipt of an error packet, the handlers alert their parent Initiator or Responder objects, which destroy the connections and handlers. The Initiator also receives a result code from the Handler, which it returns to the User in a RESPONSE packet.

1.4 STRUCTURAL REPRESENTATIONS

The idea of representing a structured system of communicating components as a set of of nested boxes connected by vectors representing communication links is a familiar and useful one. Unfortunately, this notation assumes a ***static*** system structure, thus it cannot be used to define a structural dynamic system, such as a general ISO Estelle system.

In order to define a structurally dynamic system, we introduce a notation which we call the *structure tree*. This notation is not as informative as the nested box notation as it cannot make any assumptions about the dynamic instance and communication structure. However, as it is based on generic module definition, can fully define all structural aspects of an Estelle system.

Which notation should be used depends on whether the system structure is *dynamic* or *static*. The nested box notation is more expressive than a structure tree, therefore it is recommended where the system is structurally static. At its most general, it can also usefully describe the initial structure of a dynamic system where this initial structure is statically determinable.

1.4.1 Nested boxes notation

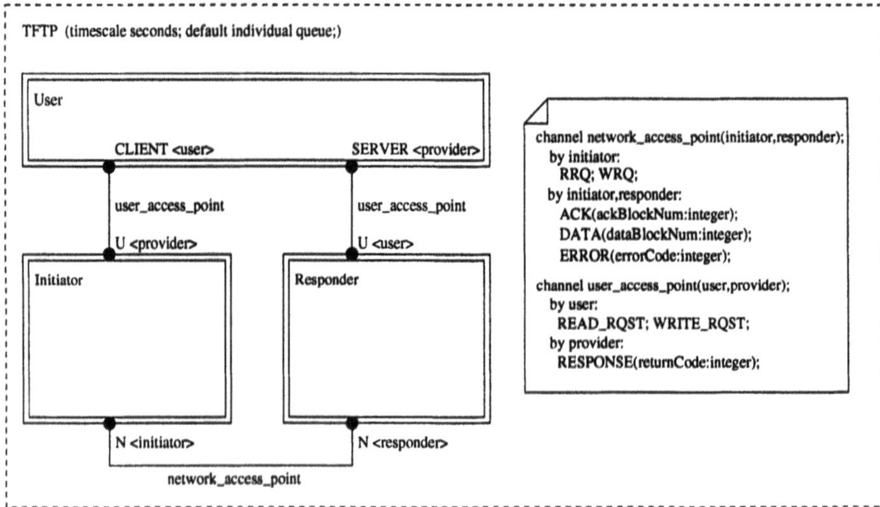

Figure 1.1 The TFTP nested box diagram (showing the initial structure of the dynamic system)

The nested box notation defines system *instances* which effectively have single, statically-determined module header and body parts. For this reason, it cannot define a dynamic system where headers have multiple associated bodies.

The nested box notation thus defines not only the system structure, but also the module body initialization parts required to initialize that structure.

Figure 1.1 shows the nested box representation of the TFTP structure. The nested box can describe the initial structure of this system, as it is statically-determinable. However, it is not *definitive*, as the full definition of the structure tree in Figure 1.3 reveals the definition of descendent modules for both the `Initiator` and `Responder` modules.

In Figure 1.1, each box defines a module instance and the graphical inclusion of one box within another indicates hierarchical substructuring. A box's shape is used to define the mode of concurrency between the included child instances. The diagram shows three systemactivity modules executing asynchronously in parallel, as their parent instance (the specification) is unattributed. The Estelle/GR uses the box shapes shown in Figure 1.2 to define concurrency between children instances.

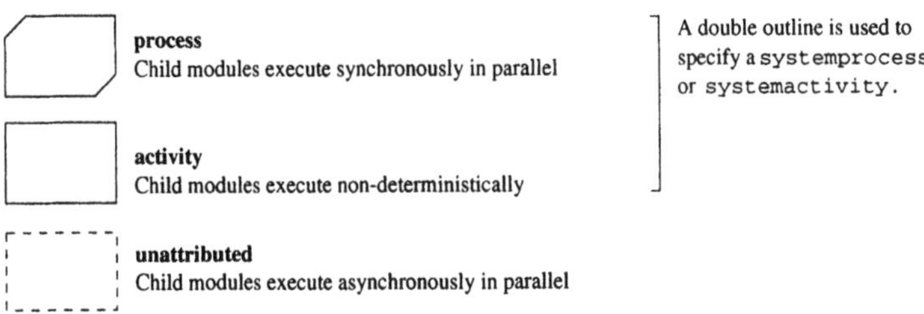

Figure 1.2 Specifying concurrency between children modules in the Estelle/GR

Child instance boxes should not be confused with channel declaration boxes, shown as a rectangle with a 'folded' corner ( ). These define channels for the containing instance.

Communication links. A major feature of the nested box notation is the inclusion of communication link information. Channels and interaction point definitions are semantically grouped by a vector terminating at two "dots". The extracted link attached between the `User` and `Responder` modules in Figure 1.1 has the following characteristics:

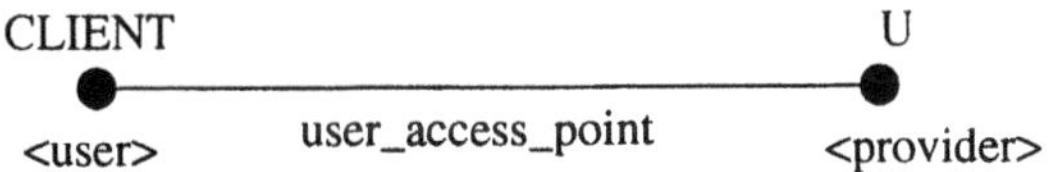

The solid '●' defines an interaction point with an individual queue (a common queue is shown by a '○'). The vector label defines the channel shared by the interaction points, and the identifier in angled brackets defines the role of each interaction point. From these labels it is possible to look up what interactions each interaction point may send in the channel declaration. For instance, interaction point `CLIENT` may send interactions `READ_RQST` and `WRITE_RQST`.

1.4.2 Structure tree notation

A structure tree is a tree of generic modules. A generic module is itself visualized as a two level tree with the module header as root and associated bodies as children. The separation of module headers and bodies makes it possible specify multiple instance sub-hierarchies which the nested box notation cannot.

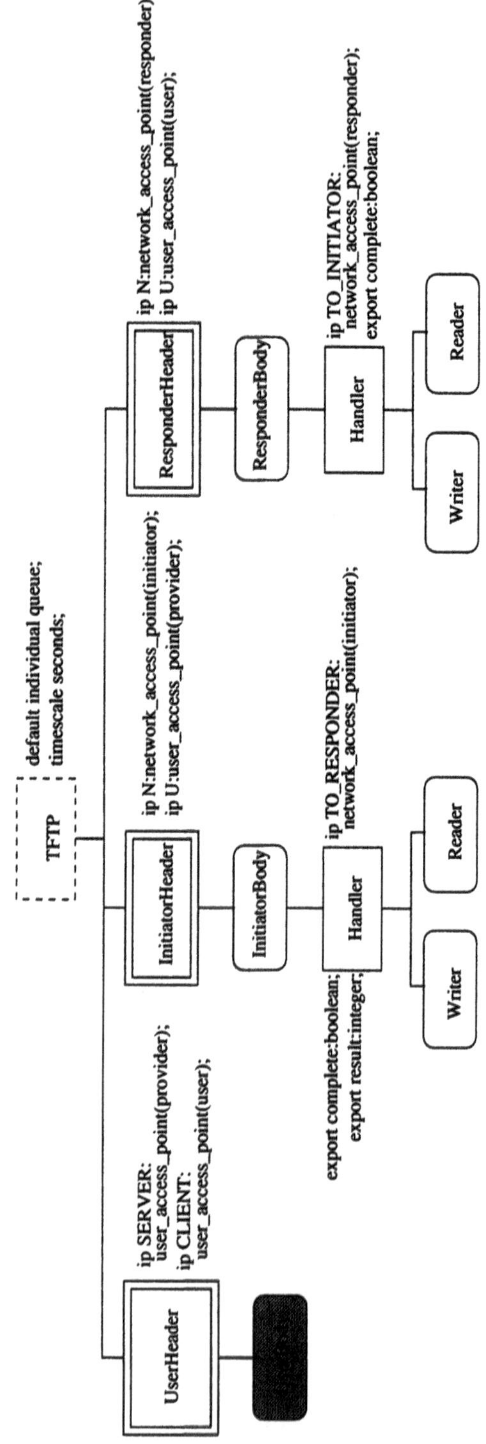

Figure 1.3 The TFTP structure tree

The tree is started using a special specification node and built downwards by adding generic modules as children of the module body in which they would be textually defined. As an example, the TFTP structure tree is illustrated in Figure 1.3.

Generic module representation. A generic module is graphically defined as follows:

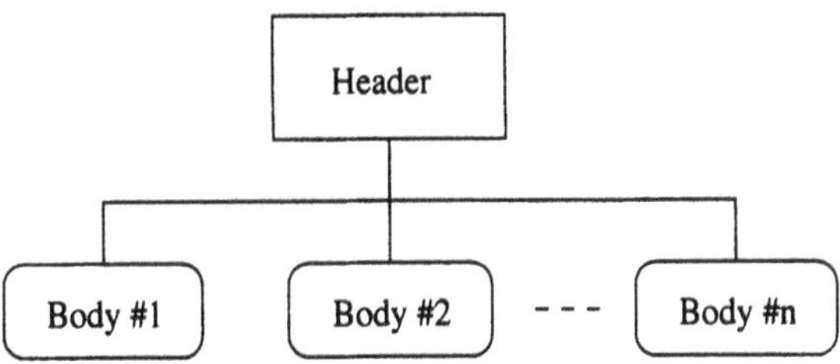

The header definition part is named `Header`, and has bodies `Body #1`, `Body #2`, ..., `Body #n` associated to it. A module instance can be created from the pairing of `Header` with any of the bodies during execution.

A module body is always represented by a rounded rectangle, but a module header may be any of the nested box shapes listed in Figure 1.2, which determine the mode of concurrency between child modules.

Communication links. The structure tree cannot represent communication links as it defines a structurally *dynamic* Estelle system, in which communication links are only created during execution

The structure tree is limited to defining the static *communication interfaces* of module headers. These consist of `interaction points`, `exported variables` and `parameters`, and are shown as textual annotations of a module header node.

This can be awkward in even a simple protocol such as TFTP, as is obvious in Figure 1.3. This is one reason the module instance diagram is recommended to be used whenever possible.

1.5 BEHAVIORAL REPRESENTATION

The Estelle/GR uses a flow diagram notation to define individual transitions in the extended finite state machine making up the internal behavior of a module instance. This notation has been widely popularised by SDL, is easy to understand, and describes all details of a transition definition.

A flow diagram defining a transition in the Estelle/GR is called a *transition tree*, and the set of transition trees making up an EFSM definition is called a *forest*. Figure 1.4 shows a small forest of transition trees.

The basic structure of a transition tree is shown in Figure 1.5. The origin and destination states, the conditions necessary to fire the transition and the

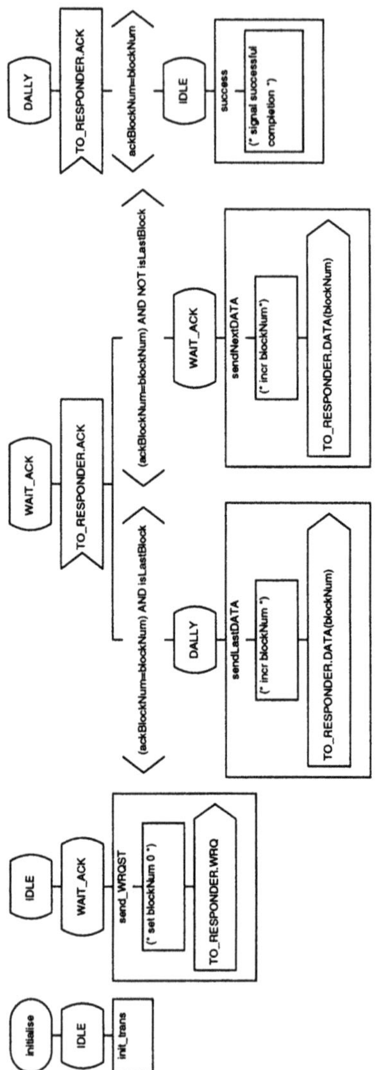

Figure 1.4 A subset of the forest of transition trees which define the `Writer` Handler of the Initiator (Figure 1.12 shows the simplified automaton of this forest)

actions taken on firing are explicitly shown as individual symbols in a flow diagram. The different parts of a transition tree are described below.

1.5.1 FROM and TO symbols

The FROM and TO symbols may be included as condition clauses in the condition part of the transition definition to allow nesting on common origin and destination states (see the section on nested transitions below). All the tran-

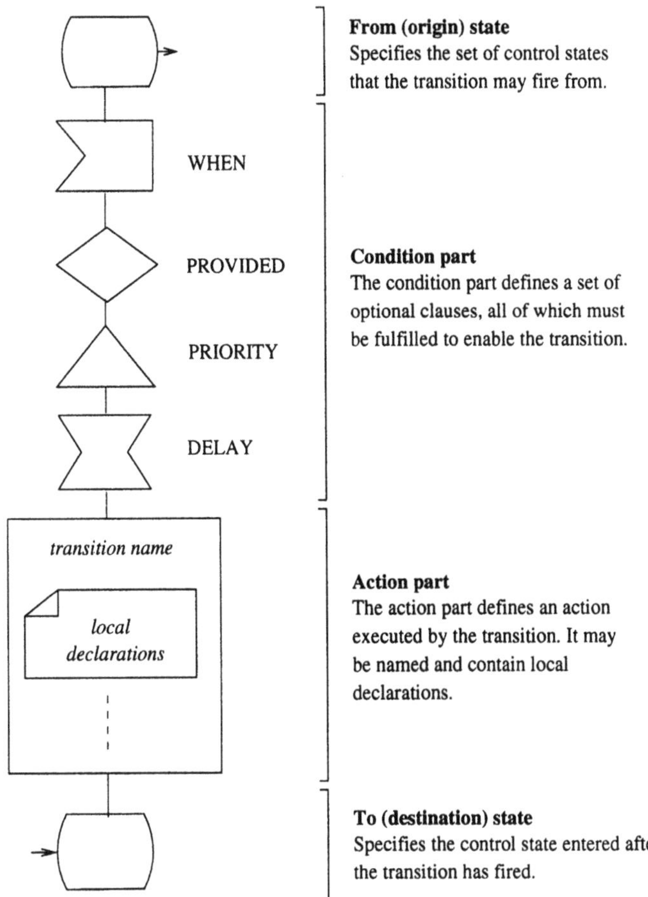

Figure 1.5 The structure of an Estelle/GR transition tree

sition examples in this section define the TO state as part of the condition part.

1.5.2 Condition part

The condition part is made up of optional condition symbols (listed in Figure 1.6) each of which defines a condition which must be satisfied before the transition may fire.

These conditions may occur in any order, but only once each. By treating the `FROM` symbol as a condition, it is possible to start the transition with any of the conditions listed in Figure 1.6. Combined with branched transition trees, this allows a number of transitions which respond to a common event to be semantically defined together (see the section on nested transitions below).

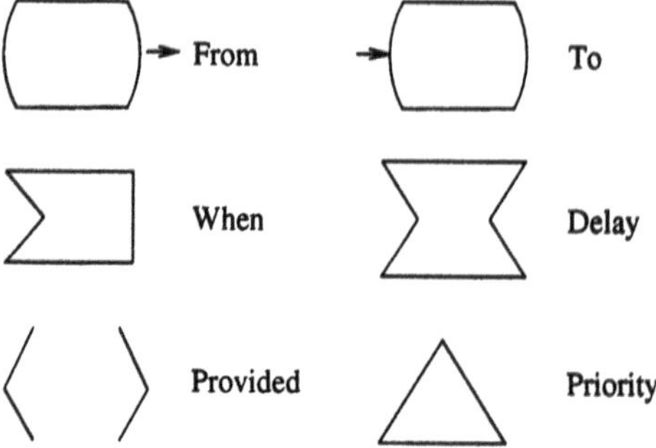

Figure 1.6 Estelle/GR condition symbols

1.5.3 Action part

The action part is separated from the condition part by a solid box, which also indicates that the enclosed actions are atomically executed. Transition naming is handled by the labeling of this box. The action part also has a scope in which declarations can be made. This is done textually within a graphical declaration symbol ( ), which is included in the action part box.

Unlike the condition part symbols, action symbols (listed in Figure 1.7), may be repeated. In the Estelle/GR we only define action symbols for Estelle-specific extensions to Pascal, and native Pascal statements are included in the graphical text symbol, . There are two exceptions to this rule:

- The Estelle `exist-one` expression has no graphical representation, as it is used as an operand in boolean expressions;

- A Pascal `procedure call` is shown explicitly, due to its value in abstracting detail.

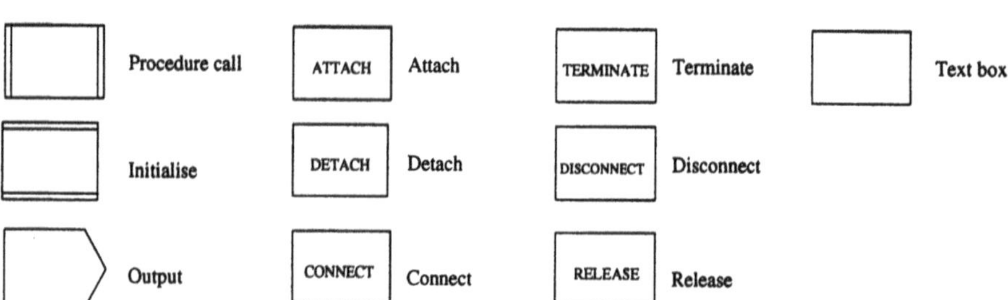

Figure 1.7 Estelle/GR action symbols

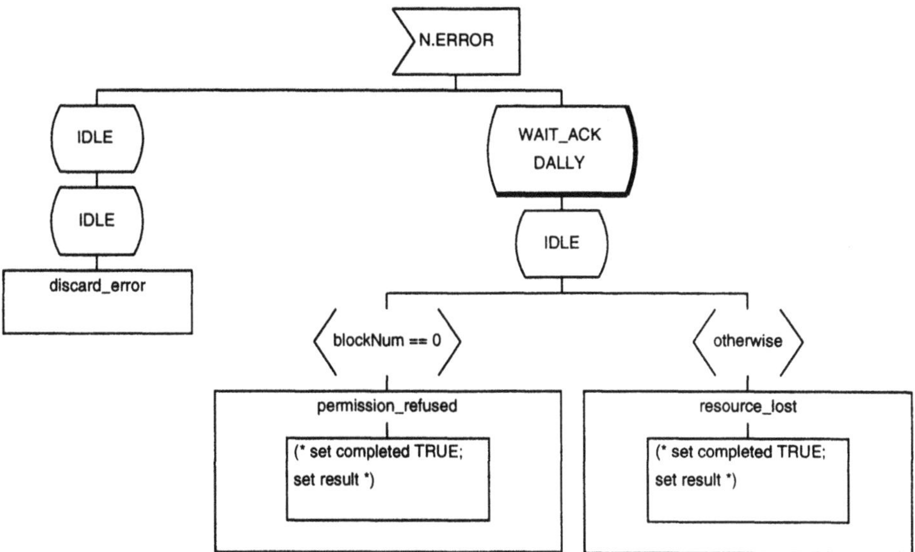

Figure 1.8 A transition nested on a WHEN condition, showing the specification of error recovery in the `Writer` handler of the `Initiator` module of the TFTP. The enboldened FROM state indicates that the transition may fire from more than one state in the EFSM

1.5.4 Nested transition trees

The term *transition tree* is used as an Estelle transition may be split into branches within the condition part. A branched transition tree defines a *nested* transition, as informally defined in the Estelle ISO standard. Within a branched transition tree, the conditions above a split apply to each branch created by the split.

The advantage of introducing branching into a transition tree is the ability to semantically group together transitions which result from a common event. For example, in the TFTP, error packets may be received at any time, in any state. Rooting a transition tree at a 'WHEN ERROR' condition, and branching the transition tree with different actions dependent on the current state, is a sensible way of specifying error-recovery behaviour. Figure 1.8 shows error recovery in the TFTP `Writer Handler` of the `Initiator`. The single transition defines three error-handling procedures depending on the state the machine is in, in a single transition tree definition. This is semantically more meaningful than defining three procedures in separate, not obviously related transitions.

1.5.5 The initialize transition

The initialize part of a module behaviour part is represented as a transition tree starting by a special initialize symbol, (initialise). The initialize transition for the TFTP specification module is shown in Figure 1.9. This initializes the

`User`, `Initiator` and `Responder` instances at the system level, and establishes the link between the `Initiator` and `Responder`.

Note that this initialization *is* statically determinable (there is only one configuration possible from the initialize transition), which is why the initial structure of the TFTP protocol can be described by an instance block diagram: the instance block diagram effectively interprets this initialization part However, when used to define a system, the instance block diagram implicitly defines this initialization part.

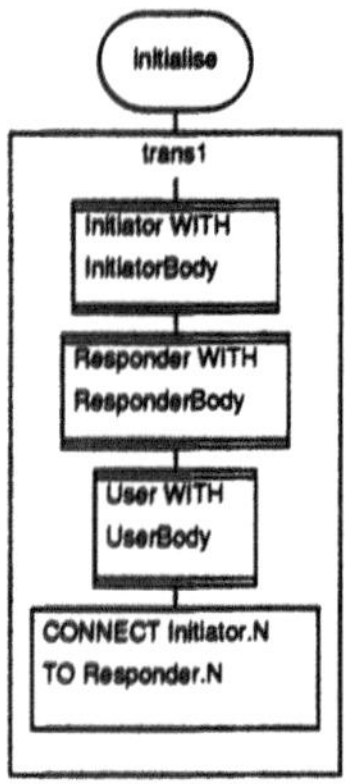

Figure 1.9 Initialize transition for the TFTP specification module

1.5.6 Special considerations

Estelle `FORONE` and `ALL` statements are represented as complex statements of other Estelle/GR symbols. Figure 1.10 show the templates for these constructs.

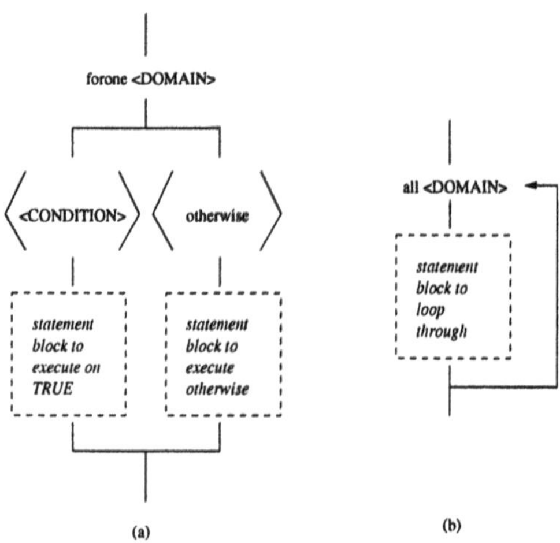

Figure 1.10 Estelle/GR complex statements: (a)`forone` and (b)`all`

The ANY macro is represented in a transition tree using a special symbol, ⏢. It defines identifiers which are available to all symbols following it in the transition tree.

1.6 THE PROTOTYPE EDITOR

The prototype editor has two main functions:

- It is a syntax-directed editor for the definition of a specification using the Estelle/GR syntax.
- It translates a graphical Estelle/GR document into Estelle/PR code.

These two functions combined enable the editor to turn the Estelle/GR into a powerful graphical technique that can make use of existing tools supporting the Estelle/PR technique.

The editor also provides a navigational aid for complex module body definitions in the form of a *simplified automaton* view.

1.6.1 Editor structure

The editor provides separate interfaces for the definition of system structure and module behaviors.

Launching the editor opens the structure editor. The structure editor opens a single specification at a time, and controls the usual functions of file handling and printing, as well as the import and export of Estelle/PR specification documents. The structure editor is used to define a specification structure in the general *structure tree* syntax of the Estelle/GR. The *instance block* notation is not supported.

For each module body of the structure tree defined, a behavior editor can be opened in which the behavior of that module body is defined using the Estelle/GR *transition tree* syntax. Multiple behavior windows can be open simultaneously, making concurrent development of bodies possible, and enabling 'cut, copy and paste' editing between two bodies. The structure tree in the permanently open structure editor is a useful navigational aid for moving between a number of module body definitions.

The structure and behavior editors are very similar in presentation, and a screenshot of the behavior editor is presented in Figure 1.11.

1.6.2 Simplified automaton view

The behavior editor can display a simplified automaton view of its behavior definition in a separate window. Figure 1.12 presents the *simplified automaton* view of the set of transition trees defined in Figure 1.4.

The automaton is *simplified* as it does not display any transition information, and aggregates all transitions between two states into a single *macro-transition*, as defined in the work of GROPE [NP93, NP90]. The simplified EFSM therefore

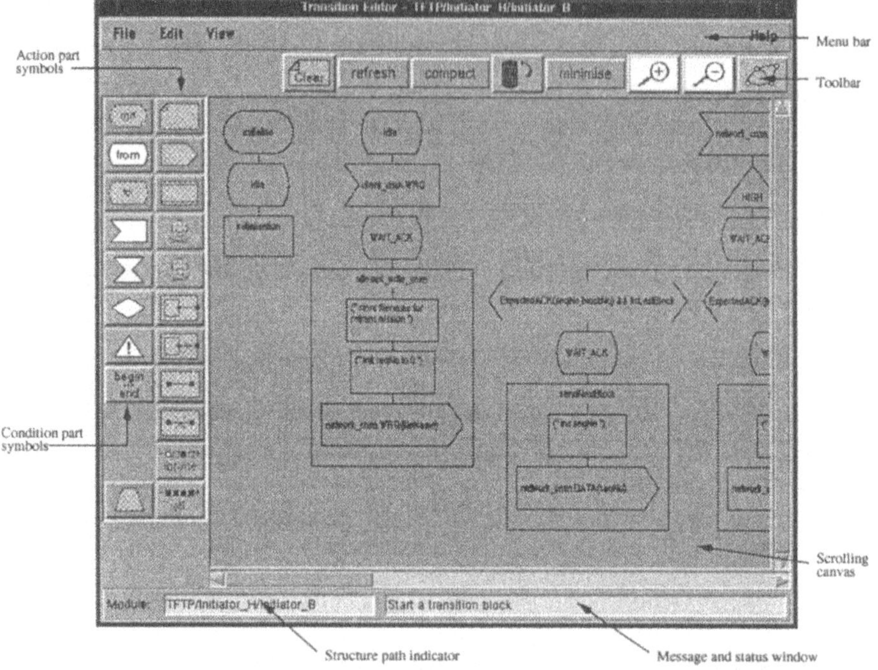

Figure 1.11 **Screenshot of the behavior editor window.**

consists of all states, but with at most one arc in each direction between each state pair, labeled with the names of the individual transitions aggregated on that arc.

The utility of the simplified automaton view is derived from the linking of each transition names in this view with its full transition tree definition in the behavior editor: When the transition name is selected in the automaton view, the transition definition is centered and highlighted in the behavior editor.

A module behavior definition may stretch over several virtual screens, whereas the simplified automaton will usually fit on one. This makes the latter a useful navigational aid when defining or exploring a module body behavior. In its simplified form it provides an overview of the transition definitions, and by way of the dynamic links, functions as an 'index' into the module body definition.

1.6.3 Ongoing work

The editor currently cannot import Estelle/PR specification documents. The instance block notation for specifying structurally static systems is not supported. These features are under development.

The editor will ultimately be integrated into the XEdt toolkit, making seamless use of the graphical technique and its set of Estelle/PR-based tools possible.

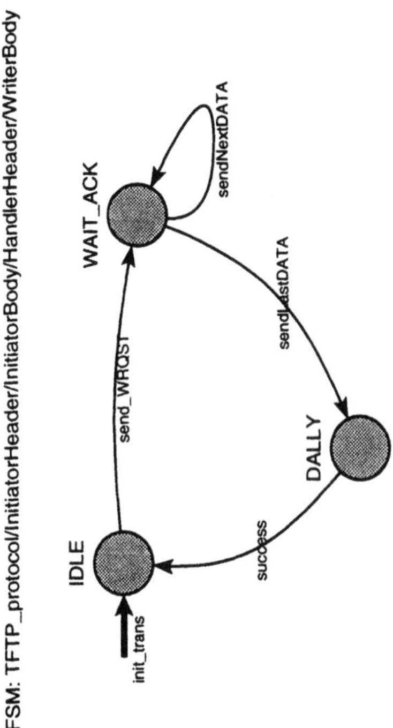

Figure 1.12 The simplified automaton view of the transition tree definitions in Figure 1.4 (Displayed sideways for easier comparison)

1.7 CONCLUSION

The recently completed MIL-STD-188-220A [FAS+97] Estelle specification is one of first Estelle descriptions to be included as an official part of a commercial formal specification. To encourage further acceptance of the Estelle technique, we believe that it is essential to make it accessible to a wider range of non-technical users. We believe that a graphical syntax and support tool is an important contribution to this process.

In the process of developing a graphical representation for Estelle, we found that it is more complex than similar techniques like SDL, as it permits a system to dynamically change its structure. This structural dynamicity means that the common "nested box" structural representation is not expressive enough to define an Estelle system, and a less expressive but more general *structure tree* notation had to be introduced. We retain the "nested box" notation in the

Estelle/GR though, for use the large subset of Estelle specifications which do not have a dynamic structure and for which it is definitive.

We believe that we have created a graphical syntax which will be immediately familiar to existing Estelle users, being built on the knowledge of existing Estelle and other FDT notations.

The functionality of the editor, and integration with existing Estelle/PR should also make the graphical technique a powerful and practically useful tool.

Acknowledgments

The authors wish to thank the anonymous reviewers for their constructive criticisms.

References

[BB87] T. Bolognesi and E. Brinksma. Introduction to the ISO specification language LOTOS. *Computer Networks and ISDN Systems*, 14(1), 1987.

[Chi94] Noel Chiappa. The TFTP protocol (revision 2). July 22, 1994.

[FAS+97] M Fecko, P Amer, A Sethi, M Uyar, T Dzik, R Menell, and M Mcmahon. Formal design and testing of mil-std 188-220 based on estelle. In *Proceedings of MILCOM '97*, Monterey, California, November 1997.

[ISO83] *ISO/IEC 7185, Information Technology - Programming Languages - PASCAL.* International Standards Organisation, 1983.

[ISO94] *ISO/IEC 7498-1, Information Technology - Open Systems Interconnection - Basic Reference Model: The Basic Model.* International Standards Organisation, 1994.

[ISO97] *ISO/IEC 9074(E), Estelle: A Formal Description Technique based on a finite state transition model.* International Standards Organisation, 1997.

[ITU89] *SDL, Specification and Description Language, Recommendation Z100.* International Telecommunications Union, Geneva, 1989.

[NP90] D. New and Amer P. Adding graphics and animation to Estelle. *Information and Software Technology*, 32(2):149–161, 1990.

[NP93] D. New and Amer P. Protocol visualisation in Estelle. *Computer Networks and ISDN Systems*, 25:741–760, 1993.

[Pet62] C.A. Petri. *Kommunikation mit Automaten.* PhD thesis, Universität Bonn, 1962.

Justin Templemore-Finlayson graduated in the degree of Bachelor of Business Science and Computer Science from the University of Cape Town in 1996. On graduation he won a scholarship to follow an internship at the Institut National des Telecommunications in Evry, where he met Stan Budkowski and Jean-luc Raffy and who introduced him to the world of Estelle. This relationship developed into a Master of Science degree to develop a formal graphical representation for Estelle.

This research is supervised by Prof PS Kritzinger of the Data Network Architectures Laboratory in South Africa, and will be completed by the end of 1998.

PART TWO

Verification 1

PROTOCOL FEATURE INTERACTIONS

Thomas F. LaPorta, David Lee, Yow-Jian Lin
and Mihalis Yannakakis
Bell Laboratories, Lucent Technologies

Abstract: Feature interaction is one of the most intriguing problems in research and development of telecommunications systems. Features may interact and result in undesirable system behavior. Today the analysis of interactions is conducted in an ad hoc manner by subject matter experts. This leads to time-consuming feature design and testing without any interaction-free guarantee.

We present a formal approach for the detection of feature interactions during feature design, specification, implementation, and integration. We use automata to model switch behaviors and services rendered to users. Given a protocol system specification, we propose detection algorithms and prove their correctness and optimality; they require minimal checking to conclude whether a given system is interaction-free. To aid with the introduction of new features to an existing protocol system, we describe efficient algorithms for the incremental validation of interaction-free conditions. Moreover, given a set of design and integration requirements, we obtain necessary and sufficient conditions to prove that an interaction-free system can be built based on the given requirements. We illustrate various concepts and techniques using telephony examples.

1 INTRODUCTION

Feature interaction is a major roadblock to the development and evolution of telecommunications systems. Typical telecommunication switches in a system support hundreds of features [1], combinations of which are marketed as services. Though some feature interactions are desirable by design, others could arise unexpectedly, resulting in user annoyance or adverse system operation. Given a complex switching system implementation, it is formidable to check for feature interactions. Hence, much effort is devoted during feature design and integration phases to ensure that unintented feature interactions do not exist. Currently subject matter experts conduct such analysis in an ad hoc manner. The task has been difficult due to large number of features on a system and complicated interaction scenarios among them. This leads to time-consuming feature design and testing, yet lack of interaction-free guarantee.

The analysis of feature interactions is a matter of determining if users' expectation of feature behaviors is consistent with that exhibited by a system. Since various combinations of features can be simultaneously in use, an interaction-free system means that the behavior of a feature or feature set must be independent of the presence or absence of any other features. Two questions follow in investigating feature interactions: (i) Given a design specification of a system that offers many features, is the system design interaction-free? (ii) Given a set of feature requirements based on users' expectations, is it possible to design an interaction-free system that meets those requirements?

To validate in an effective manner that a system is interaction-free, a major challenge is to minimize the number of feature sets to be analyzed. Similarly, in adding a new feature sets to an existing interaction-free system, the challenge is to avoid analyzing those subsets of features of which the interaction-free status has not changed.

Approaches to solving the feature-interaction problem has proliferated over the past several years [2, 3, 4]. Many of these efforts apply formal techniques, which range from languages for describing composition and alternation of features [15], to state transitions or process algebra based modeling in conjunction with detection techniques such as reachability analysis or model checking (e.g., see [5, 6, 12, 14]). Nevertheless, they mainly focus on detection mechanisms; the complexity of the above mentioned challenges remains an open issue.

This paper addresses the challenges in the context of the two raised questions. We present a formal approach to specify features from both user and network perspectives, and define conditions that guarantee a system to be free from undesirable feature interactions. Regarding the first question, we study efficient algorithms for feature interaction detection. We derive a necessary and sufficient set of combinations that need to be tested, and show that there is no need to check every possible combination of features. Furthermore, we prove that our algorithm is optimal in a sense that the number of tests conducted is minimal. Depending on how feature sets are grouped, this can significantly reduce the number of tests performed. Concerning the second question, we derive necessary and sufficient conditions for the design of an interaction-free system that satisfies feature requirements. We present an efficient algorithm and show that it is optimal. In an evolving telecommunications system, the system is augmented with new features, and the new features may interact with each other or with existing features. We thus study conditions for interactions due to introducing new features and discuss algorithms for their testing.

The remainder of the paper is organized as follows. In the next section we introduce the framework and present the basic definitions of protocol systems, features and their interaction. In Section 3 we address the issues in validating an interaction-free system. In Section 4 we illustrate our methods on a telephony example. In Section 5 we discuss the design and integration of interaction-free systems. Section 6 deals with evolving systems, followed by our concluding remarks in Section 7. Due to the space limit, we omit all the proofs and intermediate results and refer the readers to [10].

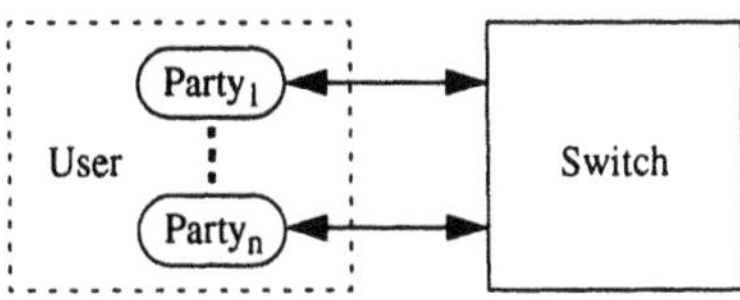

Figure 1 A reference model of a protocol system

2 PROTOCOL SYSTEMS

A communication system provides a set $\mathbf{F}$ of *services*, or *features* to its users. The feature set $\mathbf{F}$ includes the *basic feature* F_0 and a set of *value-added features* F_i, $i = 1, 2, \cdots, N$. Communication between the system and the users takes place via two sets of matching *service primitives*: $S_\mathbf{F} = \{f_1, f_2, \cdots\}$ and $\bar{S}_\mathbf{F} = \{\bar{f}_1, \bar{f}_2, \cdots\}$. The sets $S_\mathbf{F}$ and $\bar{S}_\mathbf{F}$ represent service primitives from the user's point of view and the system's (e.g., switch's or network's) point of view, respectively. Each feature F_i has two associated subsets of matching service primitives $S_{F_i} \subseteq S_\mathbf{F}$ and $\bar{S}_{F_i} \subseteq \bar{S}_\mathbf{F}$ with the property that a primitive f is in S_{F_i} if and only if its matching primitive $\bar{f}$ is in $\bar{S}_{F_i}$. We assume that $S_\mathbf{F} = \cup_i S_{F_i}$ and $\bar{S}_\mathbf{F} = \cup_i \bar{S}_{F_i}$. The switch may have additional *non-service primitives* which are used to communicate with other network entities.

We assume that for each element f in $S_\mathbf{F}$ there is one and only one matching element in $\bar{S}_\mathbf{F}$ and vice versa. The matching relation between the service primitives is for the corresponding *send* and *receive* events, denoted by $!x$ and $?x$ respectively, for a message x between a user and the switch in synchronous communication [7]. The matching relation can be extended from the service primitives to strings and languages over them. Two strings $f_{p_1} f_{p_2} \cdots f_{p_k} \in S_\mathbf{F}^*$ and $\bar{f}_{q_1} \bar{f}_{q_2} \cdots \bar{f}_{q_k} \in \bar{S}_\mathbf{F}^*$ *match* if f_{p_i} and $\bar{f}_{q_i}$ match for $i = 1, \cdots, k$. For a string x over $S_\mathbf{F}$ or $\bar{S}_\mathbf{F}$, we denote by $match(x)$ the string that matches x. For a language L over the alphabet $S_\mathbf{F}$ or $\bar{S}_\mathbf{F}$, let $match(L) = \{match(x) : x \in L\}$. Note that since the matching relation among the service primitives is one-to-one and onto, if $x' = match(x)$ then also $x = match(x')$, and, consequently, if $L' = match(L)$, then also $L = match(L')$.

Based on the network design and user demands, each user is able to use a subset of features in $\mathbf{F}$, including at least the basic feature F_0. Let $\mathcal{F} = \{F : F_0 \in F \subseteq \mathbf{F}\}$ be a class of all the possible feature sets offered by the system (this need not be the complete powerset) for a group of communicating parties; all sets include the basic feature F_0. Figure 1 depicts the relationship between network and users. In our general framework, the term "user" means a group of communicating parties (shown as a dash box in Figure 1); the features set of a user therefore refers to the combinations of features used by *all* communicating parties. We also discuss a special setting in which a feature set is the set of features available to each communicating party.

We model the behavior of each feature set F by a *user automaton* A_F with input symbols S_F. We model the switch behavior by a *switch automaton* $\bar{A}$ whose input symbols consists of the service primitives of all the features $\bar{S}_\mathbf{F}$

and possibly other non-service primitives. The automata have an initial "idle" state. All the states of the user and switch automata are considered accepting. We use $L(A)$ to denote the language accepted by an automaton A; in the case of the switch automaton, the transitions on non-service primitives are treated as τ moves, i.e., $L(\bar{A})$ is a language over the alphabet $\bar{S}_{\mathbf{F}}$.

Definition 1 A *protocol system* P on feature sets class $\mathcal{F}$ contains a switch automaton $\bar{A}$ and a set of user automata $\{A_F : F \in \mathcal{F}\}$ (one for each feature set F in $\mathcal{F}$) which communicate with $\bar{A}$ by matching service primitives.[1] □

In order to identify each transition in the switch automaton on a per feature basis, we also tag each transition with the corresponding feature(s) of which the execution traverses the transition. Multiple features may share a transition, i.e., the execution of these features may traverse the same transition labeled with either a shared primitive or a non-service primitive. In this case the transition is tagged by all the features that share the transition. A transition in the switch automaton is F_i-related if it is tagged with F_i. A transition in the switch automaton is F-related if it is F_i-related for some $F_i \in F$. Note that the feature tags are not part of the alphabet for the switch automaton.

Consider the switch behavior with respect to a feature set F in $\mathcal{F}$. If no other features are present, then the switch will traverse only F-related transitions; that is, the other transitions are as if they are missing. This gives rise to the following definition of a restriction operator.

Definition 2 Let $\bar{A}$ be an automaton whose transitions are tagged by the features. The *restriction* of $\bar{A}$ on a set of features F, denoted $\rho_F(\bar{A})$, is the automaton obtained from $\bar{A}$ by deleting all transitions that are not F-related. The remaining transitions retain in their tags only the features from F. □

To the parties that jointly use feature set F, the user automaton A_F should be consistent with the view $\rho_F(\bar{A})$ supported by the switch. Formally,

Definition 3 A protocol system $P = (\bar{A}, \{A_F\}_{\mathcal{F}})$ on feature set class $\mathcal{F}$ is *consistent* if $L(A_F) = match(L(\rho_F(\bar{A})))$ for all $F \in \mathcal{F}$. □

Note that both the switch and user automata can be nondeterministic. There are various notions of equivalence for nondeterministic automata, including language (trace) equivalence, observational equivalence etc. [13]. In this paper we will use language equivalence, mainly for reasons of simplicity; a similar development can be carried out using other equivalences.

Since the matching of service primitives of the user and switch automata is unique, given the switch automaton of a consistent protocol system, the user automata are well defined and unique in terms of language acceptance.

Example 1 Figure 2 shows a switch automaton and a user automaton as part of a protocol system from telephony. We label transitions in a format [!,?]X_Y,

[1]For clarity we omit specification P of the underlining protocol system if there are no ambiguities.

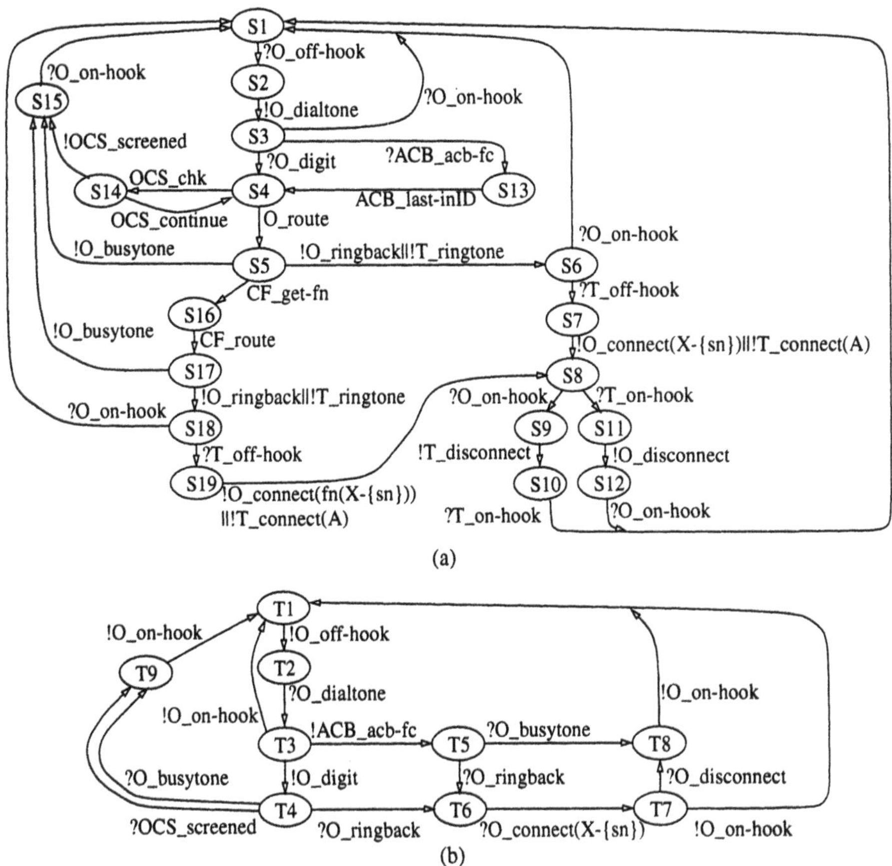

Figure 2 An example of (a) a switch automaton and (b) a user automaton

where X corresponds to a feature tag, and Y is a primitive. The labels preceded with either "!" or "?" are service primitives, and the ones without are non-service primitives. The switch is for calls involving two parties, the caller and the callee. Correspondingly, the basic feature set has two components: the originating POTS (Plain Old Telephone System), denoted O in the figure, and the terminating POTS, denoted T. The switch provides three valued-added features: Originating Call Screening (OCS), Automatic CallBack (ACB), and Call Forwarding (CF). The user automaton represents expected caller behaviors. □

Assume a set of operative features that includes a feature set F and possibly an additional set E. Consider the switch behavior as perceived by an observer who tracks only the communications involving F. We can classify transitions of the switch automaton into three categories: F-related transitions; E-related but not F-related transitions; and the rest (neither F- nor E-related) transitions. The switch now traverses transitions in the first and second categories but not the ones in the third category, as if features in $\mathbf{F} - (F \cup E)$ are not available. Consequently, we can delete the transitions in the third category from the switch automaton without affecting the apparent behavior of the switch. Furthermore, the associated semantics of transitions in the second category are irrelevant to F; hence from F's point of view we can change these transitions to τ moves. This gives rise to the following projection operator.

Definition 4 Let $\bar{A}$ be an automaton whose transitions are tagged by the features. The *projection* of $\bar{A}$ on a set of features F, denoted $\pi_F(\bar{A})$, is the automaton obtained from $\bar{A}$ by changing to τ moves all transitions that are not F-related (the τ transitions retain their tags).

For a pair of feature sets F and E, the *projection* of the automaton $\bar{A}$ on F w.r.t. (or, in the context of) E, denoted by $\pi_{F|E}(\bar{A})$, is the automaton $\pi_F(\rho_{F\cup E}(\bar{A}))$, i.e., it is the automaton obtained from $\bar{A}$ by changing the transitions as follows: (i) All F-related transitions remain unchanged; (ii) All transitions that are E- but not F-related become τ moves; and (iii) All the other transitions (those are not $(F \cup E)$-related) are deleted. □

The restriction operator ρ_F corresponds to the situation where only feature set F is available. The projection operator π_F corresponds to the situation where one is observing the system behavior only from F's viewpoint, though other features may be available and active. The operator $\pi_{F|E}$ combines the two cases.

The restriction operator ρ_F is similar to that of *protocol pruning* [11] where the transitions labeled with service primitives not in the feature set F are deleted. Furthermore, only the strongly connected component with the initial state is maintained for the protocol operations. The projection operator is different from *protocol projection* [9], which is intended for protocols with several distinguishable services. The goal there is to construct image protocols for each service that preserve safety and progress properties. In [11] the projection is done by aggregating states into blocks (and message values into blocks).

We also define our restriction and projection operators in languages.

Definition 5 Given a string $x \in S^*_{\mathbf{F}}$ ($\bar{x} \in \bar{S}^*_{\mathbf{F}}$) and a subset of features $F \in \mathcal{F}$, the *projection* of x ($\bar{x}$) onto F, denoted by $\pi_F(x)$ ($\pi_F(\bar{x})$), is obtained by deleting all the symbols in x ($\bar{x}$) that are not in S_F ($\bar{S}_F$). The definition can be extended to languages in a straightforward way: $\pi_F(L) = \{\pi_F(x) : x \in L\}$. The *restriction* of L on F, denoted $\rho_F(L)$, is the set of all strings of L that contain only symbols in $\bar{S}_F$ (or in S_F for a language on the user side). The projection of L on F w.r.t. E is $\pi_{F|E}(L) = \pi_F(\rho_{F\cup E}(L))$. □

Example 2 Figure 3 depicts the projection $\pi_{\{O\}|\{T\}}(\bar{A})$ of the switch automaton in Figure 2(a). It is the basic POTS as seen from the caller. For simplicity we have removed τ transitions and redundant states. □

We now define interactions between feature sets. The definitions of independent feature sets and interaction-free protocol systems follow.

Definition 6 A feature set $F \in \mathcal{F}$ is *independent* of (or *free of interactions* from) another feature set E, denoted as $ind(F, E)$, if $L(\pi_{F|E}(\bar{A})) = L(\rho_F(\bar{A}))$. Otherwise, we say that F has interactions from E. □

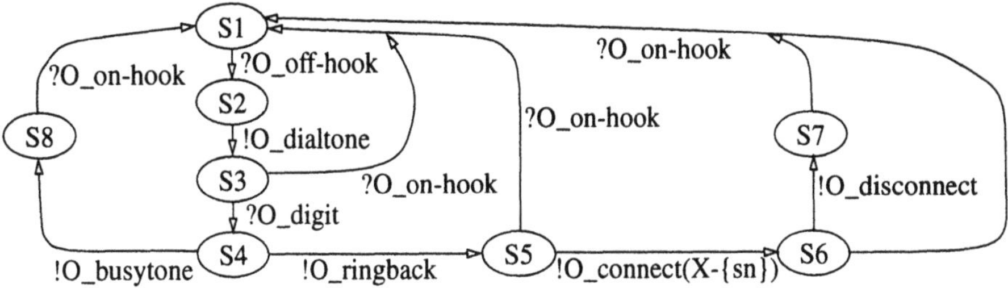

Figure 3 The projection $\pi_{\{O\}|\{T\}}(\bar{A})$ of the switch automaton in Figure 2(a)

Note that this relation is not symmetric; $ind(F, E)$ does not imply that $ind(E, F)$. Moreover, interactions between E and F need not concern us unless $(F \cup E)$ belongs to some feature set offered by the protocol system.

Definition 7 A feature set $F \in \mathcal{F}$ is *independent* (or *interaction-free*) in $\mathcal{F}$ if $ind(F, E)$ holds $\forall E$ such that $(F \cup E) \subseteq H$ for some $H \in \mathcal{F}$. A protocol system P with feature set class $\mathcal{F}$ is *interaction-free* if every set $F \in \mathcal{F}$ is interaction-free in $\mathcal{F}$. □

For a protocol system P, the design and engineering of the system is on the switch automaton $\bar{A}$. On the other hand, the services provided from the user's point of view are specified by the user automata. Hence, analyzing feature interactions means validating that P is consistent and interaction-free. In the next section we investigate the detection of feature interactions.

3 FEATURE INTERACTION DETECTION

Consider a protocol system $P = (\bar{A}, \{A_F\}_{\mathcal{F}})$ on a class $\mathcal{F}$ of feature sets. Checking for the consistency of P is straightforward from the definition: we need to check for each feature set $F \in \mathcal{F}$ that $L(A_F) = match(L(\rho_F(\bar{A})))$. The interesting part is checking for interactions.

The straightforward way to check for interaction-free is to check $ind(F, E)$ for every pair of feature sets, F, E with $(F \cup E) \in \mathcal{F}$. We seek ways to avoid checking all possible pairs. As for the basic operation of checking interactions of some pairs of feature sets, from Definition 6, the computation is reduced to checking equivalence of two automata, which has been well studied in automata theory [8]. In general, it is an expensive operation and we want to minimize the number of such operations in feature interaction detection process. For convenience, we assume that each such operation has a unit cost, and we want to design feature interaction detection algorithms that minimize the total cost.

Intuitively, if there are no interactions from a feature set then there are no interactions from its subsets as well.

Lemma 1 Suppose that F, E, E' are sets of features with $E' \subseteq E$. If $ind(F, E)$ then also $ind(F, E')$. The independence relation ind is transitive on comparable sets: for feature sets $F \subset K \subset L$, if $ind(F, K)$ and $ind(K, L)$ then also $ind(F, L)$. □

We proceed now to study how to check whether a given protocol is interaction-free. We first consider a protocol with two parties, the originating and the terminating party. The same development can be extended along the same lines to the case of multiple parties. Assume we are given a switch automaton $\bar{A}$, and that the features are partitioned into originating and terminating features, there is a class $\mathcal{F}_{\mathcal{O}}$ of originating feature sets and a class $\mathcal{F}_{\mathcal{T}}$ of terminating feature sets. A feature set in $\mathcal{F}_{\mathcal{O}}$ can pair with a feature set in $\mathcal{F}_{\mathcal{T}}$, and vice versa. There is a basic originating feature set F_0 included in all the sets of $\mathcal{F}_{\mathcal{O}}$, and a basic terminating feature set E_0 included in all the terminating feature sets. We wish to determine whether there are any feature interaction between the two parties, i.e. whether the observed behavior by one party for some feature set depends on the feature set of the other party.

Formally, a feature set $F \in \mathcal{F}_{\mathcal{O}}$ is independent of a feature set $E \in \mathcal{F}_{\mathcal{T}}$, denoted $ind(F, E)$, if $\pi_{F|E}(\bar{A}) = \pi_{F|E_0}(\bar{A})$, i.e., the originating side cannot tell the difference between the terminating side using E or its basic feature set E_0 (note that it has to use at least E_0). Similarly, E is independent of F if $\pi_{E|F}(\bar{A}) = \pi_{E|F_0}(\bar{A})$. An originating or terminating feature set is called *externally independent* if it is independent of every feature set of the other party. (Note that it may still have interactions from features of the same side). The protocol system is externally independent if all the feature sets are.

The definition of external independence asks to check two language equalities for each pair of feature sets F, E with $F \in \mathcal{F}_{\mathcal{O}}$ and $E \in \mathcal{F}_{\mathcal{T}}$. We will use the properties from the last subsection to reduce this number.

Definition 8 The *maximal subclass* of a class of feature sets $\mathcal{F}$, denoted by $\mathcal{F}^*$, is a subset of $\mathcal{F}$ such that (i) No feature set in $\mathcal{F}^*$ is a proper subset of any other one in $\mathcal{F}$; and (ii) For each feature set $G \in \mathcal{F}$ there is a feature set $F \in \mathcal{F}^*$ such that $G \subseteq F$. □

For a given class of feature sets $\mathcal{F}$, there is a unique maximal subclass $\mathcal{F}^*$; it consists of the set-theoretically maximal sets of $\mathcal{F}$. It can be easily obtained by eliminating all the elements in $\mathcal{F}$ that is a proper subset of another element. Therefore, for the two disjoint feature set classes for the two parties, $\mathcal{F}_{\mathcal{O}}$ and $\mathcal{F}_{\mathcal{T}}$, we have the two maximal subclasses, ${\mathcal{F}_{\mathcal{O}}}^*$ and ${\mathcal{F}_{\mathcal{T}}}^*$. The feature interaction procedure is summarized in Algorithm 1. Line 2 and 5 call a subroutine $ind(\cdot, \cdot)$, which checks for independence of two feature sets, as in Definition 6. It constructs two automata from the two given feature sets and checks whether they are equivalent or accept the same language with a unit cost.

The following two theorems show the correctness of Algorithm 1, i.e., checking for interaction-free with respect to the maximal feature sets is sufficient.

Theorem 1 A feature set $F \in \mathcal{F}_{\mathcal{O}}$ is externally independent if and only if $\forall E \in {\mathcal{F}_{\mathcal{T}}}^*, ind(F, E)$, i.e., F is independent of every maximal feature set E of $\mathcal{F}_{\mathcal{T}}$. Similarly, A feature set $H \in \mathcal{F}_{\mathcal{T}}$ is externally independent if and only if $\forall G \in {\mathcal{F}_{\mathcal{O}}}^*, ind(H, G)$. □

Algorithm 1 (Feature Interaction Detection - Two Parties)

Input: Feature sets $\mathcal{F}_O, \mathcal{F}_T, \mathcal{F}_O{}^*, \mathcal{F}_T{}^*$.

```
1   for (each pair (F ∈ F_O, E ∈ F_T*)) do
2       if (ind(F, E) = FALSE)
3           return "feature interaction detected";
4   for (each pair (H ∈ F_T, G ∈ F_O*)) do
5       if (ind(H, G) = FALSE)
6           return "feature interaction detected";
7   return "feature interaction free";
```

Thus, for each $F \in \mathcal{F}_O$ ($H \in \mathcal{F}_T$) the cardinality of $\mathcal{F}_T{}^*$ ($\mathcal{F}_O{}^*$) determines the number of feature sets that we must validate against in order to claim that F (H) is independent in $\mathcal{F}$. In the worst case, it is the total number of value-added features in $\mathcal{F}_O$ ($\mathcal{F}_T$); in other words, every feature set in $\mathcal{F}_O$ ($\mathcal{F}_T$) consists of exactly one value-added feature and the basic feature. In the best case, the cardinality of $\mathcal{F}_O$ ($\mathcal{F}_T$) is one, that is, there is a feature set in $\mathcal{F}_O$ ($\mathcal{F}_T$) that contains every originating (terminating) value-added features.

From Algorithm 1, the pairs of feature sets to be checked for independence is $|\mathcal{F}_O| \times |\mathcal{F}_T{}^*| + |\mathcal{F}_T| \times |\mathcal{F}_O{}^*|$ whereas a brute-force checking requires $2|\mathcal{F}_O| \times |\mathcal{F}_T|$:

Corollary 1 Given a class of feature set $\mathcal{F}$. Let n and m be the cardinality of $\mathcal{F}_O{}^*$ and $\mathcal{F}_T{}^*$, and N and M be the cardinality of $\mathcal{F}_O$ and $\mathcal{F}_T$, respectively. To check that every feature set is externally independent, Algorithm 1 reduces the number of tests from $2NM$ to $nM + Nm$. □

Can we further reduce the number of tests for external independence? The following theorem shows that Algorithm 1 is optimal in a sense that it conducts all the necessary checking and, therefore, there is no guarantee of external independence if less tests are performed.

Theorem 2 Let S be a set of pairs (F, E) with $F \in \mathcal{F}_O$, $E \in \mathcal{F}_T$. If S does not contain $(\mathcal{F}_O \times \mathcal{F}_T{}^*) \cup (\mathcal{F}_O{}^* \times \mathcal{F}_T)$, then there is a protocol system P such that $\forall (F, E) \in S$, $ind(F, E)$ holds, but P is not externally independent. □

An important special case is that there is a feature set that includes all the features: $E = \mathbf{F}_T$ and $G = \mathbf{F}_O$. In this case, $m = n = 1$, and we only need to perform a linear number of tests:

Corollary 2 Suppose that there exists a feature set $E \in \mathcal{F}_T$ and $G \in \mathcal{F}_O$ that include all the features, i.e., $E = \mathbf{F}_T$ and $G = \mathbf{F}_O$. Then all the feature sets are externally independent if $ind(F, E)$ for all $F \in \mathcal{F}_O$ and $ind(H, G)$ for all $H \in \mathcal{F}_T$. The number of tests is thus reduced to $M + N$ where and N and M be the cardinality of $\mathcal{F}_O$ and $\mathcal{F}_T$, respectively. □

We come back now to the general framework. We have a communication protocol P, given by the switch automaton $\bar{A}$, with a class $\mathcal{F}$ of feature sets. We

Algorithm 2 (Feature Interaction Detection - General Case)

Input: Feature sets $\mathcal{F}$

1 **for** (each $F \in \mathcal{F}$) **do**
2   construct interaction graph D_F;
3   **for** (each weakly connected component of D_F) **do**
4   find a sink node G;
5   **if** ($ind(F, G) = FALSE$)
6   **return** "feature interaction detected";
7 **return** "feature interaction free";

wish to check if P is interaction-free, without checking all the pairs of feature sets of $\mathcal{F}$.

First note that Lemma 1 implies that we need only perform independence tests $ind(F, E)$, where E is a maximal set of $\mathcal{F}$. For, if E is a proper subset of another set G of $\mathcal{F}$, then we can test instead $ind(F, G)$; if it holds, then also $ind(F, E)$. Recall also that independence of a feature set F requires that $ind(F, E)$ for all sets E such that $F \cup E$ is contained in some feature set G of $\mathcal{F}$. Thus, we only need to test pairs (F, G) of feature sets of $\mathcal{F}$ such that $F \subset G$ and G is a maximal set of $\mathcal{F}$. Do we need to test all these pairs? Not necessarily. We investigate below the pairs that need to be checked. Although the choice of these pairs is not necessarily unique in this case as we shall see, the number of pairs is uniquely determined.

Consider the partial order induced by set containment on the class $\mathcal{F}$ of feature sets. That is, we form a directed graph D with the feature sets as nodes and with an arc $F \to G$ if $F \subset G$. By 'connectivity' below we mean weak connectivity, i.e., the directions of the arcs are ignored and the graph is treated as undirected.

For a node (feature set) F, let D_F be the *interaction graph*, a subgraph induced by all the proper descendants of F, i.e. all nodes corresponding to sets that properly contain F. Let c_F be the number of connected components of D_F. We shall show that $\sum_{F \in \mathcal{F}} c_F$ tests are necessary and sufficient to verify that the protocol is interaction-free.

For each feature set $F \in \mathcal{F}$ and each connected component of D_F, we select a sink of the component G (i.e., G is in the component and is a maximal set of $\mathcal{F}$) to form a pair (F, G). Let S be a set of all such pairs. Note that a component may have several sinks so there is a choice for which one we pick for G to test against F. That is, S is not uniquely determined, although its cardinality is determined, namely $\sum_F c_F$. The procedure is summarized in Algorithm 2 and following theorem proves its correctness:

Theorem 3 If $ind(F, E)$ for all pairs in S, then P is interaction-free. □

We can show again that Algorithm 2 is optimal if we access the switch automaton through independence tests; if we omit any of the tests in Algorithm

2 then there is no guarantee of interaction-fee. We can assume without loss of generality that we test pairs (F, E) with $F \subset E$ feature sets of $\mathcal{F}$.

Theorem 4 Let R be a set of pairs of sets of features. Suppose that there is a feature set $F \in \mathcal{F}$ and a connected component C of D_F such that R does not contain any pair (F, E) with E in the connected component C. Then there is a protocol system P that passes all the independence tests in R but the protocol is not interaction-free. □

An important special case is that there is a feature set that includes all the features: $E = \mathbf{F}$, and it is the sink node of all connected components of the interaction graphs of all feature sets. We only need to check each feature set F against E: K

Corollary 3 Suppose that there exists a feature set $E \in \mathcal{F}$ that include all the features, i.e., $E = \mathbf{F}$. Then all the feature sets are independent if $ind(F, E)$ for all $F \in \mathcal{F}$. The number of tests is thus reduced to N where N is the cardinality of $\mathcal{F}$. □

4 EXAMPLE

Consider the switch automaton $\bar{A}$ given in Figure 2 and assume that a caller can use any combination of the two originating features, OCS and ACB, whereas a callee has the choice of using the terminating feature CF. The feature set class $\mathcal{F}$ in the setting of Section 3 is $\mathcal{F} = \{\{O, OCS, ACB\}, \{O, OCS\}, \{O, ACB\}, \{T, CF\}\}$. The following illustrates how a pair of feature sets can fail the independence checking.

Assume that a user a has the feature set $\{O, OCS, ACB\}$ and another user b has the feature set $\{T, CF\}$. Assume further that a has c on the screening list for OCS, and b has c as the forwarding number for CF. Figure 4(a) and (b) depict the projections $\pi_{\{O,OCS,ACB\}|\{T,CF\}}(\bar{A})$ and $\pi_{\{O,OCS,ACB\}|\{T\}}(\bar{A})$ respectively of the switch automaton in Figure 2(a). The two automata accept the same set of language except that the automaton in (a) also accepts a string ...O_connect(c)... through states S9 and S10 but the automaton in (b) does not. The string corresponds to the case that user a attempts to call b, who in turn forwards the call to c; hence a is connected to c even though c is on a's screening list.

Based on our results in Section 3, Table 1(a) summarizes the independence checks we need to perform before claiming that the feature set is externally interaction-free. In addition, we need to check that OCS and ACB are independent of each other. That is, we need to check for $L(\pi_{\{O,OCS\}|\{ACB,T\}}(\bar{A})) = L(\pi_{\{O,OCS\}|\{T\}}(\bar{A}))$ as well as $L(\pi_{\{O,ACB\}|\{OCS,T\}}(\bar{A})) = L(\pi_{\{O,ACB\}|\{T\}}(\bar{A}))$. According to the general framework in Section 3.3, on the other hand, the context for each independence check will be different. Note that in this setting the feature set class $\mathcal{F} = \{\{OCS, O, T\}, \{ACB, O, T\}, \{OCS, ACB, O, T\}, \{O, CF, T\}, \{OCS, O, CF, T\}, \{ACB, O, CF, T\}, \{OCS, ACB, O, CF, T\}\}$. Table 1(b) lists the six checks we derived from the interaction graph depicted

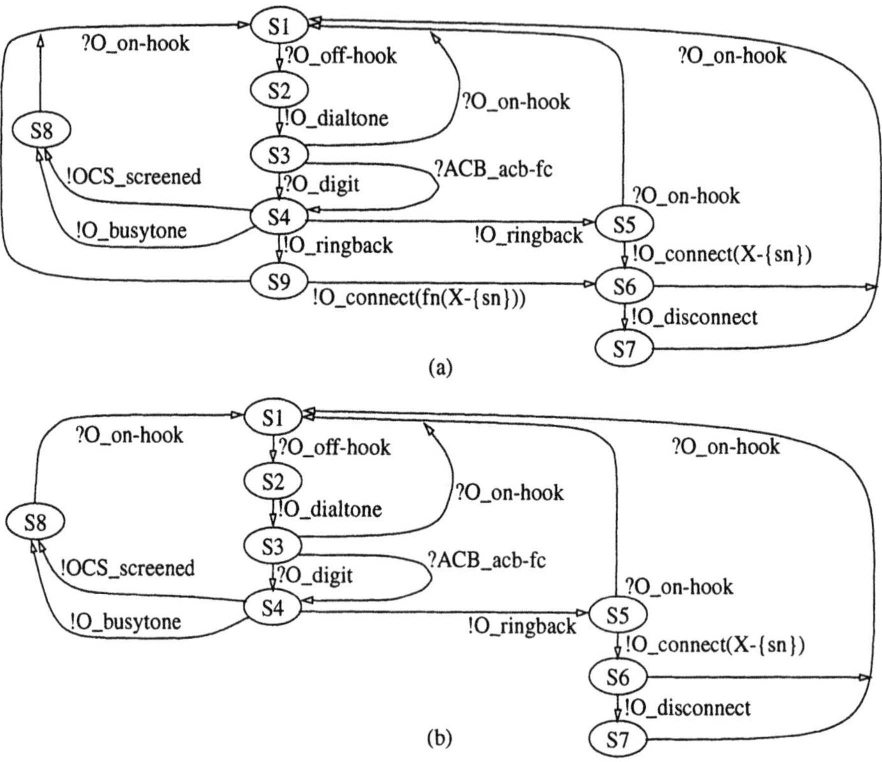

Figure 4 Projections (a) $\pi_{\{O,OCS,ACB\}|\{T,CF\}}(\bar{A})$ and (b) $\pi_{\{O,OCS,ACB\}|\{T\}}(\bar{A})$ of the switch automaton in Figure 2(a)

(a)

1	ind({O,OCS,ACB}, {T,CF})
2	ind({O,OCS}, {T,CF})
3	ind({O,ACB}, {T,CF})
4	ind({T,CF}, {O,OCS,ACB})

(b)

1	ind({OCS,ACB,O,T}, {OCS,ACB,O,CF,T})
2	ind({OCS,O,T}, {OCS,ACB,O,CF,T})
3	ind({ACB,O,T}, {OCS,ACB,O,CF,T})
4	ind({O,CF,T}, {OCS,ACB,O,CF,T})
5	ind({OCS,O,CF,T}, {OCS,ACB,O,CF,T})
6	ind({ACB,O,CF,T}, {OCS,ACB,O,CF,T})

Table 1 A summary of independence checks

in Figure 5. There is a one-to-one correspondence in the semantics of these two sets of checks.

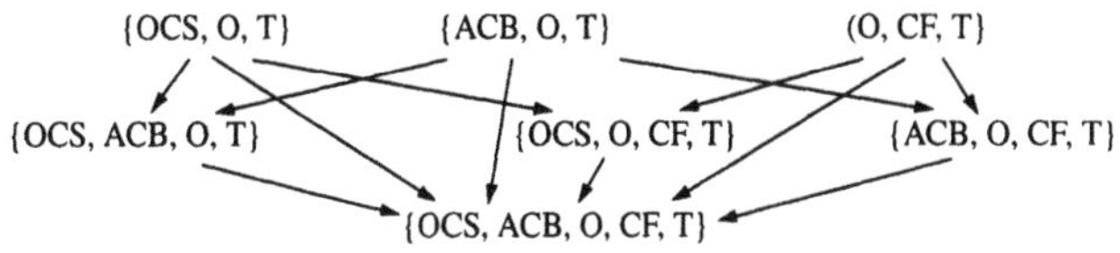

Figure 5 The interaction graph for our example

5 DESIGN AND INTEGRATION OF INTERACTION-FREE PROTOCOL SYSTEMS

In the previous development we have assumed that given a switch automaton and the user automata for the various subsets of features what the users will *actually see* in their interactions with the switch: this is basically what the consistency condition says - the language of a user automaton can be derived from that of the switch automaton. On the other hand, when new features are created, specifications for the features are first created before the features get incorporated in the switch. It is desirable to analyze feature interactions at this stage of design, and indeed typically during the generation of requirements for new features, system engineers analyze and list interactions with other features. In this point of view, we consider user automata (or languages) for various subsets of features to be specifications of what the users are *supposed to see* for different feature sets - i.e. what is desired. We wish to analyze the interactions of the features on the basis of these specifications.

A similar situation arises if we have different systems that need to be integrated. In this case the automata for different features represent the user views for different protocols that were constructed separately and we need to determine whether it is possible to combine them in a consistent manner without interactions.

We first describe a specification system that is typically constructed from design requirements. We want to check whether it is possible to build a protocol system, an implementation from the specification, that meets the design requirements and is interaction-free.

Definition 9 A *specification system* C for a set of features $\mathbf{F} = \{F_i : i = 0, \cdots, r\}$ is a finite set of user automata $\{A_F : F_0 \in F \subseteq \mathbf{F}\}$, where each automaton A_F has input alphabet S_F. The family $\mathcal{F}_C$ of feature sets for which an automaton is specified may or may not be the complete power set of $\mathbf{F} = \{F_i : i = 0, \cdots, r\}$ of all subsets that include the basic feature F_0. [2] □

Definition 10 Let $C = \{A_F : F \in \mathcal{F}_C\}$ be a specification system over a set of features $\mathbf{F} = \{F_i : i = 0, \cdots, r\}$. The system C is *interaction-free* if there is a consistent, interaction-free protocol $P = (\bar{A}, \{A_F : F \in \mathcal{F}_C\})$ whose user automata are as specified by the system C; we say that the protocol P *implements* $C = \{A_F : F \in \mathcal{F}_C\}$. □

[2]This corresponds to the case that features can be obtained only in certain combinations.

Although we did our analysis before based on the switch automaton, we can carry out a similar analysis equally well based on the user automata. That is, we can extend our methodology and algorithms to apply in the setting as well.

Theorem 5 A specification system $\mathcal{C} = \{A_F : F \in \mathcal{F}_\mathcal{C}\}$ is interaction-free if and only if $L(A_F) = \pi_F(L(A_E)) \subseteq L(A_E)$ for all feature sets $F \subset E$ of $\mathcal{F}_\mathcal{C}$. □

We can show along the same lines as Theorem 3 that it is sufficient to verify the equality $L(A_F) = \pi_F(L(A_E))$ only for the pairs (F, E) of the set S of Theorem 3; the equality then is implied for all other pairs $F \subset E$. Furthermore, clearly the containment $L(A_F) \subseteq L(A_E)$ needs to be checked only for the pairs (F, E) in the Hasse diagram (transitive reduction) of the partial order that is induced by the feature sets in $\mathcal{F}$, i.e., for the pairs $F \subset E$ in $\mathcal{F}$ such that there is no $G \in \mathcal{F}$ with $F \subset G \subset E$.

6 INCREMENTAL VALIDATION

Suppose that we have a consistent, interaction-free protocol system P with switch automaton $\bar{A}$ and user automata A_F, $F \in \mathcal{F}$. There are two possible ways of extending P: adding new features, i.e., modifying the switch automaton $\bar{A}$; or adding new feature sets to $\mathcal{F}$ only. In the following we discuss a general scenario of adding both a new feature and new feature sets.

Consider the introduction of a new feature F_{N+1} to protocol system P. It results in a new protocol system Q, where Q consists of: (i) A new switch automaton $\bar{B}$; (ii) A new class of feature sets $\mathcal{G} = \mathcal{F} \cup \{G_i : F_{N+1} \in G_i\}$; [3] (iii) A user automaton A_{G_i} for each new feature set G_i; and (iv) The old user automaton A_F for every old feature set $F \in \mathcal{F}$. We would like the new protocol to provide the same service for each one of the old features, hence the condition that the user automata for the old features remain the same in Q. To verify that Q is consistent, we need to check again for every old and new feature F that $L(A_F) = match(L(\rho_F(\bar{B}))$. Note that we have to verify again consistency for the old features since the switch automaton has changed to an arbitrary new automaton $\bar{B}$.

For feature interaction detection, we do not need to check from scratch; we can use the fact that the old protocol was interaction-free to avoid checking the old features. The procedure is summarized in Algorithm 3.

Theorem 6 Suppose that a consistent and interaction-free protocol system P of feature set $\mathcal{F}$ is incremented with new feature sets $\mathcal{F} \subset \mathcal{G}$, resulting in a new consistent protocol system Q. Algorithm 3 determines correctly whether the incremented protocol system Q is interaction-free. □

[3] We assume that all feature sets in $\mathcal{F}$ remain in $\mathcal{G}$. Adding the new feature F_{N+1} results in adding a new set of feature sets, each of which contains F_{N+1}.

Algorithm 3 (Feature Interaction Detection - Incremental Validation)

Input: Two feature sets classes $\mathcal{F}$ (old) and $\mathcal{G}$ (incremented);
$\mathcal{G} - \mathcal{F}$ is the set of new feature sets.

```
1   for (each F ∈ 𝓕) do
2       construct interaction graph D_F;
3       for (each weakly connected component of D_F) do
4           if (it does not contain any old feature set node)
5               find a sink node G;
6               if (ind(F, G) = FALSE)
7                   return "feature interaction detected";
8   for (each new feature set F in 𝓖 − 𝓕) do
9       construct interaction graph D_F;
10          for (each weakly connected component of D_F) do
11              find a sink node G;
12              if (ind(F, G) = FALSE)
13                  return "feature interaction detected";
14  return "feature interaction free";
```

7 CONCLUDING REMARKS

We have proposed a formal framework for a study of feature interaction problem. We investigated conditions and algorithms for testing features for freedom from interactions, with an eye towards minimizing the number of tests that need to be performed.

Certain features behave differently by design when other features are present. For example, the Caller ID feature will not display the ID of any caller who subscribes to the Caller ID Blocking feature. As a result, a feature set containing these features will not be independent based on our current definition. One way to address this issue is to modify our definition of independence: a feature set F is independent in $\mathcal{F}$ if the behavior of F is independent of the presence of any feature that F has no desired interactions with. Formally, let D be the set of features such that F and D have desirable interactions. $ind(F, E)$ if $L(\pi_{F|E}(\bar{A})) = L(\pi_{F|(E \cap D)}(\bar{A}))$.

In the algorithms we invoked a subroutine call $ind(\cdot, \cdot)$ to check for interactions. Also we need consistency checking. Formally, they require manipulations on the switch automaton, which is complex and often not available. Instead of explicitly constructing the switch automaton and performing projection and restriction operations for checking, we can apply conformance testing to the switch automaton, which is a "black-box" as follows. From the user automata with features of interest, we expect the structure and hence the languages accepted by the switch automaton for the interaction and consistency checking. Therefore, we can apply a conformance test to the switch automaton to confirm its structure to the expectation and conclude the consistency and the presence or absence of feature interactions.

This work is an attempt to use formal methods to model and analyze protocol feature interactions. We use automata to model both the switch and user behavior. They model the control portions of protocols well, but they are not powerful enough to succinctly model the data portions where variable values determine the system behavior and parameters are passed among system entities in messages. Extended, communicating, and parameterized finite state machines can be use to model more properly the real protocol systems. On the other hand, timers play an important role in determining protocol behaviors, and their presence further complicates the analysis of feature interactions. A lot of issues remain open in the research of feature interaction and much effort is needed before formal techniques can have a real impact on practical systems.

References

[1] 5ESS Switch Feature Handbook, AT&T, September, 1987.

[2] IEEE Communications Magazine, Feature Topic: Managing Feature Interactions in Telecommunications Systems, Ed. N.D. Griffeth, Y-J. Lin, Vol. 31, No. 8, Aug., 1993.

[3] IEEE Computer, Special Issue: Telecommunications - How many Features Can You Add?, Ed. N. D. Griffeth, Y-J. Lin, Vol. 26, No. 8, Aug., 1993.

[4] Proceedings of International Workshop on Feature Interactions in Telecommunications Networks, 1994-1997.

[5] P. K. Au and J. M. Atlee, Evaluation of a State-based Model of Feature Interactions, in [4], pp. 153-167, 1997.

[6] O. C. Dahl and E. Najm, Specification and Detection of IN Service Interference Using LOTOS, in *FORTE VI*, pp. 53–69, 1993.

[7] C. A. R. Hoare, Communicating Sequential Processes, *Prentice-Hall*, New York 1985.

[8] J. E. Hopcroft and J. D. Ullman, *Introduction to Automata Theory, Languages, and Computation,* Addison-Wesley, 1979.

[9] S. S. Lam and A. U. Shankar, Protocol verification via projections, *IEEE Trans. on Software Engineering,* vol. SE-10, no. 4, pp. 325-342, 1984.

[10] T. La Porta, D. Lee, Y.-J. Lin, and M. Yannakakis, Protocol Feature Interactions, *Tech. Mem. Bell Labs., no. 113410-980422/03*, 1998.

[11] D. Lee, A. N. Netravali and K. Sabnani, Protocol pruning, *Proceedings of The IEEE,* Vol. 83, No. 10, pp. 1357–1372, 1995.

[12] F. J. Lin and Y.-J. Lin, A Building Block Approach to Detecting and Resolving Feature Interactions, in [4], pp. 86–119, 1994.

[13] R. Milner, *Communication and Concurrency,* Prentice Hall, 1989.

[14] M. Thomas, Modelling and Analysing User Views of Telecommunications Services, in [4], pp. 168–182, 1997.

[15] K. J. Turner, An Architectural Foundation for Relating Features, in [4] pp. 226–241, 1997.

LIMITS AND POSSIBILITIES OF SEQUENTIAL HASHING WITH SUPERTRACE

Jürgen Eckerle and Thomas Lais
Institut für Informatik, Universität Freiburg,
Am Flughafen 17, Geb. 51, D-79110 Freiburg, Germany
email: {eckerle,lais}@informatik.uni-freiburg.de

Abstract

Partial search methods, like bitstate hashing or supertrace, allow formal verification techniques to be applied to problems which normally could not be solved by exhaustive verification. A high coverage (defined as the percentage of the reachable states actually explored by the verifier) is important since the higher the coverage the lower the probability that a protocol error is not detected. In literature sequential hashing is proposed to improve the coverage of supertrace (i. e. start supertrace repeatedly by using different hash functions). Since supertrace is included in many commercial and noncommercial verification tools, it is important to know where its limitations are and where there is potential for possible improvements.

We present both theoretical and experimental results to measure the quality of sequential hashing with supertrace. Tests made with the SPIN validator with different classes of hash functions show that the additional number of states reached in successive runs decreases rapidly. A coverage of close to 1 is practically impossible. While all the hash functions have the same asymptotical behaviour universal hashing seems to be best.

We present an analysis for the expected coverage of sequential hashing which explains accurately the experimental results. Our new mathematical model shows that the overlap of successive runs is much higher than predicted by former approaches. Additionally, it predicts that simple randomization strategies which in successive runs change randomly both the hash function and the traversal order of the states can increase the coverage significantily. Experimental results show that our new algorithm, *universal supertrace*, which is based on such randomization techniques, outperforms the original supertrace and allows the coverage to be increased by a factor of up to 8-10.

Keywords

Protocol verification, reachability analysis, explicit state enumeration, bitstate hashing, supertrace, sequential hashing, SPIN.

1 INTRODUCTION

In the reachability analysis protocol validation is done by state space search. Usually, the state space explosion makes a full exhaustive search impossible. Partial search methods, like bitstate hashing or supertrace (Holzmann, 1987, Holzmann, 1988), allow the time and space requirements to be reduced efficiently when compared with exhaustive search methods. As a result, the verification of protocol state spaces with orders of magnitude greater than that which can be verified with traditional methods is possible. As a result, supertrace, or bitstate hashing, is included in many tools for automated protocol validation.

Partial search methods risk that a protocol error remains undetected. Holzmann measures the quality of partial search methods by the coverage, where coverage is defined as the percentage of the reachable states actually explored by the verifier. The expected coverage equals the average probability that a particular state is not omitted. Therefore, it is imperative to reach the highest coverage possible.

A natural method of improving the coverage is to start the supertrace algorithm repeatedly with different hash functions. This method, referred to in (Holzmann, 1991) as multiple hashing and in (Holzmann, 1995) as sequential hashing, is supported by the SPIN validator of Holzmann. In theory, sequential hashing allows an expected coverage close to 1 to be achieved provided that the number of runs is sufficiently large. Up until now, it has not been determined how fast this approximation can really be done and which strategy results in the highest coverage. To get a better insight, our paper presents both theoretical and experimental results in order to measure the quality of sequential hashing.

Our tests made within the SPIN environment with different classes of hash functions show that in many cases the number of states additionally reached and checked by supertrace in successive runs decreases rapidly. While all the hash functions have the same asymptotical behaviour, universal hashing seems to be best.

We present a mathematical analysis of the combined coverage of sequential hashing which overcomes the restrictions of former approaches and is more accurate. In addition to that, we prove that the overlap of the nodes generated in the different runs is much higher than assumed in former approaches which consider the generated states as randomly chosen from the set of all reachable nodes. Our model shows that the quality of sequential hashing suffers as a result of the fixed traversal order in which supertrace generates the state space. This drawback is of principal nature and cannot be overcome by better hash functions.

Our model predicts that randomization techniques which change in each run both the hash function and the traversal order in the state space can increase the coverage by several times. We present a new algorithm, universal supertrace, which is based on such randomization techniques. Experimental results made with the SPIN validator show that randomized supertrace allows the coverage to be increased by a factor of up to 8-10 when compared with the original supertrace. We expect that additional improvements are possible using other strategies for changing the traversal order.

This paper is structured as follows. Section 2 reviews the well-known supertrace algorithm resp. bitstate/double bitstate hashing and introduces universal hashing. Our experimental results of sequential hashing with different hash functions are presented in section 3. In section 4 we mathematically analyse sequential hashing and compare our results with earlier ones. In section 5 we discuss our new algorithm, universal supertrace, its implementation and the experimental comparison with the original supertrace. Finally, the conclusion with suggestions for future research follows.

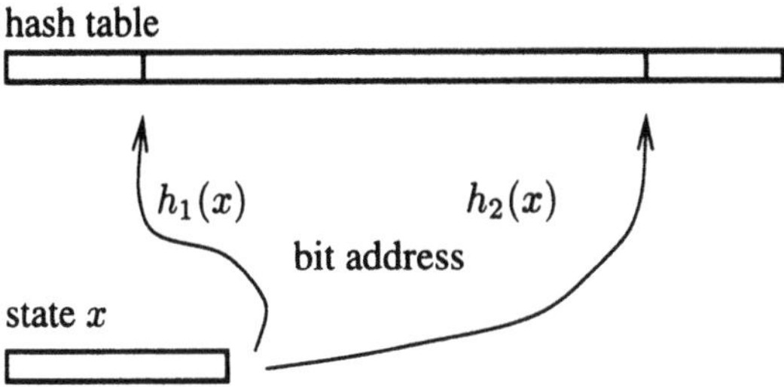

Figure 1 Double bitstate hashing.

2 BACKGROUND

In the reachability analysis the protocol is represented by an implicitly given state space, that is by a set of start states and by a set of possibly applicable transition. Protocol errors can be found by explicit enumeration of all the reachable states, that is by generating the states with either breadth-first or depth-first searches. Duplicate detection with hash tables is a widely used method of reducing the running time. The generated states are stored in a hash table which allows it to be determined whether or not a newly generated state x has been reached before and can be pruned now. Unfortunately, the number of states is often unmanageably large; this is the state explosion problem. For this reason in practice the total memory available for the hash table is often the limiting resource in verification.

2.1 Supertrace Algorithm

The bitstate hashing method or supertrace method introduced by Holzmann, 1988 aims at reducing the memory requirements for the hash table. In contrast to conventional methods where a state descriptor is stored completely, each state is represented by a bit which is addressed by the hash function h. Initially, all the bits of the table are set to 0. When a state x is inserted into the table the bit corresponding to the hash value $h(x)$ is set to 1. When the algorithm generates a new state x and finds that the bit corresponding to this state is already set to 1, it assumes that x has been visited before and prunes x without verification. If the assumption is incorrect an omission of the newly generated state x and potentially all of its successors occurs, since the search algorithm backtracks when it finds a state which has already been visited before. Usually supertrace is combined with a depth-first search with a typical threshold of $10^4 - 10^5$.

The method above is referred to as single bitstate hashing, in contrast to double bitstate hashing (Holzmann, 1991) (see figure 1). In the latter one each state corresponds to two bits calculated by two different hash functions. The probability of an omission error is reduced since a hash collision now requires a collision on both bits. Both space and time requirements are reduced efficiently by supertrace. Space is reduced, because usually the state descriptor uses about 10^3 bits which contrasts with the one or two bits if bitstate or double bitstate hashing is used. The running time is reduced, since duplicate test and insertion is done by a simple bit operation and in addition to this, hash collisions are not resolved.

2.2 Sequential Hashing

The quality of partial search techniques like bitstate hashing or supertrace can be measured by the coverage which is defined as the quotient of the number of reached states and the number of states reachable by an exhaustive search. Since the higher the coverage, the lower the average probability for an omission error, partial search methods should aim at achieving the highest possible coverage.

In (Holzmann, 1991) it is proposed that supertrace be started several times with independent hash functions. It is to be expected that different hash collisions will occur and as a result of this in each run different parts of the state space will be visited. In theory a coverage of arbitrarily close to 1 can be achieved, provided that a sufficiently large number of runs with independent hash functions are done. In the SPIN environment this method is implemented in order to increase the coverage of supertrace.

2.3 Universal Hash Functions

Since hash collisions are not resolved by bitstate hashing, it would be ideal if the used hash function could distribute the keys uniformly on the hash addresses. The key idea of universal hashing is to choose the hash function randomly from a suitable class of hash functions. If this class is carefully selected, a good average distribution can be guaranteed for an arbitrary set of hash keys. A state descriptor can be considered as a vector over the field of two elements. Let $\mathcal{H}$ be a class of hash functions from $\mathcal{A}$ to $\mathcal{B}$ where $\mathcal{A} = \{0, \ldots, 2^k - 1\}$ and $\mathcal{B} = \{0, \ldots, 2^l - 1\}$ for some integers k, l. Recall, $\mathcal{H}$ is *universal*, if $|\{h \in \mathcal{H} | h(x) = h(y)\}| \leq |\mathcal{H}|/|\mathcal{B}|$ for all x, $y \in \mathcal{A}$. That is, $\mathcal{H}$ is universal if no pair of distinct keys collide under more than $(1/|\mathcal{B}|)$-th of the functions. Universal hashing reduces the number of hash collisions on average. More precisely, if $x \in \mathcal{A}$ and $\mathcal{S}$ is any subset of $\mathcal{A}$, then the expected number of synonyms of x in $\mathcal{S}$ is lower than or equal to $|\mathcal{S}|/|\mathcal{B}|$. Note, that two keys x, y are called *synonyms* with respect to hash function h if $h(x) = h(y)$. We present some examples used in our experiments.
Class $\mathcal{H}_1$: Let the class $\mathcal{H}_1$ be defined by

$$\mathcal{H}_1 = \{f_{a,b} | f_{a,b} : x \rightarrow ((ax + b)\ mod(p))\ mod(|\mathcal{B}|), a, b \in \{0, \ldots, p-1\}, a \neq 0\}\}$$

where p is prime, $p > |\mathcal{A}|$. $\mathcal{H}_1$ is universal (Carter and Wegman, 1979). $\mathcal{H}_1$ is suitable for applications where the bit strings which represent the keys can conveniently be multiplied by the computer. If the bit vector is too long as occurs in the protocol domain, then in (Carter and Wegman, 1979) the following construction is suggested .
Class $\mathcal{H}_2$: Let the class $\mathcal{H}_2$ be defined by

$$\mathcal{H}_2 = \{h_{f_1,\ldots,f_r} | h_{f_1,\ldots,f_r} : x = (x_1, \ldots, x_r) \rightarrow f_1(x_1) \oplus \ldots \oplus f_r(x_r), f_1, \ldots, f_r \in \mathcal{H}_1\}$$

where $\oplus$ is the exclusive-or operation. If r is a power of two, then $\mathcal{H}_2$ is (nearly) universal (i. e. $1/|\mathcal{B}|$ has to be replaced by $(1/|\mathcal{B}|) \cdot (1 + (|\mathcal{B}| + 1)/(p - 1))$ in the definition above). If necessary, the state descriptor can be filled with 0's to increase the length to a power of two. In our implementation we divided the bit vector x in integer blocks and selected $p = 2^{31} - 1$ for $\mathcal{H}_1$.
Class $\mathcal{H}_3$: Let $\mathcal{M}$ be the set of Boolean matrices with l columns and k rows. Then we define

$$\mathcal{H}_3 = \{f_c | f_c : x \rightarrow x * M_c, M_c \in \mathcal{M}\}$$

where $*$ denotes the matrix multiplication. $\mathcal{H}_3$ is also universal (Carter and Wegman, 1979).

hash class	(a)	(b) ($\times 10^{-2}$)	(c)
H2	25.76	56.06	2.084
H3	19.43	46.84	1.88
HC	3.72	14.95	1.12
HP	4.57	18.30	1.13
HO	1.00	4.53	1.00

Figure 2 Measured time for hash computation of class X relative to class HO (originally in SPIN) (a), relative to the verification time with class X (b). Verification time with class X relative to verification time with class HO (c).

3 EXPERIMENTAL RESULTS

In this section we present some experimental results made with SPIN. The experimental evaluation is based on the PFPT-protocol which was also used in (Holzmann, 1995) and is provided by the SPIN environment. To evaluate sequential hashing, we measured the coverage in a sequence of tests in which the hash class and the parameters like state space size n, hash table size m and the number of runs q has been varied. In our experiments we used supertrace with the partial-order reduction methods supported by SPIN. This reduces the search effort, but has no effect on the evaluation.

The following hash classes were used: The hash functions originally included in SPIN which are based on an efficient implementation of a CRC-calculation (here denoted by **HO**) (for more details see (Holzmann, 1991)), class $\mathcal{H}_2$ (**H2**), class $\mathcal{H}_3$ (**H3**), a class based on hashing by cyclic polynomials (**HC**) (Cohen, 1997) and a class based on hashing by general polynomial division (**HP**) (Cohen, 1997). In our release of SPIN (version 2.9.5) the number of CRC-polynomials is restricted to 32. Thus, we extended SPIN in order to supply a sufficiently large set of randomly generated CRC-polynomials (**HZ**).

Figure 2 gives an overview of the running time necessary for the different hash functions. It shows that universal hashing doubles the running time when compared with CRC hashing (HO). We observed that the measured coverage of a single run differs only by a few percent between the classes, whereas when the number of runs is increased class $\mathcal{H}_2$ works significantly better than all the others, especially if $n \gg m$ (this is the relevant case in practice). In our tests n is increased by simply increasing the depth-first threshold. The improvement by $\mathcal{H}_2$ grows as (n/m) or q grows (compare fig. 3-up and fig. 3-down). For $n/m \approx 1/250$ and $q = 32$, the coverage could be increased of up to 4 (see fig. 3-up). This can be explained as follows: The higher n/m is, the more important it is that the hash functions selected in successive runs are independent, and this is best achieved by the universal class $\mathcal{H}_2$. This improvement cannot be increased indefinitely by increasing the search depth d. If n and m are fixed and d grows, the improvement first increases, but then gradually decreases (see fig. 3). Figure 5 reflects how coverage is improved by increasing m. If $n \gg m$, then doubling the hash table approximately doubles the measured coverage. This factor decreases rapidly with a decrease in n/m.

Note: In our experiments the coverage could not be increased by a simple strategy of using alternatively a high (about $10^4 - 10^5$) and a small (about $10^2 - 10^3$) threshold in successive runs with the intention of traversing different parts of the search tree.

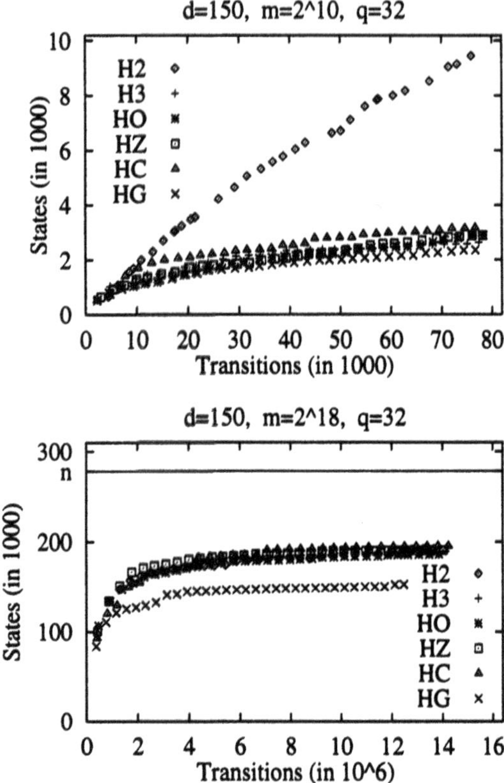

Figure 3 Measured coverage in dependence of performed transitions.

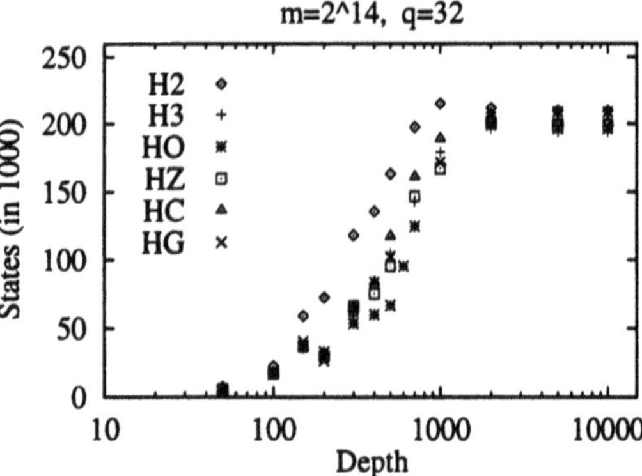

Figure 4 Measured coverage in dependence of search depth.

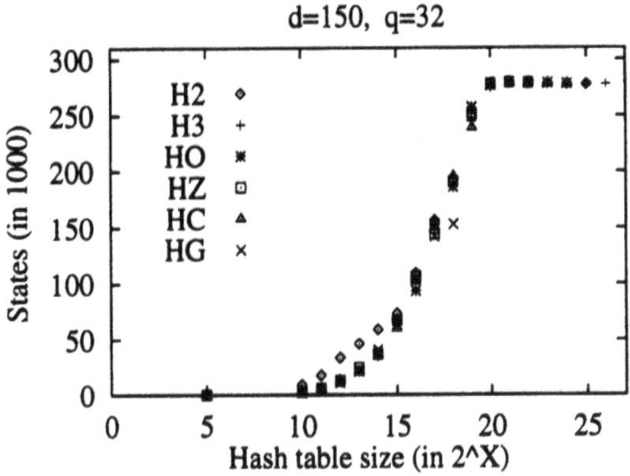

Figure 5 Measured coverage in dependence of hash table size.

4 ANALYSIS

In this section we present an analysis of bitstate and sequential hashing. Let m be the size of the hash table (measured in the number of bit addresses), q the number of runs and n the size of the state space.

4.1 Related Work

In (Holzmann, 1991) it is questioned whether sequential bitstate hashing can increase the coverage with the same system constraints. He answers with yes and gives an estimation of the coverage under certain conditions. He assumes that in each run exactly s states are generated and that the states generated in each run can be considered as randomly selected from the pool of reachable states. The combined coverage of two runs can then be computed as follows. A first run with a first hash function finds s from n states. A second run with a second hash function finds again s states, but makes a different selection. The combined coverage is greater than s, but lower than $2s$ since there is an overlap between both state sets. Under the above assumption the probability that a state x is generated in both runs is $(s/n)^2$, hence the expected overlap is $n \cdot (s/n)^2$ and the combined coverage after two runs is $2s - s^2/n$. More generally, the probability that an arbitrary state x is not generated in a selected run is $1 - s/n$, hence the probability that x is not generated in q independent runs is $(1 - s/n)^q$, and hence x is reached with probability $1 - (1 - s/n)^q$. The expected number of reached states after q runs is $n \cdot (1 - (1 - s/n)^q$ and thus the expected coverage is given by

$$1 - (1 - s/n)^q. \tag{1}$$

Equation (1) assumes that there are "ideal circumstances" which occur if the generated states can be considered as randomly chosen from the set of all reachable states. In (Holzmann, 1995) the reason for this is stated to be the fact that a graph representing a protocol state space is well-connected. We will show theoretically below that this is not sufficient. The overlap of successive runs is much higher and the combined coverage much smaller than predicted by (1).

In (Holzmann, 1995) there are presented lower and upper bounds for the expected coverage of a single run with bitstate and double bitstate hashing. These bounds are given by general constraints. Unfortunately, there are many pairs (n, m) for which either the lower bound or the upper bound is trivial or not defined. Our analysis does not suffer from such restrictions and works well even in the case $n > m$ which is the relevant case.

4.2 *Our Approach*

In this subsection we present an analysis of the expected coverage of sequential hashing based on bitstate hashing. Double bistate hashing can be analyzed in a similar way.

Equation (1) does not reflect the fact that the states are traversed in each run in depth-first order. There is an enumeration of the reachable states, say $(x_i | 1 \leq i \leq n)$, such that for each run and for all pairs i, j, $1 \leq i < j \leq n$, it is true: If both x_i and x_j are generated (reached), then x_i is generated before x_j. Note, since supertrace is a partial search algorithm it is not guaranteed that x_i and x_j are both generated. The following analysis takes into account that the states are traversed with respect to a fixed order.

In a traversal of a protocol state space, many states are generated twice. Since we compute an upper bound of the number of generated states, w. l. o. g. we may assume that duplicates have been pruned in the sequence above. Duplicates are states x_i, x_j such that $i \neq j$, but $x_i = x_j$. Such a shorter sequence is achieved by removing repeatedly in the case of two duplicates the one with the higher index. Note that this shortened sequence generates the exact same set of states as the original sequence, since only duplicates have been pruned. However, using this shortened sequence the ratio of reached states to the number of transitions is estimated too optimistically.

We assume, as in (Holzmann, 1991, Holzmann, 1995), that the hash values are distributed randomly and independently. Then we may consider the insertion operation as a random experiment which will be applied to each state with respect to the order $x_1, \ldots, x_n$. This is done in the following way. First, we compute $h(x_i)$ for state x_i. If the bit addressed by $h(x_i)$ is 0, then x_i is newly generated and inserted into the hash table by setting bit $h(x_i)$ to 1. Otherwise, x_i is considered as already generated and the hash table remains unchanged. If the insertion is successful, we say that the state has been reached, otherwise not. This reflects the fact that in the latter case state x_i is considered as a duplicate and no verification test is done. This simplification ignores the fact that in the case of $h(x_i) = 1$ (i. e. x_i is considered as a duplicate) the search tree is pruned and none of the successors of x_i are generated. As a result the expected coverage of a single run is possibly too high. This is probably also true for the combined coverage of sequential hashing.

Let $E(i, k)$ represent the event that the hash table has i entries after the k-th insertion try (of state x_k). Then event $E(i, k)$ follows from event $E(i-1, k-1)$, if x_k is inserted, and it follows from event $E(i, k-1)$, if the k-th insertion fails. This gives the following reccurence.

$$\begin{aligned} p\{E(i,k)\} &= p\{E(i,k)|E(i-1,k-1)\} \cdot p\{E(i-1,k-1)\} \\ &\quad + p\{E(i,k)|E(i,k-1)\} \cdot p\{E(i,k-1)\} \end{aligned} \qquad (2)$$

for $i \geq 0, k > 0$ and with initial values $p\{E(0,0)\} = 1$, $p\{E(i,0)\} = 0$, for $i > 0$, and $p\{E(-1,k)\} = 0$ for $k > 0$. ($p\{B|A\}$ denotes the probability for B under condition A.) It is $p\{E(i,k)|E(i-1,k-1)\} = (m-i+1)/m$ (since there are $m-(i-1)$ free hash positions) and $p\{E(i,k)|E(i,k-1)\} = i/m$ (since there are i occupied hash positions). Thus, (2) is replaced by

$$p\{E(i,k)\} \;=\; \frac{i}{m}\cdot p\{E(i,k-1)\} + \frac{m-i+1}{m}\cdot p\{E(i-1,k-1)\}. \tag{3}$$

Let $p(i,k)$ denote the probability that x_k is inserted (with the k-th insertion operation) under the precondition that the hash table has i entries after the k-th insertion. These values can be computed by

$$\begin{aligned} p(i,k) &= p\{E(i,k)|E(i-1,k-1)\}\cdot p\{E(i-1,k-1)\} \\ &= \frac{m-i+1}{m}\cdot p\{E(i-1,k-1)\}. \end{aligned} \tag{4}$$

The probability $p(k)$ that x_k is inserted is therefore given by the sum of all probabilities $p(i,k)$, that is $p(k) = \sum_{i=1}^{k} p(i,k)$. The expected number of reached elements, provided that the elements are inserted with respect to the order $(x_1,\ldots,x_n)$, is $\sum_{k=1}^{n}(1\cdot p(k))$ and the expected coverage after n insertions, denoted by $\alpha(n)$, is given by:

$$\alpha(n) \;=\; \frac{1}{n}\cdot\sum_{k=1}^{n} p(k). \tag{5}$$

A comparison between the expected coverage of a single run computed with Holzmann's equations and (5) is given in figure 6. While both Holzmann's upper bound and our estimation measure the coverage (of DTP-protocol) too optimistically (Holzmann, 1995), our estimation is more accurate. Note that our estimation predicts that bitstate hashing is better than double bitstate hashing if the state size is greater than the table size (more precisely, if $n \geq 0.965\cdot m$). This reflects quite accurately the empirical results given for DTP-protocol in (Holzmann, 1995).

Let us compute the combined coverage of q runs. The probability that state x_k is not reached in q runs is $(1-p(k))^q$ (provided that the hash values are independent). Hence, the probability that x_k is reached in q runs is given by $p_q(k) = 1-(1-p(k))^q$. As a result, the expected coverage is given by

$$\alpha_q(n) = \frac{1}{n}\sum_{k=1}^{n} p_q(k) = \frac{1}{n}\sum_{k=1}^{n}(1-(1-p(k))^q) \tag{6}$$

Of course, each state x is generated with probability $1-\epsilon$ if a large enough value of q is chosen. But in practice this can require an unacceptable amount of time.

The following diagrams summarize a comparison between the expected coverage of q runs computed with (1) and with (6). For fair comparison, we have to replace s/n in (1) by $\alpha(n)$ the expected coverage of a single run. Figure 7 and 8 shows how the number of reached states grows in dependence of the state space size and the number of runs. The coverage measured with (1) is much higher than when computed with (6). This contradicts the assumption that the states can be considered as selected randomly from the set of all reachable states. Figure 7 corresponds with figure 3 since increasing the threshold increases the number of reachable states. The figure shows that coverage grows slowly while the state space grows exponentially with the search depth. In the best case ($n \gg m$) doubling the table size doubles the coverage (as measured for class $\mathcal{H}_2$).

	m	n	$\alpha(n)$ (single bitstate hashing) Holzmann	Our approach	$\alpha(n)$ (double bitstate hashing) Holzmann	Our approach
1.	1024	256	0.8754,—	0.8851	0.9173,—	0.9419
2.	1024	511	0.7509,—	0.7872	0.6613,—	0.8321
3.	1024	1023	0.5009,—	0.6325	0.5014,—	0.6197
4.	1024	2047	—,0.5002	0.4323	0.2501,0.5002	0.3660
5.	1024	4095	—,0.2500	0.2454	0.1250,0.2500	0.1874
6.	1024	8191	—,0.1250	0.1249	0.0625,0.1250	0.0937
7.	8192	511	0.9688,—	0.9695	0.9948,—	0.9952
8.	8192	1023	0.9376,—	0.9401	0.9792,—	0.9827
9.	8192	2047	0.8751,—	0.8848	0.9168,—	0.9418
10.	8192	4095	0.7501,—	0.7870	0.6669,—	0.8319
11.	8192	8191	0.5001,—	0.6321	0.5000,—	0.6193

Figure 6 Comparison of the expected coverage of a single run with supertrace estimated with Holzmann's equations and our approach ("—" indicates when either no upper or no lower bound is defined and is to be replaced by the trivial bound 1 or 0 resp.).

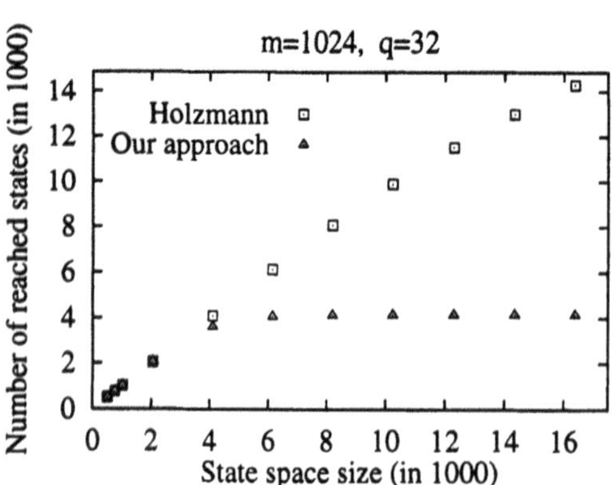

Figure 7 Holzmann's versus our approach: Expected coverage of sequential hashing in dependence of n.

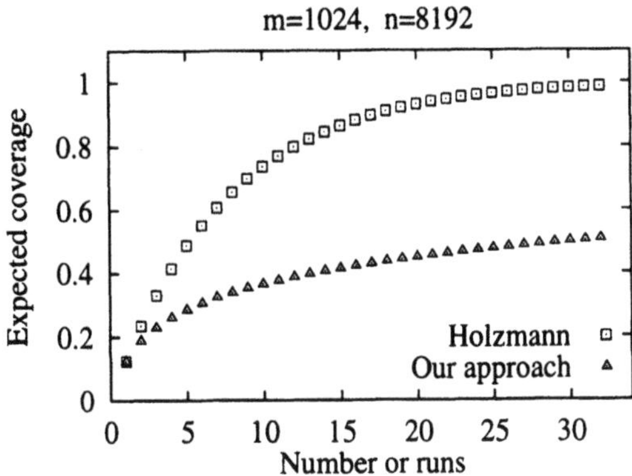

Figure 8 Holzmann's versus our approach: Expected coverage of sequential hashing in dependence of q.

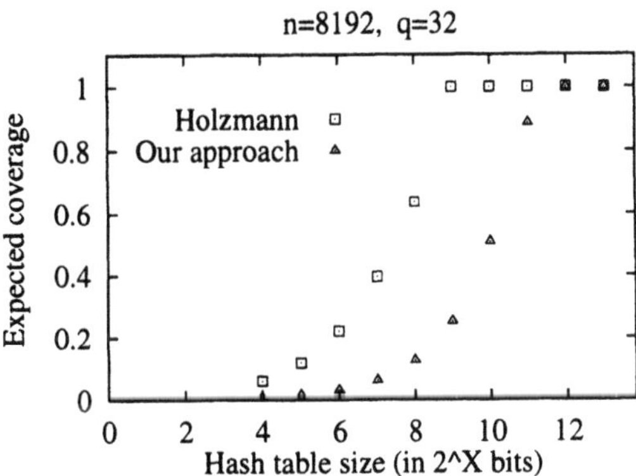

Figure 9 Holzmann's versus our approach: Expected coverage in dependence of hash table size.

5 UNIVERSAL SUPERTRACE

5.1 Theory

The hash function is used as a parameter which is changed for each run of sequential hashing with supertrace while the traversal order in which the states are generated remains unchanged. Thus, the question is: Can the combined coverage of sequential hashing be increased if in each run both the hash function and the traversal order of the states are changed. The answer is Yes.

Recall that universal hashing, for example, follows a general principle in order to get a good performance, that being a good distribution of the inserted keys. This principle suggests not to use a single algorithm for solving a class of problems, but rather a suitable set of algorithms should be provided from which one particular algorithm is chosen randomly in order to solve a given problem instance. This method is able to give a bound, which is independent of the input, for the average performance of the class of algorithms. Such a bound is valuable when it is better than the performance of any known single algorithm in that algorithm's worst case. We follow a similar principle to get a better coverage of bitstate hashing.

Let S_n be the permutation group of n elements and the sequence $x_1, \ldots, x_n$ be an explicit enumeration of all the reachable states. Let $S \subset S_n$ be a subset. Then we assign each $\sigma \in S$ a sequence A_σ: $x_{\sigma(1)}, \ldots, x_{\sigma(n)}$ which represents the order in which the states are traversed. Note that there are permutations $\sigma \in S_n$ which do not correspond to a possible traversal order of an exploration tree, since a state x is always visited before its successor states will be visited. Let $\mathcal{A}_S$ be the class of all the sequences $A_\sigma, \sigma \in S$.

A natural way of changing the traversal order is as follows: Fix a subset S and a class of hash functions $\mathcal{H}$ and for each new run randomly choose both the hash function $h \in \mathcal{H}$ and the permutation $\sigma \in S$ (with equal probabilities) and explore the state space with respect to traversal order A_σ and hash function h. We compute below the expected coverage when this strategy is applied to sequential hashing.

First, let $q = 1$: It is easy to see that for a particular run the expected coverage, denoted by $\alpha_{S,1}$, is equal to the coverage of the original supertrace. The coverage is given by the sum $\sum_{k=1}^{n} p(\sigma(k))$ provided that $\sigma \in S$ is selected. The expected coverage averaged over all permutations is

$$\alpha_{S,1}(n) \quad = \quad \frac{1}{|S|}\sum_{\sigma \in S}\frac{1}{n}\sum_{k=1}^{n} p(\sigma(k)) = \frac{1}{|S|}\sum_{\sigma \in S}\frac{1}{n}\sum_{k=1}^{n} p(k) = \frac{1}{|S|}\sum_{\sigma \in S}\alpha(n) = \alpha(n) \tag{7}$$

Note that the expected probability that the state x_k is visited is given by

$$\frac{1}{|S|}\sum_{\sigma \in S} p(\sigma(k)). \tag{8}$$

Let $q > 1$: Let us compute the expected coverage for q runs. Since we assume that the hash functions are independent, the probability that a selected state x_k is reached in at least one run, if the permutations $\sigma_1, \ldots, \sigma_q$ are selected, is given by

$$1 - (1 - p(\sigma_1(k))) \cdot (1 - p(\sigma_2(k))) \cdot \ldots \cdot (1 - p(\sigma_q(k))) \tag{9}$$

The expected coverage of q runs if the sequence $\sigma_1, \ldots, \sigma_q$ is selected is

$$\frac{1}{n}\sum_{k=1}^{n} (1 - (1 - p(\sigma_1(k))) \cdot (1 - p(\sigma_2(k))) \cdot \ldots \cdot (1 - p(\sigma_q(k)))). \tag{10}$$

The expected coverage averaged over all sequences of q permutations $(\sigma_1, \ldots, \sigma_q) \in S^q$ provided that each sequence is chosen randomly with equal probability, here denoted by $\alpha_{S,q}$, is

$$\alpha_{S,q}(n) = \frac{1}{|S|^q} \sum_{(\sigma_1,\ldots,\sigma_q)\in S^q} \frac{1}{n} \sum_{k=1}^{n} (1 - (1 - p(\sigma_1(k))) \cdot \ldots \cdot (1 - p(\sigma_q(k)))) \quad (11)$$

By changing the summation order, we obtain

$$\alpha_{S,q}(n) = \frac{1}{n} \sum_{k=1}^{n} p_{S,q}(k) \quad (12)$$

where

$$p_{S,q}(k) = \frac{1}{|S|^q} \sum_{(\sigma_1,\ldots,\sigma_q)\in S^q} (1 - (1 - p(\sigma_1(k))) \cdot \ldots \cdot (1 - p(\sigma_q(k)))) \quad (13)$$

which is the expected probability that x_k is reached after q runs. By induction over q it can be proven that

$$\alpha_{S,q}(n) = \frac{1}{n} \sum_{k=1}^{n} (1 - (1 - \frac{1}{|S|} \sum_{\sigma \in S} p(\sigma(k)))^q). \quad (14)$$

Optimization problem: The problem now is to find a suitable subset S such that the combined coverage of q runs would be maximized. An upper bound for the expected coverage is given by the following optimization problem under constraints. Consider the system

$$f(y_1, \ldots, y_n) = \frac{1}{n} (\sum_{j=1}^{n} y_j) - c \quad (15)$$

$$g(y_1, \ldots, y_n) = \frac{1}{n} \sum_{j=1}^{n} (1 - (1 - y_j)^q) \quad (16)$$

where $y_j \in [0, 1]$, $1 \leq j \leq n$. A maxima of g under the condition $f = 0$ is given by $y_j = c, 1 \leq j \leq n$. This optimization problem applied to

$$f(y_1, \ldots, y_n) = \frac{1}{n} (\sum_{j=1}^{n} y_j) - \frac{1}{n} (\sum_{j=1}^{n} p(j)) \quad (17)$$

$$g(y_1, \ldots, y_n) = \alpha_{S,q}(n) \quad (18)$$

where $y_j = (1/|S|)(\sum_{\sigma \in S} p(\sigma(j)))$, $1 \leq j \leq n$ yield a maxima for

$$y_j = (1/n)(\sum_{j=1}^{n} p(j)) = \alpha(n) = \alpha_{S,1}(n), \; 1 \leq j \leq n.$$

This maxima corresponds with the (impossible) choice $S = S_n$. Then $\alpha_{S,q}(n)$ is equal to (1). We can conclude that (1) is an upper bound for the expected coverage to be achieved by such randomization methods.

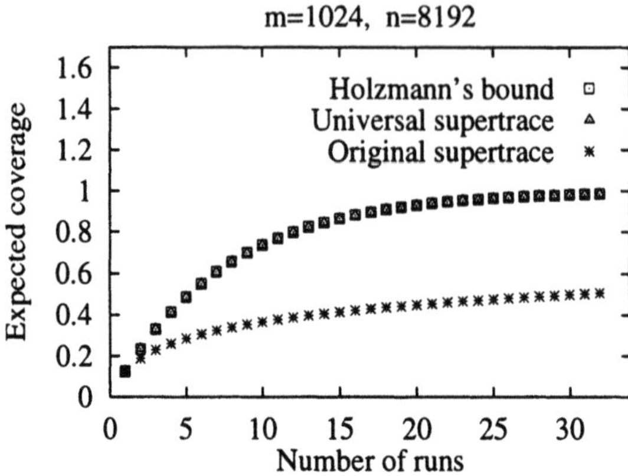

Figure 10 Expected coverage of universal supertrace versus original supertrace.

depth	measured coverage	coverage/transitions
100	0.9989	0.9849
150	2.1452	2.0614
200	4.3391	4.6981
300	8.7905	12.7175
400	8.1253	12.0768
500	8.3390	14.2580

Figure 11 Comparison of expected coverage of universal supertrace versus original supertrace (for $q = 64$, $m = 2^{16}$). The measured coverage of universal supertrace relative to that of original supertrace.

But how to select S in practice? A natural way of changing the traversal order is as follows: Keep the depth-first order, but for each reached state x change randomly the order in which the immediate successor states of x are visited. Let us call this strategy combined with (double) bitstate hashing *universal supertrace*. We have computed the expected coverage in the case that the exploration tree has a constant branching factor b (that is, that each state has exactly b immediate successor states). Surprisingly, the expected coverage comes very close to the upper bound (1). The higher the branching factor the higher the coverage, but even for $b = 2$ the coverage is very close to the upper bound (see figure 10).

5.2 Implementation and Experimental Results

We have implemented our new algorithm, *universal supertrace*, in the context of the SPIN validator. Fortunately, only slight extensions of the original supertrace were necessary. The main extension consisted of a function which allowed the order in which the immediate successor states are visited to be changed randomly. In SPIN each state refers to a chain of possibly applicable transitions. This chain reflects the order in which the successor states are generated. We included in SPIN a function which generates pointers to the transitions of the actual transition chain and changes the order randomly. In contrast to original supertrace, universal supertrace generates the successors in respect to this changed order.

It must be guaranteed that the order in which the successors of a state x are generated can be restored if the search backtracks to x in order to continue the search with the next successor of x. In the original supertrace, this order is fixed during the search. Thus it is sufficient to store only a pointer to the transition to be performed next on the depth-first stack. In contrast to this, universal supertrace changes for each state x the original order of the transitions to be applied to x. For this reason, for each state x the new transition order has to be stored on the depth-first stack. This can be done in a space efficient way. We use a random function which is called up with a randomly generated initial value for each state x on its first visit in order to generate a new order in its immediate successors: equal initial values lead to equal transition orders. It then suffices to store only those initial values with which the random function has been called up in order to generate the order of the successors to be applied to x. This leads to an additional space requirement of $O(d)$ where d is the search depth.

For evaluation, we made several tests with different parameters by using CRC-hashing without partial-order reduction. It seems that the maximal factor of improved coverage achieved with universal supertrace grows when the number of runs and the search depth are increased. For example, universal supertrace improves the coverage by a factor of up to 4.9095 compared with the original one for $q = 32$, $m = 2^{14}$ and by a factor of up to 8.7905 for $q = 64$, $m = 2^{16}$ (see figure 11). If we compare the ratio of reached states to performed transitions (coverage/transitions) as a measure of how quickly new states are generated, the improvement is still higher. For example, in the first case this factor is increased of up to 5.9739 and in the second case by up to 14.2580 (see figure 11). Similar results are obtained if additionally the partial-order reduction method in SPIN is combined with universal supertrace.

6 CONCLUSION

This paper is intended to give better insight into the quality of bitstate and sequential hashing. We have presented an analysis of the expected coverage of bitstate and sequential hashing which is more applicable and much more accurate than former approaches (Holzmann, 1991, Holzmann, 1995). Our experimental tests and our analysis shows that the quality of the original supertrace suffers as a result of the fixed order traversal used in supertrace. We have presented a new algorithm, universal supertrace, which randomly changes in successive runs both the hash function and the traversal order. Theoretical and experimental results show that universal supertrace outperforms the original supertrace (Holzmann, 1987, Holzmann, 1988) and improves the measured coverage by a factor of up to 8-10.

We expect that other strategies for changing the traversal order will have potential for additional improvement which may be the subject of our future research. The mathematical analysis developed in this paper can be applied in a similar way to other partial search methods, such as hashcompact (Stern and Dill, 1996, Stern and Dill, 1996a), and would help to clarify under which circumstances hash compact is to be preferred.

Acknowledgment

This research is supported by the DFG-project "Heuristische Suche und Anwendung auf die Protokollvalidation", No. Ot-64/11-1.

References

Carter, J. L. and Wegman, M. N. (1979). Universal classes of hash functions. *Journal of Computer and System Sciences*, 18(2):143–154.

Cohen, J. D. (1997). Recursive hashing functions for n-grams. *ACM Transactions on Information Systems*, 15(3):291–320.

Holzmann, G. J. (1987). On limits and possibilities of automated protocol analysis. In Rudin, H. and West, C., editors, *Proc. 6th Int. Conf on Protocol Specification, Testing, and Verification, INWG/IFIP*, Zurich, Sw. (Invited paper).

Holzmann, G. J. (1988). An improved protocol reachability analysis technique. *Software, Practice and Experience*, 18(2):137–161.

Holzmann, G. J. (1991). *Design and Validation of Computer Protocols*. Prentice Hall, Englewood Cliffs, NJ.

Holzmann, G. J. (1995). An analysis of bitstate hashing. In *Proc. 15th Int. Conf on Protocol Specification, Testing, and Verification, INWG/IFIP*, pages 301–314, Warsaw, Poland. Chapman & Hall.

Stern, U. and Dill, D. L. (1996). A new scheme for memory-efficient probabilistic verification. In Gotzheim, R. and Bredereke, J., editors, *Proc. 16th Int. Conf on Protocol Specification, Testing, and Verification, INWG/IFIP*, pages 333–348, Kaiserslautern, Germany. Chapman & Hall.

Stern, U. and Dill, D. L. (1996a). Combining state space caching and hash compaction. In Bernd straube and Jens Schoenherr, editors, *4. GI/ITG/GME Workshop zur Methoden des Entwurfs und der Verifikation Digitaler Systeme*, Berichte aus der Informatik, pages 81–90, Kreischa. GI/ITG/GME, Shaker Verlag, Aachen.

PART THREE

Test 1

FAULT-ORIENTED TEST GENERATION FOR MULTICAST ROUTING PROTOCOL DESIGN

Ahmed Helmy, Deborah Estrin, Sandeep Gupta

University of Southern California
Los Angeles, CA 90089
{ahelmy,estrin}@usc.edu, sandeep@boole.usc.edu

Abstract: We present a new algorithm for automatic test generation for multicast routing. Our algorithm processes a finite state machine (FSM) model of the protocol and uses a mix of forward and backward search techniques to generate the tests. The output tests include a set of topologies, protocol events and network failures, that lead to violation of protocol correctness and behavioral requirements. We target protocol robustness in specific, and do not attempt to verify other properties in this paper. We apply our method to a multicast routing protocol; PIM-DM, and investigate its behavior in the presence of selective packet loss on LANs and router crashes. Our study unveils several robustness violations in PIM-DM, for which we suggest fixes with the aid of the presented algorithm.

1.1 INTRODUCTION

Network protocol errors are often detected by application failure or performance degradation. Such errors are hardest to diagnose when the behavior is unexpected or unfamiliar. Even if a protocol is proven to be correct in isolation, its behavior may be unpredictable in an operational network, where interaction with other protocols and the presence of failures may affect its operation.

The complexity of network protocols is increased with the exponential growth of the Internet, and the introduction of new services, such as IP multicast. In addition, researchers are observing new and obscure, yet all too frequent, failure modes over the internets; such as routing anomalies [1, 2], and selective loss over LANs [3]. Such failures are becoming more frequent, mainly due to the increased heterogeneity of network components.

Many researchers have developed protocol verification methods, to ensure that certain properties of a protocol hold. Much of this work, however, was based on abstract mathematical models, with assumptions about the network conditions that may not always hold in today's Internet. To date, little effort has been exerted to formulate practical methods and tools that aid in the systematic testing of multicast protocols.

In this study, we propose a new method for automatic test generation, geared towards the study of protocol robustness in the presence of packet loss and network failures. In particular, we try to answer the question "is the protocol robust to specific failures ?" However, we do not attempt to develop a general verification method.

We borrow from well-established chip testing technologies and apply them to network protocols. We refer to our method as the fault-oriented test generation (FOTG). Starting from a given fault, the necessary topology and event sequences are established, that drive the protocol into error states, using forward and backward search techniques. As a case study, we apply FOTG to a real multicast routing protocol; PIM-DM [4], that has been deployed in parts of the Internet. We are particularly interested in multicast routing protocols, because they are vulnerable to failure modes, such as selective loss, that have not been traditionally studied in the area of protocol design [3].

The rest of this paper is organized as follows. Related work is discussed in section 1.2. Section 1.3 provides method overview and definitions. Section 1.4 describes our algorithm in detail, and summarizes the results of our case study. Conclusion is given in section 1.5.

1.2 RELATED WORK

There is a large body of literature dealing with verification of protocols. Verification systems typically address well-defined properties –such as *safety* (e.g. deadlock freedom), *liveness* (e.g. livelock freedom), and *responsiveness* (e.g. timeliness) [5]– and aim to detect violations of these properties.

In general, the two main approaches for protocol verification are theorem proving and reachability analysis [6]. Theorem proving systems define a set of axioms and relations to prove properties mathematically. Theorem proving includes *model-based* (e.g. Z [7]) and *logic-based* formalisms (e.g. Nqthm [8]). In general, however, the number of axioms and relations grows with the complexity of the protocol. We believe that these systems will be even more complex, and perhaps intractable, for multicast protocols. Moreover, these systems work with abstract specifications, and hence tend to abstract out some network dynamics that we will study; such as selective packet loss and router crashes.

Reachability analysis algorithms [9], on the other hand, try to generate and inspect all reachable protocol states. Such algorithms suffer from the 'state space explosion' problem, especially for complex protocols. To circumvent this problem, state reduction and partial search techniques [10] could be used. These algorithms, however, do not synthesize network topologies. Reduced reachability analysis has been used in the verification of cache coherence protocols [11], using a global FSM model. We adopt a similar FSM model and extend it for our approach in this study.

In [3] we have proposed a simulation-based STRESS testing method, based on heuristics and topological equivalences to reduce the number of simulated scenarios. However, we did not address automatic generation of topologies and events. Work in this paper complements our previous work, and may be integrated with the STRESS framework as part of its scenario generation.

VLSI chip testing [12] uses test vector generation to detect single-stuck faults. Test vectors may be generated based on circuit and fault models, using the fault-oriented process, that utilizes ***implication*** techniques for line justification. We adopt some implication concepts for our method, and transform them to the network protocol domain. We note that chip testing is performed for a given circuit, while a protocol must work over arbitrary and time varying topologies, adding a new dimension to the test generation problem.

1.3 METHOD OVERVIEW AND DEFINITIONS

The input to our method is the specification of a protocol, its correctness requirements, and a definition of its robustness. In general, protocol robustness is the ability to respond correctly in the face of network failures and packet loss. Usually robustness is defined in terms of network dynamics or fault models. A fault model represents various component faults; e.g. packet loss, or machine crashes. The desired output is a set of test-suites that stress the protocol mechanisms according to the robustness criteria.

Our method produces tests based on a model of the protocol. This section describes the model used and gives an overview of the case study protocol; PIM-DM.

1.3.1 System Model and Test Definition

The system consists of network and topology elements and a fault model.

Elements of the network consist of multicast capable nodes and bi-directional symmetric links. Nodes run same multicast routing, but not necessarily the same unicast routing. The topology is a N-router LAN modeled at the network level; we do not model the MAC layer.

A *fault*, is a low level (e.g. physical layer) anomalous behavior, that may affect the protocol under test. An *error* is the failure of a protocol to meet its design requirement (e.g. duplicate packet delivery).

Faults include: a) Loss of packets due to queue congestion or failure. b) Loss of state, e.g. uni/multicast tables, due to failures or crashes. Loss duration varies with the nature of the failure. c) Delays due to transmission, propagation, or queuing delays. Some delay problems may be translated into sequencing problems (see section 1.4.6).

Usually, a fault model is defined in conjunction with the protocol's robustness criteria. A design requirement for PIM is to be robust to single protocol message loss, which implies correct transitions from one stable state to another, even in the presence of single message loss. We also study the momentary loss of state. To analyze erroneous behavior, we consider single message loss per test sequence.

A test input pattern is defined as a 3-tuple '$< Topology, Events, Faults >$'; where Events is a sequence of host events; e.g. join, leave, or send packet, and the topology and faults as defined above.

Test Sequence Definition. Given sequences $T =< e_1, e_2, \ldots, e_n >$ and $T' =< e_1, e_2, \ldots, e_j, f, e_k, \ldots, e_n >$, where e_i is an event and f is a fault. Let $P(q,T)$ be the sequence of states and stimuli of protocol P under test T starting from the initial state q. T' is a test sequence if final $P(q,T')$ is incorrect; i.e. the stable state reached after the occurrence of the fault does not satisfy the protocol correctness conditions (see section 1.3.2) irrespective of $P(q,T)$. In case of a fault-free sequence, where $T = T'$, the error is attributed to a protocol design error. Whereas when $T \neq T'$, and final $P(q,T)$ is correct, the error is manifested by the fault. This definition ignores transient protocol behavior.

1.3.2 PIM-DM

As a case study, we apply our method to a version of the Protocol Independent Multicast-Dense Mode (PIM-DM) [4] protocol.

Multicast routing protocols deliver packets efficiently to group members by establishing distribution trees. PIM-DM uses broadcast-and-prune to establish the tree. In this mode of operation, a multicast packet is broadcast to all leaf subnetworks. Subnetworks with no local members send *prune* messages towards the source(s) of the packets to stop further broadcasts. Routers with new members joining the group trigger *Graft* messages towards previously pruned sources to re-establish the branches of the delivery tree. *Graft* messages are acknowledged explicitly at each hop using the *Graft-Ack* message. PIM-DM uses the underlying unicast routing tables to obtain the next-hop information, which may lead to situations where there are multiple forwarders for a LAN. The *Assert* mechanism resolves these situations and ensures there is at most one forwarder for the LAN.

In this study we target protocol robustness errors. We are interested mainly in erroneous stable (i.e. non-transient) states. We assume that protocol errors and correctness conditions are provided by the specification.

PIM Protocol Errors: A protocol error may manifest itself in one of the following ways: 1) *black holes*: consecutive packet loss between periods of packet delivery. 2) *packet looping.* 3) *packet duplication.* 4) *join latency*: excessive time taken by a receiver to start receiving packets. 5) *leave latency*: excessive time taken after a receiver leaves the group to stop the packet flow down pruned branches.

Correctness Conditions: These conditions are necessary to avoid errors during stable states in a LAN environment: 1) If there exists routers expecting packets from the LAN then there must exist a forwarder for the LAN, to prevent data loss (e.g. join latency or black holes). 2) The LAN must have at most one forwarder at a time, to prevent duplicates. 3) If there exists no routers expecting packets from the LAN there must not exist a forwarder for the LAN, to prevent leave latency. 4) The delivery tree must be loop-free. We do not consider looping in this study, as it is not a local behavior.

1.3.3 The Protocol Model

We represent the protocol by a finite state machine (FSM), and the LAN by a global FSM model (GFSM), as follows:

I. FSM model: A deterministic finite state machine modeling the behavior of a router R_i is represented by the machine $\mathcal{M}_i = (\mathcal{Q}, \Sigma_i, \delta_i)$, where: $\mathcal{Q}$ is a finite set of state symbols, Σ_i is the set of stimuli causing state transitions; and, δ_i is the state transition function $\mathcal{Q} \times \Sigma_i \to \mathcal{Q}$,

II. Global FSM model: With respect to a particular LAN, the global state is defined as the composition of individual router states. The behavior of a LAN with n routers may be described by the global FSM $\mathcal{M}_\mathcal{G} = (\mathcal{Q}_\mathcal{G}, \Sigma_\mathcal{G}, \delta_\mathcal{G})$ where: $\mathcal{Q}_\mathcal{G}$: $\mathcal{Q}_1 \times \mathcal{Q}_2 \times \cdots \times \mathcal{Q}_n$ is the global state space, $\Sigma_\mathcal{G}$: $\bigcup_{i=1}^{n} \Sigma_i$ is the set of stimuli causing the transitions; and $\delta_\mathcal{G}$: is the global state transition function $\mathcal{Q}_\mathcal{G} \times \Sigma_\mathcal{G} \to \mathcal{Q}_\mathcal{G}$,

1.4 APPLYING THE METHOD

Fault-oriented test generation (FOTG) targets specific faults. Starting from a given fault, FOTG attempts to synthesize topology(ies) that may experience an error, and a sequence of events leading to the error.

The faults studied here are single message loss, and loss of state:

1. For a given message, the algorithm uses the protocol model to identify a set of stimuli and states needed to stimulate that message, and the possible states and stimuli elicited by the message. This set of states form the global system state to be inspected. The algorithm is repeated for each message.

For loss of state, the global state is constructed from the mechanisms necessary to create the state, and the algorithm is repeated for each state.

2. Subsequent system states are obtained, through a process called *forward implication*, after the fault is included in the implication rules. Forward implication is the process of inferring subsequent states from a given state. The subsequent stable state is checked for errors.

3. If an error occurs, an attempt is made to obtain a sequence of events leading from an initial state to the error state, if such state is reachable. Such process is called *backward implication.*

Details of these algorithms are presented in section 1.4.5.

1.4.1 PIM-DM Model

Following is the model of a simplified version of PIM-DM.

I. FSM model. $\mathcal{M}_i = (\mathcal{Q}_i, \Sigma_i, \delta_i)$

For a specific source-group pair, we define the states w.r.t. a specific LAN to which the router R_i is attached. For example, a state may indicate that a router is a forwarder for (or receiver expecting packets from) the LAN.

A. System States ($\mathcal{Q}$). Possible states in which a router may exist are:

State Symbol	Meaning
F_i	Router i is a forwarder for the LAN
F_{i_Timer}	i forwarder with Timer $_{Timer}$ running
NF_i	Upstream router i is not a forwarder, but has entry
NH_i	Router i has the LAN as its next-hop
NH_{i_Timer}	same as NH_i with the Timer $_{Timer}$ running
NC_i	Router i has a negative-cache entry pointing to the LAN
EU_i	Upstream router i does not have an entry; i.e. is empty
ED_i	Downstream router i does not have an entry; i.e. is empty
M_i	Downstream leaf router with no state, and an attached member
NM_i	Downstream leaf router with no state and no attached members

The possible states for *upstream* and *downstream* routers are as follows:

$$\mathcal{Q}_i = \begin{cases} \{F_i, F_{i\_Timer}, NF_i, EU_i\} & \text{if the router is upstream,} \\ \{NH_i, NH_{i\_Timer}, NC_i, M_i, NM_i, ED_i\} & \text{if the router is downstream.} \end{cases}$$

B. Stimuli (Σ). The stimuli considered here include transmitting and receiving protocol messages, timer events, and external host events. Only stimuli leading to change of state are considered. For example, transmitting messages per se (vs. receiving messages) does not cause any change of state, except for the *Graft*, in which case the *Rtx* timer is set. Following are the stimuli considered in our study:

1. Transmitting messages: Graft transmission ($Graft_{Tx}$)

2. Receiving messages: Graft reception ($Graft_{Rcv}$), Join reception (*Join*), Prune reception (*Prune*), Graft Acknowledgement reception (*GAck*), Assert reception (*Assert*), and forwarded packets reception (*FPkt*).

3. Timer events: these events occur due to timer expiration (*Exp*) and include the Graft re-transmission timer (*Rtx*), the event of its expiration (*RtxExp*), the forwarder-deletion timer (*Del*), and the event of its expiration (*DelExp*). The expiration of a timer is implied as (*TimerImplication*) when the timer is set.

4. External host events (*Ext*): include host sending packets (*SPkt*), host joining a group (*HJoin* or *HJ*), and host leaving a group (*Leave* or *L*).

$\Sigma = \{Join, Prune, Graft_{Tx}, Graft_{Rcv}, GAck, Assert, FPkt, Rtx, Del, SPkt, HJ,$

II. Global FSM model. An example global state for a topology of 4 routers connected to a LAN, with router 1 as a forwarder, router 2 expecting packets from the LAN, and routers 3 and 4 have negative caches, is given by $\{F_1, NH_2, NC_3, NC_4\}$.

1.4.2 Transition Table

The global state transition may be represented in several ways. Here, we choose a transition table representation that emphasizes the effect of the stimuli on the system, and hence facilitates topology synthesis. The transition table describes, for each stimulus, the conditions of its occurrence. A condition is given as stimulus, and state or transition, (denoted by *stimulus.state/trans*); where the transition is given as $startState \rightarrow endState$. At least one pre-condition is necessary to trigger the stimulus. In contrast, a post-condition for a stimulus is an event and/or transition that is triggered by the stimulus, in the absence of faults (e.g. message loss). A '(p)' indicates a possible transition or stimulus, and represents a branching point in the search space. *orig*, *dst*, and *other* indicate the origin of the stimulus, its destination, and other routers, respectively. For example, a *Join* reception: a) is caused by the reception of a *Prune* from another router, with the originator of the *Join* in NH state, and b) causes *dst* to transit into F_{dst} if it exists in F_{Del} or NF state (other transitions are left out for simplicity). Following is the transition table for the global FSM discussed earlier.

Stimulus	**Pre-cond** (stimulus.state/trans)	**Post-cond** (stimulus.state/trans)
$Join$	$Prune_{other}.NH_{orig}$	$F_{dst\_Del} \rightarrow F_{dst}, NF_{dst} \rightarrow F_{dst}$
$Prune$	$L.NC, FPkt.NC$	$F_{dst} \rightarrow F_{dst\_Del}.(p)Join_{other}$
$Graft_{Tx}$	$HJ.(NC \rightarrow NH)$, $RtxExp.(NH_{\_Rtx} \rightarrow NH)$	$Graft_{Rcv}.(NH \rightarrow NH_{\_Rtx})$
$Graft_{Rcv}$	$Graft_{Tx}.(NH \rightarrow NH_{\_Rtx})$	$GAck.(NF_{dst} \rightarrow F_{dst})$
$GAck$	$Graft_{Rcv}.F$	$NH_{dst\_Rtx} \rightarrow NH_{dst}$
$Assert$	$FPkt_{other}.F_{orig}$, $Assert_{other}.F_{orig}$	(p) $F_{other} \rightarrow NF_{other}$, (p)$Assert_{other}$
$FPkt$	$Spkt.F$	$Prune.(NM \rightarrow NC)$, $ED \rightarrow NH, M \rightarrow NH$, $EU_{other} \rightarrow F_{other}$,(p) $Assert$
Rtx	RtxExp	$Graft_{Tx}.(NH_{orig\_Rtx} \rightarrow NH_{orig})$
Del	DelExp	$F_{orig\_Del} \rightarrow NF_{orig}$
$SPkt$	Ext	$FPkt.(EU_{orig} \rightarrow F_{orig})$
$HJoin$	Ext	$NM \rightarrow M, Graft_{Tx}.(NC \rightarrow NH)$
$Leave$	Ext	$M \rightarrow NM, Prune.(NH \rightarrow NC)$, $Prune.(NH_{Rtx} \rightarrow NC)$

1.4.3 State Dependency Table

To aid in test sequence synthesis through the backward implication procedure, we construct what we call a state dependency table. This table does not contain additional information about the protocol behavior to that given by the transition table, and is inferred automatically therefrom. We use this table to improve the performance of the algorithm and for illustration.

For each state, the dependency table contains the possible preceding states and the stimulus from which the state can be reached or implied. To obtain this information for a state S, we search the post-condition column of the transition table for entries where the *endState* of a transition is S. In addition, a state may be identified as an initial state (I.S.). The initial states for this study include $\{EU, ED, NM\}$.

Following is a partial state dependency table:

State	Possible Backward Implication(s)
F_i	$\xleftarrow{Fpkt_{other}} EU_i, \xleftarrow{Join} F_{i\_Del}, \xleftarrow{Join} NF_i, \xleftarrow{Graft_{Rcv}} NF_i, \xleftarrow{SPkt} EU_i$
F_{i_Del}	$\xleftarrow{Prune} F_i$
NF_i	$\xleftarrow{Del} F_{i\_Del}, \xleftarrow{Assert} F_i$
NH_i	$\xleftarrow{Rtx,GAck} NH_{i\_Rtx}, \xleftarrow{HJ} NC_i, \xleftarrow{FPkt} M_i, \xleftarrow{FPkt} ED_i$
NH_{i_Rtx}	$\xleftarrow{Graft_{Tx}} NH_i$
NC_i	$\xleftarrow{FPkt} NM_i, \xleftarrow{L} NH_{i\_Rtx}, \xleftarrow{L} NH_i$
⋮	

In some cases, multiple states need to be simultaneously implied in one backward step, otherwise an $I.S.$ may not be reached. To do this, the transitions in the post-conditions of the stimulus are traversed, and any states in the global state that are *endStates* are replaced by their corresponding *startStates*. For example, $\{M_i, NM_j, F_k\} \xleftarrow{FPkt} \{NH_i, NC_j, F_k\}$.

1.4.4 Defining stable states

As mentioned earlier, we are concerned with stable state (i.e. non-transient) behavior. To obtain erroneous stable states, we need to define the transition mechanisms between such states. We introduce the concept of transition classification and completion to distinguish between transient and stable states.

Classification of Transitions. We identify two types of transitions; *externally triggered (ET)* and *internally triggered (IT)* transitions. The former is stimulated by events external to the system (e.g. *HJoin* or *Leave*), whereas the latter is stimulated by events internal to the system (e.g. *FPkt* or *Graft*).

We note that some transitions may be triggered due to both internal and external events, depending on the scenario. For example, a *Prune* may be triggered due to forwarding packets by an upstream router *FPkt* (which is an internal event), or a *Leave* (which is an external event).

A global state is checked for correctness at the end of an externally triggered transition after completing its dependent internally triggered transitions.

Following is a table of host events, their dependent ET events and their dependent IT events:

Host Events	*SPkt*	*HJoin*	*Leave*
ET events	*FPkt*	*Graft*	*Prune*
IT events	*Assert*, *Prune*, *Join*	*GAck*	*Join*

Transition Completion. To check for the global system correctness, all stimulated internal transitions should be completed, to bring the system into a stable state. Intermediate (transient) states should not be checked for correctness (since they may violate the correctness conditions set forth for stable states, and hence may give false error indication).

The process of identifying complete transitions depends on the nature of the protocol. But, in general, we may identify a complete transition sequence, as the sequence of (all) transitions triggered due to a single external stimulus (e.g. *HJoin* or *Leave*). Therefore, we should be able to identify a transition based upon its stimuli (either external or internal).

At the end of each complete transition sequence the system exists in either a correct or erroneous stable state. Event-triggered timers (e.g. *Del*, *Rtx*) fire at the end of a complete transition, satisfying the *TimerImplication*.

Also, according to the above completion concept, the proper analysis of behavior should start from externally triggered transitions. For example, analysis should not consider a *Join* without considering the *Prune* triggering it and its effects on the system. Thus the global system state must be rolled back to the beginning of a complete transition (i.e. the previous stable state) before applying the forward implication. This will be implied in the forward implication algorithm discussed later, to simplify the discussion.

1.4.5 FOTG details

As previously mentioned, our FOTG approach consists of three phases: I) synthesis of the global state to inspect, II) forward implication, and III) backward implication. These phases are explained in more detail in this section.

FOTG starts from a given fault. The faults we address here are message and state loss.

Synthesizing the Global State.
Starting from a fault (i.e. the message or state to be lost), and using the information in the protocol model (i.e. the transition table), a global state is chosen for investigation. We refer to this state as the global-state inspected (G_I), and it is obtained for message loss as follows:

1. The global state is initially empty and the inspected message is initially set to the message to be lost.

2. For the inspected message, the state (or the *startState* of the transition) of the post-condition is obtained from the transition table. If the state does not exist in the global state, and cannot be implied therefrom, then it is added to the global state.

3. For the inspected message, the state (or the *endState* of the transition) of the pre-condition is obtained. If the state does not exist in the global state, and cannot be implied therefrom, then it is added to the global state.

4. Get the stimulus of the pre-condition of the inspected message. If this stimulus is not external (*Ext*), then set the inspected message to the stimulus, and go back to step 2.

Note that there may be several pre-conditions or post-conditions for a stimulus, in which case several choices can be made. These represent branching points in the search space.

At the end of this stage, the global state to be investigated is obtained.

For state loss, the state dependency table is used to determine the message required to create the state, and the topology constructed for that message is used for the state. This is illustrated later in this section.

Forward Implication.

The states following G_I (i.e. G_{I+i} where $i > 0$) are obtained through forward implication. We simply apply the transitions, starting from G_I, as given by the transition table, in addition to implied transitions (such as timer implication). In case of a message loss, the transition due to the lost message is not applied. If more than one state is affected by the message, then the space searched is expanded to include the various selective loss scenarios for the affected routers.

Backward Implication.

If an error occurs, backward implication attempts to obtain a sequence of events leading to G_I, from an initial state ($I.S.$), if such sequence exists; i.e. if G_I is reachable from $I.S.$

The state dependency table is used in the backward search. Backward steps are taken for the components in the global state G_I, until an initial global state (i.e. a state with all components as $I.S.$) is reached, or the termination criteria (if any) is met.

Figure 1.1 shows the above processes for a simple example of a *Join* loss. Following are the steps taken for that example:

Synthesizing the Global State
1. *Join*: *startState* of the post-condition is $NF_{dst} \Longrightarrow G_I = \{NF_k\}$
2. *Join*: state of the pre-condition is $NH_i \Longrightarrow G_I = \{NH_i, NF_k\}$, goto *Prune*
3. *Prune*: *startState* of the post-condition is F_k which can be implied from NF_k in G_I
4. *Prune*: state of the pre-condition is $NC_j \Longrightarrow G_I = \{NH_i, NF_k, NC_j\}$, goto L
5. the *startState* of the post-condition is NH which can be implied from NC in G_I

Forward implication
without loss: $G_I = \{NH_i, NF_k, NC_j\} \xrightarrow{Join} G_{I+1} = \{NH_i, F_k, NC_j\}$
loss w.r.t. R_j: $\{NH_i, NF_k, NC_j\} \longrightarrow G_{I+1} = \{NH_i, NF_k, NC_j\}$ error
Backward implication
$G_I = \{NH_i, NF_k, NC_j\} \xleftarrow{Prune} G_{I-1} = \{NH_i, F_k, NC_j\} \xleftarrow{FPkt} G_{I-2} = \{M_i, F_k, NM_j\} \xleftarrow{SPkt} G_{I-3} = \{M_i, EU_k, NM_j\} \xleftarrow{HJ_i} G_{I-4} = \{NM_i, EU_k, NM_j\} = I.S.$

Losing the *Join* by the forwarding router R_k leads to an error state where router R_i is expecting packets from the LAN, but the LAN has no forwarder.

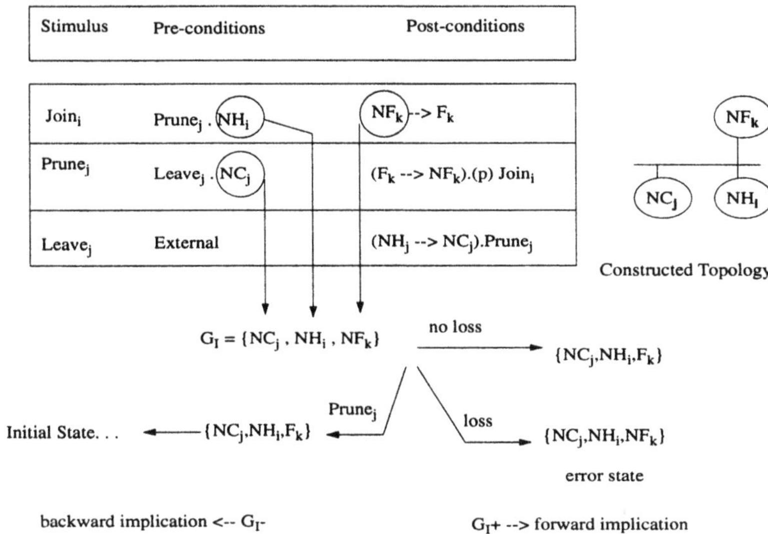

Figure 1.1 *Join* topology synthesis, forward/backward implication

1.4.6 Summary of Results

We have implemented an early version of the algorithm in the NS/VINT environment (see http://catarina.usc.edu/vint) and used it to drive detailed simulations of PIM-DM therein, to verify our findings. In this section we briefly discuss the results of applying our method to PIM-DM. The analysis is conducted for single message loss and momentary loss of state. For a detailed analysis of the results see [13].

Single message loss. We have studied single message loss scenarios for the *Join*, *Prune*, *Assert*, and *Graft* messages. For brevity, we partially discuss our

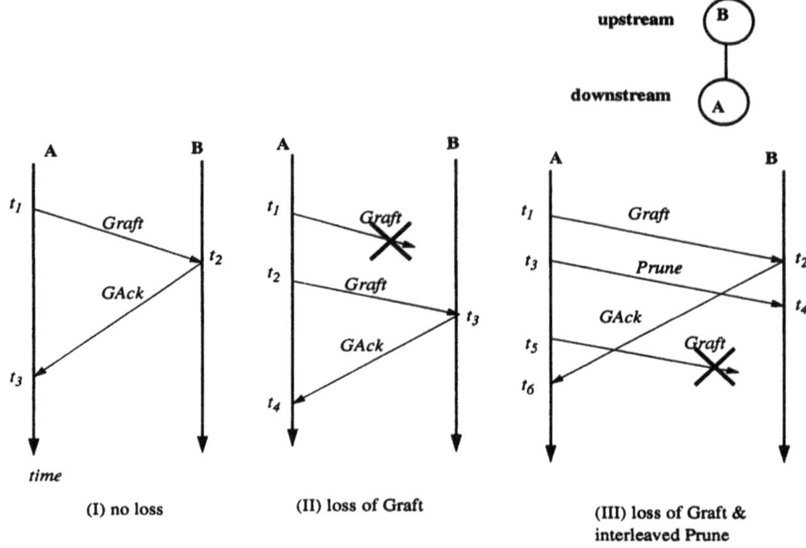

Figure 1.2 Graft event sequencing

results here. For this subsection, we consider non-interleaving external events, where the system is stimulated only once between stable states. The *Graft* message is particularly interesting, since it is acknowledged, and it raises timing and sequencing issues that we address in a later subsection, where we extend our method to consider interleaving of external events.

Join:. A scenario similar to that presented in section 1.4.5 incurred an error. In this case, the robustness violation was not allowing another chance to the downstream router to send a *Join*. A suggested fix would be to send another prune by F_{Del} before the timer expires.

Prune:. In the topology above, an error occurs when R_i loses the *Prune*, hence no *Join* is triggered. The fix suggested above takes care of this case too.

Assert:. An error in the *Assert* case occurs with no downstream routers; e.g. $G_I = \{F_i, F_j\}$. The design error is the absence of a mechanism to prevent pruning packets in this case. One suggested fix would be to have the *Assert* winner schedule a deletion timer (i.e. becomes F_{Del}) and have the downstream receiver (if any) send *Join* to the *Assert* winner.

Graft:. We did not reach an error state when the *Graft* was lost, with non-interleaving external events.

Interleaving events and Sequencing. A *Graft* message is acknowledged by *GAck*, and is robust to message loss with the use of *Rtx* timer. Adversary external conditions are interleaved during the transient states and the *Rtx* timer is cleared, such that the adverse event will not be overridden by the *Rtx* mechanism.

To clear the *Rtx* timer, a transition should be created from NH_{Rtx} to NH which is triggered by a *GAck* according to the state dependency table ($NH \xleftarrow{GAck} NH_{Rtx}$). This transition is then inserted in the event sequence, and forward and backward implications are used to obtain the overall sequence of events illustrated in figure 1.2. In the first and second scenarios (**I** and **II**) no error occurs. In the third scenario (**III**) when a *Graft* followed by a *Prune* is interleaved with the *Graft* loss, the *Rtx* timer is reset with the receipt of the *GAck* for the first *Graft*, and the systems ends up in an error state. A suggested fix is to add sequence number to *Grafts*.

Loss of State. We consider momentary loss of state in a router. A '*Crash*' stimulus transfers any state 'X' into 'EU' or 'ED'. Hence, we add the following line to the transition table:

Stimulus	**Pre-cond**	**Post-cond** (stimulus.state/trans)
Crash	Ext	$\{NM, M, NH, NC, NH_{Rtx}\} \rightarrow ED$, $\{F, F_{Del}, NF\} \rightarrow EU$

The FSM resumes function immediately after the crash (i.e. further transitions are not affected). We analyze the behavior when the crash occurs in any router state. For every state, a topology is synthesized that is necessary to create that state. We leverage the topologies previously synthesized for the messages. For example, state F_{Del} may be created from state F by receiving a *Prune* ($F_{Del} \xleftarrow{Prune} F$). Hence we may use the topologies constructed for *Prune* loss to analyze a crash for F_{Del} state.

Forward implication is then applied, and behavior after the crash is checked for correct packet delivery. To achieve this, host stimuli (i.e. $SPkt$, HJ and L) are applied, then the system state is checked for correctness.

In lots of the cases studied, the system recovered from the crash (i.e. the system state was eventually correct). The recovery is mainly due to the nature of PIM-DM; where protocol states are re-created with reception of data packets. This result is not likely to extend to protocols of other natures; e.g. PIM Sparse-Mode [14].

However, in violation with robustness requirements, there existed cases in which the system did not recover. In figure 1.3, the host joining in (II, a) did not have the sufficient state to send a *Graft* and hence gets join latency until the negative cache state times out upstream and packets are forwarded onto the LAN as in (II, b).

In figure 1.4 (II, a), the downstream router incurs join latency due to the crash of the upstream router. The state is not corrected until the periodic broadcast takes place, and packets are forwarded onto the LAN as in (II, b).

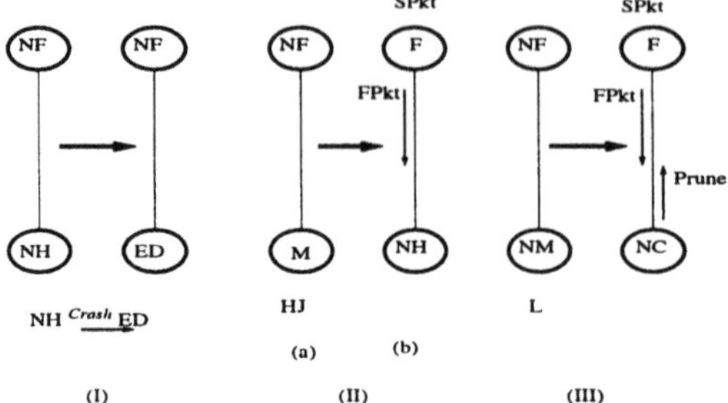

Figure 1.3 Crash leading to join latency

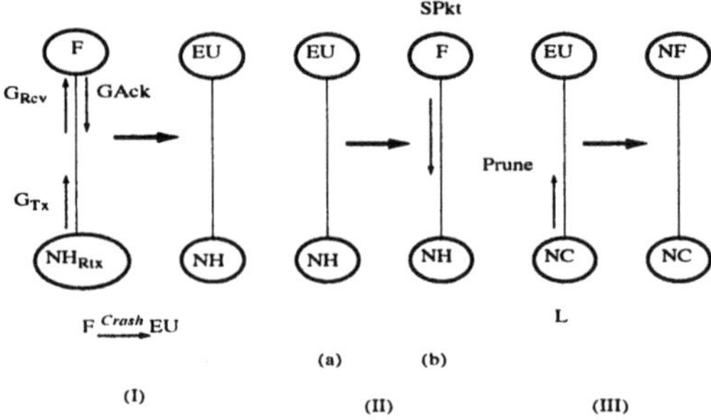

Figure 1.4 Crash leading to black holes

1.4.7 Limitations

The FOTG algorithms require a pre/post-condition global transition table, like the one presented in this paper. Generating such table from a conventional single router I/O FSM is part of our on-going work.

Currently the GFSM used in this study only models LANs. In our future work we will attempt to extend the model for regular and random topologies.

The LAN topologies constructed are inferred from the mechanisms specified by the transition table of the GFSM. The algorithm will not construct topologies resulting from non-specified mechanisms. For example, if the *Assert* mechanism was left out (due to a design error) the algorithm would not construct $\{F_i, F_j\}$ topology. So, FOTG (as presented here) may be used to evaluate behavior of specified mechanisms in the presence of network failures, but is not a general protocol verification tool.

1.5 SUMMARY AND FUTURE WORK

We have presented a new method for automating the robustness testing of multicast routing protocols, in the presence of network failures.

We do not claim nor attempt to provide mathematical proof of protocol correctness or verification. Rather, we have targeted protocol robustness and endeavored to systematize its testing and analysis for a particular domain; multicast routing.

Drawing from chip testing and FSM techniques, our method synthesizes the protocol tests automatically. These tests consist of the topology, event sequences and network faults. The following techniques were used to automate each of the test dimensions:

Topology synthesis: using the state transition table, our method synthesizes $N - router$ LAN topologies necessary to generate protocol messages or states, in terms of a global system state.

Fault investigation: from the global state, forward implication is used to test the behavior of the system in the presence of faults.

Sequence of events: if an error is found, backward implication constructs a sequence of events leading to the erroneous state, which is used to analyze the protocol behavior.

Timing problems: we have presented an example of transforming a timing problem into a sequencing problem to analyze acknowledged messages.

We have conducted a case study for PIM-DM, and found several scenarios in which the protocol behaved erroneously.

Our method may also be applicable to other protocols that can be represented by the global FSM model given in this paper.

We are in the process of conducting a quantitative analysis and evaluation of our method in terms of complexity and completeness; i.e. the number of topologies synthesized, state transitions traversed and faults covered. We are also investigating complexity reduction techniques by introducing equivalence classes of states and topologies, using counting equivalence and repetition constructors [11].

Future directions of this research include applying the FOTG method to other multicast protocols, including end-to-end performance analysis, extending the method to apply to topologies containing multiple LANs, and to include other network failures.

References

[1] V. Paxon. End-to-End Routing Behavior in the Internet. *IEEE/ACM Transactions on Networking, Vol. 5, No. 5. An earlier version appeared in Proc. ACM SIGCOMM '96, Stanford, CA*, pages 601–615, October 1997.

[2] V. Paxon. End-to-End Internet Packet Dynamics. *ACM SIGCOMM '97*, September 1997.

[3] A. Helmy and D. Estrin. Simulation-based STRESS Testing Case Study: A Multicast Routing Protocol. *Sixth International Symposium on Modeling, Analysis and Simulation of Computer and Telecommunication Systems (MASCOTS '98)*, July 1998.

[4] D. Estrin, D. Farinacci, A. Helmy, V. Jacobson, and L. Wei. Protocol Independent Multicast - Dense Mode (PIM-DM): Protocol Specification. *Proposed Experimental RFC. URL http://netweb.usc.edu/pim/pimdm/PIM-DM.{txt,ps}.gz*, September 1996.

[5] K. Saleh, I. Ahmed, K. Al-Saqabi, and A. Agarwal. A recovery approach to the design of stabilizing communication protocols. *Journal of Computer Communication, Vol. 18, No. 4*, pages 276–287, April 1995.

[6] E. Clarke and J. Wing. Formal Methods: State of the Art and Future Directions. *ACM Workshop on Strategic Directions in Computing Research, Vol. 28, No. 4*, pages 626–643, December 1996.

[7] J. Spivey. Understanding Z: a Specification Language and its Formal Semantics. *Cambridge University Press*, 1988.

[8] R. Boyer and J. Moore. A Computational Logic Handbook. *Academic Press, Boston*, 1988.

[9] F. Lin, P. Chu, and M. Liu. Protocol Verification using Reachability Analysis. *Computer Communication Review, Vol. 17, No. 5*, 1987.

[10] P. Godefroid. Using partial orders to improve automatic verification methods. *Proc. 2nd Workshop on Computer-Aided Verification, Springer Verlag, New York*, 1990.

[11] F. Pong and M. Dubois. Verification Techniques for Cache Coherence Protocols. *ACM Computing Surveys, Volume 29, No. 1*, pages 82–126, March 1996.

[12] M. Abramovici, M. Breuer, and A. Friedman. Digital Systems Testing and Testable Design. *AT & T Labs.*, 1990.

[13] A. Helmy, D. Estrin, and S. Gupta. Fault-oriented Test Generation for Multicast Routing. *USC-CS-TR 98-673, www.usc.edu/dept/cs*, March 1998.

[14] D. Estrin, D. Farinacci, A. Helmy, D. Thaler, S. Deering, M. Handley, V. Jacobson, C. Liu, P. Sharma, and L. Wei. Protocol Independent Multicast - Sparse Mode (PIM-SM): Protocol Specification. *RFC 2117. URL http://netweb.usc.edu/pim/pimsm/PIM-SMv2-Exp-RFC.{txt,ps}.gz*, March 1997.

Ahmed A-G Helmy received his M.S. ('95) from the University of Southern California, and his B.S. ('92) from Cairo University, Egypt, He is currently pursuing his Ph.D. in Computer Science at the University of Southern California. His research interests include protocol design and testing, multicast routing, and network simulation. Website: catarina.usc.edu/ahelmy.

Deborah Estrin is a Professor of Computer Science at the University of Southern California in Los Angeles where she joined the faculty in 1986. Estrin received her Ph.D. (1985) and M.S.(1982) from the Massachusetts Institute of Technology (M.I.T.) and her B.S. (1980) from U.C. Berkeley. In 1987, Estrin received the National Science Foundation, Presidential Young Investigator Award for her research in network interconnection and security. Estrin is a co-PI on the DARPA Virtual Internet Testbed (VINT) project and the NSF Routing Arbiter project at USC's Information Sciences Institute where she spends much of her time supervising doctoral student research.

Estrin is an active member of the IETF multicast routing related working groups and a long-time member of the End-to-End research group. Estrin is a member of the ACM, IEEE,and AAAS. She has served on numerous panels for The National Science Foundation, The National Academy of Engineering, and is currently a member of DARPA's Information Systems and Technology (ISAT) advisory board. Estrin has served as an editor for the ACM/IEEE Transaction on Networks and the Wiley Journal of Internetworking Research and Experience, and as a member of the program committee for many IEEE Infocom and ACM Sigcomm conferences, and she was program co-chair of ACM Sigcomm '96.

Sandeep Gupta received his Bachelors degree in Electrical Engineering from the Indian Institute of Technology, Kharagpur, India in 1985 and obtained M.S. and Ph.D. degrees in Electrical and Computer Engineering from the University of Massachusetts at Amherst in 1989 and 1991. Since 1991 he has been with the Department of Electrical Engineering – Systems at the University of Southern California, Los Angeles, where currently he is an Associate Professor. He is also a Co-Director of USC's M.S. Program in VLSI Design.

His research interests are in the area of VLSI Testing and Design and is currently involved in projects on Built-In Self-Test, delay testing and diagnosis of digital circuits, and test and validation of deep submicron, high speed circuits. He is a recipient of the National Science Foundation's Research Initiation Award (1992) and CAREER Award (1995).

ISSUES IN CONFORMANCE TESTING: MULTIPLE SEMICONTROLLABLE INTERFACES

Mariusz A. Fecko
Computer and Information Science Department
University of Delaware, Newark, DE

M. Ümit Uyar*
Department of Electrical Engineering
City College of the City University of New York

Adarshpal S. Sethi, Paul D. Amer
Computer and Information Science Department
University of Delaware, Newark, DE

Abstract

In a testing environment, where an IUT communicates with multiple entities, a tester may have differing degrees of controllability on the interactions between these entities and the IUT: directly controllable, semicontrollable, or uncontrollable. In this paper, a graph conversion algorithm is introduced that offers the testability of both the directly and semicontrollable inputs, while avoiding race conditions. Although, for the most general case, the graph conversion results in an exponentially large number of nodes, practical considerations make the converted graph size feasible. Currently, this methodology is being applied to generate tests for MIL-STD 188-220B, which increases the number of testable state transitions from approximately 200 to over 700.

*Dr. Uyar performed this research while a Visiting Associate Professor at University of Delaware.

1 INTRODUCTION

Due to increasing complexity of communication protocols, automated generation of conformance tests based on the formal descriptions has been an active research area [Ural, 1992] - [Sarikaya et al., 1987]. One problem that exists in today's conformance testing stems from the limited controllability of an Implementation Under Test (IUT), which almost always renders certain protocol features untestable. Ideally, testers should be able to generate every possible input that is defined in the Finite State Machine (FSM) modeling an IUT. Similarly, all outputs generated by an IUT should be observable by the testers.

Unfortunately, in practice, these two situations often are not possible. Testers may not have a direct access to all interface(s) in which the IUT accepts inputs. Typically, for an (N)-layer IUT, an exposed interface exists only between the IUT and the (N-1)-Service Provider [IS9646, 1991]. Interfaces with the upper layer, or with the peer entities (such as timers, etc.) are not directly accessible. In such cases, the interactions that involve these not directly controllable interfaces introduce non-determinism and/or race conditions during testing, leaving certain portions of the IUT model untestable.

Consider an (N)-layer IUT communicating with an entity FSM_i that does not have interfaces directly accessible by the tester (i.e., the tester cannot apply inputs to FSM_i). In some cases, FSM_i can be utilized to generate otherwise not directly controllable inputs to the IUT. An output from the IUT to the FSM_i can force the FSM_i to generate a response as the desired input from FSM_i to the IUT. If the tester can apply an appropriate input to the IUT, the IUT, in turn, triggers such an interaction between the IUT and FSM_i. In this case, some of the interfaces become semicontrollable (as opposed to uncontrollable).

Another difficulty that arises when there are multiple interfaces interacting with the IUT is the possible occurrence of race conditions. If an IUT moves into a state at which several inputs from different interfaces are waiting, choosing which input is consumed first may cause non-determinism.

There are many real-life protocols that possess either semicontrollable inputs, or race conditions, or both, due to a tester's limited control over the interactions between the IUT and other communicating entities. For example, in MIL-STD 188-220B [188-220B, 1998], over 70% of the transitions cannot be directly controlled. The race conditions can be shown in the IEEE 802.2 LLC Connection Component protocol [ISO8802-2, 1994].

Some problems due to the limited controllability over the IUT are addressed in the literature. Testing embedded systems [Rayner, 1987, Timohovich, 1993] where uncontrollable events may take place is discussed in [Phalippou, 1992] and [Cavalli et al., 1996]. The ferry clip testing method by [Zeng et al., 1989], and the Astride testing method by [Rafiq and Castanet, 1990] are introduced to enhance the controllability of inputs to an IUT. Deriving test sequences by combining the IUT and a single entity communicating with it into a global FSM are proposed in [Timohovich, 1993]. The testing system considered in this paper does not assume any user-defined entities within the SUT (as implied by [Zeng et al., 1989] and [Rafiq and Castanet, 1990]). Also, a global FSM is not built to avoid state explosion problem.

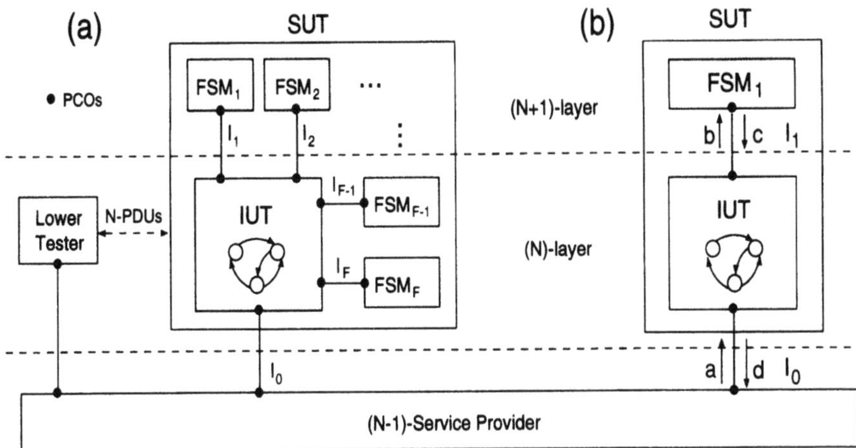

Figure 1 (a) Testing IUT with multiple interfaces; (b) Testing (N)-layer IUT with an (N+1)-layer semicontrollable interface.

This paper introduces a methodology that utilizes an IUT's semicontrollable interfaces while avoiding the race conditions. An algorithm is presented to modify an IUT's directed graph representation such that the semicontrollable portions of the IUT become directly controllable, where possible. In the most general case, such a graph conversion results in an exponentially large number of nodes. However, special considerations such as a small number of interfaces interacting with an IUT, and diagnostics considerations can make the problem size feasible for most practical cases.

The algorithm presented in this paper is being applied to MIL-STD 188-220B protocol to generate conformance tests. Initial results are promising: the number of testable transitions increased to over 700 from approximately 200 for the Class A – Type 1 Service Datalink module [Fecko et al., 1997].

This paper is organized as follows. Section 2 formally defines the *controllability problem*, which is practically motivated in Section 3 by examples from real-life protocols. Section 4 defines a system model for a testing environment with multiple interfaces with different degrees of controllability. Practical issues are introduced into this model in Section 5. An algorithm to modify an IUT's graph so that semicontrollable interfaces can be fully utilized while avoiding the race conditions is presented in Section 6. In Section 7, the application of minimum-cost test sequence generation techniques is discussed.

2 CONTROLLABILITY PROBLEM

Consider a testing environment shown in Figure 1 (a). The System Under Test (SUT) contains an IUT, which interacts with F FSMs. $FSM_1, \cdots, FSM_F$, implemented inside the SUT, interact with the IUT through interfaces $I_1, \cdots, I_F$. The points at which a testing system can apply inputs to and observe outputs from the IUT are called *points of control and observation*

(PCOs) [IS9646, 1991]. Each IUT's interface is associated with a full-duplex PCO through which inputs and outputs can be exchanged. The inputs can be of three different types:

- **directly controllable:** a tester can directly apply the inputs to the IUT through the PCO
- **semicontrollable:** a tester cannot directly apply the inputs to the IUT through the PCO. However, it is possible to utilize one of the FSMs interacting with the IUT to supply these inputs indirectly
- **uncontrollable:** the inputs may be supplied through a PCO without any explicit action of the tester. This means that inputs may be generated during testing without the tester's control

The inputs at a given PCO can belong to one or more of these three types. If a PCO has any semicontrollable inputs and does not have any uncontrollable inputs, we say that its associated interface is semicontrollable. If there are no semicontrollable or uncontrollable inputs, the interface is called directly controllable. In this paper, without loss of generality, we consider that each interface has only one type of input: either directly controllable or semicontrollable. The analysis also is applicable to interfaces with a combination of directly controllable and semicontrollable inputs (excluding uncontrollable ones).

A typical example of a directly controllable interface is a Lower Tester FSM [IS9646, 1991]. A timer FSM, whose only inputs come from an IUT (e.g., start, restart, and stop the timer), is a typical semicontrollable interface.

In the testing framework of Figure 1 (a), the tester is unable to supply inputs directly to the IUT through interfaces $I_1, \cdots, I_F$. Therefore, the interfaces $I_1, \cdots, I_F$ are only semicontrollable, provided that $FSM_1, \cdots, FSM_F$ can be utilized to supply inputs to the IUT. On the other hand, the tester can apply inputs to the IUT directly by using a Lower Tester (LT), which exchanges N-PDUs with the IUT by using the (N-1)-Service Provider. The interface I_0 between the LT and the IUT is therefore directly controllable (I_0 can be considered an interface between the LT and the IUT, because the (N-1)-Service Provider acts as a pass-through that does not alter inputs or outputs).

To test the IUT's transitions triggered by the inputs from I_i, the tester must use one of the directly controllable interfaces to force the IUT to generate outputs to I_i. These outputs are applied to FSM_i at I_i's PCO. As response to these outputs, FSM_i will send back inputs to the IUT through I_i. These inputs will trigger the appropriate transitions in the IUT.

Consider the SUT shown in Figure 1 (b). The LT, which exchanges N-PDUs with an (N)-layer IUT, represents a directly controllable interface I_0. Since the interface I_1 is not exposed in the SUT, a tester can neither directly apply inputs to the IUT nor observe the IUT's outputs to the (N+1)-layer. In this case, I_1 is at best only semicontrollable. To apply (N+1)-layer's inputs to the IUT through I_1, the LT must generate inputs to the IUT through its directly controllable interface I_0. In response, the IUT will generate outputs to FSM_1 through I_1. Subsequently, in response to these outputs, FSM_1 will generate inputs to the IUT at I_1.

This paper addresses the problem of generating optimal realizable test sequences in an environment with multiple semicontrollable interfaces. This problem will be referred to as the *controllability problem.*

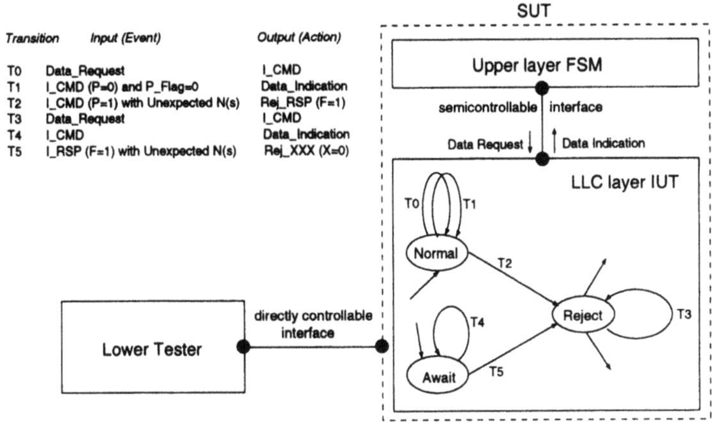

Figure 2 802.2 Type 2 Connection Component LLC Specification [ISO8802-2, 1994]

3 PRACTICAL MOTIVATION

As motivation for solving the controllability problem, two real protocols where an SUT's (N+1)-layer must be utilized indirectly to test certain transitions within the (N)-layer IUT are considered. They also illustrate an important point that must be taken into account while utilizing an (N+1)-layer indirectly – avoiding race conditions.

MIL-STD 188-220B [188-220B, 1998] is a military standard for interoperability of command, control, communications, computers, and intelligence over Combat Net Radios. In the testing framework used to test 188-220B, the upper layers cannot be directly controlled. This makes the IUT's transitions that are triggered by the inputs coming from the upper layer not directly testable.

In fact, 70% of the transitions are based on semicontrollable inputs. Without indirect testing, test coverage would be seriously reduced. However, by applying the technique introduced in this paper, almost all (>95%) transitions defined in the specification can be tested (the number of testable transitions rose to over 700 from approximately 200 for the Class A – Type 1 Service Datalink module [188-220B, 1998, Fecko et al., 1997]).

Figure 2 shows a portion of the IEEE 802.2 [ISO8802-2, 1994] LLC Type 2 Connection Component. The SUT consists of the LLC layer IUT and the implementation of the upper layer (UL), whose FSM is semicontrollable. Transitions $T1$ and $T4$ output a *Data_Indication* to the UL. Transitions $T0$ and $T3$ require input *Data_Request* from the UL.

Testing transition $T3$ requires the input *Data_Request* to be applied to the IUT. Since this input can only be applied from the UL, an indirect event triggering sequence must be used. The UL FSM contains a transition with an input of *Data_Indication* and an output of *Data_Request*. By generating *Data_Indication* from the IUT to the UL, the tester can indirectly generate *Data_Request*.

No transition in *Reject* outputs a *Data_Indication* to the UL. Therefore, a *Data_Indication* must be sent to the UL while the IUT is in another state. Then the IUT must be moved into the *Reject* before the UL outputs its *Data_Request*. For example, one straightforward solution is to use transition $T1$. The tester applies to the IUT an input *I_CMD* with the appropriate parameters and flags. The IUT outputs a *Data_Indication* to the UL. Then the tester brings the IUT to *Reject* state by applying an *I_CMD* through transition $T2$. Now transition $T3$ will be triggered by *Data_Request* that by then has arrived from the UL.

Unfortunately, this solution has a race condition: in the *Normal* state, after the *Data_Indication* is output as a result of traversing $T1$, if a *Data_Request* arrives from the UL before the tester applies $T2$'s input *I_CMD*, then the *Data_Request* will be consumed by transition $T0$ before transition $T2$ fires. To avoid this race condition after an indirectly generated *Data_Indication*, the tester must find a way of moving the IUT to the *Reject* state through states in which a *Data_Request* cannot be consumed. Consider *Await* state with transition $T4$ that can be used to generate a *Data_Indication*. There is no race condition here, because the *Await* state cannot process a *Data_Request*. A tester has sufficient time to move the IUT to *Reject* state through transition $T5$, where a *Data_Request* buffered in the interface will trigger $T3$.

As can be seen from this example, the IUT can move into a state where the IUT is forced to consume a previously buffered input. This creates a race condition if the test sequence requires that a different input be sent to the IUT by the LT. Therefore, a test sequence without such race conditions will not bring the IUT into a state where multiple inputs are pending (one from the LT, and others from the buffers). Instead, the test sequence should traverse the IUT transitions in an order that avoids these race conditions.

4 MODELING TEST SYSTEMS WITH CONTROLLABILITY ISSUES

Among the widely used specification formalisms to model protocol implementations are labelled transition systems (LTS) and Finite State Machines (FSM) [Tretmans, 1996]. In an LTS, the inputs and outputs for a system are not distinguished, but instead represented as interactions. The LTS allows an interaction to occur if both an implementation and an environment are able to perform that interaction.

An FSM, on the other hand, is a subset of an LTS where, for every state transition, an input is coupled with one or more outputs (including a *null* output). In this paper, an FSM model, which is sufficient to model protocols with finite state space and deterministic behavior, is used to represent an implementation.

Let us consider an IUT interacting with multiple semicontrollable interfaces. The goal of test generation in this environment is *to derive a set of tests exercising each transition in an IUT's FSM at least once.* Specifically, given a graph G representing an IUT's FSM, we want to find a minimum-cost tour of G such that each transition is covered at least once.

Given the graph $G(V, E)$ representing an FSM model of an IUT with multiple semicontrollable interfaces, let us define the following parameters:

- $|V|$ – number of nodes in G
- F – number of semicontrollable interfaces interacting with the IUT

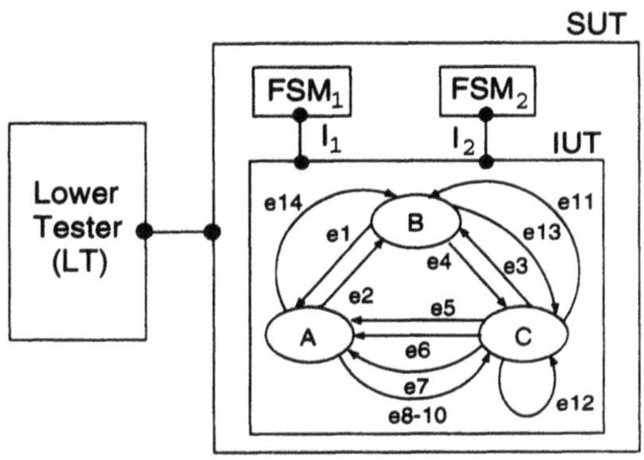

Figure 3 IUT interacting with two semicontrollable interfaces.

- $T_i \subset E$ – subset of edges in G triggered by the inputs from the i-th semicontrollable interface
- b_i – buffer size (maximum number of inputs buffered) at the i-th semicontrollable interface I_i
- A_i – set of inputs triggering transitions in T_i
- O_i – set of outputs of the IUT that force inputs in A_i to be buffered at I_i
- c_i – number of different transition classes in the IUT triggered by inputs at I_i. Two transitions t_1 and t_2 belong to the same transition class $T_{i,j} \subset T_i$ iff they both become fireable by the same input $a_{i,j} \in A_i$
- $U_{i,j} \subset E$ – set of transitions in the IUT with output $o_{i,j}$ such that, in response to $o_{i,j}$, an input $a_{i,j} \in A_i$ is buffered at I_i

Let $A_i = \{a_{i,1}, \ldots, a_{i,c_i}\}$ and $O_i = \{o_{i,1}, \ldots, o_{i,m_i}\}$. Then the sets of $T_{i,j}$, $U_{i,j}$, T_i, and U_i, are defined formally as follows:

$$T_{i,j} \stackrel{def}{=} \{e \in E : label(e) = a_{i,j}/o_{i,j}\} \qquad U_{i,j} \stackrel{def}{=} \{e \in E : output(e) = o_{i,j}\}$$

$$T_i \stackrel{def}{=} \bigcup_{j=1}^{c_i} T_{i,j} \qquad U_i \stackrel{def}{=} \bigcup_{j=1}^{c_i} U_{i,j}$$

There may be several outputs in set O_i that force input $a_{i,j}$ to be buffered at I_i. For simplicity, let $o_{i,j}$ denote any output forcing $a_{i,j}$ at I_i. Based on the above definitions, transitions triggered by the inputs from the semicontrollable interface I_i are divided into c_i classes, each corresponding to a distinct input that fires any transition within the class. No transition can belong to more than one $T_{i,j}$. Similarly, each transition can belong to only one $U_{i,j}$. In general, $T_{i,j}$ and $U_{i,j}$ may or may not be disjoint.

Example : *Consider the IUT of Figure 3 which is interacting with FSM_1 and FSM_2 through the semicontrollable interfaces I_1 and I_2, respectively. The IUT's FSM is described in Table 1. Transition e1, triggered by input x_1 from the LT, generates output $o_{1,1}$ to FSM_1. In response, FSM_1 sends back input $a_{1,1}$ which triggers transition e3. (In general, $a_{i,j}$ is the expected response to $o_{i,j}$.) e3, when*

Edge name	*Input from*	*Output to*	*Edge name*	*Input from*	*Output to*
e1	$LT?x_1$	$FSM_1!o_{1,1}$	e8	$LT?x_8$	$FSM_1!o_{1,2}$
e2	$FSM_1?a_{1,2}$	$FSM_1!o_{1,2}$	e9	$FSM_2?a_{2,1}$	$LT!y_9$
e3	$FSM_1?a_{1,1}$	$FSM_2!o_{2,1}$	e10	$LT?x_{10}$	$LT!y_{10}$
e4	$FSM_2?a_{2,1}$	$LT!y_4$	e11	$FSM_1?a_{1,2}$	$LT!y_{11}$
e5	$LT?x_5$	$LT!y_5$	e12	$LT?x_{12}$	$FSM_2!o_{2,1}$
e6	$LT?x_6$	$LT!y_6$	e13	$LT?x_{13}$	$LT!y_{13}$
e7	$LT?x_7$	$FSM_1!o_{1,2}$	e14	$LT?x_{14}$	$LT!y_{14}$

Table 1 Inputs and outputs for the edges of Figure 3. $A?x$ denotes receiving input x from A. $B!y$ denotes sending output y to B.

traversed, outputs $o_{2,1}$ to FSM_2, which responds with input $a_{2,1}$ triggering e4 or e9. $o_{2,1}$ is also output to FSM_2 by e12, which is fired by the LT's input x_{12}. Transitions e7 and e8, after being triggered by the LT's inputs x_7 and x_8, respectively, generate output $o_{1,2}$ to FSM_1. FSM_1 sends back input $a_{1,2}$, which triggers either e2 or e11. e2 outputs $o_{1,2}$ to FSM_1. Again, FSM_1 responds with input $a_{1,2}$. On the other hand, transitions e5, e6, e10, e13, and e14, can be triggered directly by the LT and do not generate outputs to the semicontrollable interfaces. For this example, we have:

- $|V| = 3$ *i.e., A, B, and C;* $F = 2$ *i.e.,* I_1 *and* I_2*;* $c_1 = 2$, $c_2 = 1$
- $T_{1,1} = \{e3\}$, $T_{1,2} = \{e2, e11\}$, $T_{2,1} = \{e4, e9\}$, $T_1 = T_{1,1} \cup T_{1,2} = \{e2, e3, e11\}$, $T_2 = T_{2,1} = \{e4, e9\}$
- $U_{1,1} = \{e1\}$, $U_{1,2} = \{e2, e7, e8\}$, $U_{2,1} = \{e3, e12\}$, $U_1 = U_{1,1} \cup U_{1,2} = \{e1, e2, e7, e8\}$, $U_2 = U_{2,1} = \{e3, e12\}$
- $A_1 = \{a_{1,1}, a_{1,2}\}$, $A_2 = \{a_{2,1}\}$; $O_1 = \{o_{1,1}, o_{1,2}\}$, $O_2 = \{o_{2,1}\}$

4.1 Buffer sizes at semicontrollable interfaces

For now, let us assume that there exists a separate FIFO buffer in the semicontrollable interface between the IUT and each interacting FSM. During testing, a buffer may be empty or store an arbitrary sequence of inputs to the IUT generated indirectly through the i-th semicontrollable interface. Then the entire system can be modeled by G (which represents the IUT's FSM) and the variables $\omega_1, \omega_2, \ldots, \omega_F$. A variable ω_i has a distinct value for each permutation of inputs that the i-th buffer can hold.

If the buffer sizes at the F semicontrollable interfaces are infinite, each variable ω_i can assume an infinite number of values. In this case, even the reachability analysis (deciding whether a given state is reachable from the initial state), which is an easier problem than finding a minimum-cost traversal of G, is undecidable [Davis et al., 1998]. Even if the buffer sizes are finite, in which case $\omega_1, \omega_2, \ldots, \omega_F$ have finite domains, the reachability analysis is PSPACE-complete for the most general case [Hopcroft and Ullman, 1979].

Given the difficulty of analyzing G and F variables, let us explore the possibility of modeling the system as an FSM, represented by $G'(V', E')$. Each vertex in V' is a tuple consisting of the original vertex in V and the set of values of variables $\omega_1, \omega_2, \ldots, \omega_F$. The maximum number of nodes $|V'|_{max}$ is

$|V'|_{max} = |V| * \prod_{i=1}^{F} B(i)$, where $B(i)$ is the maximum possible number of states of the i-th buffer defined as follows:

$$B(i) = \begin{cases} (1 - c_i^{1+b_i})/(1 - c_i) & \text{if } c_i > 1 \\ 1 + b_i & \text{if } c_i = 1 \end{cases} \quad (1)$$

In general, if each $c_i = c > 1$ and each $b_i = b$, then $|V'|_{max} = |V| * O(c^{bF})$, which indicates that the maximum number of nodes in G' grows exponentially with the number of semicontrollable interfaces F and the buffer size b. Clearly, the conversion from G to G' is not feasible for the general case. However, for the constrained environment, G' can be constructed efficiently (Section 7).

5 PRACTICAL CONSIDERATIONS FOR TEST SYSTEM

5.1 Buffering inputs at semicontrollable interfaces during testing

Although the model presented in Section 4 assumes that a semicontrollable interface consists of FIFO-type buffers, in practice this assumption may not hold true for all implementations. Typically, the interface is not part of a protocol specification. Vendors have the freedom of developing interfaces in different ways: in addition to (or instead of) FIFO-type buffers, they may have interrupt-driven mechanisms and any interface may have multiple buffers.

Therefore, test sequences generated for an IUT with only FIFO-type buffers become non-deterministic for other IUTs using different types of interfaces.

Example (cont'd): *Consider a test sequence for the IUT of Figure 3:*

$$\underline{e14, e1, e8, e3}, e4, e11, e13, e12, e5, e9, e7, e2, e13, e11, e13, e6, e10, e6 \quad (2)$$

The underlined portion of the above test sequence traverses e1, *which results in input* $a_{1,1}$ *being buffered at* I_1 *(Table 1). Subsequently, when* e8 *is executed with output* $o_{1,2}$ *to* I_1, *the buffer at* I_1 *should contain* $[a_{1,1}, a_{1,2}]$. *The IUT is expected to be in state* C *with* e3 *to be tested next. This sequence is only realizable under the assumption that inputs* $a_{1,1}$ *and* $a_{1,2}$ *are stored at* I_1 *in the FIFO order, i.e.,* $[a_{1,1}, a_{1,2}]$. *In practice, however, this may not be the case for all implementations. It is possible that, after traversing* e1 *and* e8, *the buffer will contain inputs in a different order, such as* $[a_{1,2}, a_{1,1}]$ *Then, after* e8 *is traversed,* e11 *will be triggered by* $a_{1,2}$ *instead of* e3 *being triggered by* $a_{1,1}$. e3 *will cause the IUT to fail even if implemented correctly. Clearly, the test sequence (2) is not realizable without FIFO-type buffers.*

To eliminate this type of non-determinism in test sequences, the IUT model should have a buffer size $b_i = 1$. Then, the maximum number of nodes in G' becomes $|V'|_{max} = |V| * \prod_{i=1}^{F}(1 + c_i)$. In a real testing environment, F, the number of semicontrollable interfaces, is expected to be small. For most cases, the (N)-layer IUT interacts only with an (N+1)-layer implementation and several semicontrollable timers. Typically, for each timer, the only output is the timeout, which defines c_i as 1. Therefore, for small F and c_i, the size of G' is only a small multiplicant of G.

Let us now consider the number of inputs that can be buffered simultaneously at all of an IUT's semicontrollable interfaces. When we allow inputs being buffered simultaneously at several semicontrollable interfaces, even a buffer size equal to 1 may not prevent non-deterministic behavior during testing.

Example (cont'd): *Consider a potential test sequence for Figure 3 generated for buffer sizes of 1 at I_1 and I_2:*

$$e14, e13, \underline{e12, e7, e9, e11}, e1, e10, e3, e4, e12, e5, e9, e7, e2, e13, e11, e13, e6$$

Although this sequence avoids the non-determinism due to buffer sizes shown previously in (2), it may still be non-deterministic due to the IUT interacting with multiple interfaces simultaneously. Consider the underlined portion of the above sequence. After e12 is traversed, input $a_{2,1}$ is buffered at I_2. Traversing e7 results in $a_{1,2}$ being buffered at I_1. Since $a_{2,1}$ was generated earlier than $a_{1,2}$, transition e9 is expected to be triggered instead of e2. In reality, due to the unknown response time of the interfaces, $a_{2,1}$ can be applied to the IUT earlier than, later than, or simultaneously with $a_{1,2}$. In this case, the behavior of the IUT will be non-deterministic, thereby making the test sequence invalid.

To avoid this type of non-determinism during testing, the model presented in Section 4 will be used to generate tests with the restriction that, at any time, a single input may be buffered in only one of the IUT's semicontrollable interfaces. In such case, the maximum number of nodes is $|V'|_{max} = |V| * (1 + \sum_{i=1}^{F} c_i)$.

It is important to point out that, the minimum-length test sequences generated for a restricted model will likely be longer than the minimum-length test sequences for the general model. However, restricted model tests can be used for testing implementations regardless of their interface structure. The introduced restrictions on the buffer size and on the number of buffered inputs help avoid non-deterministic behavior of the SUT during testing. Although the test sequence length is increased by these restrictions, the tests become applicable to many different SUT interface types, as further discussed in Section 7.

5.2 Diagnostic issues during testing

As presented in Section 2, during testing, an IUT typically interacts with several semicontrollable interfaces. Testing is performed under the assumption that all FSM implementations other than the IUT conform to their specifications. Otherwise, it is difficult to tell whether failure occurs in the IUT, or in the external FSM implementation, or at the semicontrollable interfaces.

Example (cont'd): *Consider* $e14, e1, e8, e3, \ldots$*, which is the beginning part of the test sequence given in (2). When this sequence is applied to the IUT, traversal of e1 should cause FSM_1 to send back input $a_{1,1}$. The IUT will move to state A with $a_{1,1}$ buffered at I_1. Suppose that a faulty implementation incorrectly contains input $a_{1,2}$ instead of $a_{1,1}$ at I_1. Then in state A transition e2 will be triggered by $a_{1,2}$, and the IUT will move to state B instead of remaining in state A after e1's traversal. This will happen even when e1 and e2 are implemented correctly. The tester cannot distinguish whether e1's or e2's implementation is faulty, or FSM_1 is not conformant to its specification, or I_1 malfunctioned.*

This practical concern of making diagnostics easier suggests the following guideline: "Test as many transitions as possible without interactions at semicontrollable interfaces." Transitions preferably should be tested when there are no inputs buffered at the semicontrollable interfaces. As a result of this approach, a minimum-cost test generation can be formulated as a Rural Chinese Postman Problem [Lenstra and Kan, 1976], as discussed in Section 7.

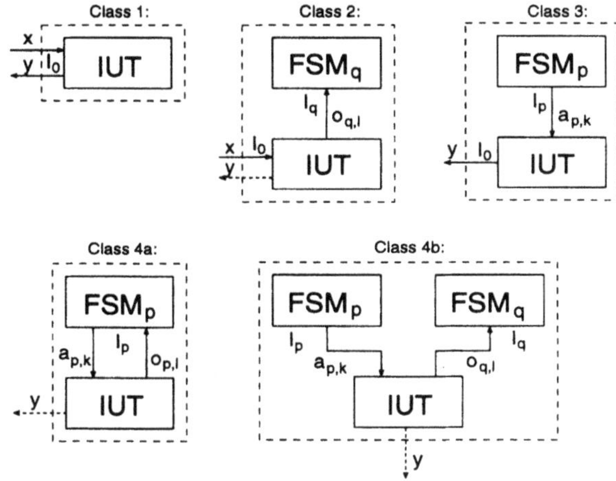

Figure 4 Classes of edges in G' (dashed-lined outputs are optional).

6 GRAPH CONVERSION FOR SEMICONTROLLABLE INTERFACES

Recall from Section 4 that an SUT, represented by variables $\omega_1, \cdots, \omega_F$, and graph G for an IUT, can be modeled as an FSM, represented by $G'(V', E')$. In this section, an algorithm sketch for converting G and $\omega_1, \cdots, \omega_F$ to G' is presented and briefly analyzed. (A detailed description of the algorithm along with its pseudocode is available in [Fecko et al., 1998]). Since the number of vertices in G' is exponential with respect to the buffer size b_i, the algorithm is most applicable for small buffers ($1 \leq b_i \leq 3$).

As discussed in Section 6, the test sequence obtained from G' constructed by the algorithm does not contain any race conditions. Recall from Section 3 that a race condition occurs when an IUT consumes previously buffered input instead of an input defined by a test sequence. Therefore, a test sequence without race conditions will never bring the IUT into a state where there is a buffered input consumable in this state and a different input applied by the LT. Such a test sequence will test the IUT's transitions in a specific order such that these race conditions will not occur. However, a test sequence of G' may still have non-determinism (Section 5). A refinement of the graph conversion algorithm to eliminate non-determinism during testing is also presented. The graph $G'(V', E')$ is built by the algorithm whose sketch is in Figure 5. Let B_i denote a sequence of inputs buffered at the i-th semicontrollable interface. Each state $v' \in V'$ has two components: the original state $v \in V$, and the current configuration of F buffers, i.e., $v' = (v, \hat{B}_1, \ldots, \hat{B}_F)$. The algorithm constructs all possible buffer configurations with up to b_i inputs buffered at I_i. Given an original edge $e = (v_{start}, v_{end}) \in E$, and the current vertex $v' = (v_{start}, \hat{B}_1, \ldots, \hat{B}_F)$, a new vertex v'_{new} is created based on a new configuration $B_1, \ldots, B_F$, i.e., $v'_{new} = (v_{end}, B_1, \ldots, B_F)$. The edge $e = (v_{start}, v_{end}) \in E$ is

$G'(V', E') =$ BUILD-EXPANDED-GRAPH $(G(V, E), r)$

1. initialize r', root of G', as $(r, \emptyset, \cdots, \emptyset)$ (root of G and configuration of empty buffers)
2. initialize E' as empty set, and V' as $\{r'\}$
3. initialize Q, queue of vertices, as V'
4. repeat until Q is empty
   (a) extract $v' = (v_{start}, \hat{B}_1, \ldots, \hat{B}_F)$ as first element from Q, where $(\hat{B}_1, \ldots, \hat{B}_F)$ is current configuration
   (b) for each outgoing edge $e = (v_{start}, v_{end}) \in E$ do
      i. determine k, index of e's class
      ii. given $(\hat{B}_1, \ldots, \hat{B}_F)$ and *Class k* definition, construct:
         - new configuration $(B_1, \ldots, B_F)$
         - new vertex $v'_{new} = (v_{end}, B_1, \ldots, B_F) \in V'$
         - new edge $e'_{new} = (v', v'_{new}) \in E'$
   (c) include new edges in E' iff inputs in $(\hat{B}_1, \ldots, \hat{B}_F)$ cannot trigger other edges outgoing from v_{start}
   (d) append to Q end vertices $v'_{new} \in V'$ of new edges included in E'
5. remove from V' all vertices from which r' cannot be reached
6. remove from E' all edges incident to such vertices

Figure 5 Algorithm sketch for converting graph G into G'.

included in E' as $e'_{new} = (v', v'_{new})$. The edge $e \in E$ belongs to one of the four classes depicted in Figure 4:

Class 1: e is triggered by an input from and generates output(s) to an LT. In this case, e's traversal does not modify the configuration $(\hat{B}_1, \ldots, \hat{B}_F)$ of e's start state. e is included in E' unconditionally.

Class 2: e is triggered by an input from an LT and generates an output $o_{q,l}$ at I_q. e is included in E'_{new} only when the number of inputs already buffered in B_q in the corresponding configuration of state $(v_{start}, \hat{B}_1, \ldots, \hat{B}_F)$ is smaller than the maximum allowable b_q. The new configuration $(B_1, \ldots, B_F)$ is obtained from $(\hat{B}_1, \ldots, \hat{B}_F)$ by appending $a_{q,l}$ to $\hat{B}_q$.

Class 3: e is triggered by $a_{p,k}$ from I_p and generates output(s) to an LT. e is included in E'_{new} only when $a_{p,k}$ is the first element buffered in B_p in the corresponding configuration of state $(v_{start}, \hat{B}_1, \ldots, \hat{B}_F)$. The new configuration $(B_1, \ldots, B_F)$ is obtained from $(\hat{B}_1, \ldots, \hat{B}_F)$ by deleting $a_{p,k}$ from $\hat{B}_p$.

Class 4: e is triggered by an input $a_{p,k}$ from I_p and generates an output $o_{q,l}$ at I_q. If e interacts with only one semicontrollable interface, i.e., $p = q$ (*Class* 4a), then it is included in E'_{new} only when e's input $a_{p,k}$ is the first element buffered in B_p in the corresponding configuration of state $(v_{start}, \hat{B}_1, \ldots, \hat{B}_F)$. Since $a_{p,k}$ is first deleted from B_p, there is always room in B_p for $a_{p,l}$. If I_p and

I_q are different, i.e., $p \neq q$ (*Class 4b*), then e is included in E'_{new} only when it satisfies the inclusion conditions for both *Classes* 3 and 2 edges, as defined above. The new configuration $(B_1, \ldots, B_F)$ is obtained from $(\hat{B}_1, \ldots, \hat{B}_F)$ by applying rules for *Classes* 3 and 2 (in this order).

More complicated cases can be modeled as a combination of these four classes. For example, the case where an input x applied at I_0 goes through IUT, FSM_p, IUT, FSM_q, and IUT, to produce an output y at I_0, can be a combination of *Classes* 2 and 3. The cases where FSM_p and FSM_q directly communicate with each other are beyond the scope of this paper.

It can be shown that $G'(V', E')$ built by the algorithm of Figure 5 is a *minimal valid* representation of the system defined by G and variables $\omega_1, \cdots, \omega_F$ [Fecko et al., 1998, Uyar et al., 1998]. The running time RT of the algorithm is shown [Fecko et al., 1998] to be:

$$RT = \begin{cases} O(c^{bF}) * \sum_{i=1}^{|V|} d_{out}(v)) = O(c^{bF} * |E|) & \text{if } c > 1 \\ O((1+b)^F) * \sum_{i=1}^{|V|} d_{out}(v)) = O((1+b)^F * |E|) & \text{if } c = 1 \end{cases}$$

Based on the practical considerations discussed in Section 5, the algorithm can be refined so that at any given point in time there could be a single input buffered in only one of the buffers B_i, which yields a linear running time of $RT_{ref} = O(c * F * |E|)$.

7 TEST GENERATION FOR PRACTICAL TESTING ENVIRONMENT

Given graphs $G(V, E)$ and $G'(V', E')$, our goal is to find a minimum-cost tour of G' such that each original edge from G is covered at least once. Recall from Section 5 that G' will likely contain multiple appearances of certain edges from the original graph G. Let $E'_c \subseteq E'$ be the subset of edges in G' such that each original edge in E is represented by at least one copy in E'_c. To build a minimum-cost tour of G' covering each original edge in E is equivalent to finding a minimum-cost tour of G' that includes each transition in E'_c (the set of *mandatory* edges) at least once, and each transition in $(E' - E'_c)$ (the set of *optional* edges) zero or more times. This problem is known as the Rural Chinese Postman Problem [Lenstra and Kan, 1976].

The issue that arises in the case of multiple appearances of certain edges in G' is which copies should be included in E'_c. Practical concerns discussed in Section 5 require that copies incident to nodes corresponding to configurations with empty buffers should be marked as mandatory. Note that when the buffer size in each semicontrollable interface is 1, each edge e that belongs to some $U_{i,j}$ or $T_{i,j}$ will appear in G' only once, and therefore will be marked as mandatory. All other edges with a copy incident to the states in V' whose second component is the configuration of empty buffers will be marked as mandatory in this copy. If no such copy exists, the edge will be marked as mandatory in all its copies with the presence of a buffered input. Therefore, each edge in E' will be marked as mandatory in one or more appearances.

[Aho et al., 1991] present a solution to the problem of finding a minimum-cost tour of $G'(V', E')$ covering subset of edges $E'_c \subseteq E'$. The sufficient condition

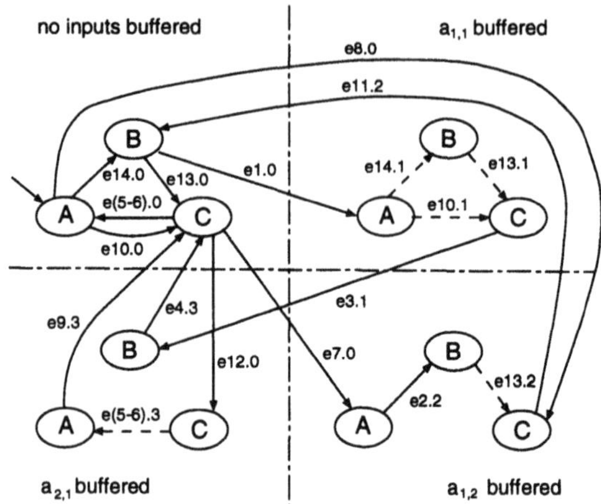

Figure 6 Graph transformation applied to the graph of Figure 3. Mandatory and optional edges appear in solid and dashed lines, respectively.

for the existence of a polynomial-time solution is that $E_c^{'}$ induce a weakly-connected spanning subgraph of $G^{'}$. It can be easily shown that $E_c^{'}$ obtained in the manner described above is a weakly-connected subset of $E^{'}$.

Example (cont'd): *Consider the graph of Figure 3. After conversion to $G^{'}$ (Figure 6), each state is replaced with at most four copies – each corresponding to the buffer configuration at a semicontrollable interface. Each edge e is annotated as e.x, where $x = 0, 1, 2, 3$, depending on the input buffered in the e.x's start state, as shown in Figure 6. Given graphs G and $G^{'}$, the sets E and $E^{'}$ are as follows:*

$$\begin{aligned} E &= \{e1, e2, e3, e4, e5, e6, e7, e8, e9, e10, e11, e12, e13, e14\} \\ E^{'} &= \{e1.0, e2.2, e3.1, e4.3, e5.0, e5.3, e6.0, e6.3, e7.0, e8.0, e9.3, \\ &\quad e10.0, e10.1, e12.0, e13.0, e13.1, e13.2, e14.0, e14.1\} \end{aligned}$$

To build the set of mandatory edges to be included in a test sequence, we adopt the approach discussed in Sections 5 and 7. In $G^{'}$, several edges appear multiple times: e5, e6, e10, e13, and e14. The edges in Figure 6 shown in solid are the mandatory edges that are incident to nodes that correspond to the case where both buffers are empty, i.e., e5.0, e6.0, e10.0, e13.0, and e14.0. The copies that can be traversed only when either buffer contains an input are shown in dashed line: e5.3, e6.3, e10.1, e13.1, e13.2, and e14.1. These are the optional edges, which will be included in the test sequence only when necessary. In this example we have:

$$E_c^{'} = \{e1.0, e2.2, e3.1, e4.3, e5.0, e6.0, e7.0, e8.0, e9.3, e10.0, e11.2, e12.0, e13.0, e14.0\}$$

Given the above sets $E^{'}$ and $E_c^{'}$, the Aho et al. technique gives the minimum-length test sequence for $G^{'}$ shown in Table 2. Steps with (→) indicate that an edge is tested in this step. Note that, for simplicity, the UIO sequences [Sabnani and Dahbura, 1988] are not included in this sequence.

Step	*Edge name*	*Input from*	*Output to*
→ 1	e14	$LT?x_{14}$	$LT!y_{14}$
→ 2	e13	$LT?x_{13}$	$LT!y_{13}$
→ 3	e5	$LT?x_5$	$LT!y_5$
→ 4	e8	$LT?x_8$	$FSM_1!o_{1,2}$
→ 5	e11	$FSM_1?a_{1,2}$	$LT!y_{11}$
→ 6	e1	$LT?x_1$	$FSM_1!o_{1,1}$
→ 7	e10	$LT?x_{10}$	$LT!y_{10}$
→ 8	e3	$FSM_1?a_{1,1}$	$FSM_2!o_{2,1}$
→ 9	e4	$FSM_2?a_{2,1}$	$LT!y_4$
→ 10	e12	$LT?x_{12}$	$FSM_2!o_{2,1}$
11	e5	$LT?x_5$	$LT!y_5$
→ 12	e9	$FSM_2?a_{2,1}$	$LT!y_9$
→ 13	e7	$LT?x_7$	$FSM_1!o_{1,2}$
→ 14	e2	$FSM_1?a_{1,2}$	$FSM_1!o_{1,2}$
15	e13	$LT?x_{13}$	$LT!y_{13}$
16	e11	$FSM_1?a_{1,2}$	$LT!y_{11}$
17	e13	$LT?x_{13}$	$LT!y_{13}$
→ 18	e6	$LT?x_6$	$LT!y_6$

Table 2 Minimum-length test sequence for the IUT of Figure 3.

8 CONCLUSION

In this paper, a testing environment is considered where the IUT communicates with multiple entities with different degrees of controllability on generating inputs. The inputs that these entities apply to the IUT may be directly controllable, semicontrollable, or uncontrollable. This paper introduces a graph conversion algorithm that takes full advantage of the semicontrollable inputs while avoiding race conditions. In a test sequence of such a modified graph, semicontrollable inputs become controllable (where possible). Although, in general, the graph conversion results in an exponentially large number of nodes, practical considerations can make the converted graph size feasible.

Currently, this methodology is being applied to generate tests for MIL-STD 188-220B. Without utilizing the semicontrollable inputs, the number of testable transitions for MIL-STD 188-220B Class the Class A – Type 1 Service Datalink module is approximately 200. With the presented methodology, initial results show that the number of testable transitions increases to over 700, a coverage of more than 95% of the transitions defined in the specification.

References

[188-220B, 1998] 188-220B (1998). *Military Standard - Interoperability Standard for Digital Message Device Subsystems (MIL-STD 188-220B).*

[Aho et al., 1991] Aho, A. V., Dahbura, A. T., Lee, D., and Uyar, M. U. (1991). An optimization technique for protocol conformance test generation based on UIO sequences and rural Chinese postman tours. *IEEE Trans. on Communications*, 39(11):1604–1615.

[Brinksma, 1988] Brinksma, E. (1988). A theory for the derivation of tests. In *Proc. IFIP Protocol Specification, Testing, and Verification, VIII*. North-Holland, Amsterdam.

[Cavalli et al., 1996] Cavalli, A. R., Favreau, J. P., and Phallippou, M. (1996). Standardization of formal methods in conformance testing of communication protocols. *Computer Networks and ISDN Systems*, 29(1):3–14.

[Chan and Vuong, 1989] Chan, W. Y. and Vuong, S. T. (1989). An improved protocol test generation procedure based on UIOs. In *Proc. ACM SIGCOMM*.

[Davis et al., 1998] Davis, M. D., Sigal, R., and Weyuker, E. J. (1998). *Computability, Complexity, and Languages : Fundamentals of Theoretical Computer Science*. Academic Press.

[Fecko et al., 1997] Fecko, M. A., Amer, P. D., Sethi, A. S., Uyar, M. U., Dzik, T., Menell, R., and McMahon, M. (1997). Formal design and testing of MIL-STD 188-220A based on Estelle. In *Proc. MILCOM'97*, Monterey, California.

[Fecko et al., 1998] Fecko, M. A., Uyar, M. U., Sethi, A. S., and Amer, P. D. (1998). Embedded testing in systems with semicontrollable interfaces. Technical Report TR-98-18, CIS Dept., University of Delaware, Newark, DE.

[Fujiwara and v Bochmann, 1992] Fujiwara, S. and v Bochmann, G. (1992). Testing nondeterministic finite state machines. In *Proc. IFIP 4th IWPTS*. North-Holland, Amsterdam.

[Hopcroft and Ullman, 1979] Hopcroft, J. E. and Ullman, J. D. (1979). *Introduction to Automata Theory, Languages, and Computation*. Addison-Wesley.

[IS9646, 1991] IS9646 (1991). *ISO International Standard 9646: Conformance Testing Methodology and Framework*. ISO, Information Technology - OSI, Geneva, Switzerland.

[ISO8802-2, 1994] ISO8802-2 (1994). *International Standard ISO/IEC 8802-2, ANSI/IEEE Std. 802.2*. ISO/IEC, 2nd edition.

[Lenstra and Kan, 1976] Lenstra, J. K. and Kan, A. H. G. R. (1976). On general routing problems. *Networks*, 6:273–280.

[Linn and Uyar, 1994] Linn, R. J. and Uyar, M. U. (1994). *Conformance Testing Methodologies and Architectures for OSI Protocols*. IEEE Comp. Soc. Press, Los Alamitos, CA.

[Miller and Paul, 1994] Miller, R. E. and Paul, S. (1994). Structural analysis of protocol specifications and generation of maximal fault coverage conformance test sequences. *IEEE/ACM Trans. on Networking*, 2(5):457–470.

[Phalippou, 1992] Phalippou, M. (1992). The limited power of testing. In *Proc. IFIP 5th IWPTS*. North-Holland, Amsterdam.

[Rafiq and Castanet, 1990] Rafiq, O. and Castanet, R. (1990). From conformance testing to interoperability testing. In *Proc. IFIP 3rd IWPTS*, pages 371–385, Washington, DC.

[Rayner, 1987] Rayner, D. (1987). OSI conformance testing. *Computer Networks and ISDN Systems*, 14(1):79–98.

[Sabnani and Dahbura, 1988] Sabnani, K. K. and Dahbura, A. T. (1988). A protocol test generation procedure. *Computer Networks and ISDN Systems*, 15:285–297.

[Sarikaya et al., 1987] Sarikaya, B., von Bochmann, G., and Cerny, E. (1987). A test design methodology for protocol testing. *IEEE Trans. Software Engineering*, 13(5):518–531.

[Timohovich, 1993] Timohovich, E. (1993). An approach to protocol entity model development for embedded testing. *Automatic Control and Computer Sciences*, 27(3):34–41.

[Tretmans, 1996] Tretmans, J. (1996). Conformance testing with labelled transitions systems: Implementation relations and test generation. *Computer Networks and ISDN Systems*, 29(1):49–79.

[Ural, 1992] Ural, H. (1992). Formal methods for test sequence generation. *Computer Communications*, 15(5):311–325.

[Uyar et al., 1998] Uyar, M. U., Fecko, M. A., Sethi, A. S., and Amer, P. D. (1998). Minimum-cost solutions for testing protocols with timers. In *Proc. IEEE Int'l Performance, Computing, and Communications Conf.*, pages 346–354, Phoenix, AZ.

[Zeng et al., 1989] Zeng, H. X., Chanson, S. T., and Smith, B. R. (1989). On ferry clip approaches in protocol testing. *Computer Networks and ISDN Systems*, 17(2):77–88.

AUTOMATIC TEST CASE GENERATION OF REAL PROTOCOLS: FRAMEWORK AND METHODOLOGY

[1]Tae-hyong Kim, Ik-soon Hwang, Chan-min Park,
[2]Jai-yong Lee, Sang-bae Lee

Dept. of Electronic Eng., Yonsei Univ.
Seodaemun-gu, Shinchon-dong 134, Seoul, 120-749, Korea
Phone: +82-2-361-2864, Fax: +82-2-312-4584
{[1]thkim,[2]jyl}@nasla.yonsei.ac.kr

Abstract: A lot of researches have been made for an effective and efficient test case generation method. However, theoretical researchers and practical researchers follow the different goals and methods each other especially in the test case generation of real protocols. In this paper, we proposed a new framework and methodology for automatic test case generation of real protocols. They bridge between the theoretical and practical researches and try to bring together the best of theory and practice in test case generation. They can cope with the change of test environment and technical state. The key idea is the distributed analysis technique. We used both simulation and model analysis according to the context dependency of the sequence. We defined test coverage selection criteria of a protocol in an EFSM(Extended Finite State Machine) model and in an EEFSM(Expanded EFSM) model. We used three techniques to reduce the complexity of simulation: the depth first search with termination conditions, the partial supplementation technique and the partitioning technique. To show the efficacy of the proposed framework and methodology, we apply them to a real communication protocol, SSCOP(Service Specific Connection Oriented Protocol) for B-ISDN protocol.

1 INTRODUCTION

Since whether a communication protocol implementation conforms to its standard specification or not is required to be tested, a great number of researches have been made in this area. A lot of studies among them are interested in test case generation method for conformance test. To generate more effective and efficient test cases, they try to generate test cases with higher fault coverage and for data flows as well as control flows in protocols. But there are many problems in those studies. One of them is to overcome the difficulties due to the complex structure and many functions of communication protocols. For the purpose, many efforts to reduce the problem size have been made such as modeling or partitioning of a protocol into a simplified structure, test selection and optimization. There is a trade-off relation between test coverage and the cost of test. Theoretical researchers, however, are more interested in the former issue. It is generally known that much time is required for the generation of test cases as well as the testing itself. Automatic test case generation is, therefore, one of the most important issues, because they will save much time.

Several automatic or semi-automatic test case generation tools such as TGV, TVEDA, TestGen and SAMsTAG have been developed and now they are under improvement(Doldi *et al.*, 1996; Grabowsky *et al.*, 1997). They try to reduce manual efforts in the course of test by the power of FDT(Formal Description Techniques) and their related tools. They also naturally focus on generation of test suite in TTCN(Tree and Tabular Combined Notation) that is directly applicable to a real test. Most of them use the simulation technique to find out the sequences corresponding to the selected test purposes. The simulation technique can also give the values to complete the constraints of the test suites. Their developers, however, do not try to apply techniques of theoretical researches to them yet such as techniques for extending fault coverage of test cases. Theoretical researchers also try to develop automatic tools to generate test cases, but their tools are mostly for validation of their researches and for simple protocol with many assumptions. So it is nearly impossible to use them directly to generate test cases for real protocols.

Like the most cases of other areas, in general, there is also a considerable gap between the theoretical approaches and the practical approaches in the area of conformance test. In industrial context, the most important goal is whether it can be applied to implementations and almost all the tests have real constraints such as deadline and the cost. In academic context, on the other hand, most of tests are free of constraints by some assumptions and the effectiveness and validity are the matter of primary concern. Their viewpoints to generation of test cases for real communication protocols show this gap well. Theoretical researchers will not use a real communication protocol as a target protocol, because they are so much complicated and difficult to model or simplify properly. Practical researchers seldom use the theoretical methods in testing real protocols because they think the methods are impractical and irrelevant for real protocols. Therefore, new test case generation framework to be able to use both the advantages of two approaches is required and we proposed a new framework for automatic test case generation and its methodology in this paper.

The paper proceeds as follows: In section 2, we show the proposed test case generation framework. The next 3 sections explain the methodology of test case generation in the framework: test coverage and test selection criteria are explained in section 3, the distributed analysis method in section 4 and reduction techniques of problem size in section 5. Section 6 shows the empirical results in applying the proposed method to B-ISDN protocol SSCOP(Ser-vice Specific Connection Oriented Protocol). We evaluate the method by comparing with related studies in Section 7. Finally conclusions and future works are given in section 8.

2 OVERALL FRAMEWORK

The characteristics of the required framework are as follows. They should be able to accept the requirements and viewpoints of the theoretical and practical approaches. They should be also able to use the advantages of the two approaches and deal with newly developed theoretical techniques, various test environment and new complex protocols. Considering the above requirements, we proposed a new framework as shown in Figure 1.

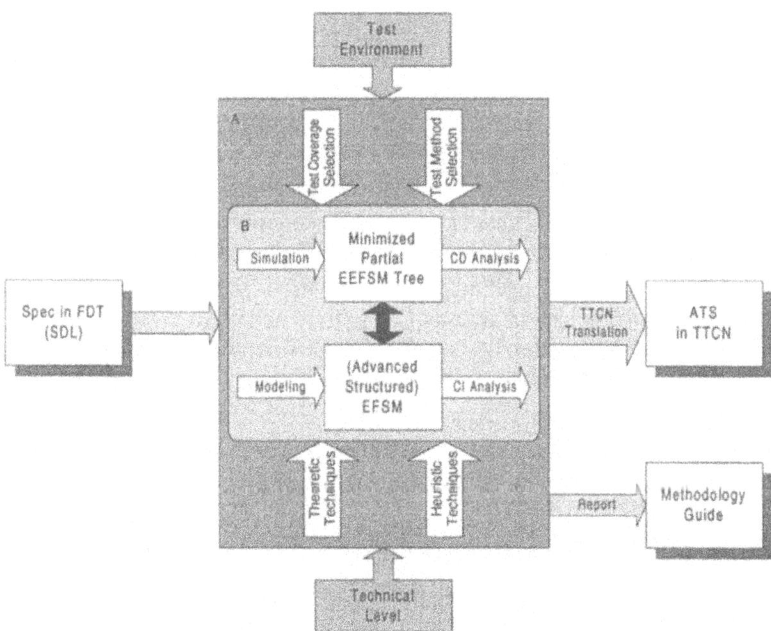

Figure 1 The proposed test case generation framework

The target protocol specification is given in FDT such as SDL. The test case generation procedure is affected by two external factors: the test environment and the technical level. The test environment factor is for several constraints of practical testing and the technical level factor is for new results of theoretical or practical researches. These two factors decide variable elements of this framework. Test constraints like cost or deadline may decide the test coverage of the test case generation method. Physical conditions of test system may select the test method. Where the PCOs(Points of Control and Observations) of the given IUT(Implementation Under Test) or SUT(Sys-tem Under Test) exist will select the specific test method. By the selected test coverage and the test method, the test purposes of the test suite to be generated are decided. On the other hand, newly developed techniques of test case generation may solve the problems existed in the existing method. Theoretical studies or a lot of test experiences can develop the techniques. The test case generation method is developed by those techniques. Thus, this rather dark region 'A' containing the above two methods can allow manual decisions or manipulations, if needed by external constraints. The test coverage selection is explained in detail in section 3 and among the techniques those for reduction of problem size are explained in section 5.

The core of the proposed framework is the distributed analysis method depicted as the region 'B' in Figure 1. In general, while the practical tools to generate test suite in TTCN use simulation to find test cases reflecting test purposes, theoretical researchers try to derive test cases by modeling the specification and its analysis. The simulation technique may take long time and the model analysis is incomplete now. Real communication protocols have several functions like a buffer management and a timer management that are difficult to model easily. The model analysis for all the protocol functions is incomplete and impractical but the analysis will be able to help to reduce the time of simulation, we think. So we used two techniques partially. The parts that are difficult for the model analysis are context dependent parts having predicate statements. Therefore the model analysis is in charge of context independent parts and the simulation technique is of context dependent parts. Section 4 explains these techniques in detail.

By the distributed analysis technique, the values of constraints in the test suite are decided and those results of the analyses are translated into an abstract test suite in TTCN. The selected test method, the techniques and other information for the generation process are reported as a methodology guide. This information is used for a new heuristic technique.

3 TEST COVERAGE AND TEST SELECTION CRITERIA

Lately as the protocol model used for test case generation of conformance test EFSM(Extended Finite State Machine) model is used in many cases, because the FSM(Finite State Machine) or LTS(Labeled Transition System) model cannot detect any error of data flows, and moreover may be impractical because of not considering executability of a test sequence. Therefore, test cases that can find the faults in data flows as well as control flows are required. In the data flow test of a protocol, Weyuker's data selection criteria based on data flow analysis of software testing are widely used as fault coverage selection criteria(Weyuker and Rapps, 1985). But her software model is different from the general EFSM model of a communication protocol; data elements like *def* or *use* are usually considered to be only in edges in an EFSM model of a protocol.

Modification of the criteria is, therefore, required. The modified the data selection criteria for protocols with the test coverage criteria of control flow test are shown in Figure 2.

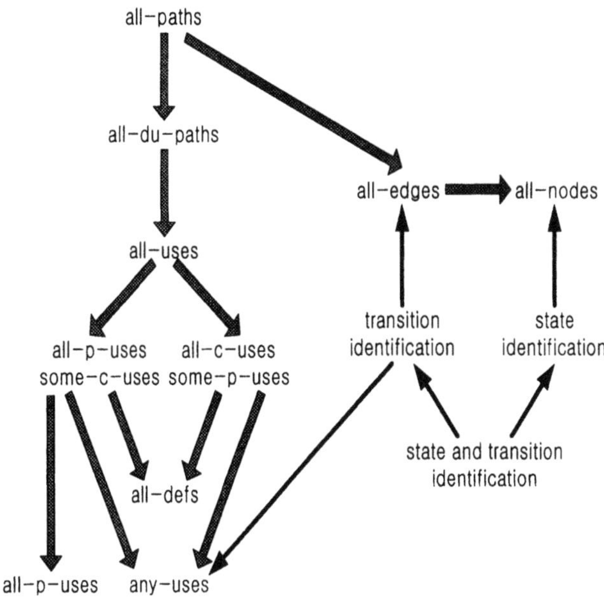

Figure 2 The test data selection criteria of EFSM

The major difference from the Weyuker's is that the test cases satisfying *all-p-uses* or *all-uses* criterion cannot cover the test cases satisfying *all-edges* criterion. There can exist edges without data element such as def or use in an EFSM model of a protocol when we do not consider major states of FSM a kind of variables as usual cases in protocol testing. All the test cases satisfying *all-uses* criterion, therfore, do not have all the edges. We added one criterion in the relations, *any-uses* criterion. The definitions of those criteria are as follows.

Definition 1 The test suite satisfies **any-uses** criterion iff for all *uses* of all the variable x whose edge number is m, there is at least one *def-clear-path* with respect to x from n to m, where n is the edge number of the *def* of the *def-clear-path.*

Any-uses criterion can be used in some practical environments. The test coverage selection criteria of a control flow test depicted together show the relations with data selection criteria. Naturally transition identification criterion covers *all-edges* criterion and *state identification* criterion covers *all-nodes* criterion. Test cases satisfying *transition identification* criterion come to have all the *uses* of the variables and their prior *defs* in natural, in other words they have at least one *def-clear paths* for each *use*. Therefore, *transition identification* criterion covers *any-uses* criterion.

As we mentioned in section 2, we use simulation for context dependent analysis. By simulation, the specification is transformed to a reachabiliy tree. The tree has normally many *homogeneous states* that were originally same states. To reduce the problem size, we use minimized EEFSM(Expanded EFSM) tree instead of the reachability graph as shown in Figure 1(Henniger *et al.*, 1995). The minimized EEFSM tree is generated by merging the homogeneous states having same data values in the reachability tree. In the minimized EFSM tree, the relations between test data selection criteria and test coverage selection criteria are changed as shown in Figure 3.

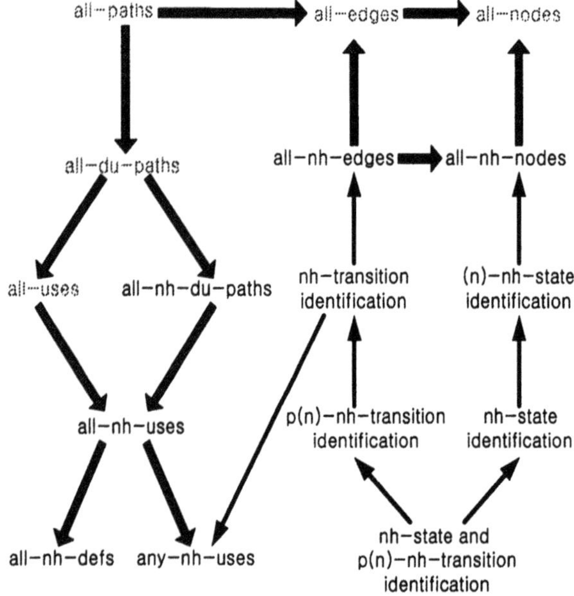

Figure 3 The test data selection criteria of expanded EFSM

In the minimized EFSM tree, there are still not a few homogeneous states, edges and data elements. Consequently, the total number of the states, edges and data elements is so large that too many test cases are required to satisfy the existing criteria. We, therefore, defined *all-nh-du-paths*, *all-nh-uses*, *all-nh-defs*, *any-nh-uses*, *all-nh-edges*, *all-nh-node*, *nh-transition identification* and *nh-state identification* criteria, where the word 'nh' means *non-homogeneous*. The defined criteria are equivalent to the criteria without 'nh' in the case of EFSM.

In test coverage selection criteria for control flow test, we added *p(n)-nh-transition identification* criterion and *(n)-nh-state identification* criterion. A series of input parameter values in a test case that make the test case executable may not be unique. Finite or infinite number of values can be valid for the test case. Which values to select in generating the test case has been an important issue in software testing or protocol testing. How to find out the exact set of values that makes the test case executable is a theoretical interest for many researchers but the problem turned out to be undecidable in normal cases. In case of simulation, it is difficult to find out the set of executable values. We get to know, by simulation, whether a series of specific values makes the test case executable or not only. But by using more than one test cases having different values of input parameters, we can generate a test suite with more test coverage. The *p(n)-nh-transition identification* criterion is satisfied by the test suite which has n test cases having different values for each input parameter. The *(n)-nh-state identification* criterion is for reduction of test case explosion by the state identification parts for all the states. The *(n)-nh-state identification* criterion is satisfied by the test suite that has the state identification parts for the specific n states.

Most test case generation tools generate a test suite satisfying the *nh-transition identification* criterion while theoretic methods use ones having more test coverage such as the *all-nh-du-paths* criterion or the *all-nh-uses* criterion. The *nh-transition identification* criterion has high test coverage to cost ratio but if a test suite having more test coverage is required, the procedure of the next section will generate them effectively.

4 DISTRIBUTED ANALYSIS METHOD

All the constraint parts of a test suite in TTCN can be completed automatically by the distributed analysis method. This method uses two well-known techniques, the simulation and the modeling, as shown in Figure 1. Why is this method needed? Using only one of two techniques, which is usual, did not produce complete results. But two techniques are complementary each other. The modeling technique can reduce the size and time of the simulation technique. On the other hand, the simulation technique can give the executable values of input parameters. Which technique to use to generate test cases for a certain sequence is decided by whether the sequence is context dependent or independent. A sequence is said to be *context independent* if the predicate of every transition in the sequence is independent of any control variable of the specification. The sequence that is not context independent is called a *context dependent* sequence. To know the executability of the sequence, simulation is required in most cases.

The test case generation procedure using the distributed analysis method is shown in Figure 4. First, the protocol specification is modeled in an EFSM for model analysis. In this paper, we confined the analysis model to a single module EFSM without local signals. In case of the protocol specification that has communicating multiple modules, each module is modeled an EFSM individually without considering concurrency between modules. Test case generation for a concurrent system is still an open problem. When the target EFSM model is built, several fundamental test components are generated by its analysis. The test components are the *preambles* to each state, the *postambles* from each state and *def-clear path candidate sets*. We first search for the preambles and

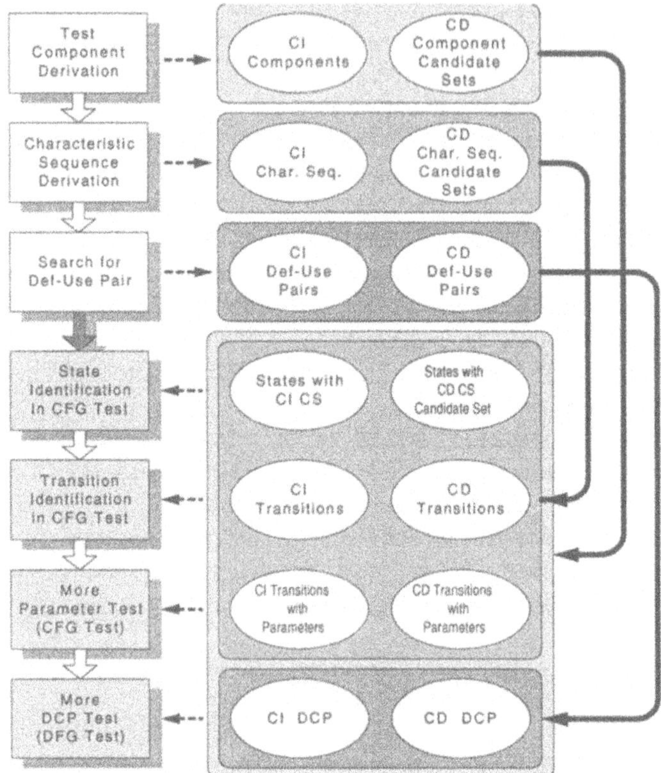

Figure 4 The test case generation procedure using the distributed analysis method

the postambles that are context independent. Only one context independent preamble or postamble for each state is found out and used. If failing to find out any context independent component for a state, we search for all the context dependent components for the state and build the context dependent component candidate set for the state. Each element in the set is a candidate for a preamble or a postamble. Then, CS(Characteristic Sequences) of states in the model are derived in the same way as test component derivation. The context independent CS are generated in the first place and the context dependent CS candidate sets are generated for the state having no context independent CS. Then, we search for *def-use pairs* for data flow test. A def-use pair is context independent if the edge of the *def* has no predicate. Otherwise, the def-use pair is context dependent.

After those basic test components are generated, main test cases for control and data flow test are derived by direct assembling or simulation. In normal cases, generation of test cases for state identification is optional while test cases for transition identification is necessary. State identification verifies the uniqueness of CS, so it increases the fault coverage of a test suite(Anido and

Cavalli, 1995). For state identification, input sequences of all CS are fed to each state to check if generated output sequences of each CS are unique. To reduce the problem size, we can select *(n)-state identification* criterion. If there are some states having context dependent component candidate sets or CS candidate sets, we can pick up the elements fit for the other given components by simulation. Then, test cases satisfying transition identification criterion are generated. As we know, these test cases are for detection of transfer errors and output errors of transitions. The test case of a transition is composed of the preamble, the transition, CS and the postamble. If you can not find out a context independent element for any element, we can pick up the proper element in the relevant candidate set by simulation. You can increase the test coverage of the test suite by generating test cases satisfying *p(n)-nh-transition identification* criterion additionally. As shown in Figure 2 and 3, *transition identification* criterion covers *any-uses* criterion. Test coverage of any uses criterion is almost lowest among data selection criteria. Test cases for data flow test has been disregarded to a test suite designer for its big size. More test coverage for data flows, however, may be needed for more complete test. Technical advance may be able to pay the expense of data flow test. In any case, if additional test cases for data flow test are required, we can select one of the data selection criteria, find out the proper def-clear paths and generate the test cases by assembling or simulation. *All-def* criterion or *all-uses* criterion will be a good selection for its efficiency, we think.

We used the DFS(Depth-First Search) method as the search method in simulation. The search method is explained in next section with other reduction techniques of problem size. The efficacy of this distributed analysis method is discussed in section 6 with the empirical results.

5 PROBLEM SIZE REDUCTION TECHNIQUES

We used three problem size reduction techniques in the proposed methodology: *the DFS with termination conditions*, *the partial supplementation technique* and *the partitioning technique* to reduce the complexity of simulation.

Normally in reachability analysis, the BFS(Breadth First Search) method is used, because the reachability graph is infinite. In this paper, however, we used the DFS method because we need the reachability graph of a specific part rather than that of the whole specification. Because we have some information about the way to the target sequences, we can find out the executable path to the target sequences faster. We used termination conditions of simulation to prevent infinite simulation. Two conditions were used for termination of simulation: the characteristics of the minimized EEFSM and the satisfaction of the selected test coverage. The minimized EEFSM is generated by merging the homogeneous states having same data values. So if the homogeneous state having same data set as the already generated states are generated by simulation, the simulation is terminated. If all the sequences satisfying the selected criterion are found out, the simulation is also terminated.

We used *the partial supplementation technique* to reinforce the fault coverage of only a specific part of the specification. We can select some states and generate test cases satisfying *(n)-state identification* criterion for the states only. We can also select some input parameterized transitions and generate test cases satisfying *p(n)-transition identification* criterion for the transitions only. In the same way, we can select some variables and generate test cases satisfying higher level of criterion among data selection criteria for the variables only.

The specification has several control variables. But some control variables may lie in a specific part in a functional group. *The partitioning technique* makes use of this feature and reduce the number of control variables partially by the functional partitioning. Each partition has, therefore, smaller control variable than the whole specification, but has full information about its control variables. If the values of the control variable of the partition change, a new state having the changed data of its control variables is created. For the control variables of other partitions but not of the present partition, each partition has the data set of the variables. If the values of one of the variables change which are not the control variables in the present partition, instead of creating a new state, only the changed value is added to the data set of the variable. Only after the execution path may go out to other partitions where the variables is the control variable, according to the elements of the data set, new states are created. This technique with the above termination conditions reduces the size of the generated EEFSM by simulation considerably.

We did not use any optimization technique such as the RCP method additionally to reduce the number of test cases. Because optimization techniques will blur the test purposes and decrease the test coverage of the generated test cases in most cases(Anido and Cavalli, 1995).

6 EMPIRICAL RESULTS

To show the efficacy of the proposed framework and methodology, we applied the methodology to the SSCOP (Service Specific Connection Oriented Protocol) used in the B-ISDN AAL(ATM Adaptation Layer). This protocol is often chosen for conformance test because its SDL specification was standardized in ITU-T and its abstract test suite in TTCN was released in ATM FORUM. Besides, several test case generation tools such as TVEDA or SAMSTAG also generated abstract test suites of the protocol.

For generation of test cases, we adopted DS(Distributed Single layer) method, because an upper interface to the IUT exists in the method, which make it possible to generate desirable test cases with high test coverage. As the test coverage of a test suite, we selected both *transition identification* criterion and *(6)-state identification* criterion for control flow test and *all-defs* criterion mainly and *all-uses* criterion partially for data flow test. The partially selected criterion increases fault coverage of the specific part. For addition of test cases satisfying *(6)-state identification* criterion, we selected the state set, { S0(Idle), S2(Outgoing Connection Pending), S10(Data Transfer Ready), S7(Outgoing

Recovery Pending), S8(Recovery Response Pending), S4(Outgoing Disconnection Pending) }. For partial addition of test cases satisfying *all-uses* criterion, we selected the data element set, { VT(S), VR(R) }.

SSCOP has 10 states and more than two hundred edges. The state, S10 is for data transfer and the edges from/to the state are rather complex. The edges have cascaded predicates and inside recursions. We cut the edges into each segment subedges. Thus we generated an EFSM model for the specification that has 10 states, 150 edges and 98 subedges. By model analysis, we generated the preamble of each state that is context independent and the context independent postamble of each state. We derived each context independent CS for 7 states but for the states, S8, S9(Incoming Recovery Pending), and S10, we generated context dependent CS candidate sets whose elements are 2, 3 and 26 each. The SDL specification of SSCOP has 12 variables, 8 PDU parameters, 5 SSCOP parameters, 7 signal parameters, 5 timers, 15 PDUs and 23 signals. We considered only variables and timers for data flow test. The numbers of *defs* to constants/*defs* of the variables, VT(PS), VT(A), VT(PA), VT(MS), VT(PD), VT(CC), VT(SQ), VR(H), VR(MR) and VR(SQ) are 17(10), 12(12), 11/11, 21/21, 17/20, 20/24, 1/17, 18/18 and 19/19 each. The numbers of the pairs *use* to predicate or output/*def-use pairs* of the variables, VT(S) and VR(R) are (10,32)/(11, 33) and (11,9)/(13,12) each.

We used *the partitioning technique* to reduce the complexity of simulation. We separated the complex data transfer part around the state S10 and partitioned it into five parts as the input. The number of the control variables for the whole specification are twelve or all the variables, but the numbers of control variables for six partitions are 2, 3, 3, 2, 2 and 5 each. This partitioning technique, therefore, could reduce the size of generated EEFSM effectively.

We generated 72 test cases satisfying *(6)-state identification* criterion for the six states and 205 test cases satisfying *transition identification* criterion for all the edges and subedges for control flow test. For data flow test we generated 159 test cases satisfying *all-def* criterion for the ten variables and 376 test cases satisfying *all-uses* criterion for the two variables. Because of not using any optimization technique, the number of the test cases for data flow test is rather large.

7 RELATED WORKS AND EVALUATION

A lot of researches has been made for automatic generation of test cases for both control and data flows(Chanson and Zhu, 1993; Kim, 1995; Ramalingom *et al.*, 1995; Bourhfir *et al.*, 1997). Those researches modeled a single module of a protocol to a single module EFSM and generated test cases satisfying transition identification criterion for control flow test and *all-du-paths* or *all-uses* criterion for data flow test. They used the techniques to give the executable values of input parameters like linear programming or CIUS(Context Independent Unique Sequences) for not considering executability of the sequences. However, the techniques are valid only in the assumptions that are too strong for the real protocols. The CIUS method is not always applicable because there are many protocols that do not have the CIUS for all states like SSCOP. In addition, the protocols used for their experiments are too simple compared with real protocols and an EFSM model is too weak to model big and complex protocols such as SSCOP.

TGV, TVEDA and SAMSTAG are automatic generation tools of test suite in TTCN(Doldi *et al.*, 1996; Grabowsky *et al.*, 1997). TGV, TVEDA V3 and SAMSTAG use simulation to search for test cases relevant to given test purposes. They generate a test suite satisfying only *all-edges* criterion. TGV uses feeds and filters to reduce the problem size. But because they exclude some messages and edges, they may not be able to find out test cases relevant to a certain test purpose. TVEDA V2 uses the syntactic techniques with some constraints and generates test purposes automatically by heuristic methods. But constraint parts of a test suite generated by TVEDA V2 are incomplete and require manual completions. TVEDA V3 overcame this weakness by using simulation but it still ignores test suite having more test coverage. SAMSTAG needs test purposes in form of MSCs, from which test cases are generated by simulation. But it fails to find out test cases for a considerable amount of test purposes. Developers of these tools are now interested in completing the tools rather than in generating more powerful test suites.

ATM Forum released an abstract test suite in September 1996(ATM FORUM, 1996). The test suite uses the verifying sequence for each state and attaches it to the transition relevant to each test purpose. The test suite, therefore, satisfies almost *transition identification* criterion. But in the test suite, a considerable amount of test cases are omitted including the ones for transitions starting from four states. And the test suite uses the same verifying sequence for each state, which may make some test cases not executable.

The proposed framework and methodology consider those weakness mentioned above. They are interested in test cases having higher test coverage as well as the generation of executable test cases for real protocols by using both model analysis and simulation.

8 CONCLUSIONS AND FUTURE WORKS

In this paper, we proposed new framework and methodology for automatic test case generation for a real protocol. The distributed analysis technique, the core of the proposed framework, makes use of the advantages of model analysis used in theoretical methods and simulation used in practical methods. We can select one of the redefined test coverage criteria for generating a test suite having higher test coverage. State space explosion by simulation can be reduced by the depth first search with termination conditions and the partitioning technique. We generated the test cases of SSCOP satisfying *(6)-state identification* criterion and *transition identification* criterion for control flow test and satisfying *all-defs* criterion and *all-uses* criterion for data flow test. The test cases were generated by a simple tool with a simple simulator partially and by manual efforts partially.

We are planning to develop a complete tool where the proposed framework and methodology are implemented. Besides, we are studying a method to cope with communicating multiple module protocols and time-relevant protocols for multimedia communication.

References

Weyuker, E. J. and Rapps, S. (1985). Selecting software test data using data flow information. *IEEE trans. on Software Engineering.*, **SE-11(4)**, 367–375.

Aho, A. V., Dahbura, A. T., Lee, David and Uyar, M. U. (1988). An optimization technique for protocol conformance test generation based on UIO sequences and Rural Chinese Postman Tours. *Protocol Specification, Testing and Verification '88.*, Atlantic City, USA, 75–86.

Lee, Do-young and Lee, Jai-yong (1991). A well-defined Estelle specification for the automatic test generation. *IEEE trans. on Computer.*, **COM-40(4)**, 526–542.

Ural, H. and Yang, B. (1991). A test sequence selection method for protocol testing. *IEEE trans. on Communications.*, **COMM-39(4)**, 514–523.

Chanson, S. T. and Zhu, Jinsong (1993). A unified approach to protocol test sequence generation. *IEEE INFOCOM '93.*, San Francisco, USA, 106–114.

Li, Xiangdong, Higashino, T., Higuchi, M. and Taniguchi, K. (1994). Automatic generation of extended UIO sequences for communication protocols in an EFSM model. *Int. Workshop on Protocol Test Systems '94.*, Tokyo, Japan, 213–228.

Anido, R. and Cavalli, A. (1995). Guaranteeing full fault coverage for the UIO-based testing methods. *Int. Workshop on Protocol Test Systems '95.*, Evry, France, 221–236.

Henniger, O., Ulrich, A. and König, H. (1995). Transformation of Estelle modules aiming at test case generation. *Protocol Test Systems 95.*, Evry, France, 45–60.

Kim, Tae-hyong (1995). Automatic generation of observation-based and length-optimized test cases for EFSM model in conformance testing. *M.S. thesis.*, Dept. of Electronic Eng., Yonsei University.

Park, J. H., Hong, J. P. and Lee, J. Y. (1995). A conformance testing framework for applying test purposes. *Protocol Test Systems 95.*, Chapman and Hall, 1995.

Phalippou, M., Guerrieri, S. and Stokar, D. (1995). INTOOL/CATG - D2: User requirements. *Document INTOOL/CATG/EC/4.*, Rev 5.

Ramalingom, T., Das, A. and Thulasiraman K. (1995). A unified test case generation method for the EFSM model using context independent unique sequences. *Protocol Test Systems 95.*, Evry, France, 289–305.

The ATM Forum Technical Committee (1996). Conformance Abstract Test Suite for the SSCOP for UNI 3.1. *af-test-0067.000.*, Sep. 1996.

Doldi, L., Encontre, V., Fernandez, J., Jeron, T., Bricquir, S., Texier, N. and Phalippou, M. (1996). Assessment of automatic generation methods of conformance test suites in an industrial context. *Testing of Communicating Systems 96.*, Chapman and Hall, 347–361.

Lai, R. (1996). How could research on testing of communicating systems become more industrially relevant?. *Testing of Communicating Systems 96.*, Chapman and Hall, 3–13.

Bourhfir, C., Dssouli, R., Aboulhamid, E. and Rico, N. (1997). Automatic executable test case genration for extended finite state machine protocols. *Testing of Communicating Systems 97.*, Chapman and Hall, 75–90.

Chin, B. M., Kim, T. H., Jang, M. S., Hwang, I. S., Lee, J. Y. and Lee, S. B. (1997). Generation of Reliable and Optimized Test Cases for Data Flow Test with a Formal Approach. *ICOIN-11.*, Taipei, Taiwan.

Grabowsky, J., Scheurer, R., Dai, Z. R. and Hogrefe, D. (1997). Applying SAMsTAG to the B-ISDN protocol SSCOP. *Testing of Communicating Systems 97.*, Chapman and Hall, 397–415.

Rayner, D. (1997). Future directions for protocol testing, learning the lessons from the pass. *Testing of Communicating Systems 97.*, Chapman and Hall, 3–17.

Kim, T. H., Hwang, I. S., Jang, M. S., Kang, S. W., Lee, J. Y. and Lee, S. B. (1998). Test Case Generation of a Protocol by a Fault Coverage Analysis. *ICOIN-12.*, Tokyo, Japan.

Invited Talk II

Quality of Service in Integrated Networks

Prof. Paul J. Kuehn, University of Stuttgart, Germany

Abstract
Communication Networks migrate towards an infrastructure, which ischaracterized by seamless interoperation of heterogeneous subnetworks with different protocol architectures, network technologies, and services. The user will not be aware of this heterogeneity, but relies on a secure, reliable and efficient operation supporting his applications. To provide such a quality of service, communication networks have to meet stringent quality requirements with respect to

- functions and features

- availability and reliability

- security

- real time performance

- speed and throughput.

For the individual requirements particular methodologies have been developed, such as specification, verification, implementation, modeling, simulation and testing techniques. These techniques reflect the particular needs and are only partially interrelated. Often, they are applied (if at all) sequentially and independently. The extended functionalities of future communication network infrastructures indicate a more integrated approach.
In the talk we will define the requirements of enhanced methodologies. Some examples will be given to elucidate the needs such as integrated function and performance testing of switching systems or feature interaction. FDTs may be rather useful but they are limited to closed subsystems. Integrated methodologies have to be supported by standardization, built-in capabilities and tools.

PART FOUR

Methodology 1

Enumeration protocol in Estelle: an exercise in stepwise development

P. Dembiñski
*Institute of Computer Science, Polish Academy of Sciences**
Ordona 21, 01-237 Warsaw, Poland
e-mail: piotrd@ipipan.waw.pl

Abstract

The paper sketches a design process that starts with a mathematical solution of the enumeration problem and ends up with a distributed, efficient, and successfully terminating protocol in the specification language Estelle. The subsequent versions of the protocol are obtained from their predecessors by well controlled and small modifications that (almost) obviously preserve correctness. The target, successful (successful with arbitrarily high probability) solution allows forgetting theoretical "impossibility results", which prove non-existence of such successful distributed solutions for some classes of networks.

Keywords

Distributed algorithms, Formal Specification, Verification, FDT Estelle.

1 INTRODUCTION

The enumeration problem is one of a number of similar problems (election of a leader, mutual exclusion, termination, etc.) that have been studied because of their importance for distributed systems and networks, i.e., systems composed

---

* Part of this work was written when the author visited INT, Evry, France

of a number of independent processes running in parallel and communicating by messages. A solution to the problem (an algorithm) is considered distributed if the processes run the same algorithm, no process is privileged, and their computation decisions are solely based on local information and information gained through messages coming from neighbour processes. A distributed algorithm in the above sense is called *protocol.*

In the enumeration problem, networks of processes are abstracted by graphs that are: finite, connected, and without self-loops. Nodes correspond to processes and the graph relation represents the communication links between them. The enumeration problem consists in finding a distributed algorithm (protocol) that terminates with all processes (nodes) of a network (graph) *properly* enumerated, i.e., each process of an N-process network gets a unique label from the set {1, 2, ... , N}.

It is known that for some graphs (*ambiguous* graphs) there is no distributed enumeration algorithm that always terminates successfully (so called, *impossibility results*, e.g., in (Litovsky at al., 1993)). Several solutions have been proposed that worked for specific classes of graphs (like, e.g., rings). In (Mazurkiewicz, 1997), an elegant protocol, that works for all unambiguous graphs, is presented and proved correct. The solution is kept at a high-level of abstraction to reduce the interference of particular realization choices to a necessary minimum.

The aim of this work has been to implement the Mazurkiewicz protocol, i.e., the protocol from (Mazurkiewicz, 1997), in the specification language Estelle (ISO, 1989) and to experiment with Estelle supporting tools EDT (Estelle Development Toolset, (Budkowski, 1992)). Also, the work seemed a good case study to show that such an implementation can be obtained in a way that preserves correctness of the original solution proved at a higher, mathematical level. This occurred feasible, but it also occurred that trying to translate, as close as possible, the main protocol ideas from (Mazurkiewicz, 1997) into environment where sharing variables is not possible (a direct availability of values of internal variables of processes in the neighbourhood of a given process, is one of the simplifying assumptions in (Mazurkiewicz, 1997)), leads to an inefficient and clumsy solution. Moreover, not all properties of this solution were exactly the same as in the original. The obtained protocol is discussed in Sec. 2 with illustrative extracts from its Estelle specification (EP1: Enumeration Protocol 1).

Being not satisfied with the obtained Estelle specification, a modification of the original protocol has been proposed that resulted in a solution which is both economic and has exactly all properties of the original. The solution (EP2) is analyzed and described in Section 3. The way it is obtained from EP1 shows again that its correctness can be deduced directly from this predecessor version, hence, indirectly from its mathematical original. In the following section the notion of ambiguity of a graph is briefly discussed in order to explain non-existence of an (always successful) distributed enumeration algorithm for ambiguous graphs. A very simple extension to EP2 (EP3) makes the protocol "always" successful, i.e., even for ambiguous graphs. The word *always* is appropriate since success can be achieved with arbitrary high probability. Again, correctness of EP3 protocol follows directly from the correctness of EP2. A full Estelle specification of EP1, EP2 and EP3 can be found in (Dembiñski, 1996). The last "probabilistic" solution led the author to a radical reformulation of the protocol into a randomized one, but this reformulation is the subject of a separate paper (Dembiñski, 1998).

2 MAZURKIEWICZ ALGORITHM AND ITS IMPLEMENTATION BY A NETWORK OF IDENTICAL COMMUNICATING PROCESSES.

The enumeration protocol in (Mazurkiewicz, 1997) is assumed to be performed by a network of sequential processes, each one for a node of the graph to be enumerated (named). Similarly as for the graph itself, we use the term *neighbour* (*neighbourhood*) of a process to mean another process (all processes) representing a node (all nodes) directly in the graph relation with the node of the process. Each process has its main local variables (we are using slightly different notation than in (Mazurkiewicz, 1997)):

own_name - storing the current name (number) assigned to the process (initially all processes have no names, i.e., *own_name* = *0*),
known_names - storing a set of names (numbers assigned to neighbour processes and currently known to the process; initially, *known_names* = ∅),
knowledge - storing a set of pairs (*name*, *set_of_names*) representing the current knowledge of names of other processes in the network and their *known_names* at the time the knowledge was acquired (initially, *knowledge* = ∅)

In the informal description of the algorithm in (Mazurkiewicz, 1997), the author explains the local behavior of each graph node by means of message exchange between neighbour nodes. In fact, such an exchange of messages is necessary if we really think of the algorithm realized in a network of sequential processes. However, in (Mazurkiewicz, 1997), for the sake of simplicity of the correctness proof of the enumeration protocol, an indivisible, atomic computation step of each node is realized as if each process had a direct read/write access to all local variables of its neighbourhood. The atomic step (identical for each process) repeated until the *termination_condition* is satisfied can be described, in Pascal-like terms, as follows:

repeat
  if *knowledge_in_the_neighbourhood_is_not_the_same* **then** *unify*;
  if *renaming_condition* **then** *rename*
until *termination_condition*

Without going into details of the conditions and procedures, it is worth to point out that the condition *knowledge_in_the_neighbourhood_is_not_the_same* and the procedures *unify* and *rename* use and change values of local variables of the neighbourhood processes. That is why the whole step is kept atomic. With such an approach elegant proofs are provided for the following results:

1. The protocol always terminates. If graph is not ambiguous (see Section 4) the protocol terminates with its nodes properly (or successfully) enumerated (*successful termination*). If graph is ambiguous the protocol may terminate in a state in which no process can continue (*unsuccessful termination*) but the graph nodes are not properly enumerated (see example in Section 4).

2. The protocol may reach any given proper enumeration for an arbitrary finite and connected graph, i.e., even for ambiguous graphs the protocol may reach a proper enumeration.

3. If the protocol terminates successfully, then each node knows about this termination. Moreover, each node has enough information to reconstruct the whole graph (i.e., the protocol solves the graph recognition problem).

4. The protocol solves the election problem (see, e.g. (Bougé, 1988)) for all unambiguous graphs (it chooses a leader-node and any node can be chosen as a leader).

5. The protocol is distributed in that, in principle, processes that do not have a common neighbour can run in parallel (the parallelism is modelled by interleaving). Otherwise, their actions must be mutually excluded.

The proofs in (Mazurkiewicz, 1997) follow from a key invariant property of the above algorithm. The property says that each computation step preserves so called *connectivity of the graph enumeration.* Assume that a graph has N nodes. An *enumeration* is a function from the graph nodes V into the set $\{1, 2, \ldots, |V|\}$. An enumeration f is called *connected* if, for every x in V such that $f(x) = n+1$, there is y in V with $f(x) = n$. The proof of invariant goes by induction. Of course, if nodes have no names (all names are set to 0), then the property holds. Each step may rename a node and therefore define a new enumeration which is increasing its maximal value. The first one is obviously connected (it gives 1 as a name to exactly one node). It is sufficient then to prove that a computation step cannot destroy this connectivity. Finally, if a node gets $|V|$ as its name, then by the connectivity argument, a proper enumeration is achieved and the algorithm terminates with each node having knowledge about the whole graph. To complete the proof it should be added that, for an unambiguous graph and for any non-terminal state, eventually a node is renamed (i.e., a new enumeration function is eventually created).

Our first goal is to specify a distributed version of the protocol, i.e., such in which local variables of a process cannot be accessed by another one. Exchange of messages can be the only way of knowing their values.

Below, we describe a distributed solution (EP1) in terms of a (well known) model assuming that each process is described by a finite set of *guarded commands*. A *guard* is a boolean condition over a space of the process *internal states* and the contents of its (unbounded) *communication mailbox* storing incoming *messages*. A *command* defines an *action* that may change an internal state, may remove some messages from the process's mailbox, and it may send messages to some other processes *connected* to the given process. A message sent to a process is (immediately) appended to its mailbox. The action defined by a process command is assumed *indivisible* (or *atomic*). The communication is *asynchronous* in that a process may always send a message or, in other words, operations of sending and receiving messages are not synchronized. A process, in each of the computation steps, performs one of the commands from among those for which their guards evaluate to *true*. The behavior of a *network of processes* is described by interleaved sequences of processes' actions.

Specifying the Mazurkiewicz protocol in the above model is fairly straightforward if one is not going into details. That is what we are trying to do in this paper: an informal description completed with some corresponding parts in Estelle to illustrate how close an Estelle specification can mimic the colloquial (though precise) language. The basic idea in this first distributed solution is to replace the direct access to the values of local variables of neighbour processes by the exchange of messages that include these values.

Enumeration Protocol (solution) 1 (EP1)

In the description below we refer to the following elements:

- messages: *request* - asking for actual contents of local variables of neighbour processes,
  response - responding to a request and containing the requested information (values of *own_name*, *known_names* and *knowledge*),
  new_knowledge - a message containing, for each neighbour, a newly created value of its *known_names* and a unified *knowledge*, for all of them; one such message is transmitted to each of the neighbours of the process that previously requested information from them.

Estelle definitions of a network node and some types of basic data stored in nodes and exchanged between them:

{ A specification of the architecture of enumeration protocol EP1 is

given, for a graph of size given by 'graph_size' constant and generated by 'graph_creation' procedure in 'initialize' part of the specification. A node of the network (graph) is defined by the module header 'node' and the module body 'node_body' with its input and output ports specified by vectors of length 'graph_size' of interaction points (ip). Ports of the instances of nodes are connected in part 'initialize' of the specification, by communication channels of type 'link(R1, R2)' and according to the edge structure of the graph. Messages circulate through channels. Allowed messages are defined within channel 'link'. Sets of names ('L_set') are defined as sets of integers from the set 'V = 1 .. graph_size', while pairs composed of a name and a set of names have the record 'K_el' definition.
Knowledge is represented by an array of 'max' length. Its actual contents are reduced to 'K_el' elements that are not equal to (max, empty). This not very elegant solution is chosen because of the Estelle restriction that does not tolerate dynamic arrays and forbids pointers in messages.
The 'systemactivity' attribute of the specification means that behavioral semantics are given by interleaving.}

```
specification EP1 systemactivity;
 default common queue;
 const
   graph_size = any integer;
   max = any integer;      {greater than graph_size, eg. graph_size^2}
 type
   zero_one = 0..1;
   V = 1..graph_size;
   neighbour_representation = array[V] of zero_one;
   graph_representation = array[V] of neighbour_representation;
   L_set = set of  0..graph_size;
   K_el = record
          ng : integer;
          known: L_set;
        end;
   K_set = array[1..max] of K_el;
channel link(R1,R2);
   by R1,R2: request(ad:V);
               new_knowledge(l:L_set; k:K_set);
               response(l:L_set; k:K_set; n:integer) ;
 module node activity (ngh: neighbour_representation; nad:V) ;
 ip s_neighbour: array[V] of link(R1);
    r_neighbour: array[V] of link(R2);
 end; {node header}
```

- local node variables *own_name, known_names,* and *knowledge* as described above,

- local node control states *idle, wait_new_knowledge, wait_response* and *termination.*

Estelle counterparts:

```
state
  idle, wait_response, wait_new_knowledge, termination;
var
  known_names : L_set;
  knowledge : K_set;
  own_name: integer;
```

- *termination_condition* that evaluates to true, in a process, if its local variable *knowledge* contains an element named by the number that equals the size of the enumerated graph.

Estelle counterpart:

```
exist i:1..max suchthat (knowledge[i].ng=graph_size)
```

- *renaming_condition* that evaluates to true, in a process, either when the node is nameless (*own_name=0*) or *knowledge* contains an element named the same as the node (i.e., the element's name equals *own_name*) but has a *stronger* (see Section 4) set of names known to it,

- *rename* procedure which assigns a new name to the node represented by the process executing the procedure; the new name is greater by 1 than all names known to neighbours; the procedure also prepares a new, unified *knowledge* for all neighbours and new *nown_names* for each of them; the new values are then communicated to neighbours.

In Estelle, both the *renaming_condition* and the *rename* procedure occur within transitions' bodies and use simple auxiliary functions. We omit them here, since their implementation has nothing specific (they are simple Pascal procedures).

A process initially has its mailbox empty, its local variables initialized as described above, and its local control state *idle.* The meaning of the guarded commands of each process is characterized as follows:

• (sending requests)
The process broadcasts *request* to all its neighbours (asking for actual values of their local variables) and then waits for their responses changing its local

state to *wait_response*, provided: its mailbox is empty, the *termination_condition* is *false* and its control state is *idle*.

Estelle counterpart:

```
from idle to wait_response
    name sending_requests:
    begin
      all i: V do
        if ngh[i]=1 then output s_neighbour[i].request(myad);
    end;
```

- (responding to requests)
  The process receiving the *request* message from a neighbour (the message is in its mailbox) responses to the requesting process sending the current values of its local variables (message *response*) provided its control state is *termination* or it is *idle* and the *termination_condition* is *false*; in the first case the control state remains the same, while in the second it changes to *wait_new_knowledge*; in any case the *request* message is removed from the mailbox.

  Estelle counterpart:

```
from idle to wait_new_knowledge
 any i: V do
  priority low
     when r_neighbour[i].request(ad)
      name response_to_request_i:
      begin
        output s_neighbour[ad].response(known_names, knowledge,
                own_name);  requesting :=ad
       end;
{ Transition priorities plus common queue organization give a possibility of
excluding the non-determinism between transitions. It is explicity excluded
in the model description by guards that are impossible to express in Estelle,
e.g., emptiness of a queue is not expressible. It is also worth to note that, in
Estelle, a message received and served by a transition is, by definition,
removed from the appropriate queue. This is not the case in the original
asynchronous model operating on a mailbox, hence it must be repeated
when needed}
```

- (receiving responses, creating and distributing knowledge)
  If *response* messages arrived to the process mailbox from all neighbours while the process's control state is *wait_response*, then the process checks the *renaming_condition* and if it is satisfied, it performs the *rename* procedure; independently of the condition value the process unifies its current *knowledge* with that included in *response* messages and then this new *knowledge*, with

possibly new *known_names,* for each neighbour (message *new_knowledge*), are transmitted back to neighbour processes; finally the process removes all *response* messages from the mailbox and changes its control state to *idle.*

Estelle counterpart:

```
from wait_response to idle
    provided count=ngh_number  {i.e., all responces already arrived }
     name creating_and_distributing_knowledge:
        begin
          count := 0;
          if (own_name=0) or (exist j:1..max suchthat
              rename_cond(knowledge[j], known_names, own_name))
            then
              begin
                maxname(maxn); new_know(maxn); own_name:=maxn+1
              end;
          all i:V do if ngh[i]=1 then add(resp_k[i], knowledge);
          all i:V do if ngh[i]=1 then {i.e., if i is a neighbour}
            output s_neighbour[i].new_knowledge(resp_l[i], knowledge);
        end;
```

- (receiving new knowledge)

The process receiving *new_knowledge* message from a neighbour (the message is in its mailbox) either ignores it, if it was already in the *termination* state (and stays in the same state) or, being in the control state *wait_new_knowledge*, it actualizes its local variables with the received values and changes its control state to *idle*; then, as always, it removes the *new_knowledge* message from the mailbox.

Estelle counterpart:

```
from termination to termination
    any i: V do
     when r_neighbour[i].new_knowledge
       name ignoring:
       begin end;
from wait_new_knowledge to idle
     when r_neighbour[requesting].new_knowledge(l, k)
       name absorbing_new_knowledge:
       begin
         known_names := l; knowledge := k; requesting := 0
       end;
```

- (terminating)

The process passes to the control state *termination* as soon as its control state is *idle* and the *termination_condition* is satisfied.

Estelle counterpart:

```
from idle to termination
  provided (exist i:1..max suchthat (knowledge[i].ng=graph_size))
    priority highest
     name terminal:
     begin end;
```

It seems completely obvious that connectivity (of enumerations determined by values of *own_names* variables) is preserved by anyone of the above transitions. Hence, (variants of) properties 1-5 could be proved by similar techniques as in (Mazurkiewicz, 1997) (of course, with all changes resulting from the change of model and differences between the chosen asynchronous model and this of Estelle). However, it is worth to point out a slight difference in properties which does exist. To prove the termination property of EP1 a global fairness in computation is needed. Without it, infinite and useless sequences of requests and responses are possible even if graph is not ambiguous. In the Mazurkiewicz protocol checking the **if** conditions do not require any actions, hence no useless actions are initiated. In case of an ambiguous graph such endless computations repeating exchange of information without further progress in enumeration are reflecting the fact that locally one cannot recognize the ambiguity. The global termination is locally known to all processes since a process that achieves its *termination* control state knows that eventually all other will end up in this state.

3 ALWAYS TERMINATING AND ECONOMIC PROTOCOL

The distributed enumeration protocol that is described in the previous section (and implemented in Estelle) has some flows that are inherited from its prototype in (Mazurkiewicz, 1997).

First of all, the amount of information that is exchanged between processes. For instance, the *knowledge* value is always included in a *response* message. The inclusion occurs because the description tries to follow, in an exact way, the original solution. The question is, however, whether it is necessary. If the answer was negative, then all protocol messages can be bounded by the graph size and not by the square of this value as it is in EP1.

Second, one would like a distributed protocol that always terminates, either successfully or not, as it is the case in the original. Moreover, this termination condition should not be dependent upon a fairness property which in its nature is of a global character. Recall, that this fairness property is hidden in the original solution by the assumption that makes value of the enabling condition *knowledge_in_the_neighbourhood_is_not_the_same* instantaneously "known" to the processes without any explicit evaluation procedure.

The protocol specified below (EP2) satisfies the above postulates and preserves all properties of the original solution.

Enumeration Protocol (solution) 2 (EP2)

All initial assumptions of EP1 remain the same for EP2 with the following exceptions and/or additions:

- two messages are added to the set of possible exchange messages (and the contents of the remaining messages are not exactly the same as before):

done - the message, spread over the network, causes a global and successful termination of the protocol,
trans_knowledge - the message transmits a name and a set of names (as also do now *new_knowledge* and *response* messages) and serves to ca newly created name of a process-node, together with names known to it,

- *termination_condition* reduces now to the equality: *own_name=graph_size,*

- *rename* procedure is reduced to produce a new name of a process.

Of course, there are many other details that are different in both solutions, i.e., EP1 and EP2, at the level of Estelle specification. Below we present the specification of types and messages of EP2 to enable the reader to grasp these minor differences. Anyway, the essential difference is in the size of messages (values of all neighbour *knowledge* variables are not transmitted) and the way new knowledge is disseminated. These differences in both solutions can be captured examining respective guarded commands of each node-processes. That is why we present guarded commands of EP2 in the style used for EP1, but Estelle transitions of EP2 are not presented here.

```
L_set = set of  0..graph_size;
K_el = record
            ng : integer;
            known: L_set;
          end;
K_set = ^KK_el;
KK_el = record
              kel: K_el;
              next: K_set
            end;
channel link(R1,R2);
 by R1,R2: done;
            request(ad:V);
            new_knowledge(el: K_el);
            response(el: K_el);
            trans_knowledge(ad:integer; el: K_el);
```

Guarded commands for EP2: (we try to keep similar wording as in EP1 to facilitate a comparison):

- (sending requests)
  The process broadcasts *request* to all its neighbours (asking for actual values of their local variables) and then waits for their responses changing its local state to *wait_response*, provided: its mailbox is empty, the *termination_condition* is *false*, the *renaming_condition* is *true*, and its control state is *idle*.

- (responding to requests)
  The process receiving *request* from a neighbour (the *request* message is in its mailbox) responses to the requesting process sending the current values of its local variables (message *response* which now includes the values of *own_name* and *known_names* only) provided its control state is *idle* and the *termination_condition* is *false*; the new control state is *wait_new_knowledge* and the *request* message is removed from the mailbox.

- (receiving responses, creating and distributing knowledge)
  If *response* messages arrived to the process mailbox from all neighbours while it is in *wait_response* control state, then the process performs the *rename* procedure and its new *own_name* together with its current value of *known_names* is broadcast to neighbour processes (*new_knowledge* message); finally the process removes all *response* messages from the mailbox and changes its control state to *idle*.

- (receiving new knowledge)
  The process, being in the control state *wait_new_knowledge* and receiving *new_knowledge* message (the message is in its mailbox) from a neighbour previously requesting the values of the process's local variables, actualizes its local variables *knowledge* and *known_names* with the received values, changes to *idle*, removes the *new_knowledge* message from the mailbox and broadcasts two *trans_knowledge* messages: one with the values as they were received in the *new_knowledge* and the second one that includes its own values *own_name* and actualised (according to the *new_knowledge*) *known_names* (of course, the first *trans_knowledge* message is not returned to the process sending the *new_knowledge* information).

- (transit of knowledge)
  The process, being in any control state and receiving a *trans_knowledge* message, checks whether the transiting knowledge is or is not known to it (the pair of values is or is not included in its *knowledge*) and if it is not, then it actualizes its own variable *knowledge* and retransmits the *trans_knowledge* message to all its neighbours but the sending one; otherwise, i.e., when it

already knows the transiting knowledge, it ignores the received information; in both cases the message is removed from the mailbox.

- (terminating)
  The process passes to the control state *termination* if only it was in the control state *idle* and either
  the *termination_condition* was satisfied or the message *done* is in its mailbox; in both cases the process sends then the message *done* to all its neighbours.

Essential change in EP2 with respect to EP1 is in that *trans_knowledge* messages are introduced and treated. In EP1, new knowledge was acquired by repeated requests to neighbours whether it was necessary or not. In EP2, new knowledge is spread over the network only if it is really new. An information included in a *trans_message* that is already known to a process is not retransmitted. That way a new information eventually reaches all processes but also its retransmission eventually ends up, i.e., there are no useless and endless loops caused by retransmissions. A *trans_message* is only created if the enumeration is successful. In that situation no other process but the terminating one may proceed. All other processes are in their idling states waiting for a message that can trigger an action. If *done* message arrives, it is retransmitted only once and the process terminates definitely. Therefore, eventually all processes terminate. Anyway, EP2 always terminates, as did the original protocol in (Mazurkiewicz, 1997), either successfully enumerating the graph (all processes are in their *termination* control state) or without a successful enumeration, with all processes waiting forever in their *idle* state.

Again, it seems obvious, that connectivity invariant holds for EP2. The proof of it, provided we've already proved EP1, is by checking whether the invariant is preserved by new transitions. The point is that those changes we are introducing result in better protocol but do not destroy the essence of our correctness proof.

4 IMPOSSIBILITY RESULTS AND ALWAYS SUCCESSFUL PROTOCOL

The reason why a distributed enumeration protocol can fail, i.e., why it can end up without a proper enumeration of a graph, is due to some graph-theoretic property of the graph itself. This property establishes a class of graphs (networks) for which so called impossibility results hold (see, e.g., (Litovsky at al., 1993)). This class of *ambiguous* graphs is defined by a notion of *graph folding* which is a graph morphism such that its restriction to any graph node and its neighbourhood is bijective. Recall, that a graph morphism is a mapping from the set of nodes of the graph into the set of nodes of another graph preserving graph relation (graph edges). Now, a graph is ambiguous if there exists a folding (from the graph nodes onto the nodes of another graph) that is not an isomorphism.

A nice characterization of the class through the notion of graph enumeration (labelling) is given and exploited in (Mazurkiewicz, 1997). Let $G = (V, R)$ be a graph. Denote by $N(x)$ the set of all neighbours of the node x in V. Recall, that any function $f: V \rightarrow \{1, 2, \ldots, |V|\}$ is called *enumeration* of the graph G (the enumeration f is *proper* or *successful*, if $f(V) = \{1, 2, \ldots, |V|\}$). Enumeration f is called *locally unique*, if for each $x, y, z \in V$ such that $f(y) = f(z)$, we have:

a. if $y, z \in N(x)$, then $y = z$
b. $f(N(y)) = f(N(z))$.

Now, a proposition from (Mazurkiewicz, 1997) says that graph $G = (V, R)$ is ambiguous if and only if there exists a locally unique and not proper enumeration of G.

In Fig. 4.1. an example of an ambiguous 6-node graph is given. A folding into a 3-node graph is depicted (dotted lines) together with a locally unique node enumeration of the bigger graph that corresponds to the folding. Both prove the ambiguity of the graph.

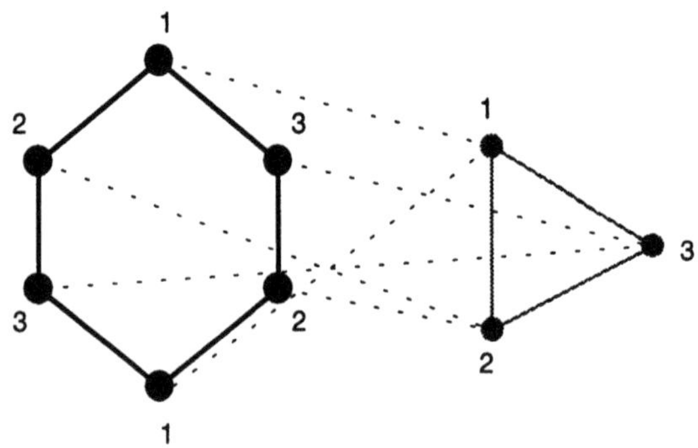

Fig. 4.1.

It is an easy exercise to get this (unsuccessful) enumeration using commands of the algorithm EP1 or EP2 in such a way that the enumeration proceeds as follows: 1, 1, 2, 2, 3, 3. It is then clear that once the above enumeration was achieved there is no hope for a successful termination. In case of EP1, only requests are possible (sending request command) but the responses are always the same, while EP2 terminates (deadlocks) in a state in which no process can continue.

The impossibility results explain that a process (node) cannot locally distinguish itself from another one that looks exactly the same in terms of attributes describing the process (node) itself and its knowledge about all neighbours. In terms of the protocol constructs, this impossibility is reflected in the formulation of the *renaming_condition.* If the attributes are equal, the *renaming_condition* remains *false* and no renaming can occur. This is exactly what happens, for the example in Fig.4.1. in the situation when the given enumeration was achieved. Existence of a linear order between information characterizing processes (nodes) of the same name is absolutely necessary to start renaming and to prevent the situation when all of them start their renaming procedure in parallel. At least one of the processes should maintain the current name to hope for a successful enumeration. In (Mazurkiewicz, 1997) this order, used in the *renaming_condition,* is of lexicographical type between the set of names known to nodes.

In view of the impossibility results, it seems hopeless to think of an always successful protocol. It is true in absolute terms. However, any computation has it own dynamics (ordered history) independent of the problem that is being solved. It may be used to add order to elements that otherwise are identical.

Imagine that every new name created for a process (node), is "stamped" with a value from an ordered set of values (e.g., integers) and that the probability of the same "stamps" to the same name is arbitrary small. Nodes that are equal in previous sense become now ordered by their "stamps". A good random number generator in each node gives a sufficient practical solution for generating of "stamps". This is the whole idea of an enumeration protocol that terminates successfully with arbitrary high probability. Hence "always " terminating in probabilistic terminology. By a slight modification of EP2 we get always successfully terminating enumeration protocol EP3. Its Estelle specification consists in a trivial adaptation of internal, primitive procedures of EP2.

Repeating the exercise performed previously with EP2 for the example in Fig.4.1, will show that with the enumeration given in this figure and achieved by the protocol EP3 there is a continuation since we have assumed, in the protocol, that stamps in such situations are always different. Therefore, there is always a node (process) for which the *renaming_condition* is satisfied. Otherwise, EP3 behaves exactly as EP2, hence all properties satisfied for EP2 are valid for EP3 as well.

5 CONCLUSIONS

We have studied the enumeration protocol given in (Mazurkiewicz, 1997) and provided a version (EP1) of it that does not use shared variables but only exchanges information by messages. This protocol is then implemented (specified) in the protocol specification language Estelle. Atomicity of actions in the new protocol is of smaller "granularity". An improved version (EP2) was then proposed which is economic in message exchange and does not assume a computation fairness implicitly present in the original solution and explicitly in EP1. This new version has exactly the same properties as the original one and solves also such problems as leader election and graph recognition.

A discussion on the source of non-existence of distributed enumeration algorithm for ambiguous graphs, shows its "static" character, i.e., it refers, roughly speaking, to impossibility of distinguishing, by local means, two graphs of different size but locally isomorphic (one graph can be "fold" into another). In any enumeration algorithm, this property translates into a possibility of reaching, from the initial state, a state in which all nodes (processes) are enumerated, but there are nodes that together with their neighbourhood are isomorphic not only with respect to the graph relation but also with respect to achieved enumeration. Such nodes cannot be distinguished, by local means, hence the enumeration procedure deadlocks with the graph not properly enumerated. However, the execution history can make the two nodes distinct even in this statically ambiguous situation. It is enough to "stamp" every newly produced name by a randomly drawn number. This simple idea has been implemented to obtain, in Estelle, a distributed enumeration algorithm that always successfully terminates, if the word *always* is understood in a probabilistic sense, i.e., the success can be assured with arbitrary high probability. The algorithm, EP3, is a simple extention of EP2 described earlier.

The paper describes an exercise (case study) of a protocol design in Estelle starting from very high level, mathematical specification and ending up with its improved and executable prototype. It shows a starting, mathematical solution makes correctness proof much easier though the solution is neither efficient nor immediately executable. Then, in small steps one can obtain an improved version of the protocol that can be almost directly mapped into such a specification language as Estelle. The gain in such design process is that the final result is correct by virtue of the original proof for a mathematical formulation and correctness preserving development steps. Protocols EP1, EP2 and EP3 are all results of such steps.

REFERENCES

Bougé L. (1988), On the Existance of Symmetric Algorithms to Find Leaders in Networks of Communicating Sequential Processes, *Acta Informatica*, **25**, 179-201.

Budkowski S. (1992), Estelle Development Toolset, *J. of Computer Networks and ISDN Systems*, **25.**

Dembiñski P. (1996), Distributed and Always Successful Enumeration Algorithm, IPI PAN Reports **812.**

Dembiñski P. (1998), Randomized Enumeration, IPI PAN Reports **848**.

ISO 9074 (1989), Information processing systems - Open Systems Interconnection - Estelle - A formal description technique based on an extended state transition model.

Litovsky I., Métivier Y., Zielonka W. (1993), The power and limitations of local computations on graphs and networks, *Lecture Notes in Com. Sci.*, **657**, 333-345.

Mazurkiewicz A. (1997), Distributed Enumeration, *Information Processing Letters* **61**, 233-239.

A DESIGN METHODOLOGY FOR THE FORMAL SPECIFICATION AND VERIFICATION OF HYPERMEDIA DOCUMENTS

C. A. S. Santos[£], J.-P. Courtiat[£]
P. de Saqui-Sannes[£,¥]

[£] LAAS-CNRS, 7 avenue du Colonel Roche,
31400 Toulouse - Cedex - France

[¥] ENSICA, 1 Place Emile Blouin,
31056 Toulouse - Cedex - France

Abstract

Hypermedia authoring tools usually suffer from a lack of validation capabilities that would make them capable of checking a document against temporal inconsistencies. The document design method proposed in the paper is meant to overcome this problem. The starting point is a document description provided in a high-level modeling technique featuring hypermedia basic concepts such as nodes (including composite nodes), anchors and links. The high-level document is then automatically translated into a RT-LOTOS formal specification on which classical reachability analysis techniques are applied, making it possible to check the temporal consistency of the document structure. A simple example is used along the text to illustrate the proposed approach.

1. Introduction

A fundamental problem in hypermedia document design is to characterize the temporal structure of such type of document. The temporal structure can be expressed by a conjunction of synchronization constraints that partly depend on user interactions (hence, the hypermedia character of the document). The quality of the document presentation heavily relies on the global consistency of these synchronization constraints.

So far, few work has been done on the identification and analysis of consistency properties for hypermedia documents (Buchanan, 1993a) (Courtiat, 1996) (Lakayda, 1995). Main reasons are twofold:

- Many models do not have any formal background to support automated proofs of consistency properties.
- The importance of these properties has often been minimized, particularly because they are sometimes obvious on models with a limited expression power (like the timeline model (Blakowski, 1996) largely used in first generation commercial authoring tools).

Hypermedia documents have been a reality for many years, but their current and already important deployment (like WEB pages and CD-ROMs) masks several limitations to be dealt with in the products of the next generation. Examples of limitations include the structure essentially static of the documents, and poor semantic models for expressing temporal synchronization, etc.

Increasing dynamism in documents together with the use of more expressive models may result in a non-consistent temporal structure (Blakowski, 1996). Hence the need for improving the design process with a technique able to check the document temporal consistency. Early results on temporal consistency for hypermedia documents have been reported in (Courtiat, 1996), based on a new temporal synchronization model. That approach was however a not general proposal since the synchronization model was directly re-engineered from the underlying formalism, namely RT-LOTOS (Courtiat, 1995), a temporal extension of LOTOS. The high level model used in (Courtiat, 1996), and consequently the hypermedia document design approach, featured three major weaknesses:

- The model was specific.
- It did not handle completely the link concept, which is classical in hypermedia systems.
- It did not distinguish a node type from a node instance; this was considered as the major limitation, since it was not easy to re-use part of a document (a node) in the design of another document.

An enhanced approach is discussed in this paper. It overcomes the previous limitations and generalizes the concept of temporal consistency. The main contribution is the definition of an hypermedia design method which includes an automatic translation from a high-level hypermedia authoring model into a RT-LOTOS specification; this authoring model relies on those objects that are usually found in hypermedia document specifications, namely nodes (including composite nodes), anchors and links. This design method is applied here to the NCM authoring model (Casanova, 1991), and it is currently being implemented within the HyperProp Authoring System (Soares, 1995).

The paper is organized as follows. Section 2 introduces the high level authoring model adopted in this work. Section 3 describes the translation procedure from the authoring model to the RT-LOTOS FDT. Section 4 surveys RT-LOTOS verification techniques, with their RTL tool support, and shows how to apply them to prove hypermedia document consistency properties by reachability analysis. This section also illustrates some verification results that have been obtained on the running example used throughout the paper. Some conclusions are finally given in Section 5.

2. Modeling hypermedia documents with temporal constraints

2.1. Hypermedia authoring systems

Hypermedia authoring systems make it possible to integrate different media types, and to define temporal and spatial[1] constraints among the document components. They offer well suited capabilities for structured document design, including navigation control and presentation monitoring. An authoring tool must be easy to use and sufficiently expressive in order to not impose a cognitive overhead to the authors. It is now admitted that a better expressiveness increases the probability to generate inconsistent documents.

Hypermedia document presentations are real-time applications. Hypermedia presentations also depend on user interactions and are therefore dynamic. For instance, a media content may, depending on some user interaction, be presented either as a text or as an audio; presentation durations are accordingly different, and synchronization constraints are to be defined at presentation time. Consistency verification becomes much more difficult.

Authoring systems usually postpone design validation to the run-time stage. This often happens too late, and the lack of well-established procedures makes the validation process uncertain. On the other hand, a complete formal approach is too complex for end-users, which prefer a high-level graphical language ranging from basic timelines to object-oriented models (Casanova, 1991) (Fraissé, 1996). The approach proposed in this paper is meant to allow the designer to use his or her favorite authoring tool. The latter's output is then automatically translated into a RT-LOTOS formal specification that is checked against design errors, using traditional verification techniques.

2.2. Hypermedia authoring models

Various informal and formal hypermedia modeling approaches have been proposed in order to describe the temporal structures of hypermedia documents. Main approaches have been classified into four categories (Blakowski, 1996):

- Interval-based specifications offer several operators on temporal intervals that characterize elementary media with minimal synchronization constraints to be satisfied in a compound document (Wahl, 1994).
- Axes-based specifications such as temporal lines are widely used in commercial tools but their lack of expressive power is notorious.

- Control Flow-based specifications include hierarchical trees, reference points as well as Petri net based models (Willrich, 1996).
- Event-based specifications are the most expressive, but the price to pay is in the complexity of edition and update functions (Buchanan, 1993b) (Casanova, 1991) (Faissé, 1996).

The hypermedia design approach presented in this paper is currently being implemented for hypermedia documents expressed in the *Nested Context Model* (NCM) the conceptual model of the HyperProp authoring system (Soares, 1995). NCM is a high-level authoring model based on the basic concepts of the hypermedia/hypertext community (such as *nodes*, *anchors* and *links*). It permits also to structure a complex hypermedia document by means of the composite node concept and belongs to the category of event-based specifications as far as the document temporal structure description is concerned. Basic features of the NCM authoring model will be introduced intuitively in the next paragraph. Details on NCM and the HyperProp system are available in (Casanova, 1991) (Rodrigues, 1997). A detailed comparison between NCM and hypermedia models used in Firefly (Buchanan, 1993b), CMIF (van Rossun, 1993), I-HTSPN (Willrich, 1996) and others, with respect to temporal and spatial synchronization aspects as well as to the semantic power of their compositions, can be found in (Rodrigues, 1997).

It is important to stress that the scope of the proposed design approach goes far beyond a particular authoring model or system, and could therefore be adapted to another one.

2.3. Hypermedia documents based on compositions

Hypermedia systems are characterized by the existence of complex relationships among the document components. Such systems also include mechanisms intended to structure the document in order to reduce the so-called *lost in the hyperspace* problem. Such a structured definition is desirable as it carries built-in concepts of modularity, encapsulation and abstraction. Conceptual models with composite nodes, like models that allow nested compositions, support such mechanisms.

Nodes are fragments of information and *links* interconnect them into networks of related nodes. *Anchors* represent an area within the content of a node that may be associated either to the source or destination of a link. The NCM model goes one step further by distinguishing two basic classes of nodes, called *content* and *composite* nodes, the latter being the central concept of the model (Soares, 1995).

Intuitively, the *content node* contains data whose internal structure, if any, is application dependent (they are the usual hypermedia nodes). The class of content nodes can be specialized into other classes (*text, video, audio, image*, etc.), as required by the applications.

A *composite node* C is a node whose content is a collection S of nodes and links. Note that a component may be included more than once in S. However, in this paper, we will assume links and nodes collection as a set, without loss of generality. An important restriction however must be done: a node cannot be recursively contained in itself. It is worth to note that composite node contains nodes and links, generalizing some models that group only nodes in compositions.

Composite nodes have several desired properties, such as:

- Compositions nesting, that is, compositions that contain other compositions.
- Grouping of the components of a document and the relationships among them independently of their types (synchronization relationships for presentation, selection relationships for usual hyperlink navigation, etc.).
- Use of the composite nodes as a new type of node, in all senses, that is:
  - They can be presented - since in a presentation, it is important to exhibit not only the data content of a document, but also its structure as it is specified in the composite node (for example, when one accesses a book chapter modeled as a composite node, besides seeing its content, one may want to visualize its section structuring).
  - Different entry points in a composition can be defined, i.e., in a composition, components may have different presentations, depending on the entry point. For instance, the duration of a composition (duration of its component exhibition) will depend not only on the duration of its components, but also on the associated entry point. In HyperProp system, descriptor objects are used to specify how and with which tool the associated node will be presented. Using distinct descriptors, one can define different presentations for the same data object. For example, a media segment can be synthesized as an audio, using descriptor D_1, or presented as a text using descriptor D_2.
  - Relations among compositions can be defined.
- Inheritance in the composition nesting, in the sense that relations can be defined in a composition *C*, referencing components recursively contained in *C*. This mechanism is extremely important in object reuse.
- Composite node presentation to help user navigation through a document - what can require the use of some filtering mechanism to present the document structure in order to lessen the user disorientation.

2.4. Example

Figure 1 depicts the composite node1 that includes 4 content nodes (node2, node3, node4 and node8), and another composite node (composite node5), which includes two content nodes (node6 and node7). The name of each node is followed by a time interval, which represents respectively the minimal and the maximal duration defined for that node. The duration of the content node may be estimated by the document author depending on the quantity of information to be presented. For a composite node, the situation is a little bit more complex since the termination of the node presentation depends on the internal logic of the node. If one does not want to put any duration constraint on the node, one specifies only $[0,\infty]$ avoiding consequently the risk to induce undesirable temporal consistencies in the node specification. The global behavior of the document is characterized by composite node1, which may informally be described as follows:

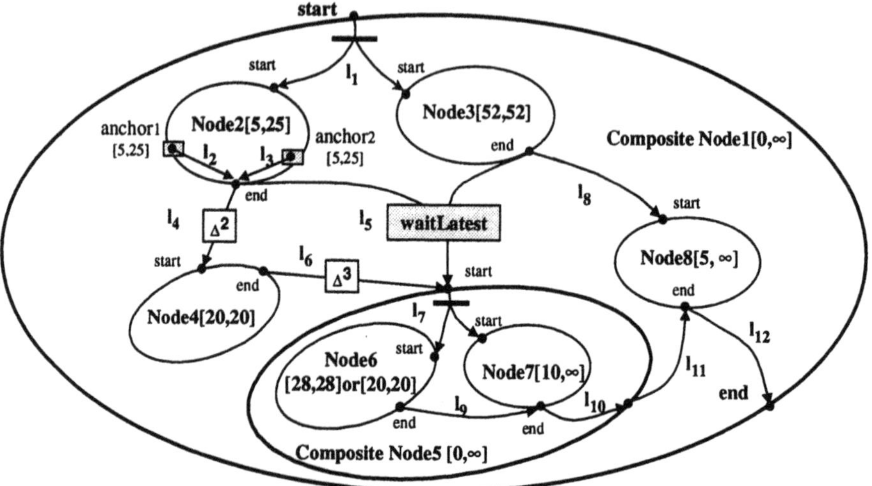

Figure 1 – Structure of the composite node1.

- The *start* of node1 leads to the immediate and concurrent *start* of node2 and node3 (see the link l_1).

- Within node2, two anchors have been defined. Node2 should terminate as soon as one of the anchors has been triggered. The selection of *anchor1* yields to *start* of node 4 starts 2s after the *end* of node2 (link l_4 with delay 2s noted Δ^2). The selection of *anchor2* yields to the immediate *start* of node5, as soon as node 3 terminated (link l_5 with a **waitLatest** logic). It is further assumed that both anchors are timed, i.e. that their activation depend on additional time constraints: they can only be triggered after *dmin* (set here to 5s) and until *dmax* (set here to 25s). Furthermore, one considers that *anchor1* is the default anchor, i.e. it triggers automatically when there is no user interaction during the time interval where both anchors are offered.

- Node3 is a content node representing an audio; its duration is initially set to 52s. The *end* of this node triggers the *start* of node8 (link l_8).

- Node4 is a content node representing a video; its duration is set to 20s.

- Node8 is a content node representing an image; its minimal duration is set to 5s, and its maximal duration is undetermined (set to infinity). In fact, the presentation of node8 will end as soon as composite node5 ends (link l_{11}); finally, the *end* of node8, causes the *end* of node1 presentation (link l_{12}).

- Composite node5 may be started 3s after the *end* of node4 or as soon as the link l_5 is triggered. As a consequence, node5 may be started with two different descriptors depending on the event which triggered its start; these descriptors have an impact on the internal structure of composite node5, i.e. the type of presentation for content node6 (either an audio or a text) and their associated duration. Node 7 contains a video, with a minimal duration of 10s.

- The *end* of node6 leads to the *end* of node7, which itself leads to the *end* of composite node5, which causes the *end* of node8.

3. Translation from the high-level model into a RT-LOTOS specification

The purpose is not to define a new formalism for the description of temporal structures in hypermedia documents. Instead, our goal is to formalize hypermedia objects defined in some high-level authoring model like NCM via a mapping of these objects onto a general-purpose formal description technique (FDT in short). The FDT is therefore completely hidden to the document authors. Having an FDT document specification, verification techniques, such as reachability analysis and model-checking, can be applied for analyzing application oriented properties of the design and consistency properties in particular.

How to perform such mapping is the prime question. The selected FDT should have the following features in order to be useful as well as efficient:

- It must have a formal semantics, and not only an intuitive one.
- It should be based on compositions, in the sense that complex document structures should be expressed by general constructs.
- It should be able to express non-deterministic behaviors, particularly interactions with the external environment (i.e., the users).
- It should have the ability to express complex time-constrained behaviors.
- It should be mature enough, and supported by software tools for automating the verification procedure.

Process algebra meet the first three requirements and LOTOS becomes a strong candidate because it is an international standard. Since standard LOTOS is not able to express time-constrained behaviors, different extension proposals have been made within the international LOTOS community, and are currently being standardized. Among these proposals, we selected RT-LOTOS (Real-Time LOTOS) proposed at LAAS-CNRS, in particular because the RTL software tool is available and operational (Courtiat, 1995). The same approach may easily be adapted to the new emerging E-LOTOS standard once stabilized with an adequate tool support.

3.1. The specification methodology

The approach developed in the paper addresses the formal semantics of the basic objects of a hypermedia document by providing a mapping between these objects and RT-LOTOS processes, and another mapping between object composition rules and RT-LOTOS process parallel compositions. The approach relies on general mapping rules, as well as on the definition of RT-LOTOS process libraries specifying the behavior of reusable basic hypermedia objects, like content nodes, links and anchors. The approach is therefore general and may possibly be fully automated within an authoring system. As previously mentioned, an experimental implementation is currently being developed for the HyperProp authoring system. Even without any full automated translation procedure currently available, various RT-LOTOS specifications of documents authored in NCM have been produced using this general translation procedure which is currently being put in practice by means of templates.

3.2. Hypermedia objects specification

The overall architecture of the specification is presented in Figure 2, where the specification of a hypermedia document (process *Hypermedia*) is composed in parallel with a process representing the environment (process *Users*).

Process *Hypermedia* is itself specified as the instantiation of a process specifying the document top level (composite) node, synchronized with a parallel composition of the processes describing the links defined for the nodes. These process models timed synchronization relationships among all components.

```
...
hide user, start, end, trigger, endRequest in
   Hypermedia[user, start, end, trigger, endRequest] |[user]| Users[user]
where
process Hypermedia[user, start, end, trigger, endRequest] : exit :=
   compositeNode[user, start, end, trigger, endRequest] (...)
       |[start, end, trigger, endRequest]|
            ( ... parallel composition of the different synchronization links ...)
endproc
process Users[user] : exit :=
            ( ... specification of the user interactions ... )
endproc
```

Figure 2 – Specification overall architecture.

The basic structure in a hierarchical composition model of documents is the Composite Node. This general structure is described by the compositeNode RT-LOTOS process definition as presented in Figure 3.

Process *compositeNode* has three formal parameters characterizing respectively the node ID, and its minimal and maximal presentation duration. It features five external gates: gates *start* and *end* associated with the start and end of the node presentation; gates *user* and *trigger* related to the capture of user interactions and to the trigger of an anchor, respectively; gate *endRequest* related to the termination of the node. All these gates are parameterized in order to identify the node ID and the media type (for gates *start*, *end* and *endRequest*), the anchor ID (for gates *user* and *trigger*).

When process *compositeNode* is instantiated with some *ID* value and other parameters, the process first synchronizes on gate *start*. Using the recursive definition of compositeNode, one may note that the composite node may be re-instantiated again. Once the synchronization on *start* is performed, the behavior of the process is essentially described by process *bodyNode* together with the specification of the termination conditions of the composite node (see process *endNode* in Figure 3).

Process *bodyNode* permits to select the body of the current composite node with respect to the node ID; assuming some generic node identifier *i*, the associated bodyNode process is *bodyNodei*; it describes the internal logical structure of composite node *i*; it is made up by the parallel composition of the RT-LOTOS processes characterizing content nodes and/or other composite nodes nested within composite node *i*.

```
process compositeNode[start, end, user, trigger, endRequest, grant]  (ID, dmin, dmax : nat) : exit :=
let composite : nat = 0 in
start !ID!composite;
 ( ( bodyNode[start, end, user, trigger, endRequest, grant](ID)
    [> endNode[endRequest,end] (ID,dmin,dmax) )
            ||| compositeNode[start, end, user, trigger, endRequest, grant] (ID, dmin, dmax) )
where
process bodyNode[start, end, user, trigger, endRequest, grant] (ID:nat) : exit :=
          ...
          [ID == i] -> bodyNodei[start, end, user, trigger, endRequest, grant]
          ...
endproc
process endNode[endRequest,end] (ID,dmin,dmax : nat) : exit :=
let composite : nat = 0 in
          delay(dmin) (endRequest!ID!composite; end!ID!composite; exit)
                    |[end]| end{dmax-dmin}!ID!composite; exit)
endproc
...
process bodyNodei[start, end, user, trigger, endRequest, grant] : exit :=
  ... ||| contentNode[start, end, user, trigger, endRequest, grant] (...)
  ||| ... ||| compositeNode[start, end, user, trigger, endRequest, grant] (...) ||| ...
endproc
process contentNode[start, end, user, trigger, grant] (ID, dmin, dmax, media : nat) : exit :=
    start!ID!media;
    grant!media?dstart:nat!true; delay(dstart) (( delay(dmin) end{dmax-dmin}!ID!media; exit )
       ||| contentNode[start, end, user, trigger, grant] (ID, dmin, dmax, media) )
endproc
```

Figure 3 - compositeNode and contentNode generic process definitions.

The general structure of a content node is described by the contentNode RT-LOTOS process presented in the bottom of Figure 3. This process is much simpler than the previous composite node specification, since there are no more nodes nested within a content node. The content node specification forces the termination of the node according to the minimal and maximal duration defined for this node. As previously for the composite node, one can note that process contentNode may be defined recursively.

As suggested by the excerpts of the formal specification described in, the approach did not try to get the most readable formal RT-LOTOS specification, but instead the most general (with respect to the expressive power of the NCM model) and generic, based on reusable components. The goal has been to completely automate the derivation of the formal specification from an NCM authoring. As a consequence, our specification methodology is based on the identification and formalization of several process libraries characterizing, basic content nodes, basic links, basic composition of nodes and links and basic termination conditions respectively. Some components of these libraries have been inherited from previous work on a simpler hypermedia synchronization model (Courtiat, 1996).

4. Consistency verification

Using formal methods within the context of a high-level authoring tool brings two main advantages. First, it provides a formal semantics to the high-level authoring model, describing without any ambiguity the behavior of a document presentation. Second, it permits to check consistency properties on the formal specification derived from the authoring model, using standard verification techniques.

4.1 Reachability analysis of RT-LOTOS specifications

The verification techniques developed and implemented for RT-LOTOS within the RTL software tool (Courtiat, 1995) consist in applying reachability analysis to the timed automaton model into which the RT-LOTOS specification is compiled. The method implemented in RTL is characterized by two important features:

- It permits to minimize the number of clocks in each control state of the timed automaton, thanks to the definition of the DTA (Dynamic Timed Automaton) model.
- Reachability analysis may be performed on the fly during the DTA generation.

The resulting graph is a minimal reachability graph. Each node of the graph represents a reachable class which may include an infinite number of elements depending on the value of the current time. Each arc of the graph corresponds either to an action occurrence (*i.e.* an event) or to a global time progression (in this case, the arc is labeled by action *time*).

RTL tool has further been interfaced with other tools: BCG for labeled transition systems display, and ALDEBARAN (Fernandez, 1996) for deriving a minimal automaton from a labeled reachability graph with respect to an equivalence relation (for instance, the observational equivalence).

4.2 Consistency properties

Hypermedia documents are expected to satisfy temporal consistency properties stating that temporal synchronization constraints to be met during the document presentation are not in conflict with one another. Depending on how these synchronization constraints are defined, there exists indeed a risk to create inconsistent situations, *i.e.* situations where different contradictory synchronization requirements cannot be satisfied together, leading to undesirable deadlocks (global or partial) during the document presentation.

By definition (Courtiat, 1996), we consider that a document presentation is consistent[2] if the action characterizing the *start* of the document presentation is necessary followed, some (finite) time later, by an action characterizing the *end* of the document presentation (Property *P1*). As a consequence, action *end* should be always reachable from the initial state of the RT-LOTOS minimal reachability graph, assuming that, by construction of the RT-LOTOS specification, action *start* is the unique action to be enabled in the initial state. The definition also implies that there is no divergence (i.e. no undesirable loop) within the reachability graph.

The above definition is related to a document presentation. Within the framework of our authoring model, the definition may also be applied to a composite node presentation.

In order to distinguish different consistency properties, let us call *P1* an *intrinsic* consistency property (Santos, 1998). It is said to be intrinsic because it is independent of any particular multimedia resource used for the document presentation. Many reasons may lead to temporal intrinsic inconsistencies, when one uses a high-level and powerful authoring model like NCM. For examples which seemed a priori not too complex, the following situations were encountered: inconsistencies between the expected duration of the nodes and the logic of the synchronization links enforcing their termination, conflicting synchronization links, bad timing of the synchronization links, omission of some default cases (for instance, what happens for a node featuring several anchors, when none of them is triggered by the user: should the node terminate or not? What the impact on other nodes?)

Another interesting consistency property (called *extrinsic* consistency property, or consistency property *P2*) to be proved when a document presentation is related to potential undesirable concurrent accesses to some multimedia presentation resources (Santos, 1998). A simple example includes an erroneous attempt to mix different audio signals on the same audio channel (sound card + loudspeakers). A complete specification of these consistency properties would probably require to define a high-level RT-LOTOS specification of the multimedia platform on which the document is supposed to be presented and to compose this specification with the RT-LOTOS specification of the document. Such a study is being carried out at LAAS-CNRS, but it is not reported in this paper. We will see however how some extrinsic consistency properties may be proven without developing any formal model of the underlying multimedia platform.

4.3 Verification approaches

Two complementary approaches have been investigated in order to prove consistency properties: global reachability analysis and node aggregation.

4.3.1. Global reachability analysis. This classical approach consists in performing the reachability analysis on the RT-LOTOS specification derived from the entire document. Its main limitation is related to the state space explosion problem. If one nevertheless succeeds in building the reachability graph, the latter can be transformed into a labeled transition system (LTS) whose observable events are appropriately selected with consistency properties verification in mind (typically, the *start* and *end* of the document or node, the *time* progression action, the *user* interactions). Other events may *de facto* be hidden. The LTS can be minimized using, *e.g.*, observational or safety equivalence relations, which are both supported by the ALDEBARAN tool.

4.3.2. Node aggregation. Validation by node aggregation is bottom up. We first select a composite node and perform reachability analysis on it. The resulting graph together with the minimized automaton output by ALDEBARAN may be used to prove, on this composite node, the required consistency property. Once the consistency of the composite node has been established, one knows, according to the consistency definition, that the occurrence of the *start* action of the composite node is followed some (finite) time later by the occurrence of the composite node *end* action. The idea is then to replace the composite node specification by a content node specification whose minimal (*dmin*) and maximal (*dmax*) duration are directly derived from the internal timing logic of the composite node. We sketched an algorithm which derives dmin and dmax information from the composite node global reachability graph. This algorithm relies on the reachability graph clock regions and the analysis of the enabled transitions, taking into account the general structure of the composite node specification. However, we did not get any proof of the algorithm correctness, but experimented successfully its "supposed" validity on different examples. Once we get this timing information, we come up with the following conjecture: any composite node may be replaced by an "abstract" content node whose *start* and *end* actions are similar to the composite node's *start* and *end* actions, and whose (possibly non deterministic) global delay is computed from the composite node's internal timing logic. The required congruency has not been formally proved, but seems to be a direct consequence of the rather particular structure of a consistent composite node (one *start* action, followed by several branching actions which are hidden within the composite node and without any divergence, leading finally to the unique *end* action). Further work is required in this respect, but the aggregation method appears to work well in practice, since it permits to analyze complex structured hypermedia documents composite note by composite node instead of globally.

4.4 Illustrations

In this section we survey some verification results achieved for the hypermedia document presented in the section 2.4.

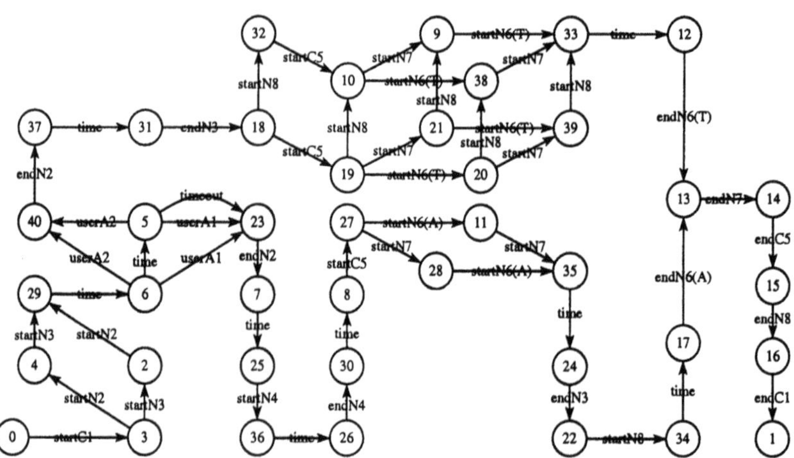

Figure 4 – Document minimal automaton.

Figure 4 presents the minimal automaton[3] of the whole document reachability graph, where only actions *start* and *end* of all nodes (composite or not), *user* (for anchors selection), *timeout* (for the automatic trigger of *anchor1*), *time* (for time progression) have been selected as visible actions. The graph shows that there is no visible loop, and that action *startC1* (i.e. the start of compositeNode1) leads eventually to *endC1*. As a consequence, the document is intrinsically consistent. The fact that there is no hidden internal loop is a direct consequence of the RT-LOTOS semantics. The invisible internal actions are indeed urgent actions, and a loop among these actions will stop the time progression; now, one can note that any *start* action is always followed by a time progression action before achieving the corresponding *end* action.

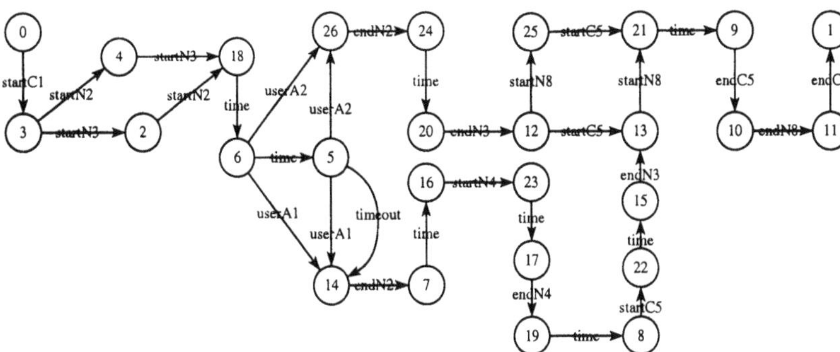

Figure 5 – Document minimal automaton (with node aggregation).

Figure 5 presents the minimal automaton of the global reachability graph which has been derived from the document RT-LOTOS specification, where composite node5 has been replaced by an abstract content node (following the aggregation method discussed previously). This graph is observationally equivalent to the minimal automaton derived from the graph of Figure 4, with actions *startN6(A)*, *startN6(T)*, *startN7*, *endA6(A)*, *endN6(T)*, and *endN7* being invisible.

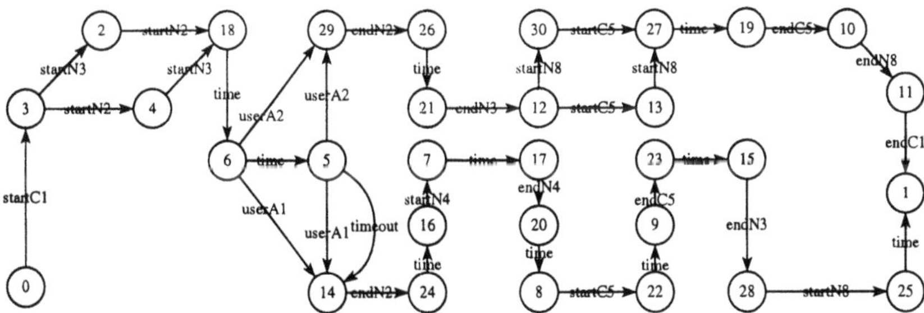

Figure 6 – Inconsistency situation (intrinsic).

Figure 6 presents the same graph as Figure 4 with the difference now that the duration of node3 has been set to 100s instead of 52s. One can see that such a specification leads to an intrinsic inconsistency since the *start* of node8 (action *startN8*) may occur after the *end* of node5 (action *endC5*); and yet, we specified that the *end* of node5 should lead immediately to the *end* of node8, hence the inconsistency since action *endC1* may be not reachable from the initial state.

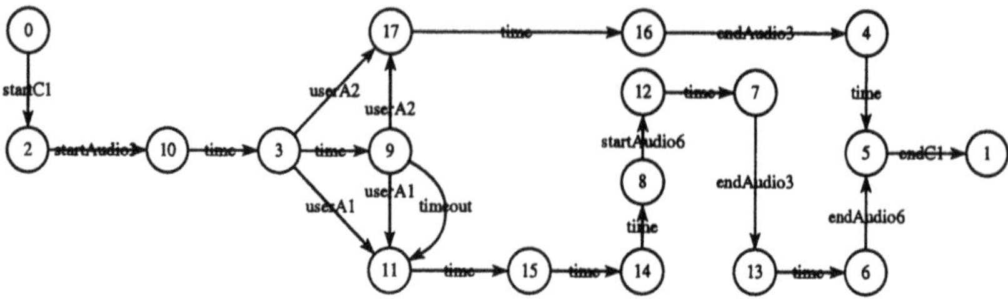

Figure 7 – Inconsistency situation (extrinsic).

Figure 7 considers the same initial case previously analyzed in Figures 4 and 5. The minimal automaton is constructed considering as visible only the actions starting and ending presentations of audio nodes (either audio content nodes or composite nodes including an audio node), plus the *user*, *timeout* and *time* actions. Analyzing this automaton, one can see that the document specification does not satisfy consistency property *P2* (see the concurrent access to an audio channel), since for some behaviors depending on the time at which the user has selected an anchor, the start of an audio presentation is followed by the start of another audio presentation, without the end of the previous presentation; hence an inconsistency, leading to an undesirable mix on the same audio channel of two audio presentations. For the construction of this automaton and for clarity purpose, *startN3* (respectively *endN3*) has been relabeled *startAudio3* (respectively *endAudio3*), and *startN6(A)* (respectively *endN6(A)*) has been relabeled *startAudio6* (respectively *endAudio6*).

This extrinsic inconsistency may be corrected by different ways: either by decreasing node3 duration, by modifying the version of node6 considering a textual presentation instead of an audio, or more probably by adding a new link in the document specification. Thus one can see that error corrections are made at the level of the authoring model, which are then translated into the RT-LOTOS formal specification, following the translation procedure of section 3.

5. Conclusions

As a conclusion, one can stress the following points. First, it is extremely easy and flexible to work with high-level models at the authoring stage. Second, it is easier for the author to handle document logical structuring than only document presentation structuring. Hence, models based on compositions allowing every type of relationships among their components become very important. Third, the use of a hidden formal description technique, like RT-LOTOS, brings two important advantages: (i) it provides a formal semantics for high level hypermedia models, permitting a better understanding of some critical behaviors that may be expressed by the model (ii) it allows applying, when possible, exhaustive verification techniques to check a document against design errors. More work is currently devoted to the study of the extrinsic properties, taking into account additional delays that may be forced by an underlying multimedia platform. The limitation of the proposed method is related to the state space explosion of the reachability graph. The definition of the aggregation method, which remains to be formalized and proved, appears to be an important step towards the verification of complex structured documents.

Endnotes

1. Only temporal constraints are considered in this paper; note however that some ideas may probably be generalized to spatial synchronization.
2. For simplification purpose, we do not consider in this definition the case where one wants, in the middle of a document presentation, restart the whole presentation; in that case, a start of a document could be followed by another start, and so on.
3. All the minimal automata referred to in this paragraph are related to the observational equivalence.

Acknowledgments

The work reported in this paper is funded by a CNRS research grant (Action Télécoms). The first author is supported by a grant from the Brazilian Government (CAPES).

References

Blakowski G.; Steinmetz R. (1996) A Media Synchronization Survey: Reference Model, Specification and Case Studies, *IEEE Journal on Selected Areas in Communications*, 24(1).

Buchanan MC.; Zellweger PT. (1993a) Automatically generating consistent schedules for multimedia documents, *Multimedia System Journal.*

Buchanan, MC; Zellweger, PT. (1993b) Automatic Temporal Layout Mechanisms. *ACM Multimedia'93.* San Francisco, USA.

Casanova, M.A. et all. (1991) The Nested Context Model for Hyperdocuments. *ACM Conference on Hypertext*, San Antonio, USA.

Courtiat, JP.; De Oliveira, RC. (1995) A Reachability Analysis of RT-LOTOS Specifications. *FORTE'95*, Montreal, Canada.

Courtiat, JP.; Oliveira, RC. (1996) Proving Temporal Consistency in a New Multimedia Synchronization Model. *ACM Multimedia'96.* Boston, USA.

Fernandez JC. et all. (1996) CADP, CAESAR/ALDEBARAN Development Package: a protocol validation and verification toolbox, *8th Conference on Computer Aided Verification*, New Brunswick, USA.

Fraissé, S.; Nanard, J.; Nanard M. (1996) Generating Hypermedia from Specifications by Sketching Multimedia Templates. *ACM Multimedia'96*, Boston, USA.

Lakaida N.; Keramane C. (1995) Maintaining temporal consistency of multimedia documents. *ACM Workshop on Effective Abstractions in Multimedia*, San Francisco, USA.

Rodrigues, R.F.; Soares, L.F.G.; Souza, G.L. (1997) Authoring and Formatting of Documents Based on Event-Driven Hypermedia Models, *IEEE Conference on Protocols for Multimedia Systems, Multimedia Networking — PROMSMmNet'97*, Santiago, Chile.

Santos, C.A.S; Soares, L.F.G; Souza; G.L; Courtiat, J.-P. (1998) Design Methodology and Formal Validation of Hypermedia Documents. *ACM Multimedia'98*, Bristol, UK. (To be published)

Soares, L.F.G.; Casanova, M.A.; Rodriguez, N.L.R. (1995) Nested Composite Nodes and Version Control in an Open Hypermedia System, *International Journal on Information Systems, Special Issue on Multimedia Information Systems.* 501-519.

Van Rossum et all. (1993) CMIFed: A Presentation Environment for Portable Hypermedia Documents, *ACM Multimedia'93*, California, USA.

Wahl T., Rothermel K. (1994) Representing time in multimedia systems. *IEEE Int. Conf. on Multimedia Computing and Systems*, Boston, USA.

Willrich, R.; Sénac, P.; Saqui-Sannes, P.; Diaz, M. (1996) Hypermedia Documents Design Using the HTSPN Model. *3rd Int. Conf. on MultiMedia Modeling - MMM'96*, Toulouse, France.

PART FIVE

Methodology 2

Frameworks by means of virtual types - exemplified by SDL

Rolv Bræk
SINTEF Telecom and Informatics, N-7034 Trondheim, Norway
and Institute for Telematics, Norwegian University for Science and Technology

Birger Møller-Pedersen
Ericsson AS, Applied Research Center,
Software Engineering Laboratory,
P.O. Box 34, N-1361 Billingstad, Norway
and Institute of Informatics, University of Oslo

Abstract: Frameworks have emerged as a very effective way to achieve reuse. A framework provides the basic structure and behaviour of a family of applications, and they simplify the task of application development. Traditionally the framework idea has not been applied to FDTs. This paper contributes to the use of SDL for building frameworks. SDL offers the opportunity to encapsulate object structures and default behaviour in frameworks defined by SDL system types. Specific types of applications are obtained by defining subtypes of framework system types and redefining virtual types. The mechanism is not specific to SDL, but can be applied to any language that supports virtual types.

Keywords: Reusable architectures, Frameworks, SDL

1 WHAT IS A FRAMEWORK?

A framework is a reusable, "semi-complete" application type that can be specialized to produce custom applications. The main property that distinguishes a library from a framework is that a framework embodies a design of a type of systems, while a library just is a collection of related classes. An elaboration of this distinction is given in Table 1.

Table 1: Library and Framework

Library	*Framework*
The application uses library classes, but the library knows nothing about the application	*The framework knows about the application and uses application classes*
No predefined system structure. The system is entirely defined in the application	*Provides structure. The system is (partially) defined in the framework*
No predefined interaction	*Defines object interaction*
No default behaviour	*Provides default behaviour*

Classical frameworks like window systems are defined as a set of related classes. The structure of these frameworks results from dynamically created objects that are kept together by object references. The event loop that dispatches mouse and keyboard events to window objects will use a list of currently active window objects. This list will typically contain different window objects, with the common property that they react on these events. The default behaviour of the framework is to call virtual procedures or callback procedures. The user of the framework redefines these procedures to what is special for the application.

The paper does not address the definition of frameworks by means of composition of components with so-called "hot-spots" where components may be exchanged dynamically. Component-based frameworks rely on a black-box approach: the user of a component only knows its interface and cannot apply specialisation in order to cover specific needs.

2 WHY MAKE FRAMEWORKS USING SDL?

Experience from SDL based development has shown that SDL systems often contains an *application specific* part and an *infrastructure specific* part (implementation specific) which are not clearly separated. The reason is that an abstract (application specific) system has to be supplemented by a large infrastructure part in order to be executable on a given platform. These two parts need to be combined for the purposes of complete system simulation and automatic code generation, but they serve different purposes and follow separate evolution patterns. Therefore an approach is needed that will allow these parts to be efficiently combined into complete systems, while keeping the parts identifiable for evolution and maintenance purposes. A solution to this *divide and combine* problem is to make a framework in SDL. Instead of making the infrastructure part again and again for each new system with the same infrastructure on the same platform, a framework that embodies both the application part and the infrastructure part is useful. The idea is that if a specific system can be made as an instance of a framework, with much of the general properties of the framework isolated in the infrastructure, then the framework will have a potential for being reused as a design. Furthermore application development may concentrate on the central application issues and disregard the infrastructure, while infrastructure development may concentrate on the central infrastructure issues and disregard the application.

In an initial development, the infrastructure aspect may not be obvious. Frameworks will often come as a result of a (successful) initial development, which is to be used as a basis for a new system. If e.g. distribution has been considered and isolated in an infrastructure part, the next system with the same infrastructure, but with a different application part can reuse this framework.The basic idea is that the infrastructure shall provide some general support to the application, and that application development may concentrate on the application parts.

3 HOW TO MAKE FRAMEWORKS IN SDL

The distinction between a library and a framework is in SDL directly reflected by the language concepts *Package* and *System Type*. An SDL specification can be either a package specification or a system specification. A Package in SDL defines a set of types, while a System (Type) defines both local types *and* a set of related instances. The interpretation of a system specification yields a system instance that is the outmost container of all instances in a system.

3.1 System types

An SDL *system* consist of *blocks*. A block contains *processes* that are extended finite state machines communicating by means of asynchronous signal exchange and remote procedure calls. Blocks are connected by *channels*, while processes are connected by *signal routes*. The connection points are called *gates*, and they are defined as part of the block- and process types. In object oriented terms, *process types* correspond to classes of "active" objects, while *block types* correspond to classes of "container" objects. Gates define simple interfaces in terms of which signals and remote procedures that are allowed. SDL have more entity kinds than system, blocks and processes, but for the purpose of this presentation only these are used. The notion of *specialisation* by means *subtyping* applies to most of the entity kinds of SDL, and also to system. A framework can therefore be defined by a *system type* in SDL, and a specific application of the framework by a *subtype* of the system type.

The framework system type will have a structure where some components may be classified as infrastructure- and some as application components, see Figure 1. The framework will have *virtual types* for those components that should be adapted to specific applications of the system type. These virtual types will be *redefined* when the framework is used as a supertype when defining the system type for a given application. In Figure 1 the system type is defined to consist of a number of blocks. The types of two of these are defined as *virtual block types,* and can as such be redefined in a subtype of the system type. The redefinitions imply that the instances (A1 and I1) specified as part of the framework system (super)type will have redefined properties. The structure specified in the supertype is inherited, while the "contents" of the instances are redefined.

The dashed boxes in Figure 1 are not part of the SDL specification, but indicates the distinction between application- and infrastructure components. The actual borderline between these may not be as sharp as indicated.

Virtual types do not distinguish between being infrastructure or application component types, so infrastructure component types can also be defined as virtual type in order to allow the infrastructure to be adapted to the specific application.

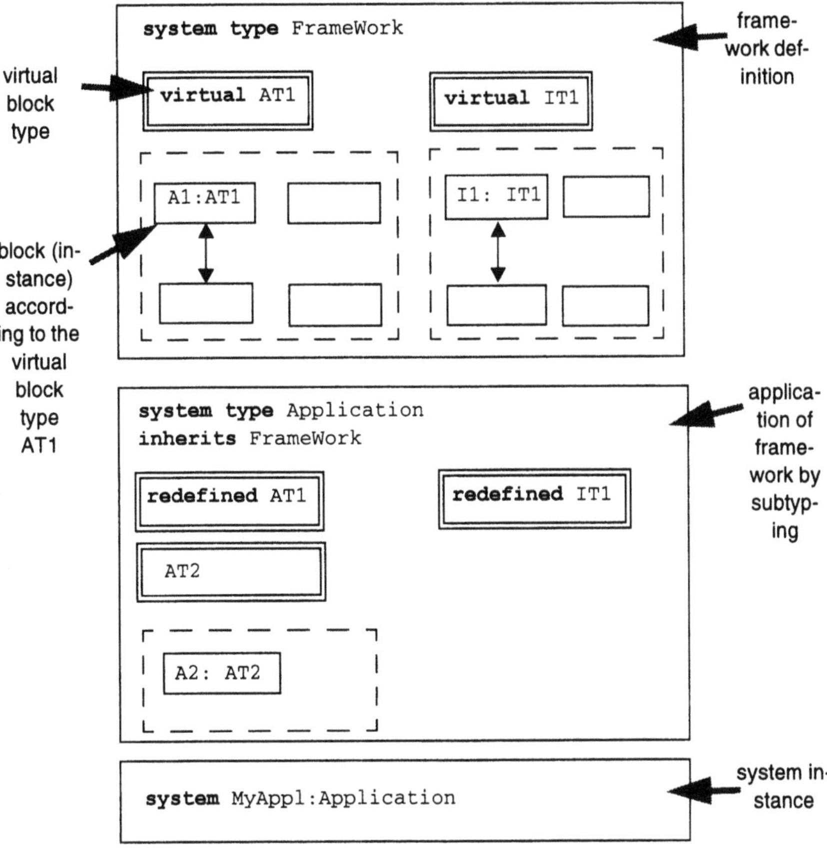

Figure 1: A framework as a System Type

3.2 Applications

In many cases the application components are blocks with some infrastructure specific parts. The block type AT1, see Figure 1, may consist of an infrastructure and an application specific part, each represented by a virtual process type. This is illustrated in Figure 2. This also illustrates that the borderline between application and infrastructure parts is not so sharp as indicated in Figure 1.

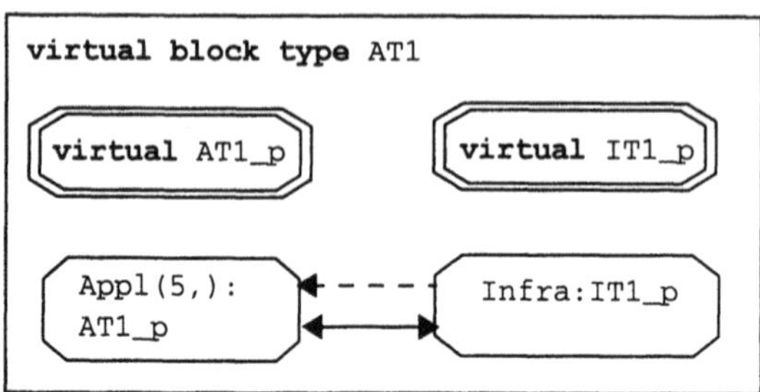

Figure 2: A virtual block type as part of framework, with application specific instances and infrastructure instances

AT1_p and IT1_p are two *virtual process types*, and Appl and Infra are processes of these types. When redefining the virtual block type AT1, the virtual process types in this may also be redefined. If the application specific part is to be adapted, then the virtual process type AT1_p is redefined. If the infrastructure part shall be adapted, then IT1_p is redefined.

3.3 Instances as part of frameworks

As described above, the way a framework is adapted to specific needs is by redefining virtual types. Redefining virtual types will, however, have no effect if the virtual types are not used to specify instances. These instances may either be specified as part of the framework or be specified as part of the specific use of the framework.

3.3.1 Application specific instances specified as part of the framework

In this case the framework defines the structure of instances with their connections in terms of channels and signal routes. This is the approach used in Figure 1 and Figure 2. SDL is special in the sense that a static structure of instances and instance sets can be specified - instances are not just created dynamically. In a specific application, the properties of the instances are provided by redefining the virtual types, while the structure and connection of instances are inherited from the framework.

The structure of instances and their connections in the framework are obtained by connecting gates of the instances with channels and signal routes. The properties of the instances may be redefined when a specific application is made, but the interconnections between gates cannot be redefined. Gates are defined as part of the virtual types, and as gates are used for connecting instances, the redefinition of these must be constrained. It is important that the framework specification can be analyzed (e.g. that the connections are valid), and that this analysis is valid for any application of the framework.

This requirement is fulfilled by the notion of *virtual type constraint*. A virtual type in SDL has a constraint (in terms of a type of the same kind), and a redefinition of the virtual type must be a subtype of the constraint. The default is that the constraint of the virtual type is the type definition itself. The gates necessary for the connections - specified as part of the framework - are properties of the virtual type constraints, and are as such guaranteed to be properties of any redefinitions.

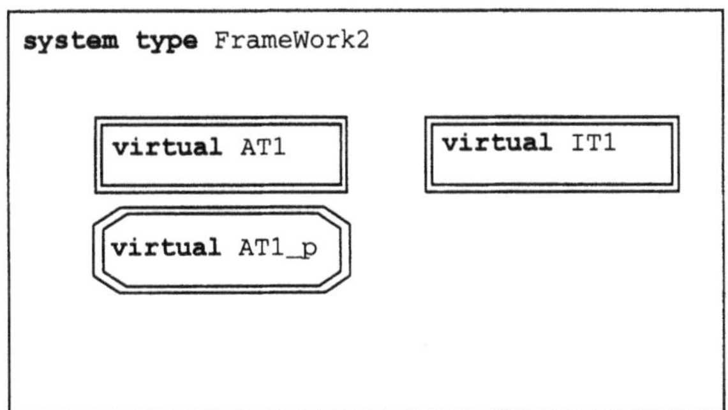

Figure 3: A framework where an application process type is common to many parts of the framework

If the structure is so that the same virtual process type is used to make process sets in many parts of the system structure, then the virtual process type should be moved to the system type, see Figure 3. In this way it can be redefined at *one* place as part of the system subtype and have implication on many parts of the system.

Application specific instances may be added either statically as part of the subtype system definition, as illustrated in Figure 1 (block A2), or they may be created dynamically. Dynamic creation of application instances are anticipated in Figure 2: the Appl specifies a set of maximum 5 processes of type AT1_p. The dashed arrow specifies that the Infra process creates processes in the Appl process set. If new process sets are added in redefinitions of the enclosing virtual type (here AT1), then the infrastructure process must be redefined accordingly to take care of their creation.

3.3.2 Application specific instances not specified as part of the framework

In this case the framework only consists of general types that may be used for the construction of the application part. The infrastructure parts may either contain instances or just be represented by types.

In this situation the creation must be anticipated by the infrastructure. SDL is special in the sense that processes are part of process sets and that creation of processes is done by referring to the name of the set and not to the name of the process type. It is therefore not enough that the infrastructure specific process types are defined in the same scope (e.g. a package or a system) as the application specific types and thereby know these types - they must have means for referring to the process sets that will be part of the specific system.

SDL provides two means for this: *context parameters* and *virtual procedures*.

Context parameters

The general types of the framework that have to create process instances according to application specific types do this through process context parameters. The actual process context parameter is the process set being part of the specific application.

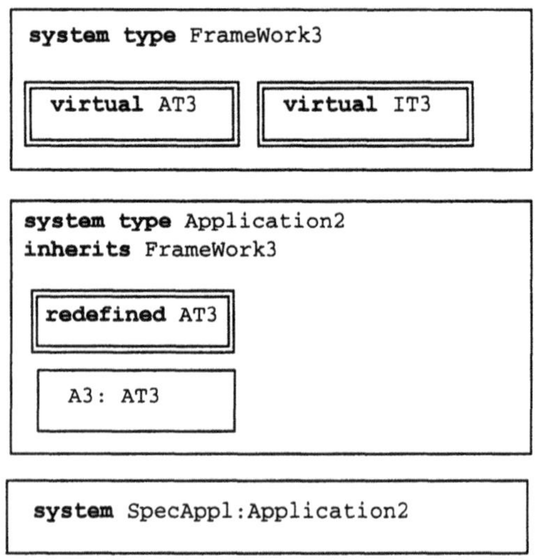

Figure 4: A framework with no instances

The scheme is illustrated in Figure 4, Figure 5, Figure 6 and Figure 7. Within the block types, there will be general process types (infPT and applicationPT) that are used as supertypes in specific systems inheriting from FrameWork3.

The infPT process type shall create instances of type applicationPT, and to this purpose it will have a process context parameter that is constrained by applicationPT, see Figure 7. The idea is that a particular system will pro-

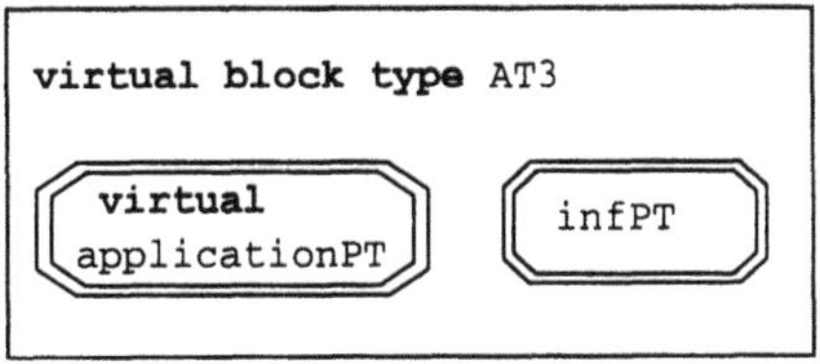

Figure 5: A virtual Application block type without instances.

vide its redefinition of applicationPT *and* its process set. By means of the context parameter, the infPT processes will be able to create instances of the redefined applicationPT without knowing the final process set.

The final system type will introduce the redefined block types with appropriate process sets, see Figure 6, and provide these as the actual context parameters.

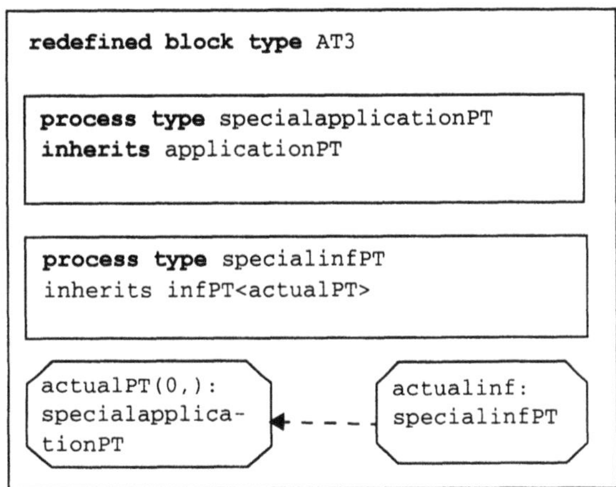

Figure 6: Redefined block type with process set and actual context parameter

The difference from the case with application specific instances as part of the framework is not so big: the process set must be foreseen (the context parameter must be defined - and correspond to a process set), but the name of it and its position in the application part is not determined. Its position will, however, be constrained by the fact that it shall be visible from the place where the actual context parameter is provided, and that it shall be created by a process in the same block.

```
process type infPT
<process cp atleast applicationPT>




 . ...create cp( . . .)
```

Figure 7: Creation of application specific processes

Another constraint with this approach is that it is not possible to specify instance sets of the infPT in the framework itself. The reason is that this process type has context parameters and therefore cannot be used for instance specification.

This approach should only be used in cases where it is important that the framework specific and application specific process types can be defined within the same enclosing block type and where the framework specific types must specify the creation of application specific processes.

Virtual creation procedures

This approach is more general in the sense that it does not have to be decided if there is one or several process sets in the final system type. In the general types where there is a need to create application specific processes, this is represented by corresponding virtual procedures. In the final system type these are redefined to create processes in the actual process sets.

This approach also has the property that instance sets of the application specific types are first introduced in the final system subtype.

In order to provide an example on this way of making frameworks, we use Figure 8. Suppose that it is a requirement that the system shall start by creating processes for each of the actual applications and that changes to the number of application processes shall be reflected while the system is running. Still we would like to define the system as a framework in the sense that it will consist of a common infrastructure and some dynamically created application instances. We assume that the division of responsibility between the two are determined, the communication is fixed and that the functionality of the infrastructure is specified - the only thing that is not specified is the types and numbers of application processes. The configuration of the system is initiated by a new signal (setUp, with appropriate parameters) coming to the infrastructure part of the system.

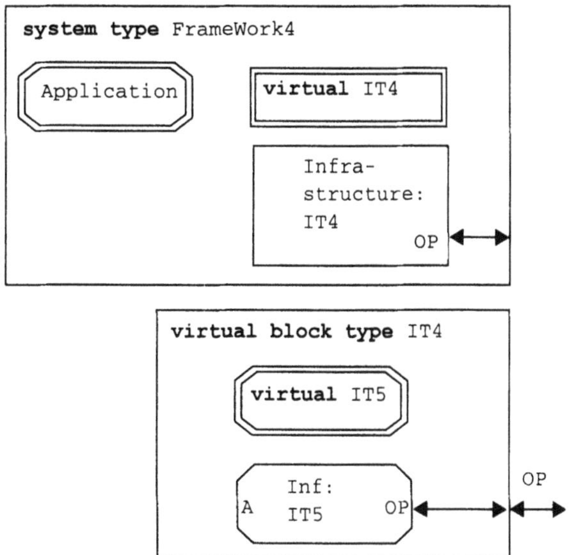

Figure 8: Framework with virtual creation procedures

The setUp signal is supposed to come to the inf process and imply the creation of application processes in the right process sets. Depending on the desired number and types of application processes, the signal will carry enough parameters for the infrastructure to create the right instances.

The infrastructure types can now be defined as before, the only difference being that they will have a virtual procedure setUp and will communicate with possible application processes via a gate that is constrained by Application - that is only process sets of Application or subtypes of Application can be connected to the gate, see Figure 9.

In addition to the normal behaviour and the creation of application processes, the Infrastructure may have behaviour that contributes to the definition of the framework. As an example there may be a limit on the total number of application processes, independent of type of application. The behaviour that ensures this will either be part of the infrastructure, e.g. some action executed each time setUp is executed, or it may be a constraint on setUp which all redefinitions will inherit.

An actual system consisting of two types of application processes is specified as a subtype of the system type Framework4, redefining the setUp procedure to cater for this and introduce the two process sets, see Figure 10.

The names of the process sets are used in the redefined setUp procedure for the specification of the creation of process instances.

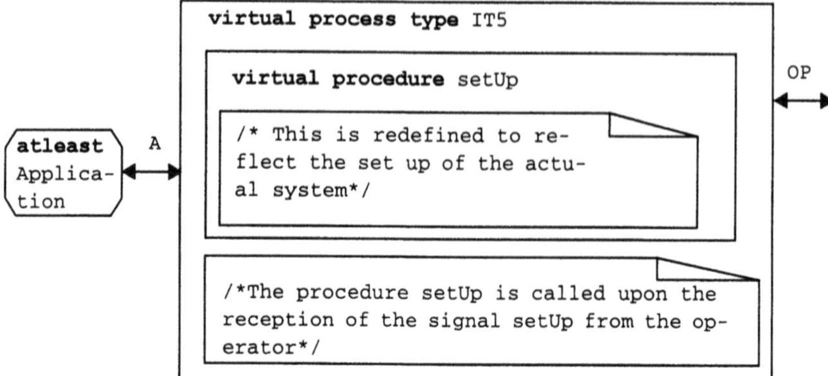

Figure 9: Setup as a virtual procedure of IT5

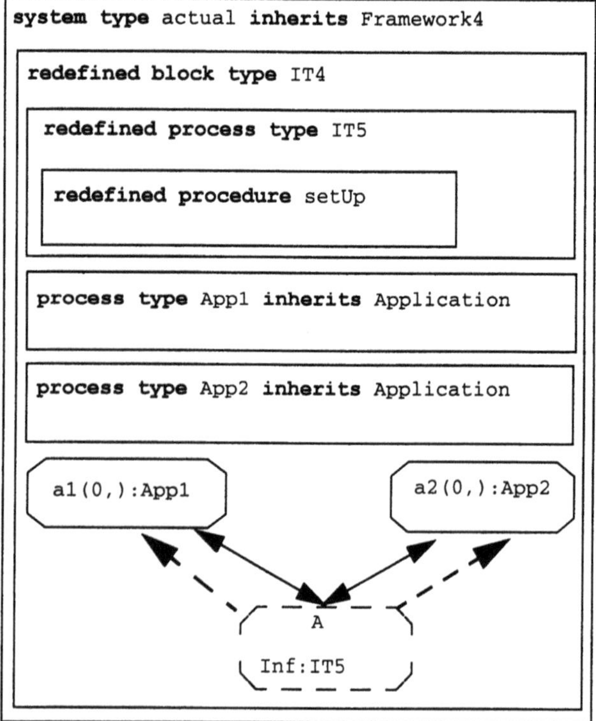

Figure 10: Dynamically created application part

4 RELATED WORK

4.1 Virtual classes/design patterns

The notion of virtual types with constraints where first introduced in BETA (Madsen, Møller-Pedersen & Nygaard, 1993) in terms of virtual patterns. Patterns in BETA is a generalisation of language concepts like class, type, procedure, etc. Virtual procedure patterns provides what is common in most object oriented languages: virtual member functions or methods that can be redefined. The use of virtual patterns to obtain virtual classes is described in (Madsen & Møller-Pedersen 1989).

In (Agerbo, 1998) virtual classes are used to express some design patterns by means of language constructs. It is argued that a design pattern should not be covered by a language construct, in order to be a design pattern, and they show that the Factory Method design pattern can be expressed by a virtual class for the class that varies with the different applications of the design pattern.

Virtual block types and virtual process types as described above for SDL corresponds to virtual classes. A requirement for using this approach to framework definition is therefore that the language has virtual classes. Most object oriented languages do not have classes within classes, with Java (Arnold &Gosling, 1996) and BETA as exceptions.

Component based software development, with solutions based on e.g. CORBA, COM or Java Beans, represents another approach to the definition of frameworks. While the approach described above relies on access to the source so that specialisation can be used to express adaptation to specific needs, component based frameworks relies on applications with so-called "hot-spots", that is components that can be exchanged with components with the same interface.

4.2 Frameworks in The Integrated Method - TIMe

TIMe is a comprehensive development methodology that uses UML , MSC and SDL as its primary languages for analysis and design. It emphasises that SDL is used to make models that are both readable, formal and sufficiently complete for extensive simulation and automatic code generation. TIME recommends that design is carried out in three main steps: application design, architecture design and framework design, where application design and framework design are abstract models expressed primarily in

SDL and MSC (possibly using UML for parts where SDL is not well suited).

4.2.1 Application design: where the service functionality is designed

The first purpose of an application design model is to describe the system behaviour at an abstraction level, where it can be understood and analysed independently of a particular implementation.

The second purpose is to be a firm foundation for designing an optimum implementation satisfying both the functional and non-functional requirements.

4.2.2 Architecture Design: where the implementation architecture is decided

The purpose of architecture design is to design an implementation architecture that will behave as defined in the application design model and satisfy the non-functional properties. The purpose of architecture design is to answer *how* the system is going to be realised. This is expressed using Architecture descriptions that show:

- the overall architecture of hardware and software;
- how frameworks and applications are mapped to the architecture.

While the application and the framework has focus on functional properties and behaviour, the architecture has focus on non-functional properties and physical structures. The purpose is to give a unified overview over the implementation and to document the major implementation design decisions.

Architecture design determines critical architectural issues such as physical distribution, global addressing schemes and fault handling. Some of these may subsequently be reflected in the framework model in order to describe the complete system behaviour.

During normal application evolution, the architecture will be stable, and system evolution can take place mainly at the Application level.

In an initial development, architecture design will come before framework/infrastructure design.

4.2.3 Framework Design: from Infrastructure to Framework

The purpose of framework design is to describe the complete system behaviour taking the underlying implementation as defined in the architecture design into account. In this step the implementation dependent infrastructure functionality is taken into account, e.g. distribution support, error handling and configuration. The infrastructure part of a system contains additional behaviour needed to fully understand what the system does (i.e. the complete system behaviour). Here we find objects and parts of objects that support distribution, system administration and other facilities not directly related to user services (applications). Whenever practical, the application and the infrastructure should be put together in a framework that serves to simplify the definition of new systems and separate the evolution of applications from the evolution of infrastructures. The framework model defines a system type or a block type, with predefined structure so that a specific application system only has to provide the specific "contents" of part of this structure using the approaches explained earlier.

An application system designed before the infrastructure was known,

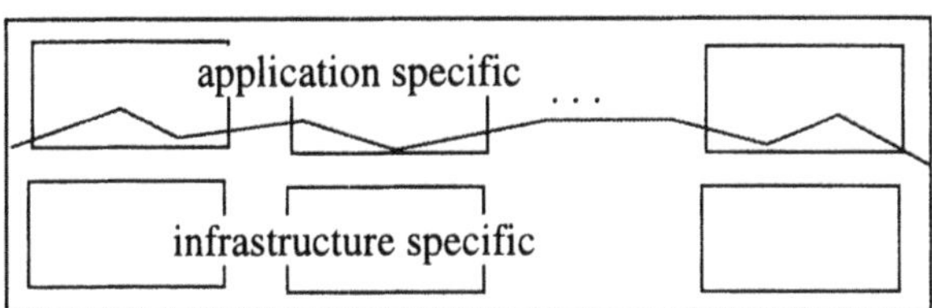

Figure 11: A framework with application and infrastructure specific parts of systems

may have to be redesigned somewhat in order to satisfy the infrastructure interfaces. Restructuring does not mean that everything has to be redefined, only those parts that depend on the infrastructure. When the infrastructure is known, new application types may be defined right away to comply with the infrastructure interfaces.

In general it will be an advantage if the application design has been done by means of types that are as general as possible.

When the framework has been defined, making a specific system instance is a matter of exchanging the application types of the framework with either improved versions or new application types with e.g. new functionality.

5 CONCLUSIONS

The notion of a framework has a potential to be very useful when making abstract system models using FDTs, especially when the models are complete and used as basis for system evolution and code generation. This paper has only considered the possibility of making frameworks using SDL. However, the underlying needs that are addressed and the potential benefits of using frameworks are independent of the FDT actually used and by no means restricted to SDL. The important thing is that the language of the FDT is able to support frameworks in an efficient way. As has been demonstrated in this paper, SDL provide some basic features that make frameworks possible. Using these, a divide-and-combine approach to application and infrastructure development is possible that enable a high degree of reuse, and simplify system evolution.

6 REFERENCES

Steve Sparks, Kevin Benner, Chris Faris; *Managing Object-Oriented Framework Reuse;* IEEE Computer, September 1996.

Mohammed E. Fayad, Douglas C. Schmidt; *Object-Oriented Application Frameworks*; Communications of the ACM, October 1997.

Z.100; *CCITT Specification and Description Language (SDL)*, ITU-T, June 1994

ITU (1993d) Z.120 Message Sequence Charts (MSC), ITU-T, September 1994, 36 p ("MSC-92")

Rolv Bræk and Øystein Haugen (1993); *Engineering Real Time Systems*. Hemel Hempstead: Prentice Hall, 1993.

TIMe - The Integrated Method. SINTEF Telecom and informatics 1997, see http://www.sintef.no/time.

Booch, G., Jacobsen, I. and Rumbaugh, J. (1997b). *The Unified Modeling Language*, Version 1.1, Rational Software Corporation, http://www.rational.com (September 1997)

Ole Lehrmann Madsen, Birger Møller-Pedersen and Kristen Nygaard; *Object Oriented Programming in the BETA Programming Language*, Addidon-Wesley 1993

Ole Lehrmann Madsen and Birger Møller-Pedersen; *Virtual Classes - A powerful Mechanism in Object-Oriented Programming*, OOPSLA 1989.

Ellen Agerbo and Aino Cornils: *How to preserve benefits of Design Patterns*, OOPSLA 1998.

K. Arnold and J. Gosling: *The Java Programming Language*, Addison-Wesley 1996

SDL-Pattern based Development of a Communication Subsystem for CAN *

B. Geppert, A. Kühlmeyer, F. Rößler, and M. Schneider
University of Kaiserslautern, Computer Science Department
P.O. Box 3049, 67653 Kaiserslautern, Germany
{geppert, akuehlm, roessler, schneider}@informatik.uni-kl.de

Abstract

SDL patterns are reusable software artifacts. They represent generic solutions for recurring design problems with SDL as applied design language. We have developed a construction set of protocol building blocks consisting of a pool of SDL patterns and an accompanying methodology for the incremental design of communication protocols. In this paper, we present a case study on the specification of communication protocols based on SDL patterns. The case study is part of a more comprehensive project, where a real-time communication subsystem was developed on top of a Controller Area Network (CAN) installation. We demonstrate how the protocols supporting user communication and certain management tasks were configured. We also applied the SDT Cadvanced code generator to implement the resulting design specification on a PC cluster. Generally, it turned out that SDL-pattern based configuring of communication protocols yields more systematic designs, i.e., readability and maintainability is improved and less design errors occur, since the design decisions are well founded and documented.

1. Introduction

The reuse of predesigned solutions for recurring design problems is of major concern in object-oriented software development. During the past few years, design patterns have emerged as an especially fruitful approach to software reuse [2] [5]. Contrary to the traditional paradigm of class and function libraries, which is solely concerned with code reuse, design patterns aim to focus on the invariant parts of a design solution and offer by far more flexibility for adaptation to the embedding context. That is, the potential of reuse is substantially increased. There are several advantages commonly attributed to design patterns: patterns capture solutions, which have evolved over time and serve as an elegant way to make designs more flexible, modular, reusable, and understandable. They reflect experiences gained in prior developments and therefore help designers to reuse successful designs and architectures. As a consequence, the design process becomes faster and the number of design errors decreases.

* This work is supported by the German Science Foundation (DFG) as part of the Sonderforschungsbereich SFB 501 *Development of Large Systems with Generic Methods.*

In [6] [7] we present the SDL-pattern approach that integrates the design patterns concept with SDL [11]. Generally speaking, SDL patterns describe generic solutions for recurring design problems, which can be customized for a particular context. While conventional design patterns are specified independently from a possible design language, it is assumed that the target language for SDL pattern instantiation is SDL. Thereby we benefit from the formal basis provided by SDL, so that SDL patterns actually characterize as formalized design patterns. Instead of specifying and applying the patterns rather informally, a formal target language such as SDL offers the possibility to precisely specify how to apply a specific pattern, under which assumptions this will be allowed, and what properties result for the embedding context. This is a major improvement compared to conventional design patterns, which mainly rely on natural language based pattern description and still have to leave pattern application to a large degree to the personal skills of the system designer. However, we do not deal with formalizing design patterns in general. Instead of formalizing reuse concepts we aim to support reusability within the formal methods area.

In this paper, we present a case study on the specification of communication protocols based on SDL patterns. The case study is part of a more comprehensive project, where a real-time communication subsystem was developed on top of a Controller Area Network (CAN) [3] installation. We demonstrate how the protocols supporting user communication and certain management tasks were configured. We also applied the SDT Cadvanced code generator to implement the resulting design specification on a PC cluster running under the real-time operating system QNX.

The remainder of the paper is organized as follows: Section 2 introduces the SDL-pattern approach. In Section 3 we demonstrate how a real working communication subsystem for CAN was configured using SDL patterns. The development steps for (semi-)automatic code generation of the resulting SDL design are outlined in Section 4. Finally, we summarize the results in Section 5.

2. The construction set of protocol building blocks

This section summarizes concepts for a construction set of protocol building blocks, from which a protocol designer can select SDL patterns and configure them to a customized, formal protocol specification.

2.1. SDL patterns

An SDL pattern describes a generic solution for a recurring, context-specific design problem. It is assumed that the target language for pattern instantiation is SDL. Though the concept is not restricted to a specific application domain, we are mainly concerned with communication protocols.

For the specification of SDL patterns we have defined a standard description template with the main items sketched in the following. The mere syntactical part of the design solution is defined by a generic *SDL fragment*, which has to be instantiated and textually embedded into the context specification when applying the pattern. SDL fragments represent context invariant parts of the design solution. Instantiation and embedding of SDL fragments is prescribed in terms of *syntactical embedding rules*, which, e.g., guide renaming of generic identifiers or specialization of embedding design elements. Usually pattern semantics is not completely captured by an SDL fragment. Due to language constraints this would otherwise result in an overspecification of the design solution and reduce potential of reuse. Thus additional *semantic properties* are included specifying preconditions for pattern application as well as behavioral changes of the embedding context. Though semantic properties are currently stated in natural language, it is possible to express them precisely in a temporal logic. Also, restrictions on the *refinement* of pattern instances are specified in order to prevent a pattern's intent from being destroyed by subsequent development steps. A more detailed discussion of the SDL-pattern description template and a comparison to existing description templates of conventional design patterns is given in [6].

The current pool of protocol building blocks contains SDL patterns that deal with interaction behavior of distributed objects, error control (lost or duplicated messages), lower layer interfacing, or dynamic creation of protocol entities. To further illustrate the functional scope of SDL patterns we shortly introduce some examples. Note that the SDL patterns below are not completely specified. We basically summarize a pattern's intent and skip the description items explained above.

- **BlockingRequestReply:** The *BlockingRequestReply* pattern introduces a confirmed interaction (two-way handshake) between two given automata. After a trigger from the embedding context, the first automaton sends a request and is blocked until receiving a reply. The request is eventually received by the second automaton, which replies and finally releases the first automaton from its waiting state.
- **DynamicEntitySet:** Consider a given server entity that provides its service exactly one time and terminates thereafter. In order to offer this service several times (e.g., to more than one client), the *DynamicEntitySet* pattern is applied. For each client a new server entity is dynamically created by a special entity administrator. Subsequently, the administrator acts as a proxy from the perspective of the clients, which forwards service requests to the corresponding server entity.
- **Codex:** The *Codex* pattern provides mechanisms to allow two (or more) entities, which interact directly through SDL channels, to cooperate by the means of a given communication service. In general, the introduction of a basic service involves many specialities. Among others, these are segmentation, reassembly, upgrade of basic service quality (e.g., in case of loss, disruption or duplication of messages), lower layer connection setup, or routing decisions. The *Codex* pattern is only concerned about a minimal subset of these functionalities, namely interfacing with the basic service by the means of service primitives. That is, *Codex* essentially provides a mapping from protocol data units to basic service primitives and vice versa.

2.2. Configuration process

For the design of SDL protocol specifications we have defined a configuration process supporting the reuse of protocol building blocks represented as SDL patterns (Figure 1). The configuration process suggests an incremental protocol design, where the whole set of communication requirements is first decomposed, i.e., partitioned and (where appropriate) simplified. Decomposition classifies as an analysis task that identifies separate protocol functionalities. Thereby it is possible to consider a protocol functionality under different assumptions. For instance, interaction sequences for connection establishment are less complex on top of a reliable basic service instead of an unreliable basic service. Experience has shown that protocol functionalities can often be specified one after the other and - in addition - be stepwise completed (e.g., adapted to the non-ideal properties of an underlying basic service). This suggests that we perform an individual development step in order to incorporate an additional protocol functionality or relax a corresponding simplification. Thereby each development step divides into analysis, design, and validation and yields an executable SDL design specification. In the following the different activities within a development step are sketched.

First, an object-oriented analysis of the current protocol functionality is performed. This results in an updated analysis model from the previous development step. It is suggested to provide an UML [1] object model and an MSC [12] use case model, which together identify participating objects and typical interaction scenarios.

The analysis model is implemented in the following design activity. Here, SDL patterns come into place. Starting point is the context SDL specification, i.e., the SDL design specification obtained from the previous development step. This may, e.g., be a protocol specification, which relies on a reliable basic service. Hence, the design problem (stated in the analysis model) could then be to suit the protocol to an unreliable basic service. In order to meet the new requirements a number of design steps are performed that apply separate SDL patterns to the context specification. Note that for some design problems the pool of predefined protocol building blocks may not contain an adequate solution, so that an ad hoc solution must be found. The selection of an SDL pattern is supported by several items of the SDL pattern description template, namely *intent*, *motivation*, *structure*, *message scenario*, *semantic properties* and *cooperative usage*. As patterns represent generic design solutions, the corresponding SDL fragment has to be adapted in order to seamlessly fit the embedding context. This is instructed by the *renaming* parts of the *syntactical embedding rules* and the *refinement rules*. The resulting pattern instance finally has to be composed with the embedding context, which is prescribed by the *composition* part of the syntactical embedding rules and also by the refinement rules of embedding pattern instances.

The result of this design activity is an intermediate SDL design specification, which is subsequently validated against the analysis use case model. Also, the correctness of the SDL specification concerning general properties such as freedom from deadlocks is checked. If any faults are discovered, a return to one of the previous development or design steps is needed (not shown in Figure 1). Otherwise the validated specification serves as the context specification for the next development step. If all simplifications are eliminated and all requirement subsets are implemented the final design specification is given by the validated design specification of the last development step.

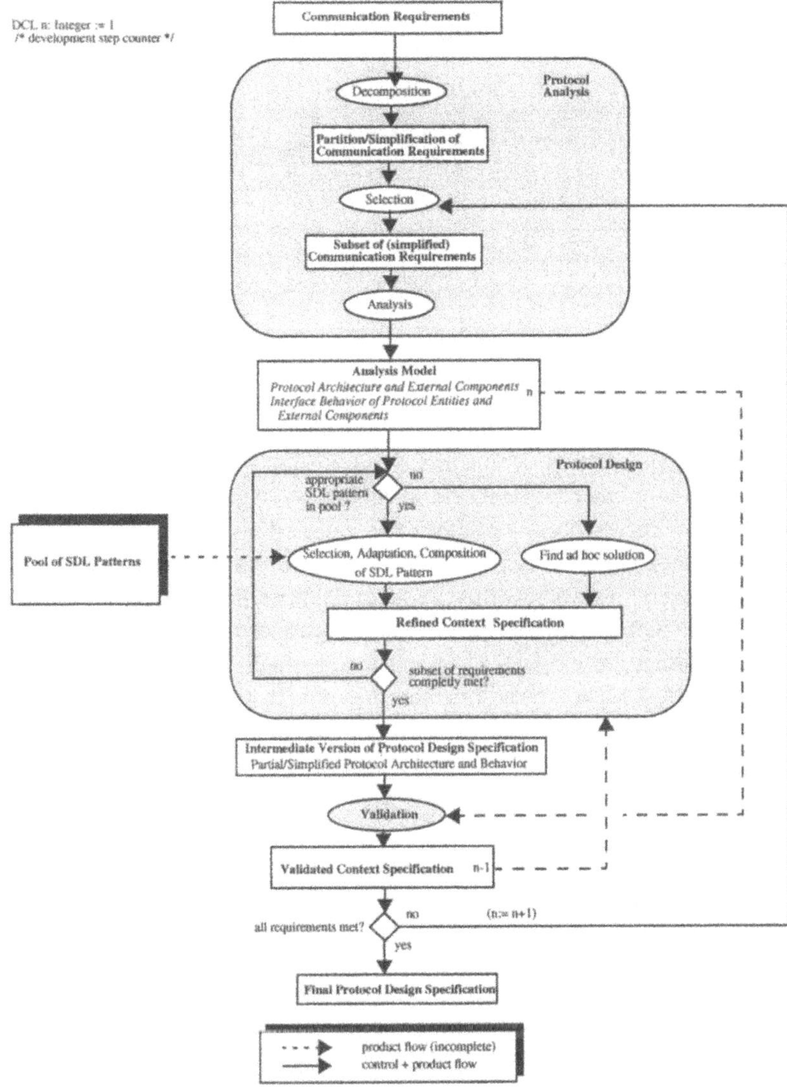

Figure 1. Configuration process (excerpt)

3. Configuring a communication subsystem for CAN

Controller Area Network (CAN) is a field-bus standard originally developed for mobile systems. However, due to exceptional features such as high data rates and fault tolerance, CAN became also established in other areas of the field-bus domain [3]. We have configured a real-time communication subsystem for CAN using SDL patterns. Note that the presentation below follows the configuration process introduced in Section 2.2.

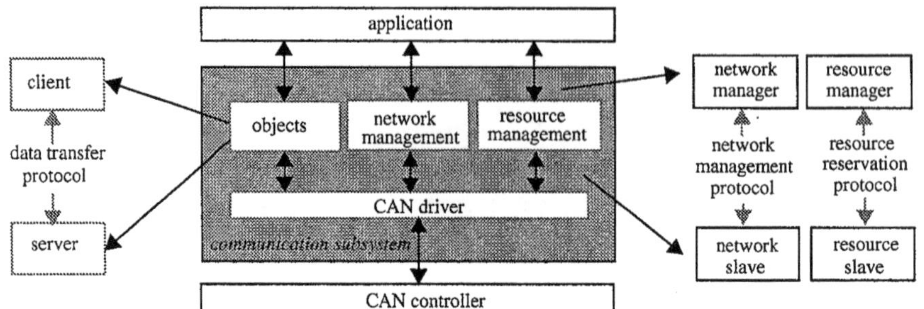

Figure 2. CAN subsystem architecture

3.1. Communication requirements

Communication between intelligent sensors, actuators, and controller components imposes stringent requirements on timeliness and predictability. Thus CAN employs bitwise bus arbitration based on message identifiers to support deterministic media access. Message identifiers are to be interpreted as priorities and therefore a global, priority-driven scheduling of message transfers is realized. As timeliness and predictability strongly depend on worst case traffic load, field-bus installation normally precedes a configuration step, where communicating peers as well as traffic loads are identified and priorities are assigned accordingly. This step is carried out prior to system operation and works fine for static settings. Nevertheless, the operator or owner of a field-bus installation may change communicating devices or applications dynamically (e.g., in building automation). Though dynamic insertion and removal of communicating devices is generally supported by CAN, timeliness guarantees have to be concerned from scratch.

Therefore a communication subsystem shall be developed on top of CAN that performs admission control, priority assignment, and traffic policing *automatically during system operation.* This can be seen as a kind of "plug and play" functionality for timeliness guarantees, i.e., when, for instance, a new communicating device is inserted, the system automatically handles priority assignment with respect to timeliness requirements and under consideration of the already existing traffic load. If there are not enough network resources left, the system shall deny additional communicating peers.

3.2. Decomposition

The real-time CAN subsystem (Figure 2) is based on a generic QoS architectural framework [8]. This framework was also previously instantiated for a real-time communication subsystem on top of a Token-Ring network [9]. Essentially, the subsystem for CAN contains protocol functionalities for initialization of CAN modules (network management), resource management, and object-based user communication (e.g., access to remote read-only-variables). With the additional simplification that communicating peers exchange protocol data units (PDU) directly, i.e., without using an underlying basic service, we get the decomposition of Figure 3. The chronological order of the performed development steps is also illustrated.

3.3. Development steps

3.3.1. Development step I: Initializing CAN modules, directly connected peers. Before communication can start, some CAN modules must be initialized (e.g., programs or parameters are downloaded). This process would consume many CAN identifiers that are permanently assigned to individual modules. In order to save resources, it is suggested to share CAN identifiers for initialization purposes. As a consequence, a special control protocol for medium access is needed, which was specified in the first development step. The protocol is implemented by a network slave on each CAN module and a central network manager.

The interaction between the network manager and a network slave is structured into different phases. When the slave requests access to the CAN bus, the corresponding request message contains no data in order to avoid collisions. Thus, the network manager must identify the slave before it can be addressed. This is done by a binary search that broadcasts identification messages, which contain an interval of valid module identifiers. Only those slaves with a valid module identifier are allowed to reply.

A slave enters the next phase, when an initialization message with its own module identifier is received. As already mentioned, the initialization protocol employs CAN identifiers that are shared among all slaves. In order to avoid collisions on the CAN bus their use is coordinated by sequencing the initialization phases of individual slaves.

The interaction scenario is implemented by cooperative application of several SDL patterns such as *SendReceive*, *BlockingRequestReply* or *SingleRequestMultipleReplies*. For instance, the exchange of the request and initialization message follows the *BlockingRequestReply* pattern, while the binary search for identification of network slaves applies *SingleRequestMultipleReplies*. Because of space limitations we skip further details here.

3.3.2. Development step II: Resource reservation, directly connected peers. Before communication can take place a certain amount of network resources must be reserved in order to guarantee quality of service (QoS). Figure 4 illustrates the interaction between a resource slave and the central resource manager. The application asks its local resource slave to establish a connection with certain QoS (*SP_id.req*) and is

step	basic service	protocol functionalities
I	directly connected	initialization of CAN modules
II		resource reservation
III		object-based user communication: •read-only-object •write-only-object •uncontrolled-event-object •stored-event-object •read-write-object •domain-object
IV	CAN basic service	

Figure 3. Decomposition and chronological order of selection

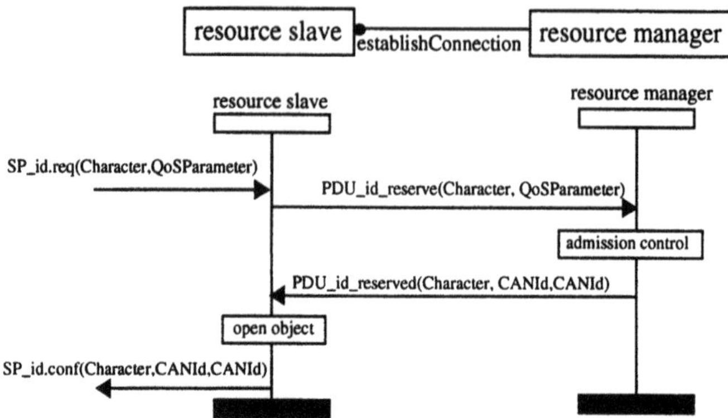

Figure 4. Analysis model of development step II (excerpt)

informed about the result by a corresponding reply (*SP_id.conf*). Embedded into this interaction is a two-way handshake between the slave and the resource manager. The slave sends a *PDU_id_reserve* message with a flow specification containing all relevant QoS parameters. The resource manager then decides, if the connection will be established and assigns unique CAN identifiers to the communication object that will be used for data transfer. The identifiers are returned within the *PDU_id_reserved* message. The interaction is realized by a cascaded application of the *BlockingRequest-Reply* pattern. The shaded parts of Figure 5 determine the renamed SDL fragments of the embedded pattern instances.

3.3.3. Development step III: Object-based communication, directly connected. The CAN subsystem supports object-based user communication. That is, a connection is established by common access to the same communication object, which, e.g., represents a physical quantity or event of a technical process. Access to remote objects is realized according to the Client/Server paradigm. Thereby the server side is responsible for updating the logical communication object with its physical counterpart, while the client side offers remote access. We defined six types of communication objects with different access characteristics, such as read-only or write-only objects (Figure 6). An object consists of a certain number of clients and servers, which are usually located on different nodes. In case of a read-write object (RWobject) the object consists of one server and a maximum of one clients (Figure 6). The MSC diagrams describe interactions between server and client for opening, reading, writing, and closing such an object. In the following, we will only consider RWobjects. It should be noted that communication based on other objects can be specified independently from each other.

Design of object-based communication divides into two major steps. At first, it is assumed that no more than one server and one client of each type of communication object exists per node. Figure 7 shows a possible architecture. Two processes are added to the context specification that resulted from development step II (Figure 5), one for the client and one for the server of a RWobject. The client is assumed to be located on node *Com1* and the corresponding server on node *Com3*. For definition of the behavior SDL patterns are applied. In the following, we will concentrate on the

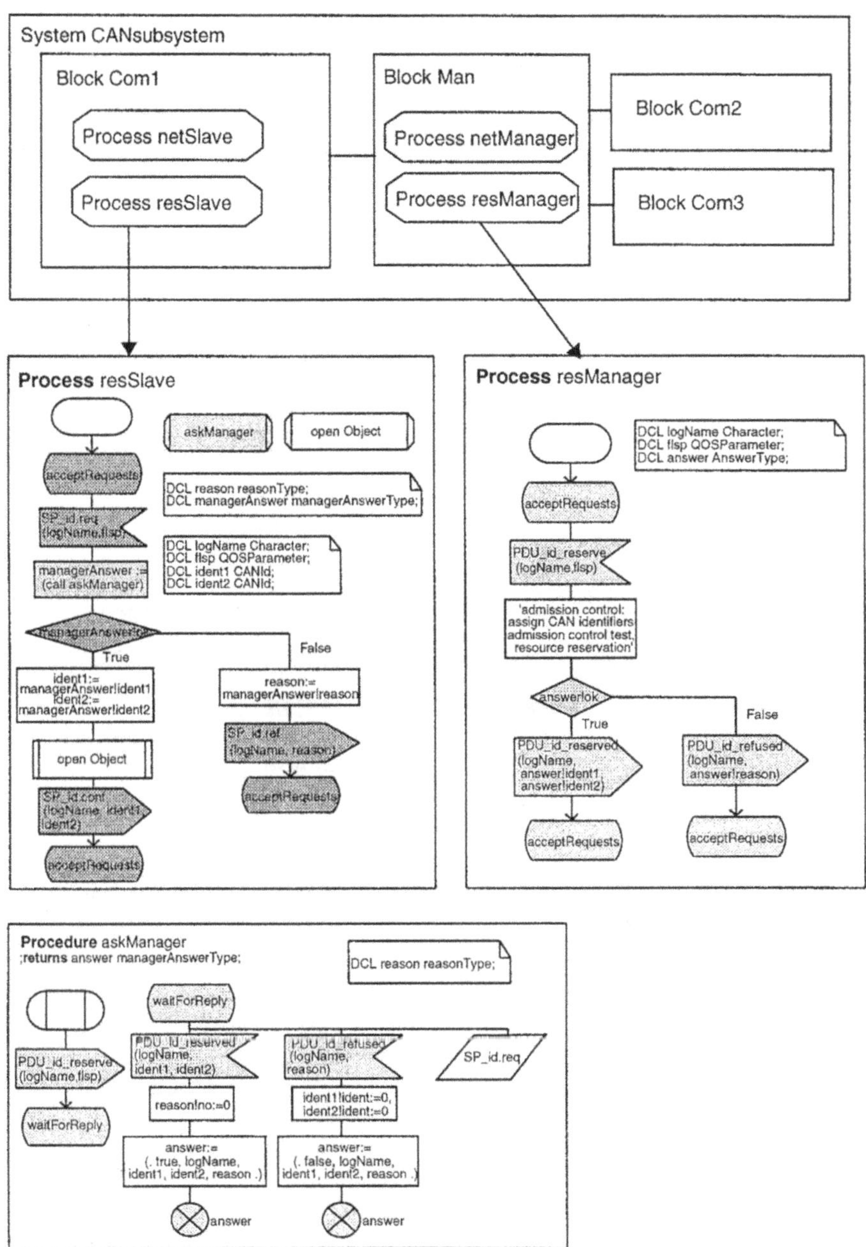

User communication (BlockingRequestReply, replier)

Communication between protocol instances (BlockingRequestReply, requester & replier)

Figure 5. Design specification of development step II (excerpt)

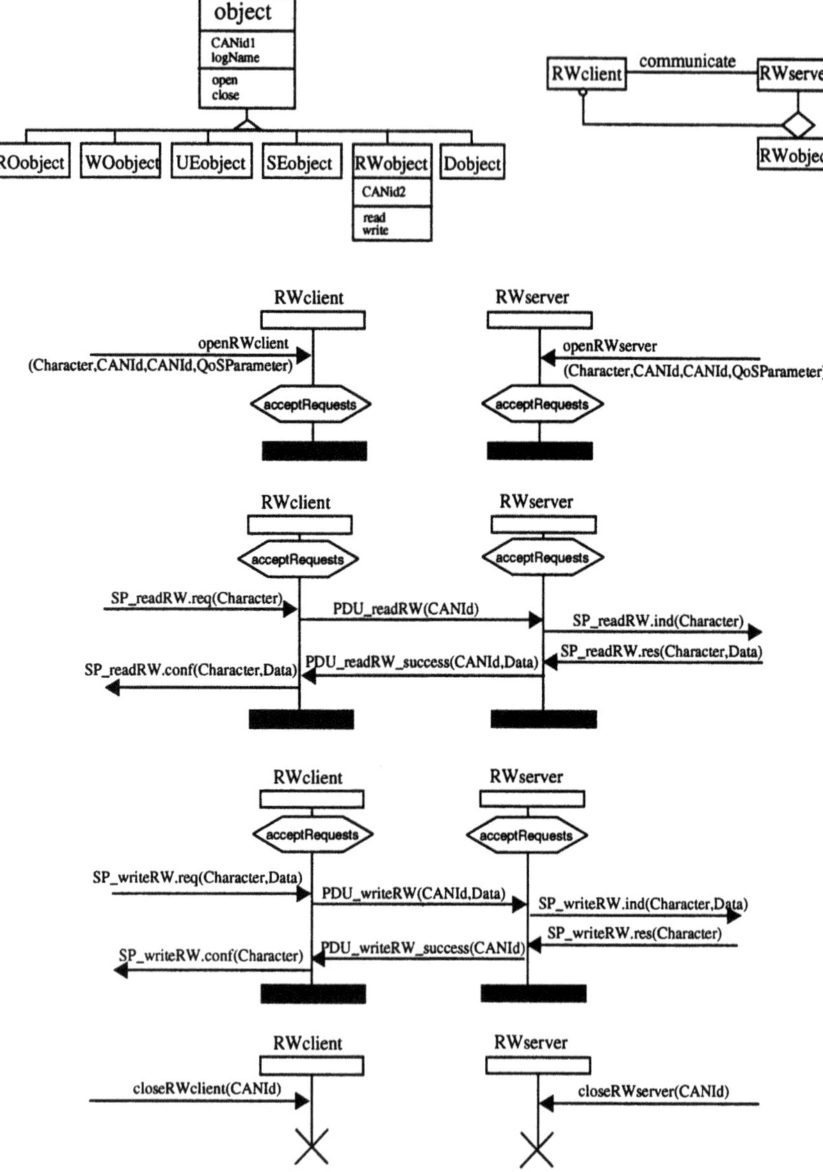

Figure 6. Analysis model of development step III (excerpt)

interactions for reading the object. Write access is specified accordingly. It follows from the corresponding MSC of Figure 6 that the interactions actually describe three cascaded two-way handshakes between the application and *RWclient*, *RWclient* and *RWserver*, and finally between *RWserver* and the application. An SDL pattern generating a two-way handshake is *BlockingRequestReply*. We applied the pattern three times, resulting in the chained *BlockingRequestReply* instances of Figure 7. For the first pattern instance, which is shaded ▒ only the replier is specified, because the request-

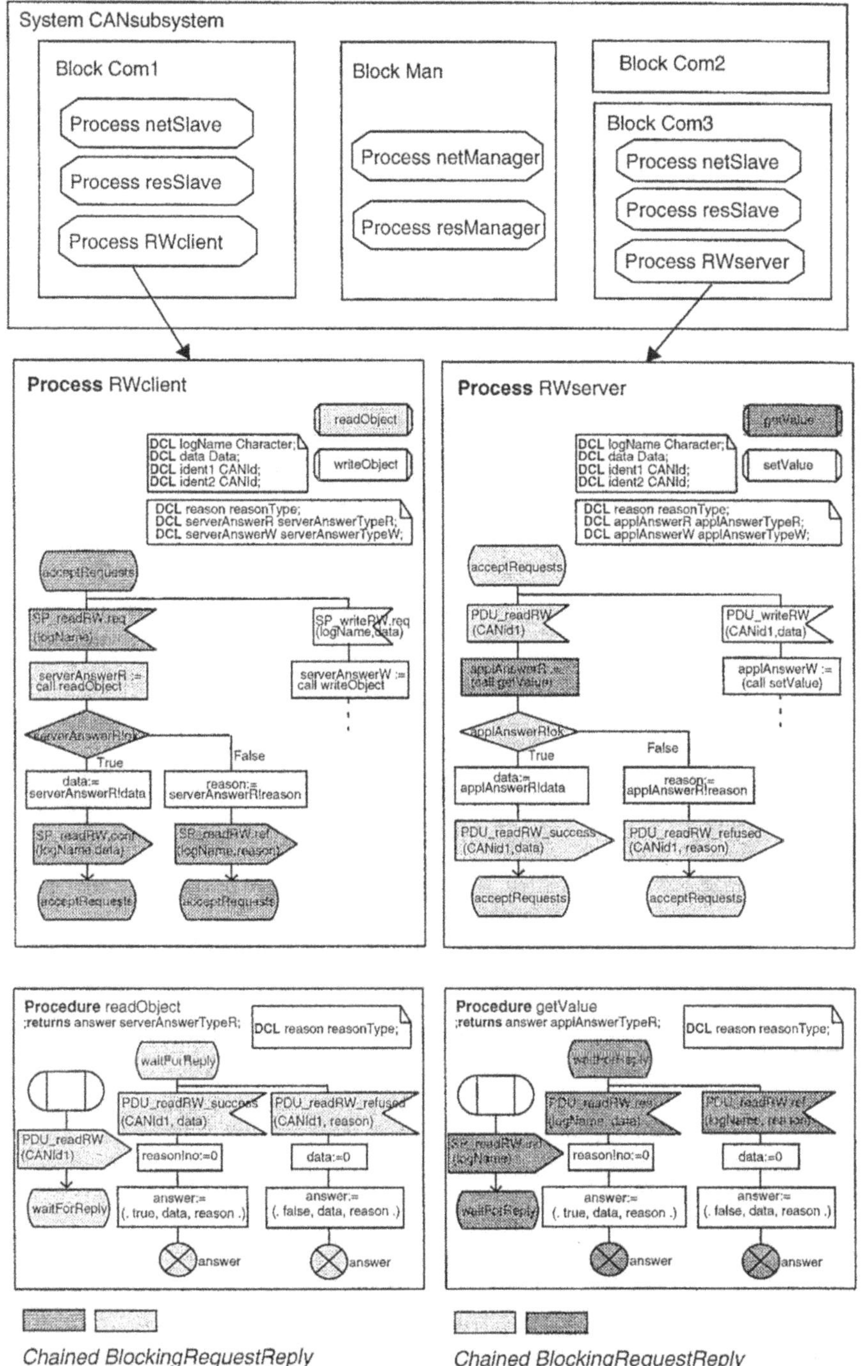

Figure 7. Design specification of development step III: Part 1 (excerpt)

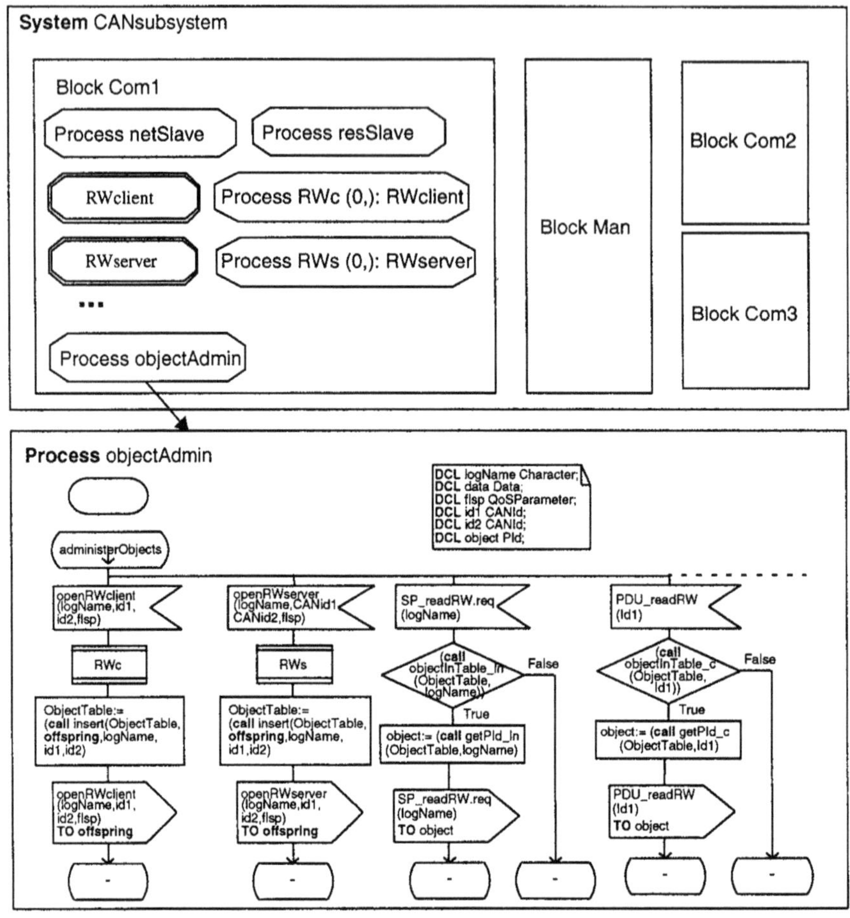

Figure 8. Design specification of development step III: Part 2 (excerpt)

ing application is part of the environment. The requester of the second pattern instance (shaded) is embedded in the first one and defined by the procedure *readObject*. The corresponding replier is located in *RWserver* and also shaded . The requester of the third *BlockingRequestReply* instance is given by the procedure *get-Value* (shaded). The corresponding replier is part of the application on the server side and therefore not explicitly specified.

In the second major design step we relax the assumption that no more than one server or client of each object type be allowed per node. In order to realize this we applied the *DynamicEntitySet* pattern. According to the syntactical embedding rules the server and client processes are replaced by process sets with corresponding process types. For instance, the process *RWserver* is replaced by a process type of the same name and a corresponding process set *RWs* (Figure 8). Additionally, an administrator process *objectAdmin* is introduced, which is responsible for dynamically creating the processes, whenever a new object is opened. Subsequently, *objectAdmin* forwards incoming signals to the right instance of the process set. As a consequence, certain signal routes must be re-routed, i.e., disconnected from the client and server processes and connected to the administrator, which itself must be connected to the process sets. For reasons of readability, channels and signal routes are not shown in Figure 8.

3.3.4. Development step IV: Introducing the CAN basic service. Finally, we replace the direct connections between the communicating peers by the CAN basic service. We therefore have to map PDUs to service primitives offered by CAN. A pattern dealing with that problem is *Codex* (Section 2.1). Its application results in a new SDL process for each communication node. We skip further details here.

4. Generating the implementation

The design specification must finally be transformed into executable code. Note that this is actually not part of the configuration process discussed in Section 2.2, but is added to cover the whole development cycle. Implementation can be done manually or using a code generator such as the SDT Cadvanced Compiler. For our case study, we followed the implementation activity of the SOMT method [10] belonging to the SDT development tool set. That is, we first partitioned the SDL design to define the different run-time modules according to the light integration for the real-time operating system QNX. Environment functions are to be implemented, which are responsible for interfacing the generated code with its physical environment. Next the implementation must be generated using the SDT Cadvanced compiler with its run-time library for QNX. As the development tool set does not run under QNX but is implemented on our SUN cluster running Solaris, the generated C files must be downloaded and compiled on the QNX-PCs.

4.1. Partitioning and light integration

Cadvanced allows two types of integration into other operating systems, which mainly differ in the way SDL processes are mapped to operating system (OS) processes. The tight integration creates an OS process for every SDL process, while the light integration creates a single OS process that handles all SDL processes of the SDL partition. We decided to use light integration, because at the time of implementation only the light integration supported SDL services, which our specification heavily relied on.

We prepared two different partitions. One that covers the protocol functionalities of a slave node (where only network and resource slaves reside, Figure 2) and one for the manager node (where additionally the network and resource managers reside, Figure 2). The hardware-specific low-level CAN driver was hand-implemented. For performance reasons, we also decided to integrate the communication objects (Figure 2) into the low-level driver.

4.2. Interfacing with the environment

The generated modules communicate with the environment by the environment functions xInEnv and xOutEnv. xInEnv provides the functionality to somehow receive messages from the environment and transmit the messages as signals to the generated module. xOutEnv is the counterpart of xInEnv. It forwards SDL signals from the specification to the environment.

4.2.1. Interface to the low-level driver. The low-level driver provides the functionality to transmit messages of up to 8 bytes (the maximum length of a CAN frame). A second part of the interface relates to the object-based user communication. We focus on the basic routines for sending and receiving simple messages, because the generated modules only communicate this way. The send routine has three parameters: the CAN identifier of the message type, a pointer to a buffer, where the message is stored, and the size of the message. The send procedure blocks the calling process until the message is delivered. The receive function has the same parameters, while it is non-blocking.

4.2.2. Interface to the applications. The SDL design defines signals to indicate the initialization phase and to start communication. To realize this we provide two functions that block the application until the corresponding signal is received. Additionally, we define a function that calls the local reservation slave via QNX specific IPC mechanisms in order to establish a communication object.

4.3. Environment functions

For every signal that enters or leaves the specification, we define a C data type named by the signal. For example, for the signal PDU_iden_rem, specified as

```
SIGNAL PDU_iden_rem( integer, integer )
```

we define the data type

```
typedef struct {
      char sigtype;
      short int1, int2;
} PDU_iden_rem_T;
```

`sigtype` is a unique number to identify the signals. Instances of these types are then transferred via the CAN bus.

The environment function *xOutEnv*. There are two types of signals: the signals that have to be transferred via the CAN bus and the signals that realize the communication to the application.

The signals to the low-level driver are implemented by simply allocating an instance of the corresponding data type, filling it with the supplied arguments of the SDL signal, and sending it by calling the low-level send function. For example, the code for the PDU_iden_rem signal looks as follows:

```
...
} else if( (*S)->NameNode == PDU_iden_rem ) {
     PDU_iden_rem_T data;
     data.sigtype = SIG_PDU_iden_rem;
     data.int1 = ((yPDef_PDU_iden_rem *)(*S))->Param2;
     data.int2 = ((yPDef_PDU_iden_rem *)(*S))->Param3;
     send( ((yPDef_PDU_iden_rem *)(*S))->Param1.ident,
          /* ID */ &data, sizeof( data ) );
} else if ...
```

`SIG_PDU_iden_rem` is a unique number for this signal. Expressions such as `((yPDef_PDU_iden_rem *)(*S))->Param1` are the Cadvanced standard method to provide signal parameters to the environment.

Communication with the application is realized differently: an application requests a service via IPC. When the request is handled, the protocol entity first stores the application's process identifier. If the answer is available, the stored process identifier will then be used to reply the answer to the right requester.

The environment function *xInEnv*. Again there are two types of signals: requests from applications and messages from the low-level driver.

To receive signals from the low-level driver, one has to poll the driver by calling its receive function. For every message type there normally exists a different CAN identifier. So, the xInEnv function calls the receive function sequentially for every possible CAN identifier of incoming messages. A typical piece of code looks as follows:

```
if( receive( ID_PDU_iden_rem, data, sizeof(PDU_iden_rem_T)) > 0 ) {
     S = xGetSignal( PDU_iden_rem, xNotDefPId, xEnv );
     ((yPDef_PDU_iden_rem *)(S))->Param1.ident = ID_PDU_iden_rem;
     ((yPDef_PDU_iden_rem *)(S))->Param2 =
         ((PDU_iden_rem_T *)(data))->int1;
     ((yPDef_PDU_iden_rem *)(S))->Param3 =
         ((PDU_iden_rem_T *)(data))->int2;
     SDL_Output( S, xSigPrioPar(xdefaultPrioSignal) (xIdNode *)0 );
}
```

The code inside the `if` block is the Cadvanced standard method for creating a signal instance, defining the parameters, and sending it to the generated module. The code for receiving a request from the application is basically the same, only that the message is received via the QNX specific "Creceive" call and the signals are differentiated by the first byte (the signal identifier sigtype).

5. Conclusion

We have presented a case study on the SDL-pattern based configuring of communication protocols. SDL patterns characterize as formalized design patterns. However, we do not deal with formalizing design patterns in general. That is, instead of formalizing reuse concepts we aim to increase reusability within the formal methods area. Thereby we naturally benefit from the formal basis provided by SDL. SDL-pattern based protocol configuring leads to formal specifications, which can be further used for validation and code generation. This is supported by existing SDL development tools.

As compared to SDL specifications that have been developed the usual way, SDL-pattern based configuring leads to a more systematic design. This is due to the fact that design decisions are well founded and documented. As a consequence, readability of the specification and communication among team members is improved. Also confidence in the correctness of the resulting product increases. It is worth mentioning, that a very large portion of the final specification resulted from SDL patterns, where each pattern has been applied several times. This provides some evidence that the predesigned patterns have been well chosen. From these observations we infer that our approach has the potential of substantially reducing the effort for customizing and maintaining communication protocols, which seems to be a prerequisite for developing protocols that support applications in the best possible way. Though these statements result from experience of several test projects, we intend to validate them in a stronger sense. In [4] we present an approach for the experimental evaluation of empirical properties of SDL patterns.

References

[1] G. Booch, J. Rumbaugh, and I. Jacobson, *Unified Modeling Language*, Version 1.0, Rational Software Corporation, 1997

[2] F. Buschmann, R. Meunier, H. Rohnert, P. Sommerlad, and M. Stal, *Pattern-Oriented Software Architecture - A System of Patterns*, John Wiley & Sons, 1996

[3] K. Etschberger, *CAN Controller-Area-Network - Basics, Protocols, Building Blocks, Applications* (in German), Hanser Verlag, 1994

[4] R. L. Feldmann, B. Geppert, and F. Rößler, *Towards an Experimental Evaluation of SDL-Pattern based Protocol Design*, SFB 501 Report 04/98, Computer Science Department, University of Kaiserslautern, Germany, 1998

[5] E. Gamma, R. Helm, R. Johnson, and J. Vlissides, *Design Patterns - Elements of Reusable Object-Oriented Software*, Addison-Wesley, 1995

[6] B. Geppert, R. Gotzhein, and F. Rößler, *Configuring Communication Protocols Using SDL Patterns*, SDL'97 Time for Testing - SDL, MSC and Trends, Proceedings of the 8th SDL Forum, Paris/Evry, France, 1997

[7] B. Geppert and F. Rößler, *Generic Engineering of Communication Protocols - Current Experience and Future Issues*, Proceedings of the 1st IEEE International Conference on Formal Engineering Methods, ICFEM'97, Hiroshima, Japan, 1997

[8] F. Rößler and B. Geppert, *Applying Quality of Service Architectures to the Field-Bus Domain*, Proceedings of the 2nd IEEE International Workshop on Factory Communication Systems, WFCS'97, Barcelona, Spain, 1997

[9] F. Rößler, A. Kühlmeyer, *Implementing Real-Time Communication on a Token-Ring Network*, Proceedings of the 6th Open Workshop on High Speed Networks, Stuttgart, October, 1997

[10] Telelogic, *TAU 3.3 Methodology Guidelines – Part1: The SOMT Method*, Telelogic, Sweden, 1998

[11] Z.100 *CCITT Specification and Description Language (SDL)*, ITU-T, 1996

[12] Z.120 *Message Sequence Chart (MSC)*, ITU-T, 1996

PART SIX

Verification 2

"ON-THE-FLY INSTANTIATION" OF VALUE-PASSING PROCESSES

Huimin Lin

Laboratory for Computer Science
Institute of Software, Chinese Academy of Sciences
lhm@ox.ios.ac.cn

Abstract: The traditional approach to automatic verification of value-passing processes is to translate them into pure processes, by instantiating input variables over all possible values, then submit the resulting pure processes to verification tools. The main disadvantage of this approach is that problems with infinite value domain can not be dealt with, even if these values are never used by the program, as in the case of communication protocols. In this paper we propose an algorithm which works directly on value-passing processes, represented as *symbolic transition graphs with assignment*. The key idea is to instantiate input variables "on-the-fly", while checking bisimulation, making it possible not to store the entire concrete transition system either before or during the verification process. As a consequence problems with large, even infinite, state spaces can be handled. The algorithm has been implemented and tested on some typical verification problems.

1 INTRODUCTION

The issue addressed in this paper is deciding bisimulation equivalences for value-passing concurrent systems. By "value-passing" we mean data values can be transmitted via communication channels between processes running in parallel. A typical class of value-passing processes is communication protocols. A characteristic feature of value-passing processes is that they can per-

Supported by research grants from the Chinese Academy of Sciences, the National Science Foundation of China, and EU KIT 119 project SYMSEM.

form input and output actions of the forms $c?x$ and $c!e$, where x is a data variable and e a data expression. In contrast, "pure" processes can only perform atomic actions which essentially allow processes to *synchronise* with each other, instead of exchange data. While algorithms and verification tools for pure processes have received considerable attention in the past two decades ([CPS93, SV89, GLZ89], among others), the area of automatic verification of value-passing processes remains almost unexplored.

The traditional approach to value-passing processes is to translate them into pure processes. A key step in such translation is to instantiate input variables with all possible values from the data domains which the variables range over. For example, the following defines a very simple value-passing process (in CCS syntax)

$$P = c?x.d!x.P$$

where x is a variable of type integer. All P can do is to repeatedly receive an integer along channel c then transmit it along channel d. Simple as it looks, so far there are no automatic verification tools that can handle it. To use existing tools one has to first translate P into pure process P':

$$P' = \ldots + c?{-1}.d!{-1}.P' + c?0.d!0.P' + c?1.d!1.P' + \ldots$$

which is also beyond the scope of the existing tools because P' has infinitely many transitions.

For value-passing processes there are two sensible notions of bisimulation, namely *early* and *late*, depending on when input variables get instantiated: during inferring transitions (early), or during matching for bisimulation (late) ([MPW92, HL95]). The translation-based approach can only handle early bisimulation.

In this paper we propose a new approach to value-passing systems without resorting to such translation. The approach is applicable to both late and early bisimulation equivalences. The main novelty is that it combines the *symbolic* framework with "on-the-fly" instantiation technique. As it has been recognised, in many practical value-passing systems (with communication protocols in particular) a distinction can often be made between *control* and *data* variables: the control variables may participate in complicated computations affecting the control flow of the systems, while the data variables just hold values "passively" without being tested nor computed. It is often the case that control variables take values from finite sets and data variables may assume values from infinite domains. For instance, in Alternating Bit Protocols the variables for the boolean flags are control variables, while those holding the values being transmitted are data variables.

We present an algorithm for deciding bisimulation equivalences between such value-passing processes. Instead of instantiating input variables before checking bisimulation, the algorithm instantiates these variables "on-the-fly", during the verification process. To see how it works, suppose we are comparing the following two processes for bisimulation:

$$c?x.P \qquad \text{and} \qquad c?y.Q$$

where x and y range over $\{1,2,3\}$. At this stage we know that the two processes can only mimic each other when x and y are instantiated with *the same value.* Therefore we can take each value v from the set $\{1,2,3\}$ in turn, substituting it for x in P and y in Q, then go on to compare $P[v/x]$ and $Q[v/y]$. There are only three pairs of residuals, namely $(P[1/x],\ Q[1/y])$, $(P[2/x],\ Q[2/y])$ and $(P[3/x],\ Q[3/y])$,to be compared at the next stage. In contrast, the translation-based approach would first translate them into

$$c?1.P[1/x]+c?2.P[2/x]+c?3.P[3/x] \text{ and } c?1.Q[1/x]+c?2.Q[2/x]+c?3.Q[3/x]$$

which will then require nine comparisons, with six definitely fail. If x and y are data variables, then we adopt the symbolic bisimulation technique by instantiating them with the least unused *symbolic values.*

To incorporate the on-the-fly instantiation strategy, as illustrated by the above example, the bisimulation checking algorithm should work on the process structures in "top-down" fashion. Such an algorithm is also called "on-the-fly" since, when working top-down, it is possible to generate the transition graphs piece-by-piece while checking bisimulation, instead of creating the whole transition graphs before verification starts [FM91]. For this idea to work the key issue is how to *efficiently* generate transitions "on-the-fly" at each stage. Direct calculating from process syntax using the operational semantics is too costly, because each time the same term is encountered the same calculation has to be repeated again, and due to recursion such repetition is unavoidable. Using the technique of "transition caching" may easy the pain of repeated calculations, but the caching table may take up a considerable amount of memory, a resource already in short supply.

The solution proposed in this paper is to use *symbolic transition graphs with assignment* (STGA for short). It is shown in [Lin96] that STGAs correspond exactly to *regular* value-passing processes, and moreover, they are closed under parallel composition, hence capable of representing *networks of parallel processes.* An example STGA is shown in Figure 1.1. As can be seen there, each node of a STGA is associated with a set of free data variables, and each edge carries three pieces of information: a boolean condition, a multiple assignment statement, and an abstract action. A *state* over a STGA, representing a *closed term*, then consists of a node together with a vector of data assigning values to the free variables at the node. These values can then be used to evaluate outgoing edges at the node, to produce concrete transitions, in the following manner: first the boolean condition is evaluated, if the result is *false* then this edge does not contribute to the real transition from this concrete state; Otherwise the assignment is evaluated and the result is used to update the data

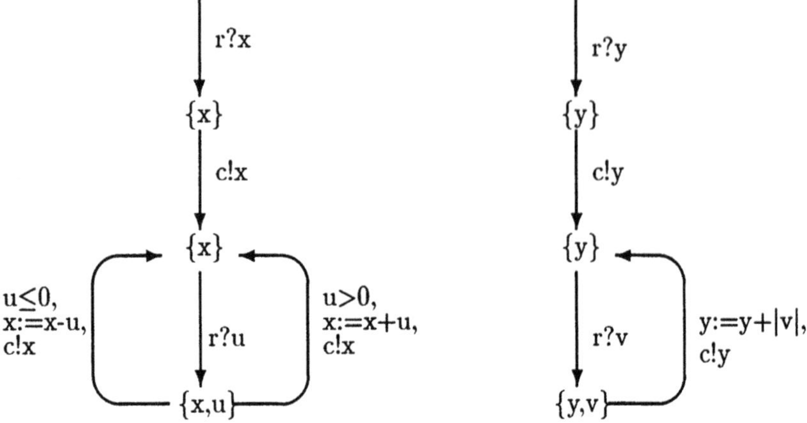

Figure 1.1 Example STGAs

values, resulting in a new vector of data used to evaluate the action if it is an output action (input actions are instantiated on-the-fly as described above).

STGAs can be generated from process terms using *symbolic transition semantics.* For regular processes the sizes of the graphs are linear to the length of the terms. In practice both the time used to generate STGAs and the amount of memory used to store them are negligible. STGAs can be regarded as "half-cooked" syntax: all process dynamics has been represented graphically, while data and boolean expressions remain "uncooked" and are carried in the edges. The task of deriving transitions from a closed process term is reduced to evaluating outgoing edges from a state.

A node in a STGA denotes a subterm, and edges leading to a common node represent sharing a common substructure. Thus STGAs allow to avoid not only direct manipulations on syntactical terms, but also repeated calculations due to recurrence of the same term.

Based on STGAs we have developed an algorithm exploiting the "on-the-fly instantiation" idea. Starting from the root nodes of two STGAs with initial value vectors, the algorithm tries to build the *smallest* bisimulation including the initial states. At each stage transitions are derived as described above. These transitions are then compared for bisimulation. The algorithm only stores pairs of state which have been compared. In particular, it does not store any concrete transitions (which requires more space than states). As a consequence it can handle problems with much larger state spaces than the partition algorithm which works on complete transition graphs. The algorithm has been implemented and tested on some typical verification problems, and the early results are very encouraging.

An important issue in designing verification algorithms and tools is the generation of informative diagnosis when verification fails, because producing just a "no" answer has little help in debugging the input. As mentioned before it is difficult for the the translation-based approach to generate helpful diagnosis information, due to the fact that most structuring information presented in the original value-passing processes gets lost in the translation. In this respect the on-the-fly instantiation algorithm has its advantage. When the two input value-passing processes are not bisimilar, it can generate diagnosis information in the form of a sequence of state pairs witnessing the failure. As a state consists of an (open) subterm in the process definition together with a list of values for the free data variables of the subterm, it is easy to debug the original design using such "high level" diagnosis information.

The rest of the paper is organised as follows: The notion of symbolic transition graphs with assignment is briefly reviewed in the next section. The algorithm is then presented in Section 1.4, with its correctness discussed and complexity analysed. Section 4 deals with early bisimulation. The paper is concluded with Section 5.

Related work The on-the-fly algorithm for deciding bisimulation for pure processes first appeared in [FM91] (to the author's knowledge). The Mobile Workbench also implemented a version of on-the-fly algorithm for open bisimulation in the context of π-calculus [VM94]. The main contribution of the present paper is twofold: *(1)* Lifting the on-the-fly bisimulation checking algorithm to handle value-passing processes, by instantiating input variables "on-the-fly" over symbolic transition graphs with assignment; *(2)* Combining the symbolic bisimulation and "on-the-fly instantiation" techniques so that a class of infinite-state problems can be dealt with.

2 AN ABSTRACT MODEL FOR VALUE-PASSING SYSTEMS

In this section we will only give a brief and informal introduction to *symbolic transition graphs with assignment*, and refer the readers to [Lin96] for a detailed account on this subject.

We assume the existence of the following sets: *Val* (data values), ranged over by $v, u, \ldots$; *Var* (data variables), ranged over by $x, y, z, \ldots$; *DExp* (data expressions), ranged over by $e, e', \ldots$; *BExp* (boolean expressions), ranged overy by $b, b', \ldots$; *Chan* (channel names), ranged overy by $c, c', \ldots$. We will use the customary notations $\overline{x} := \overline{e}$ for multiple assignment, $[\overline{e}/\overline{x}]$ for substitution, and sometime identify $\overline{x} := \overline{e}$ with $[\overline{e}/\overline{x}]$. It is assumed that $Var \cup Val \subseteq DExp$, and $e = e' \in BExp$ for any e, $e' \in DExp$. We also assume that data expressions are equipped with reasonable notions of free and bound variables, and write $fv(e)/bv(e)$ for the sets of free/bound variables of e.

The set of abstract actions, $\{\, \tau, c?x, c!e \mid c \in Chan, x \in Var, e \in DExp \,\}$, is ranged over by α. The sets of free/bound variables of abstract actions are

$$\frac{m \overset{b,\theta,\tau}{\longmapsto} n}{m_\rho \overset{\tau}{\longrightarrow} n_{\theta\rho}} \; \rho \models b \qquad \frac{m \overset{b,\theta,c!e}{\longmapsto} n}{m_\rho \overset{c!\rho(e\theta)}{\longrightarrow} n_{\theta\rho}} \; \rho \models b \qquad \frac{m \overset{b,\theta,c?y}{\longmapsto} n}{m_\rho \overset{c?y}{\longrightarrow} n_{\theta\rho}} \; \rho \models b$$

Figure 1.2 Late Operational Semantics

defined by $fv(c!e) = fv(e)$, $bv(c?x) = \{x\}$, and $fv(\alpha) = bv(\alpha) = \emptyset$ for any other action α.

An *evaluation* $\rho \in Eval$ is a type-respecting mapping from *Var* to *Val* and we use the standard notation $\rho\{x \mapsto v\}$ to denote the evaluation which differs from ρ only in that it maps x to v. An application of ρ to a data expression e, denoted $\rho(e)$, always yields a value from *Val* and similarly for boolean expressions; $\rho(b)$ is either true or false. We will write $\rho \models b$ to indicate that $\rho(b) = true$. If $\sigma \equiv [\overline{e}/\overline{x}]$ then the application of σ to ρ is denoted by $\sigma\rho = \rho\{\overline{x} \mapsto \rho(\overline{e})\}$.

Definition 2.1 A symbolic transition graph with assignments (STGA for short) is a rooted directed graph where each node n is associated with a finite set of free variables $fv(n)$ and each edge is labeled by a triple $(b, \overline{x} := \overline{e}, \alpha)$. A STGA is *well-formed* if for each edge from n to m, written $n \overset{b,\overline{x}:=\overline{e},\alpha}{\longmapsto} m$, it holds that $fv(b, \overline{e}) \subseteq fv(n)$, $fv(\alpha) \subseteq \{\overline{x}\}$, and $fv(m) \subseteq \{\overline{x}\} \cup bv(\alpha)$. □

We will often simply write $n \overset{\theta,\alpha}{\longmapsto} m$ for $n \overset{true,\theta,\alpha}{\longmapsto} m$, and will also omit θ when it is the identity assignment on $fv(n)$.

Traditionally (pure) processes are modeled by *labeled transition systems* (LTSs for short) which are directed graphs in which arcs are labeled with atomic actions. A vertex in a LTS represents a state and the outgoing arcs show the possible actions the process at the state can perform to evolve into new states. Existing verification algorithms and automatic tools for process algebras are all based on LTSs ([CPS93, SV89, GLZ89]).

STGAs are graphic representations for value-passing processes, just as traditional transition graphs for pure processes. However the STGA representations are at a very abstract level as most of the characteristic structures of value-passing processes, namely data expressions, boolean conditions and input/output actions, are retained in the edges.

A STGA $\mathcal{G}$ can be "expanded" into a "partial" concrete labeled transition system whose vertices are pairs of (n, ρ) with n a node of $\mathcal{G}$ and ρ an evaluation mapping free variables at n to values. Transitions are generated according to the operational semantics given in Figure 1.2. Transition systems so generated are called "partial" because input actions are not yet instantiated. The instantiation will be carried out while graphs are compared for bisimulation.

Definition 2.2 A late bisimulation is a symmetric relation R over states such that: if $(m_\rho, n_\varrho) \in R$ then

1. $m_\rho \xrightarrow{c?y} m'_{\rho'}$ implies there exists $n_\varrho \xrightarrow{c?z} n'_{\varrho'}$ and for all $v \in Val$ $(m'_{\rho'\{y\mapsto v\}}, n'_{\varrho'\{z\mapsto v\}}) \in R$.

2. for any other actions $m_\rho \xrightarrow{a} m'_{\rho'}$ implies there exists $n_\varrho \xrightarrow{a} n'_{\varrho'}$ and $(m'_{\rho'}, n'_{\varrho'}) \in R$.

We write $m_\rho \sim n_\varrho$ if there is a late bisimulation R such that $(m_\rho, n_\varrho) \in R$. Two graphs $\mathcal{G}$ and $\mathcal{G}'$ are bisimilar with respect to ρ and ρ' if $r_\rho \sim r'_{\rho'}$ where r and r' are the roots of the two graphs, respectively. □

As an example of how to generate symbolic graphs from a process description language let us consider regular value–passing *CCS* given by the following BNF grammar:

$$t \ ::= \ \mathbf{0} \mid \alpha.t \mid b \to t \mid t+t \mid P(\overline{e})$$

where for each process identifier P there is a definition $P(\overline{x}) \Leftarrow t$ which satisfies $fv(t) \subseteq \{\overline{x}\}$ and is *guarded*, *i.e.* every identifier in t is within the scope of an action prefixing α. .. Given a term t in this language a STGA can be generated using the following rules:

$$\frac{}{\alpha.t \overset{true,\emptyset,\alpha}{\longmapsto} t} \qquad \frac{t \overset{b,\theta,\alpha}{\longmapsto} t'}{t+u \overset{b,\theta,\alpha}{\longmapsto} t'} \qquad \frac{t \overset{b,\theta,\alpha}{\longmapsto} t'}{u+t \overset{b,\theta,\alpha}{\longmapsto} t'}$$

$$\frac{t \overset{b,\theta,\alpha}{\longmapsto} t'}{b' \to t \overset{b\wedge b',\theta,\alpha}{\longmapsto} t'} \qquad \frac{t \overset{b,\theta,\alpha}{\longmapsto} t'}{P(\overline{e}) \overset{b[\overline{e}/\overline{x}],\theta[\overline{x}:=\overline{e}],\alpha}{\longmapsto} t'} \quad P(\overline{x}) \Leftarrow t \text{ is a definition}$$

Two STGAs can be composed using parallel composition operator $\|$ and restriction operator $\backslash$. When composing two STGAs it is required that they use disjoint name spaces for data variables (so the only way for them to cooperate is through communication). The nodes of the parallel composition $(\mathcal{G}\|\mathcal{H})\backslash R$, with R a set of channel names, are pairs of those of $\mathcal{G}$ and $\mathcal{H}$ with $fv(<n,m>) = fv(n) \cup fv(m)$, and the edges are created by the following rules (where the symmetric rules of par and com have been omitted):

$$\text{par} \ \frac{n \overset{b,\overline{x}:=\overline{e},\alpha}{\longmapsto} n'}{<n,m> \overset{b,\overline{x},\overline{y}:=\overline{e},\overline{y},\alpha}{\longmapsto} <n',m>} \quad \begin{array}{l} chan(\alpha) \cap R = \emptyset \\ fv(m) = \{\overline{y}\} \end{array}$$

$$\text{com} \ \frac{n \overset{b_1,\overline{x}:=\overline{e}_1,c?z}{\longmapsto} n', \quad m \overset{b_2,\overline{y}:=\overline{e}_2,c!e}{\longmapsto} m'}{<n,m> \overset{b_1\wedge b_2,\overline{x},\overline{y},z:=\overline{e}_1,\overline{e}_2,e[\overline{e}_2/\overline{y}],\tau}{\longmapsto} <n',m'>}$$

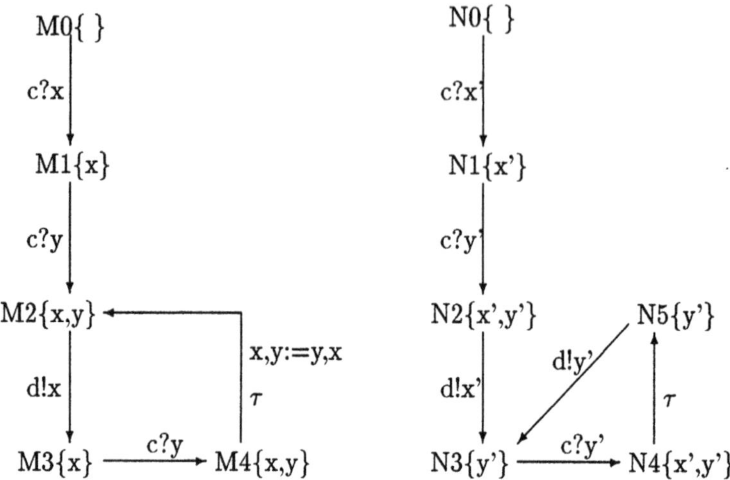

Figure 1.3

It is easy to see that parallel composition is symmetric and associative. This is CCS-style "handshaking" communication where communication happens between two processes capable of performing complementary (input and output) actions, yielding an "invisible" τ action. Similarly one can define parallel composition adopting "broadcasting" communication, as in CSP and LOTOS.

3 THE ALGORITHM

Given a symbolic graph, a variable is "data-independent" if it does not appear in the boolean condition part of any edge label, nor occur as a proper subterm of any data expression. Thus data variables are not subject to any operation other than assignment. The algorithm presented in this section works on symbolic graphs where all variables are either data-independent or with finite domains.

As observed in [JP92], it is sufficient to "instantiate" each data variable with a fresh *schematic variable* when deciding bisimulation. The approach proposed in [JP92] is based on the (global) partition algorithm and requires *a prior* construction of complete transition systems. Here we incorporate the idea by adopting the "symbolic bisimulation" technique [HL95]. We assume a countable set of "symbolic values" which are totally ordered, and the function *nextSVal*(s, t) always returns the smallest symbolic value that does not appear in the value vectors at states s and t. To illustrate how symbolic values are generated, we use the example shown in Figure 1.3 where all variables are data-independent. Assume the symbolic values are $v_1, v_2, v_3, \ldots$ (in that order). We start with the root nodes. At the state pair

$(M0\{\}, N0\{\})$ no symbolic value is used, hence $nextSVal(M0\{\}, N0\{\}) = v_1$ which is associated to x and x'. Now at the new state pair $(M1\{x = v_1\}, N1\{x' = v_1\})$ we have $nextSVal(M1\{x = v_1\}, N1\{x' = v_1\}) = v_2$, so the next state pair is $(M2\{x = v_1, y = v_2\}, N2\{x' = v_1, y' = v_2\})$. After the output transitions the new state pair is $(M3\{x = v_1\}, N3\{y' = v_2\})$. Since $nextSVal(M3\{x = v_1\}, N3\{y' = v_2\}) = v_3$, the next state pair is $(M4\{x = v_1, y = v_3\}, N3\{y' = v_3\})$. Passing through the τ and output transitions we go back to to nodes $M3$ and $N3$ again, with the state pair $(M3\{x = v_3\}, N3\{y' = v_3\})$ (note that x and y got transposed after τ move). Now we have $nextSVal(M3\{x = v_3\}, N3\{y' = v_3\}) = v_1$.

The algorithm for deciding late bisimulation for symbolic transition graphs with assignment is presented in Figure 1.4. It operates on a pair of STGAs $\mathcal{G}$ and $\mathcal{H}$. The function $bisim(s, t)$ will always terminate with answer *true* or *false*. It starts with the initial states pair (s, t), trying to find the smallest bisimulation containing the pair by matching transitions from each pair of states it reaches. While searching the symbolic graphs, at each pair of nodes the algorithm uses the values for the state variables to evaluate the edges, according to the operational semantics in Figure 1.2, producing concrete transitions and next states. The transitions are then matched for bisimulation, and the algorithm goes on to the new state pairs if the matches are successful. However when input edges are encountered the evaluation can not be fully carried out because the values for the input variables at the next pair of nodes are not yet known. In this case the algorithm behaves as follows: if the input variables are data variables then the function $nextSVal$(s, t) is called and the returned symbolic value is used for the input variables; otherwise each element is taken from the value domain associated with the input variables, one by one, to instantiate the input variables "on-the-fly". The algorithm then moves to the new states.

Three auxiliary data structures are used:

- *NotBisim*: List of state pairs which have already been detected as not bisimilar.

- *Visited*: List of state pairs which have already been visited.

- *Assumed*: List of state pairs which have already been visited and assumed to be bisimilar.

The core function, *match*, is called from function *bis*inside the main function *bisim*. It performs a depth-first search on the product of the two concrete transition graphs which are never fully created. Instead states and transitions of the concrete graphs are only generated from the symbolic graphs on-the-fly, when needed. Whenever a new pair of states is encountered it is inserted into *Visited*. If two states fail to match each other's transitions then they are not

$bisim(\mathrm{s},\ \mathrm{t}) =$ {
  $NotBisim := \{\ \}$
  fun $bis(\mathrm{s},\ \mathrm{t}) =$ {
    $Visited := \{\ \}$
    $Assumed := \{\ \}$
    $match(\mathrm{s},\ \mathrm{t})$}
  } handle $WrongAssumption \Rightarrow bis(\mathrm{s},\ \mathrm{t})$
  return($bis(\mathrm{s},\ \mathrm{t})$)

$match(\mathrm{s},\ \mathrm{t}) =$
  $Visited := \{(s,\ t)\} \cup Visited$
  $\mathrm{b} = \bigwedge_{a \in \{\tau, c!v, c?\}} match_a(s,\ t)$
  if $\mathrm{b} = false$ then {
    $NotBisim := \{(s,\ t)\} \cup NotBisim$
    if $(s,\ t) \in Assumed$ then raise $WrongAssumption$}
  return(b)

$match_{\{\,a | a \in \{\tau, c!v\}\,\}}(\mathrm{s},\ \mathrm{t}) =$
  for each $s \xrightarrow{a} s_i$, for each $t \xrightarrow{a} t_j$
    $b_{ij} = close(s_i,\ t_j)$
  return$((\bigwedge_i(\bigvee_j b_{ij})) \wedge (\bigwedge_j(\bigvee_i b_{ij})))$

$match_{c?}(\mathrm{s},\ \mathrm{t}) =$
  for each $s \xrightarrow{c?x} s_i$, for each $t \xrightarrow{c?y} t_j$
    if x and y are data-independent then
      let $z = nextSVal(s,\ t)$
        $b_{ij} = close(s_i[x \mapsto z],\ t_j[y \mapsto z])$
      return$((\bigwedge_i(\bigvee_j \forall z b_{ij})) \wedge (\bigwedge_j(\bigvee_i \forall z b_{ij})))$
    else for each $s \xrightarrow{c?x} s_i$, for each $t \xrightarrow{c?y} t_j$
      for each $v \in Val$
        $b^v_{ij} = close(s_i[x \mapsto v],\ t_j[y \mapsto v])$
      return$((\bigwedge_i(\bigvee_j \bigwedge_{v \in Val} b^v_{ij})) \wedge (\bigwedge_j(\bigvee_i \bigwedge_{v \in Val} b^v_{ij})))$

$close(s,\ t) =$
  if $(s,\ t) \in NotBisim$ then return($false$)
  else if $(s,\ t) \in Visited$ then {
    $Assumed := \{(s,\ t)\} \cup Assumed$
    return($true$)}
  else return($match(\mathrm{s},\ \mathrm{t})$)

Figure 1.4 The algorithm

bisimilar and the pair is inserted into *NotBisim*. In case the current state pair has been visited before, we first check if it is in *NotBisim* and return *false* if so. Otherwise a loop has been detected and we make assumption that the two states are bisimilar, by inserting the pair into *Assumed*, and return *true*. If we find the two states are not bisimilar after finishing searching the loop, then we know that the assumption is wrong, so we first add the pair into *NotBisim* and then raise the exception *WrongAssumption* which forces a rerun of *bis*, with one piece of new information, namely the two states in this pair are not bisimilar.

The correctness of the algorithm is not difficult to justify. Suppose there are k data-independent variables at nodes n and m, then, at most $k+1$ different schematic values will be used in any state pair (s, t) at (n, m). So termination is guaranteed. For these variables the only places they can be used are simple output actions of the form $c!x$ where x is data-independent. When the algorithm returns *true* any pair of such output actions successfully passed the comparison in $match_{c!}$ must have their variables instantiated with the same schematic value, *i.e.* they are instantiated at the same time, when the two input actions which bind the output variables are compared in $match_{c?}$. Therefore if the input variables are instantiated with the same "real" value, the two output actions will still match each other.

Each call of *bis*(s,t) performs a depth-first search in the product graph of the two concrete transition graphs (which are "generated" piece by piece from the symbolic graphs during the search). *bis*(s,t) is only recalled when the exception *WrongAssumption* has been raised (by *match*, as explained above), in this case the size of *NotBisim* has been increased by at least one. Hence *bis*(s,t) can only be called for finitely many times. Therefore *bisim*(s,t) will always terminate. If *bisim*(s,t) terminates with *true*, then the set (*Visited* − *NotBisim*) constitutes a bisimulation containing (s,t):

Theorem 1 *Suppose G and H are finite symbolic graphs with roots n and m and initial values ρ and ρ', respectively, and the value domains for the control variables are finite. Function bisim$((n,\rho),(m,\rho'))$ always terminates, and if it returns true then (n,ρ) and (m,ρ') are bisimilar.*

On the other hand, when the algorithm returns *false* it is not necessary that the two graphs are not bisimilar when data-independent variables are interpreted over *any* domain. This can be illustrated by a very simple example: Running the algorithm on the two processes

$$c?x.c?y.d!x.d!y.\mathbf{0} \quad \text{and} \quad c?x.c?y.d!y.d!x.\mathbf{0}$$

will return *false*, but they are bisimilar when x, y are only allowed to take value from a singleton set. Intuitively this is because the data domain involved does not have enough elements to distinguish between different data-independent variables. To give a precise characterisation let $DIN_{\mathcal{G}}$ be the largest number of data-independent variables at any node of graph $\mathcal{G}$, and $DIN_{\{\mathcal{G},\mathcal{H}\}}$ the minimum of $DIN_{\mathcal{G}}$ and $DIN_{\mathcal{H}}$. Then

Proposition 3.1 *If the algorithm returns false then the two input processes, with symbolic graphs $\mathcal{G}$ and $\mathcal{H}$, are not bisimilar when their data-independent variables take values over any data domain with cardinality no less than $DIN_{\{\mathcal{G},\mathcal{H}\}}$.*

To see the complexity of the algorithm, let N be the number of nodes in the production graph of the two concrete transition graphs. The time required for the first call of *bis* is (at most) $O(N)$, for the second call is $O(N-1)$, ... Hence the worst case complexity of *bisim* is $O(N+(N-1)+...+1) = O(N^2)$. However experiences show that the square fact does not appear in practical applications: in most cases the algorithm terminates with only one or two calls of *bis*.

The algorithm has been implemented in Standard ML, and tested on a number of examples. In the Appendix is the input file for the alternating-bit protocol problem which allows the media to lose data (adapted from an example in pure CCS included in the Concurrency Workbench distribution). In the file a type named `message` is defined as a subtype of integer with range `1 ... 10`. Then all process identifiers, channel names and variables are declared with appropriate types (`Bool` is a built-in type), followed by the conjecture to prove. Finally, recursive definitions are provided for the process identifiers (the `where` part). The right hand-side of the conjecture is the specification of the protocol, while the left hand-side is the implementation. Both the specification and the implementation are compiled into STGAs before subject to the algorithm. The STGA for the specification has only 2 nodes and 2 edges, while the one for the implementation has 32 nodes and 80 edges.

As comparison we also run the same examples using the Concurrency Workbench of North Carolina [CS96] which implements the partition algorithm for deciding bisimulation [KS83]. As the examples are written in full CCS Bruns' value-passing front-end [Bru96] is used to translate them into pure CCS before fed into the Workbench. Both systems are written in New Jersey ML version 109, and all results are obtained on a Sparc station with 60MHZ twin-CPU and 64MB memory. The first example is the alternating-bit protocol mentioned above. The second is the "triple-modular redundancy" problem taken from [Bru96]. By increasing the sizes of data domains bigger and bigger transition graphs can be achieved. When the domains grow up to certain limits (100 in the case of alternating-bit protocol and 20 in the case of triple modular redundancy problem) the Workbench runs out of memory and stops. On the other hand the on-the-fly algorithm can handle much larger domains: 350 for the alternating-bit protocol (with 32204 states) and 140 for the triple modular redundancy problem (with 140561 states). Moreover, the variables holding the messages being transmitted in the alternating-bit protocol are data independent. When it is declared as such (by changing the definition of type `message` to `message = data` in the input file) the verification takes less than a second to finish with answer *true*, which means the protocol is correct over even infinite data domains.

When the input processes are not bisimilar, the algorithm terminates with *false*. In this case informative diagnosis information can be easily computed from *NotBisim*. It consists of a sequence of matching transition and ends with a pair of states, one of which can do a transition that the other can not. For example, if we change the definition of process R(rb) in the AB-protocol example in the Appendix to

```
R(rb) = tau.sack!(not(rb)).R(rb) +
        r?(ra,rm).if ra==rb then receive!rm.sack!ra.R(rb)
                  else sack!ra.R(rb)
```

i.e., when the newly received frame has expected boolean flag (`ra==rb`), the message body is transmitted (`receive!rm`) and an acknowledge is sent back (`sack!ra`), but the flag is wrongly remained unswitched (`R(rb)` instead of `R(not(rb))` as in the original description). Now feed the revised file to the algorithm, we will get the following diagnosis:

```
<R(rb)|Mlossy|S(sb){rb=false,sb=false},Spec{}>==send?sm,send?m=>
<R(rb)|Mlossy|S1(sb,sm){rb=false,sb=false,sm=1},
   receive!m.Spec{m=1}>==receive!1,receive!1=>
<R(rb)|Mlossy|if sa==sb then S(not(sb)) else S1(sb,sm)
   {rb=false,sa=false,sb=false,sm=1},Spec{}>==send?sm,send?m=>
<R(rb)|Mlossy|S1(sb,sm){rb=false,sb=true,sm=1},
   receive!m.Spec{m=1}>

now <receive!m.Spec{m=1}>==receive!1=><Spec{}>
but <R(rb)|Mlossy|S1(sb,sm){rb=false,sb=true,sm=1}> has no
matching move
```

The last two lines read: the specification is at a state where a transmission is possible (`receive!1`), while the implementation is at the state `R(rb)|Mlossy|S1(sb,sm)` with boolean flags `rb=false, sb=true` and can not perform such a transition (in fact it is ready to do a `send?sm` action).

Such diagnosis information points directly to the source-level process description, hence is very helpful for debugging purpose.

4 THE EARLY CASE

The bisimulation equivalence discussed above is called "late" because input variables are instantiated when input actions are matched for bisimulation, instead of when concrete transitions are inferred from process terms (i.e. when input edges are evaluated in our case) which takes place before bisimulation matching. In late bisimulation each abstract input action of the form $c?x$ from one process has to be matched by a *single* corresponding action from the other, for all possible values. If this restriction is relaxed by allowing an input action to be matched by several input actions, possibly instantiated with different values, from the other process, the resulting relation is called "early bisimulation".

Early bisimulation can be obtained by swapping the quantifications "for all $v \in Val$" and "there exists" in the first clause of Definition 2.2, so for different values the input move from m_ρ is allowed to be matched by different input moved from n_ϱ:

1. $m_\rho \xrightarrow{c?y} m'_{\rho'}$ implies for all $v \in Val$ there exists $n_\varrho \xrightarrow{c?z} n'_{\varrho'}$ such that $(m'_{\rho'\{y\mapsto v\}}, n'_{\varrho'\{z\mapsto v\}}) \in R$.

Concerning the algorithm only the $match_{c?}$ function needs modifying:

$$
\begin{aligned}
match^E_{c?}(\mathrm{s},\ \mathrm{t}) = \ & \text{for each } s \xrightarrow{c?x} s_i \\
& \quad \text{if } x \text{ is data-independent then} \\
& \quad\quad \text{let } z = nextSVal(s,\ t) \\
& \quad\quad\quad \text{for each } t \xrightarrow{c?y} t_j \\
& \quad\quad\quad\quad b_{ij} = close(s_i[x \mapsto z],\ t_j[y \mapsto z]) \\
& \quad\quad \text{return}((\bigwedge_i(\forall z \bigvee_j b_{ij})) \wedge (\bigwedge_j(\forall z \bigvee_i b_{ij}))) \\
& \quad \text{else for each } v \in Val \\
& \quad\quad\quad \text{for each } t \xrightarrow{c?y} t_j \\
& \quad\quad\quad\quad b^v_{ij} = close(s_i[x \mapsto v],\ t_j[y \mapsto v]) \\
& \quad\quad \text{return}((\bigwedge_i(\bigwedge_{v\in Val}(\bigvee_j b^v_{ij}))) \wedge (\bigwedge_j(\bigwedge_{v\in Val}(\bigvee_i b^v_{ij}))))
\end{aligned}
$$

5 CONCLUSIONS

We have presented new algorithms for deciding both late and early bisimulation equivalences for value-passing processes. The main novelty of our approach is that input actions are instantiated on the fly, making it possible to check bisimulation for value-passing processes without first expanding them into pure processes. Process terms are compiled into *symbolic transition graphs with assignment.* During verification concrete transitions are derived from the symbolic graphs hence there is no need to store entire concrete transition systems, in particular transitions are never stored. The technique of symbolic bisimulation has been incorporated so that data-independent variables ranging over potentially infinite value domains can also be dealt with. The algorithm has been implemented and tested on some examples, and the results show that it can handle much larger problems than the translation-based approach. When the input graphs are not bisimilar an informative ("source-level") diagnosis information can be generated, which is very helpful for debugging purpose.

It should be pointed out that the on-the-fly instantiation method does not rely on any particular properties of the data domain involved, hence any techniques for reducing data domain size can be incorporated.

As further work we are adapting the "on-the-fly instantiation" algorithm to check early and late bisimulations for the π-calculus, which can be viewed as complementary to the *Mobile Workbench* [VM94], a verification tool dedicated to checking *open bisimulation* in the π-calculus.

References

[Bru96] G. Bruns. *Distributed Systems Analysis with CCS.* Prentice Hall, 1996.

[CPS93] R. Cleaveland, J. Parrow, and B. Steffen. A semantics based verification tool for the verification of concurrent systems. *ACM Trans. Program. Lang. Syst.*, 15(1):36–72, 1993.

[CS96] R. Cleaveland and S. Sims. The NCSU Concurrency Workbench. In *CAV'96*, number 1102 in LNCS, pages 394 – 397. Springer–Verlag, 1996.

[FM91] J.-C. Fernandez and L. Mounier. On the fly verification of behavioural equivalences and preorders. In *CAV'91*, volume 575 of *LNCS*, pages 181–191. Springer–Verlag, 1991.

[GLZ89] J. Godskesen, K. Larsen, and M. Zeeberg. Tav user manual. Report R89-19, Aalborg University, 1989.

[HL95] M. Hennessy and H. Lin. Symbolic bisimulations. *Theoretical Computer Science*, 138:353 – 389, 1995.

[JP92] B. Jonsson and J. Parrow. Deciding bisimulation equivalences for a class of non-finite-state programs. *Information and Computation*, 1992.

[KS83] P. Kanellakis and S. Smolka. CCS expressions, finite state processes, and three problems of equivalence. In *Proc. 2nd Ann. ACM Symposium on Principles of Distributed Computing*, pages 228–240, 1983.

[Lin96] H. Lin. Symbolic transition graph with assignment. In *CONCUR'96*, volume 1119 of *LNCS*, pages 50 – 65. Springer–Verlag, 1996.

[MPW92] R. Milner, J. Parrow, and D. Walker. A calculus of mobile proceses, part I,II. *Information and Computation*, 100:1–77, 1992.

[SV89] R. De Simone and D. Vergamimi. Aboard auto. Report RT111, INRIA, 1989.

[VM94] B. Victor and F. Moller. The mobility workbench – a tool for the π-calculus. In *CAV'94*, number 818 in LNCS, pages 428 – 440. Springer–Verlag, 1994.

Appendix: An Example Input File

```
type
        message = 1 ... 10
process
        S : Bool
        S1 : Bool message
        S2 : Bool message
        Mlossy :
        Msafe :
        R : Bool

        Spec :
channel
        send : message
        receive : message
        r : Bool message
        s : Bool message
        rack : Bool
        sack : Bool
variable
        sb,sa,rb,ra,ma : Bool
        sm,rm,mm,m : message
conjecture

        (R(false) | Mlossy | S(false))\{r,s,rack,sack} = Spec
where

 S(sb) = send?sm.S1(sb,sm)
 S1(sb,sm) = s!(sb,sm).S2(sb,sm)
 S2(sb,sm) = tau.S1(sb,sm) + rack?sa.(if sa==sb then S(not(sb))
                                      else S1(sb,sm))

 R(rb) = tau.sack!(not(rb)).R(rb) +
         r?(ra,rm).if ra==rb then receive!rm.sack!ra.R(not(rb))
                   else sack!ra.R(rb)

 Mlossy = s?(ma,mm).(r!(ma,mm).Mlossy + Mlossy) +
          sack?ma.(rack!ma.Mlossy + Mlossy)

 Spec = send?m.receive!m.Spec

end
```

Solving asynchronous equations

Alexandre Petrenko [1] *and Nina Yevtushenko* [2]
1 *CRIM, Centre de Recherche Informatique de Montréal, 550, Sherbrooke St., Suite 100 Montréal, H3A 1B9, Canada, petrenko@crim.ca*

2 *Tomsk State University, 36 Lenin St., Tomsk, 634050, Russia, yevtushenko@elefot.tsu.tomsk.su*

Key words: I/O FSMs, nondeterminism, equation solving, verification, submodule construction, controller design, machine factorization, livelocks

Abstract: The paper addresses the problem of solving an equation formulated in terms of the input/output finite state machine model. The relation used in equations is either FSM equivalence or reduction, i.e. trace inclusion. The composition operator corresponds to asynchronous communications between nondeterministic input/output machines. These types of FSM equations are called asynchronous. A procedure for determining the largest potential solution to a given asynchronous equation is proposed. To verify whether or not the largest potential solution satisfies the equation, the absence of livelocks needs to be established. A solution to the livelock problem is also suggested.

1. INTRODUCTION

Designing complex systems usually involves an important step of dividing the whole system into a number of separate components which interact in some well-defined way. In this context, a question arises how to design a component that combined with a known part of the system, called the context, satisfies a given overall specification. The problem is also known as the problem of submodule construction [Mer83], component redesign [Kim72], [Rho91], [Wat93], controller design [Azi95], equation solving [Par89], [Lar90], [Qin91], [Che96] or machine factorization [Qin91]. This problem has many important applications, ranging from hardware optimization to protocol design. In the case of compound systems, one may encounter it at the level of validation [Hee95] and test derivation [Petr93], [Petr96].

Equation solving has been extensively studied within the process algebra, however, not much work has yet been done to solve equations formulated in terms of input/output finite state machines, I/O FSMs. In the context of hardware optimization [Rho91], one usually assumes strictly synchronous communications between components, as several input signals should be processed simultaneously, see [Yev91], [Wat93], [Azi95]. In many applications, including distributed systems, e.g. protocols, we are obliged to consider more loose communications when components interact asynchronously via message buffers or queues. While the problem of composing asynchronously communicating I/O FSM has been addressed in a number of works in various contexts (mainly within the reachability analysis) [Lee93], [Pet93], [Pet94], we are not aware of any result directly related to the problem of decomposing a nondeterministic I/O machine into asynchronously communicating machines, i.e. equation solving.

This paper addresses the problem of equation solving for a system of I/O possibly nondeterministic FSMs that communicate asynchronously via bounded input queues where actions are stored, assuming that the system has a single message in transit. Under this assumption, any output of one component is immediately consumed by another one as its input, so one may say that communications are performed by rendezvous. However, we cannot directly use the results obtained in process algebra to solve an equation in terms of the FSM model. The problem is that the set of actions of an FSM is divided into two sets, inputs and outputs, and any transition is an atomic transition labeled by an I/O pair. An I/O FSM can, in fact, be transformed into a labeled transition system, but a solution (in trace semantics) to an equation can not always be translated back into an I/O FSM. Another approach has to be used to ensure that the obtained solution to the equation remains in the domain of I/O FSMs. In this paper, we propose a procedure for determining the largest potential solution to the equation, in the sense that any other potential solution is trace included in the obtained machine, i.e. it is a reduction of the largest potential solution. A potential solution becomes a solution if there are no livelocks when it is combined with the given part of the system. When the largest solution to the equation is determined an optimal implementation of the component of interest can be obtained as an appropriate reduction of the largest solution. For example, an optimal implementation should have minimal number of states [Dam94], [Wat94] or transitions. In the case where potential solutions exist and the largest solution does not, it remains an open question whether a solution to the equation exists or not. We demonstrate in this paper that the largest solution may sometimes not exist. We also consider an equation where composed machines are allowed to communicate internally a certain number of times. The idea is to avoid livelocks solving the equation.

The paper is organized as follows. In Section 2, we give necessary basic notions and define a composition operator for I/O nondeterministic finite state machines through a parallel composition operator for labeled transition systems. A method for FSM decomposition in terms of equation solving is presented in Section 3. In Section 4, we propose some enhancements to our method which allow us, solving the equation, to avoid livelocks. Section 5 presents possible future work.

2. PRELIMINARIES

2.1 Finite state machines and labeled transition systems

A finite state machine (FSM), often simply called a machine throughout this paper, is a completely specified, initialized, (possibly nondeterministic) observable Mealy machine which can be formally defined as follows. A *finite state machine* is a quintuple (S, X, Y, h, s_0), where S is a set of states with s_0 as the initial state; X - a finite nonempty set of input symbols; Y - a finite nonempty set of output symbols; and h - a behavior function h: $S \times X \rightarrow \wp(S \times Y) \backslash \varnothing$, where $\wp(S \times Y)$ is the powerset of $S \times Y$ and $|\{s' | (s',y) \in h(s,x)\}| \leq 1$ holds for all $(s,x) \in S \times X$ and all $y \in Y$, i.e. the machines considered in this paper are observable [Star72]. The machine becomes *deterministic* when $|h(s,x)|=1$ for all $(s,x) \in S \times X$. In a deterministic FSM, instead of the behavior function which is required to represent a nondeterministic behavior, we use two functions: the next state function, and the output function. An FSM $B = (S', X, Y', h', s_0)$ is said to be a *submachine* of an FSM $A = (S, X, Y, h, s_0)$ if $S' \subseteq S$, $Y' \subseteq Y$ and $h'(s,x) \subseteq h(s,x)$ for all $(s,x) \in S' \times X$. We extend the behavior function to a function on the set X^* of all input sequences containing an empty sequence ε, i.e., h: $S \times X^* \rightarrow \wp(S \times Y^*) \backslash \varnothing$. For convenience, we use the same notation h for the extended function, as well. Assume $h(s,\varepsilon) = \{(s,\varepsilon)\}$ for all $s \in S$, and suppose that $h(s,\beta)$ is already specified. Then $h(s,\beta x) = \{ (s',\gamma y) \mid \exists s'' \in S\ [(s'',\gamma) \in h(s,\beta)\ \&\ (s',y) \in h(s'',x)] \}$. The function h^1, often called the next state function, is the first projection of h, while h^2, often called the output function of A, is the second projection of h, i.e., $h^1(s,\alpha) = \{ s' \mid \exists\ \beta \in Y^*\ [(s',\beta) \in h(s,\alpha)] \}$, $h^2(s,\alpha) = \{\beta \mid \exists s' \in S\ [(s',\beta) \in h(s,\alpha)] \}$ for all $\alpha \in X^*$.

Given alphabets X and Y such that $X \cap Y = \varnothing$ and sequences $\alpha = x_1 \ldots x_k \in X^*$ and $\beta = y_1 \ldots y_k \in Y^*$, the sequence $\alpha \mathbin{¤} \beta = x_1 y_1 \ldots x_k y_k$ is called a *trace* over alphabets X and Y. Given state $s \in S$ of the FSM $A = (S, X, Y, h, s_0)$, the trace $\alpha \mathbin{¤} \beta$ is called a trace of the FSM A at the state s if $\beta \in h^2(s,\alpha)$ or simply a trace of the FSM A if s coincides with the initial state s_0. Given an input alphabet X and output alphabet Y, there exists a special nondeterministic

FSM such that any trace over these alphabets is a trace of this machine. We call such a machine a *chaos* machine over alphabets X and Y. In particular, a single-state chaos machine is an FSM $(\{p\}, X, Y, H, p)$, where $H(p,x) = \{p\} \times Y$ for all $x \in X$.

Given two states s of the FSM A and r of the FSM $B= (T, X, Y, H, t_0)$, state r is said to be a *reduction* of s, written $r \leq s$, if for any input sequence the condition $H^2(r,\alpha) \subseteq h^2(s,\alpha)$ holds; otherwise, r is not a reduction of s. States s and r are *equivalent* states, written $s \cong r$, iff $s \leq r$ and $r \leq s$. On the class of deterministic machines, the above relations coincide.

Given two machines, A and B, over the same input alphabet, B is a reduction of A, written $B \leq A$, if the initial state of B is a reduction of the initial state of A. Similarly, the equivalence relation between machines is defined. $B \cong A$, iff $B \leq A$ and $A \leq B$.

A *labeled transition system* (LTS) is a quadruple (S, L, T, s_0), where S is a finite set of states with s_0 as the initial state, L is a finite nonempty set of observable actions and $T \subseteq S \times (L \cup \{\tau\}) \times S$ is a transition relation, where $\tau \notin L$ is a nonobservable action. For state $s \in S$, we denote $in(s)$ and $out(s)$ subsets of actions labeling incoming and outgoing transitions at the state s, respectively. As usually, the set of traces of the LTS I is denoted by $Tr(I)$. The LTS I is said to be *deterministic* if for each state $s \in S$, it holds that $\tau \notin out(s)$ and for each pair (s,a), $s \in S$, $a \in out(s)$, there exists at most one state s' such that $(s,a,s') \in T$.

If the set L of actions of the LTS I is partitioned into sets X and Y, $L = X \cup Y$, $X \cap Y = \emptyset$, then for a sequence σ over the alphabet L, the *X-projection* of σ is a sequence obtained by deleting all the actions $y \in Y$ from the sequence σ, and the *X-projection* of the LTS I [Che96] is an LTS that is obtained by hiding all the actions from Y in I and determinizing the resulting LTS by means of a subset construction [Hop79].

Given an FSM $A = (S, X, Y, h, s_0)$, a minimal deterministic LTS $I = (Q, X \cup Y, T, q_0)$ is said *to correspond* to A, if $Tr(I) = \{\alpha ¤ \beta \mid \alpha \in X^* \ \& \ \beta \in h^2(s_0,\alpha)\}$. We denote I_A an LTS corresponding to A. An LTS corresponding to an FSM can be obtained through the following steps:

- unfold every atomic transition *s-x/y->s'* into two consecutive transitions *s-x->s''-y->s'*;
- determinize the obtained LTS by a subset construction;
- minimize the resulting LTS in trace semantics.

2.2 Composition of LTSs

Given two deterministic LTSs $C = (Q, L_1, H, q_0)$ and $D= (T, L_2, G, t_0)$, the *composition* of C and D is the LTS $(Q \times T, L_1 \cup L_2, F, q_0t_0)$, denoted $C \| D$, such that for all states $(q,t)\in Q\times T$ and $a\in L_1\cup L_2$, a triple $(qt, a, q't')$ is a transition of the LTS $C\|D$ if one of the following conditions holds

- $a\in L_1\cap L_2$, $(q, a, q')\in H$ and $(t, a, t')\in G$;
- $a\in L_1\backslash L_1\cap L_2$, $t'=t$ and $(q, a, q')\in H$;
- $a\in L_2\backslash L_1\cap L_2$, $q'=q$ and $(t, a, t')\in G$.

We usually consider the initially connected part of the composition and trim off states that are not reachable from the initial state and their outgoing transitions.

2.3 Composition of FSMs

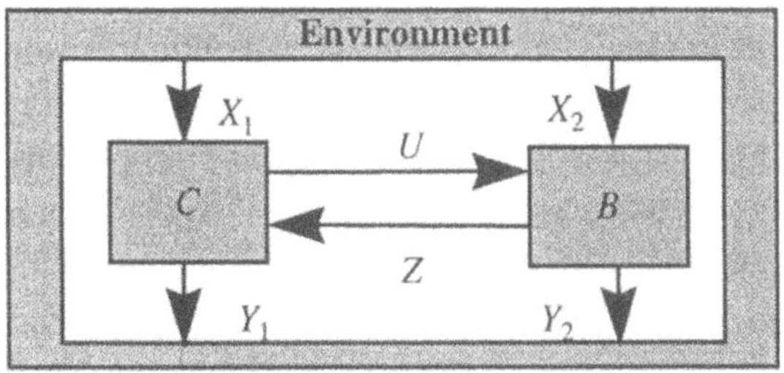

Figure 1. A closed system

We consider a system of two components which are connected as shown in Figure 1 and communicate via bounded input queues, where actions are stored, assuming that the system has a single message in transit. Let the FSM C be defined over alphabets I_1, O_1 and the FSM B over alphabets I_2, O_2 such that $I_1\cap I_2 = \varnothing$ and $O_1\cap O_2 = \varnothing$. The following relationships hold between the alphabets: $X_1 = I_1\backslash O_2$, $Y_1 = O_1\backslash I_2$, $X_2 = I_2\backslash O_1$, $Y_2 = O_2\backslash I_1$, $Z = I_1\cap O_2$ and $U = I_2\cap O_1$. We call symbols of the alphabet $X_1\cup X_2 = X$ *external* inputs, symbols of the alphabet $Y_1\cup Y_2 = Y$ - *external* outputs, while symbols of the alphabet $Z\cup U$ - *internal actions*. We further assume that at least one of the sets X_1, X_2 and at least one of the sets Y_1, Y_2 are not empty. Otherwise, i.e. when the system has no inputs or no outputs, it can not be represented by an FSM. As the system has a single message in transit, at any moment either the components exchange messages or one of them communicates with its environment which submits a next external input to the system only after the system has produced an external output in response to the previous input. In other words, we consider a closed system, where the environment is described by the LTS L_E with two states (a total chaos system) shown in Figure 2. The behavior of the closed system can be described by the composition $L_C\|L_B\|L_E$ where L_C and L_B are LTSs corresponding to FSMs C and B.

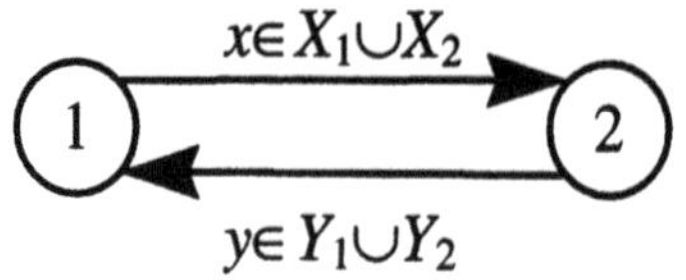

Figure 2. The environment LTS

If the system $L_C || L_B || L_E$ has no livelocks then it can be transformed into an FSM over the external inputs X and external outputs Y. The system of the FSMs C and B is said to have *a livelock w.r.t. an external input sequence* $\alpha \in X^*$ if there exists a trace β of the LTS $L_C || L_B || L_E$ such that α is the X-projection of β and the LTS has a path labeled with α containing a cycle labeled only with internal actions in $Z \cup U$. A livelock is said to be an *output deadlock* if no state s with $out(s) \cap Y \neq \varnothing$ is reachable from the cycle. The system in an output deadlock will never produce an external output even in the case of a non-catastrophic interpretation of livelocks when it is assumed that a system may get out of a livelock. We assume here a catastrophic interpretation of livelocks (a divergent LTS) and leave the composition operator for FSMs undefined. Formally, we define the composition operator $\lozenge$ for two FSMs as follows.

Given the LTS $L_C || L_B || L_E$ without livelocks, we determine $(X \cup Y)$-projection of the LTS $L_C || L_B || L_E$ and transform the projection into an FSM, denoted $C \lozenge B$, by pairing each input with a subsequent output and replacing the pair of corresponding transitions by a single transition. We are usually interested in the initially connected part of $C \lozenge B$, called a *composite* machine of C and B. We keep the same notation $C \lozenge B$ for a composite machine.

Compared with LTSs, the definition of the composition operator on FSMs is slightly more involved, and opposed to the composition of LTSs, the operator is not defined in case of livelocks. Another difference is that the environment is required to alternate between inputs and outputs, enabling the extraction of the composite FSM $C \lozenge B$ from the LTS which describes the closed system.

Our definition of the composition operator is, of course, applicable to two FSMs communicating without feedback. If, for example, $U = \varnothing$ or $Z = \varnothing$ we have a well-known *loop-free* composition of FSMs [Cec86]. The case when both sets are empty corresponds to a *parallel* (independent) composition of two FSMs. If the sets U, X_1 and Y_2 or Z, X_2 and Y_1 are empty then the system is a *sequential* composition of two FSMs studied, e.g. in [Kim72], [Petr94]. All these compositions are constrained versions of the one considered in this paper.

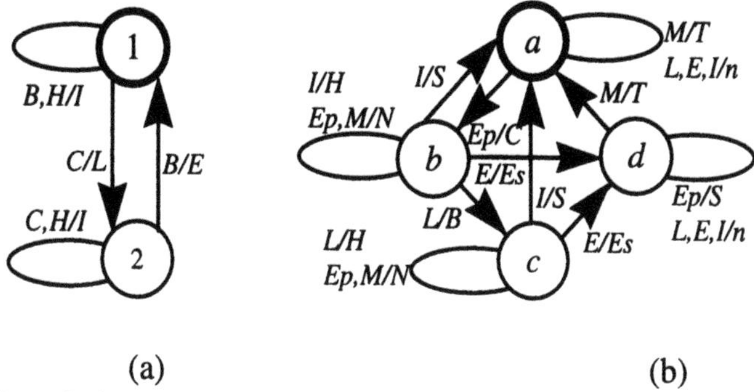

(a) (b)

Figure 3. An espresso machine, the FSM B (a), and waiter, the FSM C (b)

Example. Consider a coffee shop which has a machine that produces a cup of espresso, it is the FSM C with the inputs **C**oin, **B**utton, and **H**it, and the outputs **L**amp, **E**spresso, **I**dle, shown in Figure 3 (a) and a waiter, the FSM B that accepts two inputs, **E**spresso-**p**lease and **M**oney, from the environment (customer) and three inputs, **L**amp, **E**spresso and **I**dle, from the coffee machine, the outputs are **E**spresso-**s**erved, **T**hanks, **S**orry, **N**o action (to the environment); **C**oin, **B**utton, **H**it (to the espresso machine) or **n**o action at all, shown in Figure 3 (b). Note that the waiter behaves nondeterministically, when the machine, having accepted a coin, idles she/he either apologizes to the customer or hits the machine. The customer submits an input **E**spresso-**p**lease or **M**oney after one of the outputs has been received from the coffee shop in response to the previous input. Such behavior is modeled by the LTS L_E shown in Figure 2.

The detailed behavior of each machine becomes clear from the LTS $L_C||L_B||L_E$, Figure 4 (a). The obtained LTS has no livelock, so a (composite) FSM over the external alphabets can be extracted. The composite FSM $C\Diamond B$ is shown in Figure 4 (b).

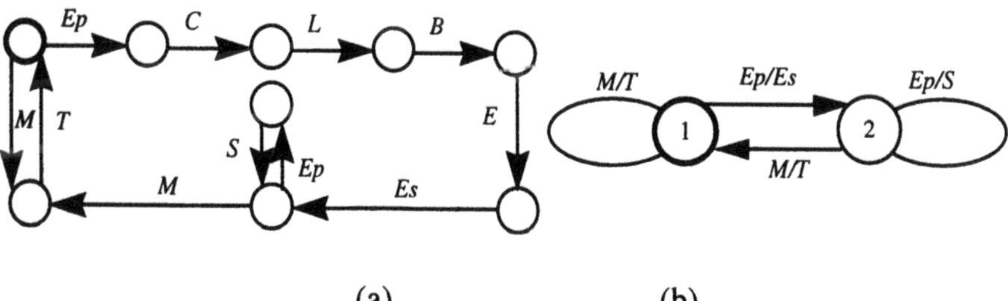

(a) (b)

Figure 4. The LTS $L_C||L_B||L_E$ (a) and the composite FSM $C\Diamond B$ (b)

Note that our definition of the composition operator on FSMs essentially differs from that used in the area of hardware, see for example, [Wat93], [Lin95], [Azi95]. Synchronous circuits are usually constructed in such way that several input signals to the same component are synchronized. It means that in the synchronous case, for example, the FSM B in Figure 1 should have a compound input alphabet $X_2 \times U$ modeling such synchronization. In the asynchronous case, the input alphabet $X_2 \cup U$ corresponds to interleaved, i.e. asynchronous, communication considered in our setting. The two definitions of the operator come to the same only when each component has just one input. Moreover, to exclude oscillations, hardware components must participate in just a single interaction in response to external input. Opposed to the synchronous case, asynchronously communicating FSMs for different external input sequences may involve into a various number of internal interactions before one of them produces an external output. As a result, no external output might be produced when the system falls into livelock. Thus, the problem of livelocks becomes crucial for solving equations in the asynchronous case.

3. FSM DECOMPOSITION

3.1 Asynchronous FSM equations

The decomposition problem can be formulated by means of an equation in terms of FSMs, similar to that considered within process algebra. Given two FSMs, A over alphabets X and Y (the overall specification) and C over alphabets $X_1 \cup Z$ and $Y_1 \cup U$ (the context), we are required to determine an FSM B such that the composition operator $\Diamond$ is defined for the FSMs C and B and a relation, reduction or equivalence, holds for the FSMs $C \Diamond B$ and A. For nondeterministic FSMs, we have the following three (asynchronous) equations

$$C \Diamond W \le A \text{ (1)}; \quad A \le C \Diamond W \text{ (2)}; \quad C \Diamond W \cong A \text{ (3)},$$

where W is the unknown FSM to be found (if possible). Intuitively, the equation, for example, (2) corresponds to the case where the system to be designed in given alphabets is allowed to do more, but no less than a given specification.

An FSM B over alphabets $X_2 \cup U$ and $Y_2 \cup Z$ is said to be a *solution to the equation* (1), (2) or (3) if

- the composition operator is defined for the FSMs C and B and
- $C \Diamond B \le A$, $A \le C \Diamond B$, or $C \Diamond B \cong A$, respectively.

A solution B is called the *largest* solution to an equation if any solution to the equation is a reduction of B. Once the largest solution is found, one can then reduce it to a solution that is optimal according to a chosen criterion, for example, a machine with a minimal number of states and transitions [Dam94], [Wat94]. Similarly, one may define the smallest solution to an equation.

Generally speaking, one may treat each equation separately. On the other hand, given the largest solution to (1) and the smallest solution to (2), the product of these solutions is obviously a solution to (3). Hereinafter, we consider the equation $C\Diamond W \leq A$ only, as it seems most interesting for practice. Solving this equation, we solve, in fact, the FSM equation $C\Diamond W \cong A$ in the deterministic setting.

3.2 The largest potential solution to the equation $C\Diamond W \leq A$

Any potential solution to the equation $C\Diamond W \leq A$ is a reduction of the single-state chaos machine over the alphabets $X_2 \cup U$ and $Y_2 \cup Z$, denoted *Ch*. This observation suggests that one may solve the equation by deleting certain traces from the chaos machine. To this end, we first state the following obvious property of the largest solution.

Proposition. Let an FSM $M \leq Ch$ be the largest solution to the equation $C\Diamond W \leq A$. Then any FSM $B \leq Ch$ such that at least one of the three following conditions is satisfied

a) $C\Diamond B \leq A$ does not hold,

b) the LTS $L_C \| L_B \| L_E$ has an output deadlock,

c) the LTS $L_C \| L_B \| L_E$ has a livelock,

is not a reduction of *P*.

The statement indicates the cases when a certain trace of the chaos machine should be deleted to reduce it to the largest solution or at least to some solution may the largest solution not exist. We will proceed in two steps. We first define and construct a so-called largest potential solution by weakening the three conditions to the first two and then we check whether the composition operator is defined for the context and obtained machine (the condition c). The last step can be performed by constructing a composite machine, as explained in Section 3.1. This motivates the following definition.

An FSM $P \leq Ch$ is said to be the *largest potential solution* to the equation $C\Diamond W \leq A$ if any FSM $B \leq Ch$ such that

a) $C\Diamond B \leq A$ does not hold, or

b) the LTS $L_C \| L_B \| L_E$ has an output deadlock

is not a reduction of *P*.

By definition, the largest potential solution *P* composed with the context *C* has no non-conforming external behavior, but may yet have livelock and the composite machine might not be defined. In other words, the largest solution (if it exists) is a reduction of the largest potential solution, and the latter becomes the largest solution to the equation when the composition operator is defined. Then it remains to determine how the largest potential solution can be constructed.

We first present a scheme that generalizes existing procedures for equation solving in terms of the LTSs and the FSM (synchronous case) in trace semantics.

Step 1. Based on a chosen definition of the composition operator, derive a global machine describing all the sequences of actions that may occur in the composition.

Step 2. Construct the acceptor whose designated dead-state* accepts all the traces of the global machine which result in an external behavior not conforming to A (according to a given conformance relation).

Step 3. Construct the projection of the acceptor into the alphabets of solutions to the equation. Replace every state of the projection, comprising the dead-state, by the dead-state.

Step 4. Construct the complement of the obtained projection (and in case of an FSM equation, convert the obtained complement into an FSM).

These steps use various procedures depending on the model, conformance relation, and composition operator. Adapted to our setting, the scheme can be implemented in the following algorithm.

Algorithm 3.1. Construction of the largest potential solution to the equation $C \Diamond W \leq A$.

Input. The FSM A and context C.

Output. The largest potential solution $[[A,C]]$ (if it exists).

Step 1. Transform FSMs C and Ch into corresponding LTSs L_C and L_{Ch}. Construct the LTS $L_C \| L_{Ch} \| L_E$ where L_E is the environment LTS.

Step 2.

a) Transform the FSM A into the corresponding LTS L_A. Construct the LTS $L_{\hat{A}}$ by adding transitions to a designated dead-state labeled with all $y \in (Y \cup Y_1) \backslash out(s)$ from each state s of the LTS L_A such that $out(s) \cap (Y \cup Y_1) \neq \varnothing$.

b) Compose the LTSs $L_C \| L_{Ch} \| L_E$ and $L_{\hat{A}}$ and replace all global states of the composed LTS, containing a component dead-state, as well as all states from which no state s such that $out(s) \cap Y \neq \varnothing$ is reachable, by the (global) dead-state. Let $L_{A,C}$ be the obtained LTS.

Step 3. Construct the $(X_2 \cup U \cup Y_2 \cup Z)$-projection of the LTS $L_{A,C}$ by a subset construction. Replace every state of the $(X_2 \cup U \cup Y_2 \cup Z)$-projection, which comprises a component dead-state, by a dead-state. Let $\hat{L}$ be the resulting LTS.

Step 4. Convert the LTS $\hat{L}$ into an FSM $[A,C]$. Complete the FSM $[A,C]$ by adding a *don't care* state and its incoming transitions from each state with an undefined transition for an input in $X_2 \cup U$ and label them with a corresponding input and each output in $Y_2 \cup Z$. Delete the dead-state with its

* The dead state in an LTS has no transitions, but in an FSM, it has self-loops labeled with each I/O pair over the given alphabets.

incoming transitions and consecutively all states whose transition becomes undefined for some input until the initial state is deleted or no more state can be deleted. In the first case, the equation has no solution and in the second case, the largest potential solution [[*A*,*C*]] is returned.

We immediately notice that since the LTS $L_{\hat{A}}$ is trace included into the LTS L_E, one can directly construct $L_C \| L_{Ch} \| L_{\hat{A}}$ instead of $L_C \| L_{Ch} \| L_E \| L_{\hat{A}}$. We first illustrate the algorithm using our example considered above and then prove its validity.

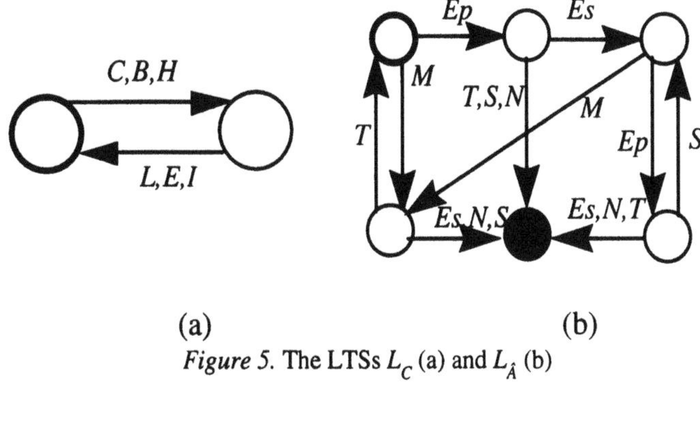

(a) (b)

Figure 5. The LTSs L_C (a) and $L_{\hat{A}}$ (b)

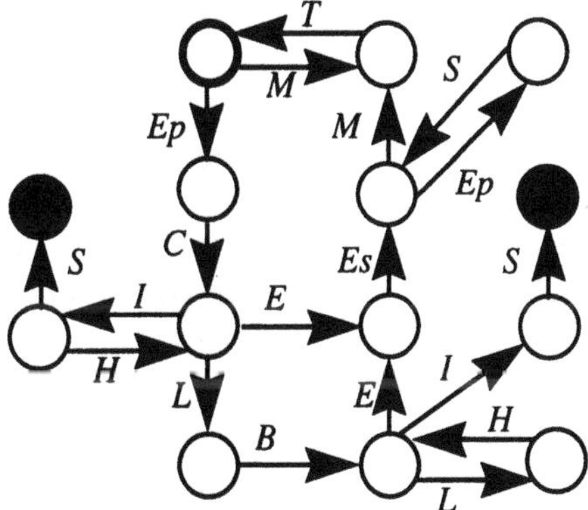

Figure 6. The LTS $L_{A,C}$ obtained at Step 2

Example. We apply the above algorithm to construct a coffee machine which can be used in the coffee shop, the FSM *A* (Figure 4b) with the waiter, the FSM *C* (Figure 3b). Figure 5 shows the LTS corresponding to the chaos machine over the alphabets of the coffee machine and the LTS $L_{\hat{A}}$, where a black hole represents a dead-state. Figure 6 shows the LTS $L_{A,C}$ obtained at Step 2, where black holes represent a global dead-state. Figure 7 shows the last intermediate result, the LTS $\widehat{L}$ obtained at Step 3 (a) and the

FSM $[[A,C]]$ (b), where dashed lines represent don't care transitions to a don't care state. The largest potential solution $[[A,C]]$ is a nondeterministic machine. Based on this machine, one can construct a few (deterministic) coffee machines, including the simplest machine which dispenses a cup of espresso once a coin is inserted.

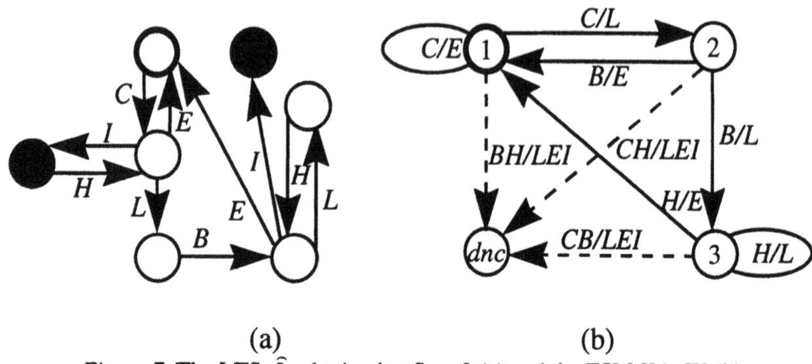

Figure 7. The LTS $\hat{L}$ obtained at Step 3 (a) and the FSM $[[A,C]]$ (b)

Theorem 1. Given two FSMs, A over alphabets $X_1 \cup X_2$ and $Y_1 \cup Y_2$ and C over alphabets $X_1 \cup Z$ and $Y_1 \cup U$, if Algorithm 3.1 does not return an FSM the equation $C \Diamond W \leq A$ has no solution, and if the composition operator is defined for the FSMs C and $[[A,C]]$ derived by Algorithm 3.1, then the FSM $[[A,C]]$ is the largest solution to $C \Diamond W \leq A$. Moreover, if $C \Diamond [[A,C]]$ is not equivalent to A then the equation $C \Diamond W \cong A$ has no solution.

Proof. According to Algorithm 3.1, a trace β¤γ over alphabets $X_2 \cup U$ and $Y_2 \cup Z$ is not a trace of the resulting FSM $[[A,C]]$ if and only if one of the following conditions holds.

(1) There exists a trace of the LTS $L_{A,C}$ (Step 2b) with the $(X_2 \cup U \cup Y_2 \cup Z)$-projection β¤γ resulting in the external behavior different from that of A.

(2) There exists a trace of the LTS $L_{A,C}$ (Step 2b) with the $(X_2 \cup U \cup Y_2 \cup Z)$-projection β¤γ resulting in an output deadlock of the composed system (Step 2).

(3) Trace β¤γ cannot be a trace of a potential solution (Step 4), because there exists $v \in X_2 \cup U$ such that any trace βv¤$\gamma\delta$ over alphabets $X_2 \cup U$ and $Y_2 \cup Z$ takes the LTS $\hat{L}$ obtained at Step 3 to the dead-state.

If Algorithm 3.1 does not return an FSM then there exists a sequence $\beta \in U^*$ such that for each trace β¤γ over alphabets $X_2 \cup U$ and $Y_2 \cup Z$ one of the above conditions 1, 2, or 3 holds while each FSM B over alphabets $X_2 \cup U$ and $Y_2 \cup Z$ has a trace β¤γ for an appropriate $\gamma \in U^*$. If B has trace β¤γ such that condition 3 holds then there exists its prolongation βv such that for each trace βv¤$\gamma\delta$ over alphabets $X_2 \cup U$ and $Y_2 \cup Z$ the condition 1 or 2 holds. In other words, without loss of generality, we assume that B has a trace β¤γ

for which the condition 1 or 2 holds. In this case, there exists an external input sequence α such that the composition of the context C and FSM B exhibits the behavior different from that of the FSM A or has an output deadlock when α is submitted, i.e. B cannot be a solution to the equation $C\Diamond W \leq A$ and therefore cannot be a solution to the equation $C\Diamond W \cong A$.

Suppose now that Algorithm 3.1 returns an FSM [[A,C]]. Trace β¤γ is a trace of [[A,C]] if and only if none of the above conditions 1, 2 and 3 holds, i.e. only a reduction of [[A,C]] may be a solution to the equations $C\Diamond W \leq A$ and $C\Diamond W \cong A$. Thus, if the composition operator is defined for the FSMs C and [[A,C]] derived by Algorithm 3.1, but $C\Diamond$[[A,C]] $\cong A$ does not hold then the equation $C\Diamond W \cong A$ has no solution. On the other hand, any reduction of a solution to the equation $C\Diamond W \leq A$ is also a solution, i.e. if the composition operator is defined for the FSMs C and [[A,C]] derived by Algorithm 3.1 then [[A,C]] is the largest solution to the equation $C\Diamond W \leq A$.

❑

If the machine [[A,C]] is not a solution to the equation $C\Diamond$[[A,C]] $\cong A$ then the question as to whether there exists a solution to both equations remains open. This is the case in our example, where the composition operator is not defined for the FSMs C and [[A,C]] due to livelock. The problem is that the waiter may repeatedly hit the coffee machine which continues to light the lamp, so the customer may no longer see the waiter (state 3 in Figure 7b).

Concluding this section, we notice that the largest solution to the equation $C\Diamond W \leq A$ may not exist even when the equation has a largest potential solution. Consider the FSMs A and C in Figure 8 (a) and (b). A solution to the equation is to be found in the alphabets $U = \{u_1, u_2\}$ and $Z = \{z_1, z_2\}$, as $X_2 = Y_2 = \varnothing$.

The LTS $L_{A,C}$, obtained at the second step of Algorithm 3.1, has no dead-state (Figure 9). Therefore, the largest potential solution [[A,C]] is a single-state chaos machine over the alphabets $\{u_1, u_2\}$ and $\{z_1, z_2\}$, however, it is not a solution to the equation $C\Diamond W \leq A$. At the same time, each FSM trace over alphabets $X_2 \cup U$ and $Y_2 \cup Z$ is a trace of some solution to the equation. The FSM [[A,C]] is not a solution to the equation and cannot be reduced without losing some solutions. In fact, by direct inspection of the LTS in Figure 9, one can verify that an FSM B is a solution if and only if the set of traces of B includes the regular set

$$L = [u_2(z_1u_1)^*z_2 \vee (u_1z_2)^*u_1z_1]^*$$

Apparently, there exists a number of such FSMs. Thus, for each trace β¤$\gamma \in L$ there exists a solution to the equation $C\Diamond W \leq A$ with this trace. Consider now trace β¤γ over alphabets $X_2 \cup U$ and $Y_2 \cup Z$ such that β¤$\gamma \notin L$. Then, by definition of L, the maximal prefix of β¤γ that is in the set L is the empty sequence or is terminated with an appropriate $z \in Z$, i.e. the regular set

$L\cup\{\beta\alpha\gamma\}$ also is in the set of traces of an appropriate FSM that is a solution to the equation.

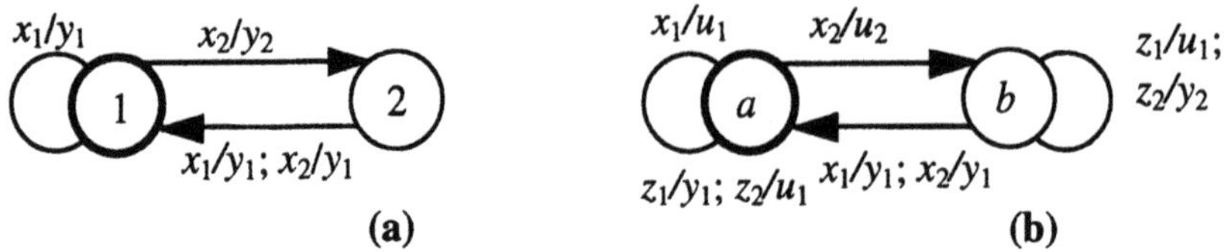

Figure 8. The FSMs A (a) and C (b)

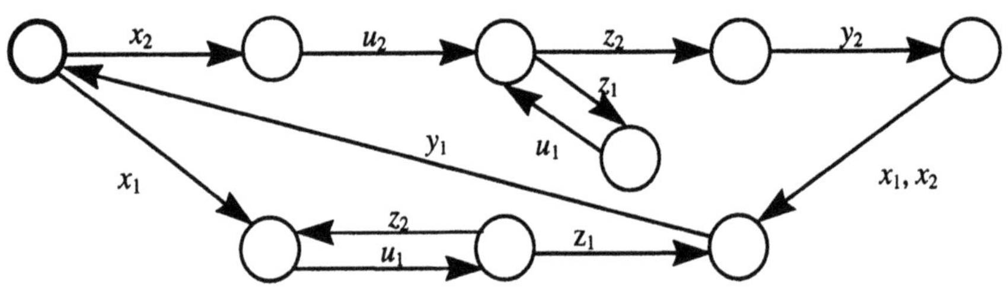

Figure 9. The LTS $L_{A,C}$

4. AVOIDING LIVELOCKS

As discussed in the previous section, the largest potential solution is not always a solution to the equation due to livelocks. Livelock corresponds to a situation when the component and its context in response to external input involve into unbounded internal interactions. To avoid livelocks we may thus bound the maximal number of internal actions executed by the system in response to any external input. One may also impose such a restriction attempting at finding a solution such that the response time of the overall system does not exceed a certain limit.

We have to find a solution to the equation $C\Diamond W \leq A$ such that in the resulting composition, the number of internal actions successively executed between any external input and a resulting output never exceeds a given threshold l, $l\geq 1$. We first introduce the l-restricted composition operator $\Diamond_l$. Given two FSMs C and B, the FSM $C\Diamond_l B$ coincides with the FSM $C\Diamond B$ if the LTS $L_C \| L_B \| L_E$ has no path labeled with more than l consecutive internal actions. Otherwise, i.e. when there exists a path with more than l consecutive internal actions, the operator $\Diamond_l$ is not defined for the FSMs C and B. Next, we demonstrate how the results of Section 3 can be further extended to solve the equation $C\Diamond_l W \leq A$.

Given two FSMs, A over alphabets $X=X_1\cup X_2$ and $Y=Y_1\cup Y_2$ and C over alphabets $X_1\cup Z$ and $Y_1\cup U$, an FSM B over alphabets $X_2\cup U$ and $Y_2\cup Z$ is said to be *a solution to the equation* $C\Diamond_l W \leq A$ if the l-restricted composition operator is defined for the FSMs C and B and $C\Diamond_l B \leq A$.

The idea behind our approach to solving such an equation is to replace a total chaos system L_{Ch} over alphabets $X_2\cup U$ and $Y_2\cup Z$ by a so-called *l-restricted* chaos system $L_{l\text{-}Ch}$ that can be obtained from the LTS L_{Ch} by redirecting internal traces of length exceeding l to a designated dead-state. Since the appearance of external output terminates any internal trace we add at appropriate states of the LTS $L_{l\text{-}Ch}$ corresponding transitions resetting the machine back to the initial state. Figure 10 (a) gives an example of an l-restricted chaos system $L_{l\text{-}Ch}$ for the coffee shop, where l=2.

Replacing the LTS L_{Ch} by the LTS $L_{l\text{-}Ch}$, we can use Algorithm 3.1 to find a solution. A small adjustment is required, though. In Step 4, if at state s of the LTS $\widehat{L}$ there is a transition labeled with $a\in U$ to dead-state then in the corresponding FSM, denoted $[A,C,l]$, there is a transition from s to dead-state labeled with a/b for each $b\in Y_2\cup Z$. Such transitions may exist since the l-restricted chaos system has transitions to dead-state labeled with internal inputs. We denote $[[A,C,l]]$ the resulting machine.

Theorem 2. Given two FSMs, A and C, and an integer $l\geq 1$, if Algorithm 3.1 does not return an FSM then the equation $C\Diamond_l B \leq A$ has no solution. If it results in an FSM $[[A,C,l]]$ then $[[A,C,l]]$ is the largest solution to the equation $C\Diamond_l W \leq A$.

The proof of the theorem is similar to that of Theorem 1. As the FSMs C and $[[A,C,l]]$ may execute at most l internal actions before producing an external output, the composition operator for the FSMs C and $[[A,C,l]]$ is always defined; thus, the FSM $[[A,C,l]]$ is the largest solution to the equation $C\Diamond_l W\leq A$.

Example. For the coffee shop, we solve the equation $C\Diamond_l W\leq A$ for l=2 and 4. Figure 10 shows the FSM $[[A,C,2]]$ (b) and the FSM $[[A,C,4]]$ (c). Note that for l=3, no solution can be found. The submodule has only internal inputs and outputs, so the number of internal interactions is always even.

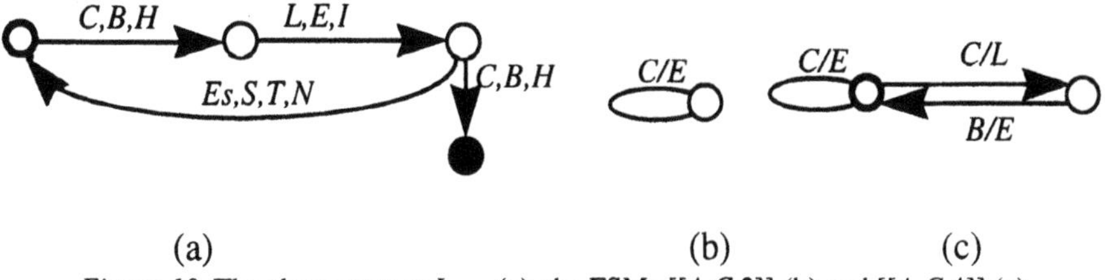

Figure 10. The chaos system $L_{2\text{-}Ch}$ (a), the FSMs $[[A,C,2]]$ (b) and $[[A,C,4]]$ (c)

5. CONCLUSION

We considered the problem of equation solving for asynchronously communicating nondeterministic finite state machines. It was demonstrated that opposed to the case of labeled transition systems, equation solving is complicated by potential livelocks arising when two machines interact via feedback. Due to livelocks, a given equation may have no solution. By the same reason, asynchronous equations differ also from synchronous ones considered in the context of sequential circuit optimization. An algorithm for finding the largest potential solution to the asynchronous equation for trace inclusion, i.e. reduction relation, was presented. The largest potential solution is the largest solution when the resulting composition has no livelock. We also demonstrated that the largest solution may sometimes not exist. To resolve the problem of livelocks we proposed to use a special chaos system that prevents the composition from falling into livelock by imposing a predefined limit on the possible number of interactions inside the system. This approach could also be useful when one needs to decompose a given system and to keep its response time within a certain bound. The results of this paper could be used in a number of applications, such as protocol design, controller design, system validation, and test derivation for embedded testing.

Future work in this direction is to generalize the presented approach to incompletely specified nondeterministic finite state machines. It would also be interesting to study the relation between the complexity of a solution (e.g. in terms of the number of transitions) and the number of internal interactions. Whether the simplest submodule also needs a minimal number of internal interactions in the resulting composition (the system with a minimal response time) remains an open question.

ACKNOWLEDGMENTS

The first author acknowledges the support of the NSERC grant OGP0194381. The authors would like to thank G. v. Bochmann for stimulating discussions on the problem of submodule construction.

REFERENCES

[Azi95] A. Aziz, F. Balarin, R. K. Brayton, M. D. DiBenedetto, and A. Saldanha, *Supervisory Control of Finite State Machines*, Proceedings of the 7th International Conference, CAV'95, Belgium, pp. 279-292, 1995.

[Cec86] F. Cecseg, *Products of Automata*, Monographs in Theoretical Computer Science. Springer Verlag, 1986.

[Che96] S. C. Cheung and J. Kramer. *Context constraints for compositional reachability analysis*, ACM Trans. on Software Engineering and Methodology, Vol. 5, N 4, 1996, pp. 334-377.

[Dam94] M. Damiani, *Nondeterministic finite state machines and sequential don't cares*, Proceedings of the Euro Design and Test Conference, 1994.

[Hee95] L. Heerink and E. Brinksma, Validation in context, *Proceedings of the 15th IFIP WG6.1 International Symposium on Protocol Specification, Testing, and Verification*, Chapman & Hall (1995), pp 221-236.

[Hop79] J. E. Hopcroft and J. D. Ullman (1979) *Introduction to automata theory, languages, and computation*, Addison-Wesley, New York.

[Kim72] J. Kim and M. Newborn. *The simplification of sequential machines with input restrictions*, IEEE Trans. on Computers, C-20, pp. 1440-1443, 1972.

[Lar90] K. Larsen and L. Xinxin, *Compositionality through an operational semantics of contexts*, Lecture Notes in Computer Science, Vol 443, 1990.

[Lin95] B. Lin, G. de Jong, and T. Kolks, *Hierarchical optimization of asynchronous circuits*, Proceedings of the 32nd Design Automation Conference, 1995, pp. 712-717.

[Lee93] D. Lee, K. Sabnani, D. Krispot, S. Paul, and M. Uyar, *Conformance testing of protocols specified as communicating fsms*, INFOCOM'93.

[Mer83] P. Merlin and G. v. Bochmann, *On the Construction of Submodule Specifications and Communication Protocols*, ACM Trans. on Programming Languages and Systems, Vol. 5, No. 1, Jan. 1983, pp. 1-25.

[Qin91] H. Qin and P. Lewis, *Factorisation of Finite State Machines under Strong and Observational Equivalences*, Journal of Formal Aspects of Computing, Vol. 3, July-Sept. 1991, pp. 284-307.

[Par89] J. Parrow. *Submodule construction as equation solving in ccs,* Theoretical Computer Science, 68, pp. 175-202, 1989.

[Pet93] A. Petrenko, N. Yevtushenko, A. Lebedev, and A. Das, *Nondeterministic state machines in protocol conformance testing*, Proceedings of the IFIP Sixth International Workshop on Protocol Test Systems, pp. 363-378, 1993.

[Pet94] A. Petrenko, N. Yevtushenko, and R. Dssouli, *Testing strategies for communicating FSMs,* Proceedings of the 7th IWTCS, (1994) pp. 193-208.

[Pet96] A. Petrenko, N. Yevtushenko, G. v. Bochmann, and R. Dssouli. *Testing in context: framework and test derivation*, Computer Communications, Vol. 19, pp. 1236-1249, 1996.

[Rho91] J.-K. Rho, G. Hachtel, and F. Somentzi. *Don't care sequences and the optimization of interacting finite state machines*, Proceedings of the IEEE Conference on Computer-Aided Design, Santa Clara, 1991, pp. 414-421.

[Sta72] P. H. Starke, *Abstract automata*, American Elsevier Publishing Company, Inc. New York, 1972.

[Wat93] Y. Watanabe, and R. K. Brayton. *The maximal set of permissible behaviors for fsm networks*, Proceedings of the IEEE/ACM International Conference on Computer-Aided Design, pp. 316-320, 1993.

[Wat94] Y. Watanabe and R. K. Brayton. *State minimization of pseudo nondeterministic FSMs*, Proceedings of Euro Design and Test Conference, pp. 184-191, 1994.

[Yev91] N. Yevtushenko and A. Matrosova. *On one approach to automata networks checking sequences construction*, Automatic Control and Computer Sciences, Allerton Press, NY, Vol. 25, N 2, pp. 3-7, 1991.

Shared Variables and Efficient Synchronization Primitives for Synchronous Symbolic Verifiers

S. V. Campos[1]
Univ. Federal de Minas Gerais, DCC
Belo Horizonte — MG — Brasil
scampos@dcc.ufmg.br

Abstract

In spite of the increasingly frequent application of formal verification tools today they still are difficult to use for non trivial verification tasks. One reason is that the specification languages used by these systems are still quite different than the programming languages designers use. One significant difference is the absence of shared variables for synchronous verifiers. Many systems use shared variables in their implementation, but most synchronous formal verification tools do not model them. Instead the user must use additional variables to maintain the illusion of shared memory and model the correct behavior. Unfortunately this can cause a significant overhead making the final model much larger than necessary. This work proposes a new technique called *BDD markings* to overcome this problem. BDD markings provide a way of keeping track of control flow information and have allowed us to implement shared variables in Verus, a synchronous symbolic verifier. Using the new method we have been able to verify a variant of a mutual exclusion algorithm used in the Mach operating system and to determine the existence of priority inversion in a complex real-time system. Priority inversion is a serious and subtle problem that can cause real-time systems to fail unexpectedly. Its importance has been highlighted recently when it was discovered that priority inversion has caused system shutdowns in the Pathfinder spacecraft soon after it started collecting Martian meteorological data. This example has been modeled with and without shared variables. The new method has allowed verification to be performed up to 10 times faster and using up to 20 times less memory, demonstrating the efficiency of the technique proposed.

Keywords

Symbolic model checking, synchronization variables, BDD, priority inversion

---

1. This research was sponsored in part by Conselho Nacional de Pesquisa e Desenvolvimento, CNPq.

1 Introduction

Formal verification tools are used more frequently today than ever before and their application is increasing even more. However, using such systems is still not as simple as using most other tools such as compilers or debuggers. Several reasons make it difficult to employ formal verification tools in practice such as the complexity of the verification algorithms: for example, careless coding can easily lead to state explosion. Another reason is that most languages used to model systems are quite different than the languages used to implement them. Languages such as Esterel (Berry, 1992), Promela (Holzmann, 1996) and Modechart (Jahanian, 1988) have very different syntaxes or semantics than regular programming languages making it difficult to translate a program from C or a similar language to one of these languages.

We have previously addressed this problem by defining a new language called *Verus*, integrated into the Verus verification system (Campos, 1996a and 1995c). The Verus language is an imperative 'C-like' language that has been designed to allow straightforward modeling of systems being verified. Using this tool, the system being verified is compiled into a state-transition graph. The model uses a discrete model of time, in which each transition corresponds to one time unit. A symbolic model checker allows the verification of untimed properties expressed in CTL (Clarke, 1986 and McMillan, 1992) and of time bounded properties using RTCTL model checking (Emerson, 1990). Moreover, algorithms derived from symbolic model checking are used to compute *quantitative* information about the model (Campos, 1996a; 1995a and 1995c). The information produced allows the user to check the temporal correctness of the model as well as determine reaction times to events and several other system parameters. This information provides insight into the behavior of the system and in many cases it can help identify inefficiencies and suggest optimizations to the design. Verus has been used to verify several large complex systems such as aircraft (Campos, 1994), robotics (Campos, 1995c) and medical controllers (Campos, 1995a), as well as the PCI local bus (Campos, 1995b).

However, a Verus program still behaves quite differently than a similar C program. Some differences are needed in order to allow an efficient verification, such as the granularity of discrete time used in Verus. Others have been artificially imposed, and make programming in Verus more difficult than it should be. One of these differences is the lack of shared variables. In Verus, as well as most other synchronous verification tools, multiple processes can be defined. The values of variables can be read by all processes, but they are "owned" by one process, that is, variables can be modified only by that process. The original release of Verus as well as other verifiers such as SMV (McMillan, 1992), CV (Deharbe) and VIS (Brayton, 1996) have this restriction when being used in synchronous mode. This restriction is a significant hurdle for many applications because shared variables are extensively used in the actual code. Implementing an application that uses shared variables in a system that does not allow them is often a complex task. Moreover, it not only increases modeling time but also makes the final model much larger, since additional variables (and even processes) may be needed to model the correct behavior. Shared variables are present in asynchronous verifiers, but certain types of systems such as real-time systems cannot be easily verified by such tools.

This work describes a new technique used to implement shared variables in Verus, *BDD Markings*. In the new method BDD variables mark specific points in the control flow of the program in order to identify where variable assignments from multiple processes can potentially conflict. An important advantage of the method is that once the transition relation for the model has been constructed, the variables introduced to mark the control flow can be existentially quantified out. In this way the technique does not impose an overhead in the model because the new variables are removed before the transition relation is constructed.

BDD markings also allow the determination of which transitions in the state-transition graph correspond to given points in the program. This can be used to implement a new feature which allows properties to refer directly to specific statements in the source code. Properties often refer to points in the control flow where some computation is performed. Without BDD markings additional variables have to be declared to keep track of this information. These variables make the final model larger and more expensive to verify. This overhead can be avoided using the method proposed.

We have used the new method to verify two examples, a mutual exclusion and the priority inversion example described in (Campos, 1996b). The mutual exclusion algorithm is particularly appropriate since it relies on shared variables for its correctness. A variant of this algorithm has been used in the Mach operating system kernel (Rashid, 1989). The priority inversion example has previously been implemented without shared variables. Its importance has been highlighted recently when it was shown that priority inversion caused the sporadic system shutdowns experienced by the Pathfinder spacecraft soon after it started collecting Martian meteorological data (Wilner, 1997). By revisiting the example we have been able to compare both alternatives and to see that the shared variable model has performed significantly better. In some cases the new model executes 10 times faster using 20 times less memory!

The paper is organized as follows: we first briefly introduce Verus in section 2. Section 3 discusses how a Verus program can be represented using BDDs. In section 4 we present BDD markings and in section 5 we show how they can be used to implement shared variables. Finally sections 6 and 7 describe the examples implemented using the new method, and section 8 concludes the paper.

2 The Verus tool

Verus is a formal verification tool used to verify finite-state real-time systems that can also be applied to systems without real-time constraints. The application being verified is modeled in the Verus language and compiled into a state-transition graph using symbolic algorithms and represented using BDDs. Model checking and quantitative timing algorithms are then applied to the model in order to determine its correctness as well as its timing characteristics. Verus is particularly suited to model and verify reactive systems defined by a set of concurrent processes. Even though shared variables are used to model communication between processes, this does not restrict the application of the tool. Distributed systems can be efficiently verified by modeling messages as shared variables and the exchange of messages as changes in the values of the corresponding variables.

The Verus Language

The main goal of the Verus language is to allow engineers and designers to describe real-time systems easily and efficiently. It is an imperative language with a syntax resembling that of C. Special primitives are provided for the expression of timing aspects such as deadlines, priorities, and time delays. These primitives make timing assumptions explicit, leading to clearer and more complete specifications. The data types allowed in Verus are fixed-width integer and boolean. Nondeterminism is supported, which allows partial specifications to be described. Further details about the Verus language including its formal semantics can be found in (Campos, 1996a).

A fragment of a simple real-time program is used to give an overview of the language. This program implements a solution for the producer-consumer problem by bounding the time delays of its processes. The code for the `producer` process is shown below. Variable `p` is a pointer to the buffer in which data is stored. The `producer` initializes its pointer `p` to 0 and the `produce` variable to *false*. It then enters a nonterminating loop in which items are produced at a certain rate. Line 7 introduces a time delay of 3 units, after which an item will be produced. Line 8 marks the production of an item by asserting `produce`. In line 9 the pointer is updated appropriately. Line 10 makes sure that the event `produce` is observed. It is needed because the state of a Verus program can only be observed at `wait` statements, since all other statements execute in time zero. This allows a more accurate control of time, eliminating the possibility of implicit delays influencing verification results. It also generates smaller models, since contiguous statements are collapsed into one transition.

```
 1  producer(p)
 2  {
 3    boolean produce;
 4    p = 0;
 5    produce = false;
 6    while(!stop) {
 7      wait(3);
 8      produce = true;
 9      p = p + 1;
10      wait(1);
11      produce = false;
12    };
13  }
```

Figure 1 Producer code.

```
16  consumer(p, c)
17  {
18    boolean consume;
19    c = 0;
20    consume = false;
21    while (!stop) {
22      wait(1);
23      if (p != c) {
24        consume = true;
25        c = c + 1;
26        wait(1);
27        consume = false;
28      };
29    };
30  }
```

Figure 2 Consumer code

The consumer code shown above is similar to the producer. the main differences are that it must check to be sure that there are items to be consumed (`p != c`), and that it takes less time to execute than the producer. The fact that the consumer is faster than the producer ensures that the buffer will not overflow, as will be verified later.

```
28  main()
29  {
30    int p, c;
31
32    process prod producer(p),
33            cons consumer(p, c);
34
35    spec MIN[prod.produce, cons.consume]
36         MAX[prod.produce, cons.consume]
37         AG(prod.produce -> AF cons.consume)
38  }
```

Figure 3 Producer/consumer main function

As in the C language, `main` has a special function in Verus. In this function all processes are instantiated, and global variables can be declared. The variables `p` and `c` (used as pointers in the buffer) are declared and the `producer` and `consumer` processes are instantiated in the `main` function of the example code.

Process instantiation in Verus follows a synchronous model. All processes execute in lock step, with one step in any process corresponding to one step in the other processes. Asynchronous behavior can be modeled by using *stuttering*, as described in (Campos, 1996a). An implicit instantiation of the `main` module is assumed, where the code in `main` executes as another synchronous module.

Specifications can also follow the code as can be seen. The specifications above compute the minimum and maximum time between producing an item and consuming it, as well as checking that a `produce` is always followed by a `consume`.

CTL and RTCTL Model Checking

Verus allows the verification of untimed properties expressed as CTL formulas (Clarke, 1986 and McMillan 1992) as well as of timed properties expressed as RTCTL formulas (Emerson, 1990). CTL formulas allow the verification of properties such "*p* will eventually occur", or "*p* will never be asserted". However, it is not possible to express bounded properties such as "*p* will occur in less than 10ms" directly. RTCTL model checking overcomes this restriction by allowing bounds on all CTL operators to be specified.

Quantitative Algorithms

Most verification algorithms assume that timing constraints are given explicitly. Typically, the designer provides a constraint on response time for some operation, and the verifier automatically determines if it is satisfied or not. Unfortunately, these techniques do not provide any information about how much a system deviates from its expected performance, although this information can be extremely useful in fine-tuning the behavior of the system.

Verus implements algorithms that determine the minimum and maximum length of all paths leading from a set of starting states to a set of final states. It also has algorithms that calculate the minimum and the maximum number of times a specified condition can hold on a path from a set of starting states to a set of final states. Our algorithms provide insight into *how well* a system works, rather than just determining whether it works at all. They enable a designer to determine the timing characteristics of a complex system given the timing parameters of its components. This information is especially useful in the early phases of system design, when it can be used to establish how changes in a parameter affect the global system behavior.

Several types of information can be produced by this method. Response time to events is computed by making the set of starting states correspond to the event, and the set of final states correspond to the response. Schedulability analysis can be done by computing the response time of each process in the system, and comparing it to the process deadline. Performance can be determined in a similar way. The algorithms have been used to verify several real-time and non real-time systems.

3 FROM A VERUS PROGRAM TO A STATE-TRANSITION GRAPH

Symbolic Representation

States are defined by the assignment of values to program variables, where each possible assignment to the program variables is a state. Boolean formulas over program variables can be true or not in a given state. The value of a boolean formula in a state is obtained by substituting into the formula the values of the variables in that state. In general, sets of states can be represented by boolean formulas, where each formula represents the set of states in which the formula is true.

Transitions can also be represented by boolean formulas. A transition $s \rightarrow s'$ is represented using two sets of variables, one for the current state and another for the next state. If state s is represented by the formula f_s over the current state variables, and state s' is represented by formula $f_{s'}$ over the next state variables, then the transition $s \rightarrow s'$ is represented by the formula $f_s \wedge f_{s'}$. The transition relation of a graph is the disjunction of all transitions in the graph. The meaning of the formula representing the transition relation is the following: there exists a transition from s to s' iff the substitution of the variable values for s in the current state variables and s' in the next state variables of the transition relation yields *true*.

Tracking the Control Flow — Wait Graphs

The execution of a Verus statement may change the value of one or more program variables. In general, it changes a given state s into another state s'. Executing a sequence of statements in a given state then generates a sequence of states. However, in Verus not all of those states are observable. The state of the program can only be observed at `wait` statements. When a `wait` is executed all changes caused by the execution of the block of statements since the previous `wait` take effect at the same time. Transitions in the graph occur only when `wait` statements are executed. Each transition in the graph corresponds to time elapsing by one unit. The statement `wait(1)` corresponds to one transition in the graph. Longer waits are modeled by a sequence of unit transitions.

It is easier to understand the behavior of a Verus program by concentrating on its `wait` statements. This can be accomplished by translating the program into a *wait graph*. The wait graph corresponding to a Verus program is a graph in which the states are the `wait` statements in the program. It corresponds to an intermediate representation between the Verus program and the corresponding state-transition graph. It is used only to illustrate how this translation occurs and is not actually constructed.

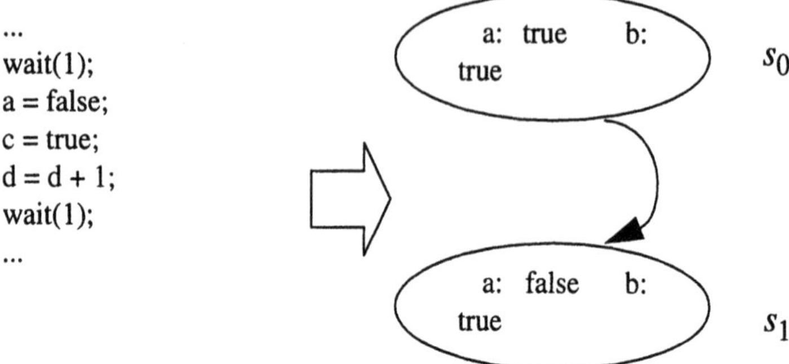

Figure 4 If s_0 is the current state at the first `wait`, s_1 will be the current state at the second.

Transitions between `waits` are defined as follows. A transition between $\texttt{wait}_i$ and $\texttt{wait}_j$ ($\texttt{wait}_i$ represents the ith occurrence of a `wait` statement in the program) exists iff $\texttt{wait}_j$ can be reached from $\texttt{wait}_i$ in the control flow of the program without going through intermediate `waits`. The edges of the wait graph are labelled by a relation T_{ij} between any two states in the state-transition graph. This relation describes the state changes caused by the execution of the statements between the corresponding `waits`. It is determined during compilation in the following way:

- When a `wait` statement is parsed an initial relation is set up. It basically states that no variables change value (variables only change value when assigned to): $(v = v')$; and that the next transitions that will be generated will start on this `wait` statement (a wait counter variable is introduced to maintain this information): $(wc = i)$.
- Each statement parsed changes the intermediate relation computed by the previous statement according to what the statement does, and produces a new intermediate relation.

The construction of this relation is described in detail in (Campos, 1996a). Intuitively, given two states s and s', $T_{ij}(s, s')$ means that if the execution of the program is in $\texttt{wait}_i$ and the current state is s, then there exists a path in the control flow leading to $\texttt{wait}_j$ (without intermediate waits) and the execution of the statements on this path will change state s into state s'. The construction of the relation between wait statements is complex and its explanation is beyond the scope of this paper. However, figure 5 can give an intuition on how these relations are constructed.

4 BDD MARKINGS

Let's assume we want to implement a primitive called `assert`. The statement `assert(`*name*`)` can be inserted at any point in the program. Its semantics correspond to creating a variable *name* in the model such that *name* is true in exactly those states reached immediately after executing the `assert` statement.

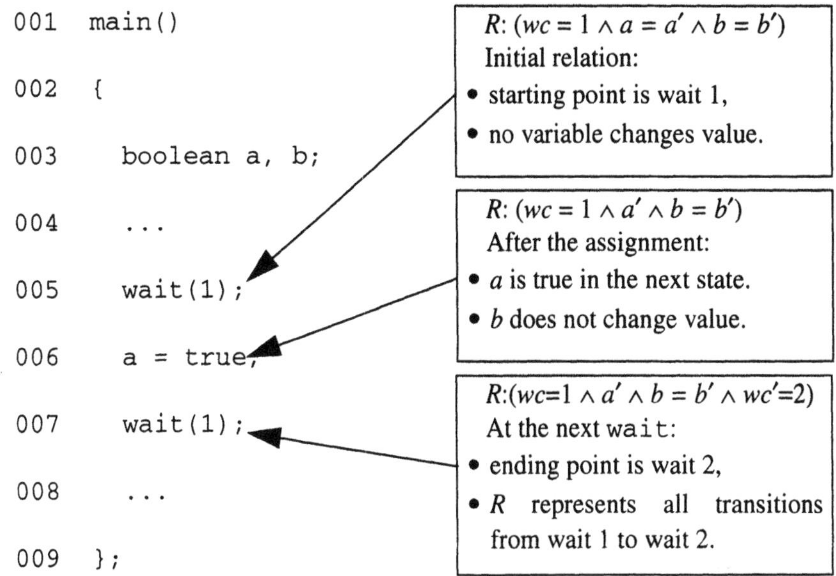

Figure 5 Constructing the transition relation for a Verus program.

Implementing `assert` requires a method to keep track of the control flow of the program. This is not always an easy task. It is sometimes difficult to determine if (or when) a program executes specific statements. Traditionally a "printf" is inserted in the program, and if the control flow goes through that point a message shows up on the screen. Unfortunately it is not possible to use this technique in Verus, since it does not follow a single execution path, but rather explores all of them. A possible solution in Verus is to insert a `wait` statement at the point of interest in the program and check to see if that `wait` statement can be reached. However, this technique affects the time delays in the program and can change its behavior. BDD markings provide a solution for this problem. A BDD variable is used to mark the transitions that correspond to the execution of `assert` in the following way:

1. Create a BDD variable *name* and make it false in every transition by default. This is accomplished by inserting the clause $\neg name$ in the initial transition relation created at the `wait` statement.
2. When the statement `assert(`*name*`)` is found, assign the value *true* to *name* in the current relation, overriding the clause $\neg name$.
3. Whenever a `wait` statement is parsed, check to see if there are subsets of the current relation that satisfy *name*. Those transitions are the ones that traverse `assert` in the source code.

Step 2 ensures that all transitions that execute `assert` are marked with the variable *name*. Step 1 guarantees that the transitions that do not execute `assert` are not marked. Step 3 is necessary because transitions in the model are created only at `wait` statements. Before reaching the next `wait` statement the current relation describes a partial transition, whose end state may never be reached because the statements following `assert` may change the relation and consequently its end state. So, only when reaching the next `wait` statement we can be sure to have determined a transition that executes `assert`.

The transitions mentioned in step 3 above represent the execution of `assert` in the program. The subset of R satisfying *name* at that point, R', is a formula that expresses when `assert` is traversed. We can then save this formula and associate with it the symbol *name*. Then, the BDD variable *name* can be quantified out of R', since it is no longer needed. References to *name* will point to the formula that represents it. In this way we can determine when a statement is executed in the program without adding extra BDD variables to the model.

The primitive `assert` can be implemented in this way, and it can be used to express properties about the control flow of the program. This is more efficient than the method used otherwise, which is to create additional program variables and assign values to them according to the point in the program that is being executed. The same method is also used to implement shared variables as described in the next section.

5 IMPLEMENTING SHARED VARIABLES

In most languages used by synchronous formal verification tools variables are "owned" by processes, that is, they can be read by all processes, but can only be changed by a single process. However, this is an unrealistic assumption, "shared" variables that can be modified by all processes are very frequently used. In fact, many applications rely on shared variables in their implementations, for example, to synchronize processes. But their behavior is difficult to model in symbolic verifiers, because concurrent assignments are hard to identify and model correctly using synchronous composition. Imperative languages in most cases pose yet another problem. The reason is that in those languages when a variable is not assigned to it maintains its value. Unfortunately this means that there is an implicit assignment in every transition that does not mention that variable. Because of this, shared variables are usually not modeled in symbolic verifiers. Unfortunately, modeling a system that uses shared variables in a language that does not allow them is frequently difficult and usually creates a significant overhead because additional variables (and sometimes processes) have to be created, making the final model significantly larger than it needs to be.

Both types of conflict described above are present in Verus. The implicit assignment problem is caused by the clause $v = v'$ inserted in all transitions that do not assign values to v. It maintains the value of variables that have not been assigned to. However, this is no longer valid when using shared variables because a transition that does assign a value to v from another process may be executed concurrently, causing a conflict. In this case the conflict eliminates the global transition from the transition relation, because synchronous composition conjuncts transitions from concurrent processes (Campos, 1996a and McMillan 1992), and the conjunction of the formulas corresponding to the conflicting assignments yields *false*.

To avoid this problem we create a BDD variable $assigned_v$ for each shared variable v. This variable is used to mark where assignments to v occur. We proceed by implicitly inserting an `assert` ($assigned_v$) everywhere assignments to v are made. The call to `assert` does not actually exist in the source code, but it is modeled in the compiler. At the `wait` statements we "collect" the transitions that assign values to v by disjuncting them into a partial transition relation TR_v. At the same time, we eliminate the clause $v = v'$ from all transitions that do not assign values to v. After compiling all processes we compute the global transition relation (without the $v = v'$ clause) and in the end conjunct the clause ($v = v' \vee TR_v$) to the global transition relation. This clause makes sure that for transitions that do not assign values to v from any process (TR_v is *false*) the value of v is maintained ($v = v'$). For transitions that do assign values to v (TR_v true), the clause $v = v'$ does not necessarily apply and the new value does not conflict with the previous one.

The method described above implements shared variables in Verus, but it does not handle explicit concurrent assignments. They still generate a conflict, eliminating the transition from the final model. Consequently, conflicting assignments potentially generate states that have no outgoing transition. Verus checks for the presence of these states and prints a counterexample showing how conflicting assignments can be reached. Racing conditions in the system can be identified this way.

However, sometimes conflicting assignments are necessary. They can be modeled by noticing the following: In most cases when a process writes to a shared variable it must foresee the possibility of not succeeding, that is, another process may overwrite the value written. We can model this behavior by allowing the possibility of *not* writing to the variable and therefore avoiding the conflict. As shown in figure 6 we can model conflicting accesses using a variable `random` (an `extern` variable which changes value nondeterministically (Campos, 1996a)) to allow the process to avoid writing.

6 EXAMPLE: FAST MUTUAL EXCLUSION

We have used the method described above to verify Lamport's fast mutual exclusion algorithm (Lamport, 1987) shown in figure 7 below. A variant of this algorithm is used in the Mach operating system kernel (Rashid, 1989) to provide mutual exclusion in those systems that have no hardware assistance to this end. In this algorithm two variables are used to achieve mutual exclusion. Processes try to write their own value on both variables. A process gains access to the `mutex` area when it succeeds in writing to both variables. If it does not succeed, it tries again. The advantage of this algorithm is that it provides a simple and efficient way of implementing mutual exclusion without hardware assistance. The disadvantage is that it may cause starvation. We have been able to show that the algorithm is correct and that it indeed can cause starvation, as foreseen. However, we have also been able to determine a property that was not described anywhere else. Starvation only occurs when more than two processes compete. For a two process system there is no starvation.

Original program:

```
...
while (M != 1) {
  M = 1;
  delay();
};
...
```

Verus program:

```
...
while (M != 1) {
  if (random) M = 1;
  wait(1);
};
...
```

Figure 6 Modeling conflicting accesses in Verus

```
start:  x = i;
        if (y != 0) goto start;
        y = i;
        if (x != i)
          delay();
          if (y != i) goto start;
        mutex();
        y = 0;
```

Figure 7 Lamport's fast mutual exclusion algorithm

7 EXAMPLE: PRIORITY INVERSION

We introduce a hypothetical air-traffic control system to illustrate how priority inversion can affect a real-time system and cause it to become unschedulable. It concentrates on two of the processes in the system. The first, called *sensor*, reads airplane position data from radars, sets alarms on catastrophic conditions (conditions that cannot wait for a detailed analysis), and puts the data into shared memory. The other process, the *reporter*, reads the data collected by the *sensor*, and updates the traffic controller screens. The *sensor* is a high priority process since it processes urgent events, and must not be blocked by other processes. The *reporter* on the other hand, is a low priority process. Since it doesn't process urgent events, it may be delayed by other tasks. The *sensor* and the *reporter* share data, accessed in mutual exclusion.

However, this may result in priority inversion, as shown in the following scenario. Suppose a third process, called the *analyzer* is added to the system. This process reads data generated by other components of the air-traffic controller and processes it. The *analyzer* is less important than the *sensor* and has a lower priority. But it is more important than the *reporter*, since urgent conditions may arise as the result of the analysis and handling them is more important than updating the screens. Consider now the same scenario as above, with the *reporter* inside the critical section, and the *sensor* waiting on the mutex. At this point, the *analyzer* starts executing. It will block the *reporter*, since it has higher priority. However, the *sensor* is waiting for the *reporter* (and therefore also for the *analyzer*). Since the *analyzer* doesn't know the relation between the *reporter* and the *sensor*, it may execute for an unbounded amount of time and delay the *sensor* indefinitely. If a catastrophic event occurs, it will go unnoticed, because the *sensor* is blocked. The behavior of the system becomes unpredictable. This unbounded priority inversion can be seen in figure 8.

Priority inheritance protocols are one way of preventing unbounded priority inversions (Rajkumar, 1989). A typical protocol might work in the following manner. As soon as a high priority process is blocked by a low priority one, the low priority process is temporarily given the priority of the blocked process. In our example, while inside the critical section the *sensor* is trying to access, the *reporter* will execute at high priority. When the *reporter* exits the critical section, it will be restored to its original priority. In this way, the *analyzer* will not be able to interrupt the *reporter* when the *sensor* is waiting.

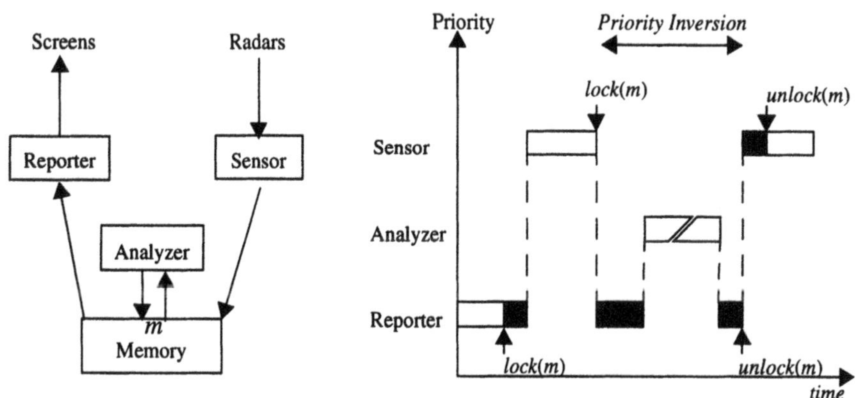

Figure 8 Unbounded priority inversion

In the original model mutual exclusion is achieved using an additional process that controls the mutual exclusion variable. This process receives requests for mutual exclusion, decides among competing requests and grants mutex to one process at a time. A high overhead is imposed in this model. In addition to the mutual exclusion variable we need one request variable for each process and a new process to control the variable. In the new model we only use the mutual exclusion variable, and accesses to mutual exclusion are modeled similarly to figure 6, except that we must wait until there is no process in mutex (signalled by `M == 0`) until we try to write to `M`. Priorities are modeled by forcing high priority processes to write to the variable (removing the `random` clause). In this case conflicts are avoided by the low priority process.

```
...
while (M != 1) {
   if (random && M == 0) M = 1;
 wait(1);
};
...
```

Figure 9 Accessing shared variables in the priority inversion example.

This example has been described in (Campos, 1996b). It has originally been implemented without shared variables. We have been able to reproduce the results previously obtained, but with a significantly smaller model. The old model has 20 boolean variables more than the new one (a savings of about 30%), and verification has been performed up to 10 times faster using up to 20 times less memory.

8 CONCLUSIONS

This work describes a new technique for keeping track of control flow information in synchronous symbolic verifiers. It has been used to implement shared variables in the Verus verification system. We have used the new feature to verify a mutual exclusion algorithm and to determine the existence of priority inversion in a real-time system. Using shared variables we have been able to perform the verification up to 10 times faster and using up to 20 times less memory than the same verification without shared variables. These gains demonstrate the efficacy of the new implementation and the importance of efficient synchronization primitives in symbolic verifiers. They show that the new method can help to make possible the verification of even larger systems than previously manageable.

Acknowledgment

The author would like to thank Edmund Clarke for the support and many helpful discussions during the time when part of this work was done at CMU.

9 REFERENCES

Berry, G. and Gonthier, G. (1992) The Esterel synchronous programming language: design, semantics, implementation. *Science of Computer Programming*, **19**.

Brayton, R. et. al. (1996) VIS: a system for verification and synthesis. *8th Conference on Computer-Aided Verification, LNCS,* **1102**. Springer-Verlag.

Campos, S. (1996a) *A quantitative approach to the formal verification of real-time systems*. Ph.D. thesis, SCS, Carnegie Mellon University.

Campos, S. et. al. (1996b) Analysis of real-time systems using symbolic techniques. *Trends in Software — Formal Methods for Real-Time Computing* (ed. C. Heitmeyer and D. Mandrioli). John Wiley & Sons Ltd.

Campos, S. et. al. (1995a) Timing analysis of industrial real-time systems. *Workshop on Industrial-strength Formal specification Techniques.*

Campos, S. et. al. (1995b) Verifying the performance of the PCI local bus using symbolic techniques. *ICCD.*

Campos, S. et. al. (1995c) Verus: a tool for quantitative analysis of finite-state real-time systems. *Workshop on Languages, Compilers and Tools for Real-Time Systems.*

Campos, S. et. al. (1994) Computing quantitative characteristics of finite-state real-time systems. In *IEEE Real-Time Systems Symposium*, 1994.

Clarke, E. et. al. (1986) Automatic verification of finite-state concurrent systems using temporal logic specifications. *ACM TOPLAS*, **8**(2):244-263.

Deharbe D. *CV, a VHDL model checker.*
`http://www.cs.cmu.edu/~modelcheck/cv/project.html`

Emerson, E. et. al. (1990) Quantitative temporal reasoning. *Conference on Computer Aided Verification.* Springer-Verlag.

Holzmann G. and Peled D. (1996) The state of Spin. *8th Conference on Computer-Aided Verification, LNCS* **1102**. Springer-Verlag.

Jahanian, F. and Stuart, D. (1988) A method for verifying properties of modechart specifications. *IEEE Real-Time Systems Symposium.*

Lamport, L. (1987) A fast mutual exclusion algorithm. *ACM TOCS.* **5**(11).

McMillan, K. (1992) *Symbolic model checking — an approach to the state explosion problem.* Ph.D. thesis, SCS, Carnegie Mellon University.

Rajkumar, R. (1989) *Task synchronization in real-time systems.* Ph.D. thesis, ECE, Carnegie Mellon University.

Rashid, R. et. al. (1989) Mach: a foundation for open systems (operating systems). *Workshop on Workstation Operating Systems.*

Wilner, D. (1997) Keynote Speech. *IEEE Real-Time Systems Symposium.*

10 BIOGRAPHY

Sérgio Vale Campos has received the bachelors degree in computer science from Universidade Federal de Minas Gerais (Belo Horizonte — Brasil) in 1986, the masters degree in computer science from the same university in 1990 and the Ph.D. degree in Computer Science from Carnegie Mellon University in 1996. He has spent one year as a post-doctoral researcher at Carnegie Mellon and is now an associate professor at the Universidade Federal de Minas Gerais. His interests include the design, analysis and verification of real-time, parallel and distributed systems.

PART SEVEN

Case Studies

SDL SPECIFICATION AND VERIFICATION OF UNIVERSAL PERSONAL COMPUTING WITH OBJECTGEODE

M. Törö, J. Zhu, V. C. M. Leung
Dept of Electrical and Computer Engineering
The University of British Columbia
Vancouver, BC, Canada V6T 1Z4

Abstract
Support of nomadic computing on the Internet requires the design of new protocols handling issues of user mobility. Our efforts target the universal personal computing (UPC) system to provide a continuous personal computing environment for mobile users. We selected SDL as a specification language for this new service, which allowed us to use Verilog's SDL tool: ObjectGEODE, for design verification. In the paper, we discuss the main principles of UPC, and how the simulator and verifier of ObjectGEODE were used during the stepwise system design. We illustrate the types of errors detected by the verifier, and show that the consistent application of the SDL methodology and the tool has increased our confidence in the correctness of our specification. The specification is the basis of our further work on UPC prototyping.

1. Introduction

Widespread use of laptop, palm-top computers, PDAs and the possibility of wireless connection to communication and computer networks anticipate the ubiquitous support of nomadic computing on the Internet. One of the main goals of our IWIN (Internetworking Wireless Data Networks for Mobile Computing) project involves the design and prototyping of a universal personal computing (UPC) system (Li, Y. and Leung, V.C.M., 1997a, b), which allows nomadic users of the Internet to work in their *familiar personal computing environment* regardless of their locations, or means of access. We envision that the UPC service will be provided at any location where Internet access is available, so that users everywhere can benefit from it, even during migration, through any available terminals or computers owned by the user or provided by the service provider or a third party on location.

To achieve our goals several new techniques are employed and integrated into our system. We selected SDL (Specification and Description Language) of ITU (ITU Z.100, 1992) and the related methodology to reinforce the design and development process with formalism.

Our design of UPC relies on the following emerging techniques/technologies:

- The Internet Engineering Task Force approved in October 1996 the IP Mobility Support (MIP) protocol (RFC 2002, 1996), which enables terminal mobility on the Internet that is transparent to the TCP level.
- We achieve a homogenous computing environment by applying CORBA (Common Object Request Broker Architecture, 1997) technology in UPC. The IDL (Interface Definition Language) of CORBA provides a common basis to design interchangeable components. CORBA services and facilities are pre-defined building blocks for new systems.
- Java's (Arnold, K. and Gosling, J., 1997) platform independence allows the portability of components. In the case of the CORBA based UPC, the client objects are brought to the nomadic user's site to deliver the familiar computing environment.

We started our work with the investigation of the IP Mobility Support protocol. We specified it in SDL. This was also an exercise of the specification of mobility with SDL. We reported this work in (Törö, M., 1997).

Currently we are working on the SDL design of the UPC system. We have chosen a stepwise approach: after specifying the basic design, new features are added gradually. After each addition or modification of the design we verify the correctness of the design. Besides, between the major steps we also validate that the new version of our design still satisfies the original requirements. This work essentially differs from the previous one, since it is not based on an existing standard or specification. For the design of the UPC system we apply Verilog's SDL tool: ObjectGEODE (ObjectGEODE, 1997). Beside SDL, ObjectGEODE supports MSC (Message Sequence Chart) of ITU (ITU Z.120, 1992), OMT of (Derr, K.W., 1995), and state and entity relationship diagrams to help the design and documentation process. It provides tools for the entire system life cycle. We present this work in the following sections of the paper.

Before going to the details of our results we give a short summary on SDL and ObjectGEODE, the tool we used in our work. In the third section we discuss the UPC system and its specification in SDL. The fourth section brings up the verification results and our experience with the tool. We draw our conclusions in the fifth section.

2. SDL and SDL tool support

SDL is an asynchronous extended finite-state machine based language standardized by the ITU for specification and description of communication protocols. It was designed for interactive, distributed, real-time systems.
The most important novelty of the 1992 version of SDL was the introduction of object-oriented methodology for the structural elements of the specification. It also introduced the remote procedure call as a new way of synchronous communications between processes.
There are several SDL based tools mainly utilized in the telecommunication industry. Most of them are in-house products and not commercially available. There are two major commercial SDL tools: SDT by Telelogic (Sweden) and ObjectGEODE by Verilog (France). None of these tools support the full Z.100 standard. Only a third, recently available commercial tool: Cinderella by Cinderella (Denmark) claims the full support of the 1992 and 1996 version of the language.
As we mentioned, in our current work we have employed ObjectGEODE.

ObjectGEODE

ObjectGEODE integrates the SDL related methodology with others such as OMT (Object Modeling Technique) to assist the protocol life cycle from the very early design up to the implementation and conformance testing stages. It offers tools for automatic documentation and version maintenance.
ObjectGEODE provides a graphical user interface for specification using the graphical representation of SDL. From the entered specification executable code is generated for both simulation/verification, and implementation. The simulator has the options of interactive or automatic random simulation. The verifier, which is integrated with the simulator, implements breadth-first, depth-first, or supertrace reachability analysis, and liveness analysis. To verify certain properties one can choose between the construction of observers and the specification of the properties in MSC. Afterwards the verifier automatically checks if there is any inconsistency between the specified property and the system behavior, or verifies the presence of (un)required properties observed by the observer. The verification results are presented in several reports: a summary of the detected problems, number of reached states and fired transitions, is listed after each run. The transition and state coverage reports are presented to the user on demand. During verification problematic scenarios are logged, so that one can replay them to analyze the problem.
MSCs are also used in ObjectGEODE to specify test purposes for conformance test suite generation. For each MSC scenario a test case is generated by finding the appropriate transitions in the reachability tree and generating the alternative transitions and the preambles and postambles. The generated test suite is specified in

TTCN-MP (Tree and Tabular Combined Notation, Machine Processable format, IS-9646 Part 3, 1992)

3. Universal Personal Computing

3.1. Informal description of UPC

UPC is designed to provide a transparent continuous computing environment for mobile users on the Internet. The key issue of the system is the representation of the user's computing environment and its delivery to the current user location. For economic and performance reasons we also require the system to provide the requested services locally whenever it is possible or desired by the user. Our design includes the case where the terminal or computer through which the user accesses the network is a 'foreign' one, e.g., one leased from a third party at the visited location.

As we mentioned in the introduction the emerging CORBA architecture provides a virtually uniform platform for object-oriented technology. In addition Java provides the portability of the required modules or objects. The CORBA architecture allows us to interpret the user's personal computing environment as the set of objects the user usually uses, which we call the user profile. Since the interface between the client and the server part of an application is described in standard IDL, any other object having the same interface can provide a requested service. This way, certain services may be obtained at the user's actual location through negotiation with the local Trader Service, instead of being delivered remotely from the user's home domain. Only if a requested service could not be obtained at a visited location would the service be invoked remotely.

According to the CORBA architecture, the client part of an application delivers the familiar user interface to the user. Therefore it is enough to migrate or copy only the client part to the user's location. It is possible that the required client part can also be obtained at the visited location, but still the user specific preferences should be considered. We define the user profile as the set of clients and the related environment objects. Java provides the portability of the client objects.

Users are identified in the system by their unique logical user identifier (LUI). They login to the network by providing this LUI and the password.

We define a user agent for each user, located in the user's home domain. It is the object in the system to which the LUI is bound, hence it can be found via the Naming Service. The user agent authenticates the user whenever it is contacted from anywhere on the Internet. The user agent also maintains the user profile and delivers it to the current user location after the user has been authorized by the system. Since a given terminal may not support every feature specified in a user profile, the capabilities of each terminal are described in a terminal profile, which is checked against the user profile and only the supported subset of the user profile is transmitted to the user location.

Since the user agent is always aware of the availability and the current location of the user, its functions are extended to include forwarding of incoming messages to the user, and acting on the user's behalf when the user is off-line.

To maintain a given user-terminal binding and to provide the personal computing environment at a terminal, we define the user-terminal agent. It is started up after the reception of the user profile, based on the features specified in the profile. The user-terminal agent takes care of the user requests to start different services. At such request it either starts the negotiation with the local Trader or obtains the required object according to the user profile. Since after a migration of the mobile user some of the local services may be obsolete, the user-terminal agent maintains also a list of these services. Whenever a new location is reached it automatically initiates a new negotiation with the new local Trader to replace the obsolete services. CORBA calls rely on TCP connections, at which level terminal migration is hidden by the MIP protocol. Only the MIP level is aware of which subnetwork the mobile node is currently attached to, therefore we define a signaling of the motion from the MIP level to the user-terminal agent. It is the user-terminal agent's responsibility to decide whether the migration has resulted in a change of the serving object request broker (ORB) as well.

3.2. SDL specification of UPC

At the UPC level we can distinguish three groups of entities as entities at the user's home domain, entities at the visited domain, and entities in the user's terminal. We group these entities in three blocks. Since we require CORBA in all parts of our system, in all of these blocks as a minimum requirement a CORBA Naming Service process needs to be defined. In addition, at the terminal and at the visited domain we define Trader Services. As part of the system initialization procedure the Naming Services exchange their process identifiers and name bindings to become known everywhere in the system.

In CORBA, objects are located on the basis of their object references. For this purpose we used the SDL process identifiers in our specification. Therefore a Naming Service registers the name and process identifier bindings in a given block, and for other Naming Services in the system. The Naming Service process contains two states and two procedures with no states. It accepts five input signals.
The Trader Service, on the other hand, maintains a database of the properties of different available services. At request the Trader Service matches them with the required service properties and returns the appropriate object, which in our case is a process identifier. At the current state, this is only an imaginary process identifier, since we have not yet specified the server objects. However, for the comparison of each of the defined properties we have to specify the appropriate operators. The Trader Service is specified as a one-state process with one procedure. We specify only the look up CORBA method; therefore the process has a single valid input signal.

As we mentioned earlier, SDL is an asynchronous language, and the remote procedure call has been added lately to provide synchronous communications. However, our tool does not support this feature yet. CORBA, on the other hand, defines synchronous and one-way method invocation through the static invocation interface, and deferred invocation via the dynamic invocation interface. SDL's asynchronous communication is adequate for the one-way and the deferred invocation. To provide synchronous calls in our UPC specification, whenever a synchronous invocation is assumed, we introduce a blocking state, from which only the inputs received in response to the invocation lead out. Any other input is saved for this state. Since we have not specified the object request brokers separately (the SDL interpreter provides their functions of signal distribution), these blocking states are also guarded with timers from indefinite blocking of the system.

Beside the CORBA components we have the following processes in each block as shown in Figure 1:

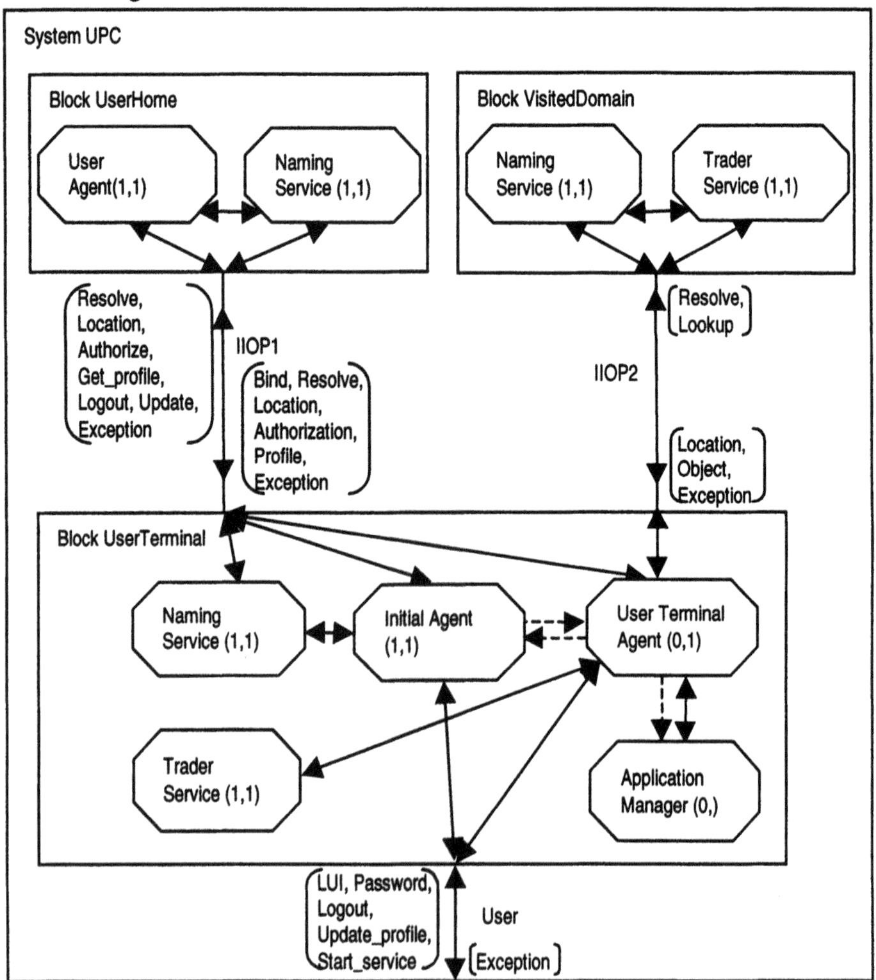

Figure 1 The simplified SDL structure diagram of UPC

- User's home: User Agent process

- User Terminal: Initial Agent process, User-Terminal Agent process, Application Manager process
- Visited domain: none

The User Agent process is created at the system start and it remains until system termination. Its behavior is straightforward: after initialization, which includes registration of the user's name and process identifier binding with the Naming Service, and initialization of the user profile data structure, the User Agent reaches the off-line state, and remains there until a login is initiated from the User Terminal. The User Agent authenticates the user based on the submitted password. If the authentication is successful, the User Agent moves to the online state, where it stays until a logout signal is received. In the online state the User Agent can send the user profile, update the profile, or deal with exceptions. Accordingly, in the User Agents process, two states and three no-state procedures are specified. It accepts five input signals.

The Initial Agent of the User Terminal accepts the login information: the user's LUI and the password. With the LUI, it finds the User Agent through the Naming Services, and verifies the user provided password with the User Agent. If the authentication is successful, it sends the terminal profile to the User Agent and receives the user profile. Upon the receipt of the user profile, the Initial Agent creates the appropriate User-Terminal Agent for the user, and then terminates. The user login and authorization are realized by seven states and six input signals. The type of the created User-Terminal Agent depends on the characteristics of the terminal and information contained in the user profile.

The User-Terminal Agent represents a given user-terminal binding. It takes care of the actions the user may request: updating the user profile, logout, or starting a service. To start up a service, it first checks in the user profile where to look for the service. If the service should be started at the user's home, the User-Terminal Agent simply creates a new Application Manager and passes the actual parameters required to adjust the service to the user preferences. If the service can be provided locally, it checks first with the User Terminal Trader whether the service is available, and if not it contacts the Trader of the Visited Domain. After receiving the information about a potential server from any of the Traders, this information and the related data from the user profile are passed to a newly created Application Manager. Services provided by the visited domain are restarted any time the user moves to a new domain. The User-Terminal Agent's functionality is specified via four states and seven procedures. Four of the procedures have no states while the rest have 1, 2 and 3 states. The process accepts three inputs from the user and five inputs from other processes of the system. The signaling of a change of the terminal location is specified as a spontaneous transition.

The Application Manager administers a given service for the user. It has two states and accepts two inputs sent by the User-Terminal Agent. At service termination it terminates and notifies the User-Terminal Agent about the termination. The termination is specified as a spontaneous transition in the process. If the user requests for an update of the user profile, the User-Terminal Agent collects the necessary information from the Application Managers and sends it to the User Agent. At user logout the User-Terminal Agent sends a termination request (one-way request) to the Application Managers, notifies the User Agent about the logout, and after creation of a new Initial Agent instance the User-Terminal Agent also terminates.

Table 1 summarizes the number of process instances, states, input signals and spontaneous transitions in the system.

Table 1 Summary on the UPC system specification

Process	*Naming Service*	*Trader Service*	*User Agent*	*Initial Agent*	*User Terminal*	*Application A. Manager*
Number of:						
Process instances	3	2	1	1 or 0	0 or 1	0 to 20
States	2	1	2	7	4	2
Inputs	5	1	5	6	5	2
Variables						
simple	11	2	5	12	12	3
complex	1	2	3	2	4	1
States in procedures	0	0	0	0	6	0
Spontaneous transitions	0	0	0	0	1	1

Although our design would benefit from the object-oriented extension of SDL (for example for the processes providing CORBA services), we could not use this feature because of restrictions in the tool; i.e., ObjectGEODE does not support context parameters or external synonyms.

4. Specification and verification of UPC in progress

4.1. Specification with the assistance of the verifier

We started the specification of UPC directly with SDL. We did not use other options offered by ObjectGEODE, since we already had a basic design at the time the tool was available to us. We use the stepwise approach to specification, i.e., starting with a basic design, to add new features and changes gradually. After each modification we verify that the system is free from deadlocks or exceptional situations. The following discussion demonstrates this approach.

As the first sketch of the design in SDL, we specified only the main parts of the UPC system and the basic communications between them. SDL allows informal parts in the specification: in tasks and decisions. The ObjectGEODE's simulator allows running such specifications, although it obviously cannot interpret the informal parts. The simulation at this stage helped us in the general understanding of the basic characteristics of the system. The verification could not give any useful results at this stage.

As the next step, we replaced the informal parts related to the user profile, i.e., we worked out a detailed data structure and operations on it. This modification allows us to use the verifier. A typical report is shown in Table 2(a).

The verification summary gives information about the type of executed verification, the number of reached states and transitions covered, the maximum depth reached during the exploration and execution time. The detected errors are reported in three categories: an *exception* refers to a dynamic error during execution; a *deadlock* is a system state when no transition can be fired in the system; a *stop condition* is a condition defined by the user, usually related to a given feature under investigation. Additionally the percentages of covered states and transitions are given. The liveness mode also reports the number of loops found with no success states. This method requires observers to identify the success/error states (ObjectGEODE, 1997).

In the second version of our specification we replaced the signal outputs representing synchronous CORBA calls, with the construction of going into a blocking state and setting a timer. In the blocking state all but the response for the call messages are saved. When we made this change, in one of the processes we forgot to specify the transition for the 'exception' signal. This immediately resulted in the report of Table 2(b) about a deadlock situation in the system.

Note that the deadlock situation was reported only by the breadth-first analysis. Although the depth-first method reported a similar coverage as shown in Table 2(c), it did not detect the deadlock for the same specification.

When we thought that the deadlock was corrected we restarted the verification and received the report of Table 2(d) with a number of exceptions (i.e., dynamic errors) found. Again only the breath-first analysis reported the exceptions; none of the other methods found any error in the specification.

The dynamic errors were the result of only partially correcting the deadlock, by adding the 'exception' input in the blocking 'wait_location' state of the User Terminal Agent, but forgetting the appropriate value assignment to the 'local' variable, which contained the reference (process identifier) to the visited domain's Trader Service. Since the 'exception' signal meant that there was no Trader available at the visited domain, the variable should not be used subsequently for the service lookup call. However, because of the obsolete value, this call was issued, and no recipient was found in the system.

Table 2 Summary of verification results

	(a)	*(b)*	*(c)*	*(d)*	*(e)*
Method:	supertrace	breadth	depth	breadth	breadth
Number of states:	162030	262142	262142	262142	524286
Number of transitions:	433021	684944	719355	723606	1566044
Maximum depth reached:	30088	37	1000 (truncated)	35	35
Maximum breadth reached:	n/a	90506	n/a	86714	148866
Duration:	11mn 28s	6mn 16s	17mn 50s	13mn 56s	18mn 28s
Number of exceptions:	0	0	0	18991	0
Number of deadlocks:	0	1	0	0	0
Number of stop conditions:	0	0	0	0	0
Transitions coverage rate:	46.15	62.94	62.94	59.21	56.60
transitions not covered	77	53	53	62	69
States coverage rate:	76.00	94.00	94.00	88.68	88.68
states not covered	12	3	3	6	6

Note that the current version of SDL would not consider this situation an error; however ObjectGEODE reports them at verification and interrupts the simulation in interactive or random mode. To avoid exceptions and the consequent interruption of the simulation in our dynamically changing system configuration, we use timers to delay the process termination whenever we expect such a situation.

After correction of these errors we ran all verifications again, and no new problem was found. However, the request-response pairs could not be matched in the system. The message identification was lacking. To correct it, we added a sequence number field to the messages, so that an old response could not be accepted any more for a newer request. We also guarded all such calls with timers. After all these changes we ran the verifications again, and this time the verifier reported no errors by any of the methods. We show only the report of the breadth-first analysis in Table 2(e).

A nice feature of ObjectGEODE is that the new verification results update the previous results and so does the simulation. This is important because different methods capture different parts of the system state space. This was the reason why the depth-first method did not find the deadlock, and this is the explanation of the following experiment:

During the verification we tried different parameter settings for the depth-first analysis. Originally we defined the truncation level of the reachability tree at depth 1000, which normally resulted in a poor state and transition coverage compared to runs of the breadth-first analysis, where the number of explored states and transitions were similar. By defining the truncation level closer (500, 100, 50) to the reached maximum depth of breadth analysis (35), we got similar results with both methods, since the smaller truncation level forced the verifier at the depth-first analysis to go through the similar part of the reachability tree. The higher numbers of truncation kept the analysis in the left most part of the reachability tree.

Currently we are refining the third version of the UPC specification. To check the consistency of a new version of the specification with an older one, we generated a number of message sequence charts from the older version and ran the new version against them. This serves to verify that a refinement retains the properties of a previous version. For such verification ObjectGEODE generates observers from a MSC document. We found no inconsistency between the different versions of the design so far.

4.2. State and transition coverage

We ran the verifications without any limitation set on the exploration size, so all the runs used the maximum memory available, and stopped when they reached this limit. Eventually, as seen in Table 2(e), we reached approximately half a million system states, and one and a half million transitions were investigated. There would be no other way of verifying such a number of states and transitions except with the use of an appropriate tool.

One can see from the report in Table 2, that there were still uncovered states and transitions. To find out more about these we used the state and the transition coverage reports of ObjectGEODE.

The state coverage report lists all the states for all the processes and procedures and for each of them gives the number of hits cumulatively for all instances. Beside the process and procedure state symbols (unlike in the verification results, in the coverage report the data part is not considered) the tool takes into account the process and procedure start symbols as states. The coverage rate is calculated as the proportion of states visited at least once to the number of states of the process or procedure. Table 3 presents an excerpt of a state coverage report.

Table 3 Excerpt from state coverage report

```
states coverage of userhome!namingservice : rate 66.67
        start : 1
        listen : 72731
        wait : 0

states coverage of userhome!namingservice!find : rate 100.00
        start : 64512

states coverage of useragent : rate 100.00
        start : 1
        online : 87694
        offline : 34967
        states coverage of check_profile : rate 100.00
        start : 95
        states coverage of update_profile : rate 100.00
        start : 87336
        states coverage of useragent!create_profile : rate 100.00
        start : 29491
```

From the state coverage list we can easily determine the states which have not been checked during the verification. In this case the unchecked states are the states of the Naming Services, which have been introduced for compatibility with the CORBA standard but are not used by the UPC system. In other words, we have covered all the reachable states in the system; hence there is no need of further verification from this point of view.

The transition coverage report presents the cumulative statistics on transitions (state-input pair) fired in each process and procedure. It also gives the proportion of transitions fired at least once to the number of transitions in the process or procedure. Table 4 shows an example of transition coverage report.

One can see that ObjectGEODE lists as transitions the procedure calls and automatically adds a 'discard' transition for each process or procedure, representing the SDL implicit transition (i.e., an input for which there is no transition specified explicitly so it is discarded in a given state). The tool also adds a time-out counter for each timer of a process. These automatic additions and the timer handling are the features of ObjectGEODE we found arguable.

The simulator and the verifier of ObjectGEODE handle differently the signals entered from the system environment (composing the feed-list) and the internal signals with respect to the time-out signals. If a time-out signal's alternative is from the feed-list (i.e., from the environment), the time-out signal is encountered as an alternative, otherwise not. As a result, transitions which are triggered by time-outs and have alternative internal signals are never executed during automatic verification, which in our case includes most of the time-outs in the system. We could check these transitions only by interactive simulation, i.e., by stepping forward the system's timer, so the time-out would happen, and the alternative became available. At this point, when we had to do the job manually, we really appreciated the convenience of the automatic verification for all the other cases.

Table 4 Excerpt from transition coverage report

transitions coverage of userhome!namingservice: rate 50.00
start: 9336
from_wait_input_location: 0
from_wait_input_exception: 0
from_wait_input_reply: 0
from_listen_input_unbind: 5465
from_listen_input_bind: 29692
from_listen_input_exception: 0
from_listen_input_resolve: 10540
call_delete: 5465
call_find_1: 29692
call_find_2: 10540
call_find_3: 0
discard: 0
timeout_reply: 0

Furthermore, in the current way of interpretation, a time-out would be encountered several times at the calculation of the transition coverage as the report of Table 4 shows. It would be taken into account once when the specification encounters the time-out signal (e.g. from_wait_input_reply), and once at the automatically added 'timeout_reply' transition.

We also debate with the suitability of two other automatically added items on the report, i.e., the introduced 'discard' transition and the procedure calls (e.g. call_find_1). In the case of the discard transition, our problem is that the tool does not check whether or not all of the inputs are handled explicitly by the specification. This addition is justified only when some of the inputs are not handled by the specification. Otherwise this addition falsifies the transition coverage rate. It is further questionable whether it is better to add a 'discard' transition for each state or state-input pair, or one for the whole process. They will give different coverage rates. In any case, the main issue for the designer is to be aware of the fact that a signal has been discarded by the system via an implicit transition. For this purpose any of the solutions is helpful. Regarding the inclusion of the procedure calls in the process transition coverage again the question is whether or not this figure is important for the designer. Since the transition coverage is listed for the procedures similarly as for the processes we do not see the need of the inclusion of these calls. In contrary, encountering each of them as a transition corrupt the transition coverage report for the caller process or procedure. In reality these calls might not represent different transitions at all; for example if they were subsequent calls in the same transition. They might not represent different state transitions at all, if the procedures contain no states, or on the opposite, they might hide several transitions if a procedure contains a state with multiple possible inputs.

It is important to be aware of how ObjectGEODE interprets these characteristics, so one can compare the reported results with others calculated by another approach.

In our case after analysing the results and taking into account all these, it can be seen that we have reached the maximum possible transition coverage for the specification. The recalculated state coverage after excluding the start transitions was 96.27. Uncovered states were only those in the Naming Service introduced for compatibility. The transition coverage was 73.11 after excluding the procedure calls, time-out counters and 'discard' transitions where the automaton was fully specified. Uncovered transitions were related to the uncovered states and timers or error handling, which would not happen during automatic verification. They had to be checked manually.

5. Conclusions

We have found very useful and effective the application and the assistance of ObjectGEODE in the design of UPC. During this work the static and dynamic checkers of the tool brought our attention to any possible inconsistency in our specification. In many cases these were only warnings, but many times they helped to spot errors, which would be extremely difficult to find without this assistance. Most of the errors were result of inattentiveness. It was trivial to correct them, as we showed in our example, once we knew that there was an error in our specification. However, even at the early stage of the work it was difficult to take into account all possible states and behaviors of our system even with the help of the simulator, because of its distributed and interactive feature. The simulation and the subsequent verification gave us confidence of the correctness and completeness of the specification.

Selecting a standard language like SDL gave us the advantage of:

- The application of the methodology worked out for the language based on the experience of many researchers and technicians around the world, and
- A powerful tool, ObjectGEODE, which was introduced to the market several years ago and is gradually finding its way to acceptance by the industry.

In spite of some short-comings, ObjectGEODE's capabilities are overwhelming in comparison with the in house tools we used earlier. We are convinced that to design a complex system, which is typical for modern communication systems, such design help and assistance as provided by ObjectGEODE are imperative from the very beginning of the work. ObjectGEODE is also easy to learn and use. There was some awkwardness in the earlier version of the tool, but most of it was corrected in the newversion 3.2. The new HTML based manual is more convenient, though addition of some more links would be a great help for the user.

We are continuing our work on the UPC system with further refinements of the specification. Some of the features we plan to add include allowing multiple user-terminal bindings, specification of the type hierarchy of the user-terminal agents, and refinement of the application manager.

Beside the specification, we are also working on prototyping the system, based on the specification. This work was started before the final specification, since the prototype will not implement all the options specified in the final version of the design. At the same time, practical implementation issues bring up new questions which help us in the refinement of the specification. For prototyping we could use the C code generated by the design tool. Unfortunately, as we mentioned earlier, many of the system components require Java implementation for portability, which is not yet supported in ObjectGEODE or any other SDL based tool we know.

Acknowledgment

This work was supported by a grant from Motorola Wireless Data Systems Division, Richmond, B.C., Canada.

References

1. Arnold, K. and Gosling, J. (1997) The Java programming language, Addison-Wesley.
2. CORBA (1997) Common Object Request Broker Architecture documentation, Object Management Group, http://www.omg.org
3. Derr, K.W.(1995) Applying OMT: A Practical Step-By-Step Guide to Using the Object Modeling Technique, SIGS Books & Multimedia.
4. IS-9646 Part 3 (1992) Tree and Tabular Combined Notation
5. ITU Z.100 (1992) Specification and Description Language (SDL).
6. ITU Z.120 (1992) Message Sequence Chart (MSC).
7. Li, Y. and Leung, V.C.M.(1997a) Supporting personal mobility for nomadic computing over Internet, *ACM Mobile Computing and Communications Review*, Vol.1, No.1, 22-31.
8. Li, Y. and Leung, V.C.M.(1997b) Protocol architecture for universal personal computing, *IEEE Journal on Selected Areas in Communication*, October 1997
9. ObjectGEODE (1997) documentation, Verilog, Inc.
10. RFC 2002 (1996) IP Mobility Support, Internet Engineering Task Force.
11. Törö, M. (1997) Simulation of IP Mobility Support: An experiment in mobile protocol specification with SDL, in *SDL, MSC and Trends* Proceedings of the Eighth SDL Forum (ed. A. Cavalli and A. Sarma), Elsevier Science.

BIOGRAPHY

Mária Törö received her Ph.D. in Computer Science from the Technical University of Budapest in 1995. Her main field of expertise is formal description techniques and their application for specification and conformance testing of communication protocols. Currently she works on the SDL design and specification of the UPC system in the IWIN project at the University of British Columbia. Her previous experience includes formal specification, analysis and testing of the mobile IP, X.25 and other protocols, and development of tools for protocol engineering at UBC and at the Research Institute for Measurement and Computing Techniques (Hungary).

Jinsong Zhu received his B.Sc. and M.Sc. in Computer Science from Tsinghua University, China in 1985 and 1987 respectively. He was a faculty member at Tsinghua University from 1987 to 1991. From 1995 to 1996, he was a Senior Software Engineer at Infonet Software Solutions, Inc., Canada. He is currently working towards his Ph.D. degree at the Computer Science Department of the University of British Columbia. His research interests include protocol testing and verification, high speed networks, mobile computing and distributed object computing technology.

Victor C. M. Leung received the B.A.Sc. and Ph.D. degrees in Electrical Engineering from the University of British Columbia (U.B.C.) in 1977 and 1981, respectively. From 1981 to 1987, he was a Member of Technical Staff at MPR Teltech Ltd., Burnaby, B.C., specializing in the design and analysis of satellite systems. He lectured in the Department of Electronics at the Chinese University of Hong Kong in 1988. Since 1989, he has been a faculty member with the Department of Electrical and Computer Engineering at U.B.C., where he is currently an Associate Professor. His research interests are in telecommunications and computer networking. Dr. Leung is a senior member of the IEEE, and a voting member of the ACM.

Development of a Validation System based on Formal Description Techniques: Application to an Ada Run Time System

G. Huecas[1], T. Robles[1], L.M. González[2], J. Serrano[3]

[1] *DIT-UPM, Technical University of Madrid, Ciudad Universitaria s/n, 28040 Madrid SPAIN, E-mail :* gabriel@dit.upm.es

[2] *GMV S.A., c/Isaac Newton, 11, 28760 Madrid SPAIN, E-mail :* lmgonzalez@gmv.es

[3] *CRISA S.A., c/Torres Quevedo, 9, 28760 Madrid SPAIN, E-mail :* serrano@crisa.es

Abstract: In December 1996, a project called LVARTS was finished and delivered to the ESA. The goal was to validate a real system, namely ATAC, an ADA co-processor chip, running on a real board. The system was big enough to develop specific methodologies and tools, which are described in this paper. LOTOS was chosen to formally specify ATAC. The formal specification was used to produce test cases that were executed against the chip, after a completion process to obtain executable test cases.

1. INTRODUCTION

This paper is based on the experience gained from the Project "LOTOS Validation of an ADA Runtime System", ESA contract No. 10797/94/NL/FM(SC), carried out by CRISA, GMV and DIT-UPM (hereby, the Consortium) and started in April 1994 and finished in December 1996.

The objective was to apply rigorous validation strategies on a Complex Hardware System (the ADA Tasking Co-processor, ATAC) based on Formal Description Techniques (FDTs). This implies the definition and developing of theoretical and practical procedures, concepts, tools and whatever was needed for validating the real, big, complex system working in a real environment.

The project was divided in four major phases, as the chip was available and it was not needed to develop it:

1. Production of the User Requirement Document (URD) of ATAC.
2. Specification of ATAC in LOTOS, including the validation of the formal specification against the URD.
3. To develop a Validation Environment and to derive Conformance Test Cases (through a Conformance Test Tool also generated within the project) from the formal specification.
4. To validate a real ATAC implementation that consisted of a board using the MA31750 μP and the co-processor ATAC.

This paper focused on point 3, as it was intended by the Consortium that both the methodology and the tools were as general as possible, so it could be applied to other systems, even using other FDTs. The article layout is as follows: section 2 describes different Validation strategies considered by the Consortium and their uses applying Formal Methods are looked at. Section 3 focuses on the Validation Environment, comprising hardware and software. In Section 4, conclusions are exposed.

2. VALIDATION METHODOLOGY

Current day conformance testing is based on the standard ISO-9646 "OSI Conformance Testing Methodology and Framework" [ISO:9646]. ISO-9646 is a five-part standard that defines a methodology and framework for conformance testing, mainly oriented to protocols. However, much of that standard applies to conformance of any product, not specifically communication protocols.

2.1 Validation and Conformance Testing

Validation and Verification (V&V) is a collection of analysis and techniques activities across the full life cycle [Wall89] of a product, oriented to determine that it performs its intended functions correctly, to ensure that it performs no unintended functions, and to measure its quality and reliability.

Verification involves evaluating the product during each life-cycle phase to ensure that it meets the requirements set forth in the previous phase. It answers the question “are we constructing the product correctly?” Validation involves testing the product against its requirements. It tries to answer “Are we constructing the correct product?”

A particular case of validation is when a formal specification of a system is available. In that case, the implementation of the system is tested against that specification called Reference Specification. Conformance is to establish if an implementation conforms to its specification. Let define Conformance formally in next section.

Let us define Conformance formally. It is said that an implementation I conforms to a specification S if and only if 1) for any behavior imposed by S, I is able to perform it, and 2) for any behavior rejected by I, it is foreseen in S this rejection. In other words, the implementation only behaves as indicated by S, and it is only able to reject something if it is considered in the specification. This definition includes several aspects:

1) Conformance testing only takes into account allowed behavior in the specification.
2) The specification only imposes some minimum conditions and behavior.
3) The specification may have optional aspects: there is no obligation to follow them, but if it is the case, the implementation should follow the specification.

Usually, it is not possible to test conformance at 100%. The problem may be mathematically exposed, but undecidable in practice: all possible behaviors, and data values should be tested. However both are usually infinite, so it should be fixed a sensible, reachable limit, that defines certain degree of tested or covered specification. These are coverage metrics, defined in the field of conformance against what part of the specification has been tested. The more tests executed on the product, higher values of coverage will be obtained.

The use of formal techniques in the conformance process implies a number of advantages, discussed in next section. For the conformance process, it supposes that a formal, computer-treated specification is available, from which test cases may be automatically generated. Conformance process is represented in figure 1.

The first step is to generate a test suite for the product. This is the test generation or derivation phase. The test suite is divided in test groups, organized hierarchically, divided functionally. A single test is called test cases, composed by test events. Test cases are organized by test purposes.

A test purpose is a description of the goal of the test precisely defined, oriented to a conformance requirement, accordingly to the system specification.

Test cases are functionally composed by a preamble, a test body and a postamble. The preamble is a preliminary phase that drives the Implementation Under Test (IUT) into the desired state in which the test will start. In simple tests, this preamble is empty. The test body is the sequence of test events that ensures the accomplishment of the test purposes. The postamble is the sequence of actions that conduct the specification to a stable state. Sometimes it is as simple as a reset button.

The second step in the conformance process is to make the test cases able to be executed. This is called Test Case implementation, and the result is to obtain the Executable Test Cases (ETC), in order to be applied to the IUT. ETCs are particular for every product. In this paper, we will use Test Procedure (TProc) as a synonymous for ETC.

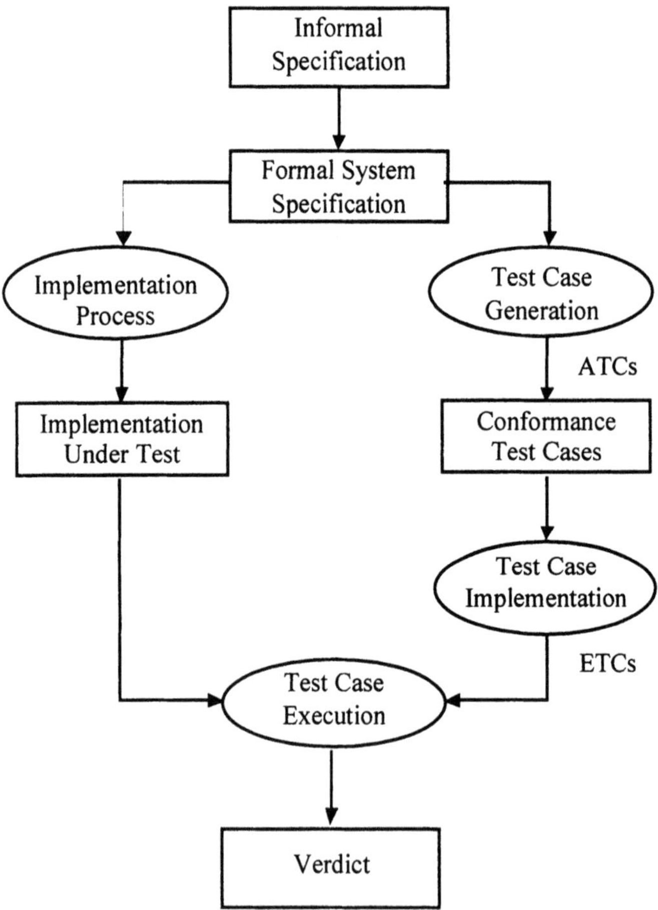

Figure 1: Conformance Process

The last phase is Test Execution. The ETCs are executed on the IUT. The result is a conformance verdict: PASS, if the product conforms to the specification, FAIL if the product does not conforms and INCONCLUSIVE, if neither a PASS or FAIL verdict may be emitted.

2.2 Formal Description Techniques

The growing complexity of real systems nowadays has originated the need for well-defined, complete, unambiguous and precise specifications. From this point Formal Description Techniques (FDTs) appeared to cover these problems. By using these techniques, systems are described with languages that have a formal syntax and semantics. FDTs are based on well-defined mathematical models, which allows a formal analysis of the system. Currently, there exists FDTs based on EFSM (Extended Finite State Machine) as SDL, with a wide impact on industry and based on Process Algebra as LOTOS, with a growing field of application and use.

LOTOS (Language Of Temporal Ordering Specification) was designed for describing open distributed systems, and is applied to concurrent and/or distributed systems in general. The basis of LOTOS is the description of a system by means of the definition of the timing relation, which constitute the observable behavior of a system. In LOTOS, a system may be specified by the composition of the different subcomponents that concurrently performs the desired functionality. In that sense, LOTOS is able to capture the architectural composition of the system.

2.3 Different approaches

There are a number of theories that pretend to generate, more or less automatically, test cases from a formal specification written in LOTOS. In this section, an overview of them is presented. Also a couple of test execution methods may be identified, and here they are discussed.

Brinksma's Canonical Tester [Bri89] uses a process model (failure trees) that identifies processes test equivalents. This work includes an algorithm to develop such Canonical Tester. The Canonical Tester is the tree of all test traces for the specification. Although it model a lot of basic concepts, as testing relations, etc, the Canonical Tester is complete, in the sense of it test all the behavior at once. This makes impossible to develop the Correspondent ETS and there is no possibility of organization by requirements, purposes, etc. As exhaustive testing is not possible in practice, this approach should be given up.

Wezeman [Wez88], starting from Brinksma's Canonical Tester, developed another method called CO-OP. It tries to produce tests step by step, by identifying COmpulsory and OPtional set of events. However, the method ends obtaining the Canonical Tester, result that validates the process but makes unfeasible for industrial application.

In 1991, Robles, Mañas and Huecas [Rob90] developed a Test Conformance Derivation and Execution methodology based on ISO-9646.

The key concept is the combination of Test Purposes and the Formal Specification, which allows driving a trace extraction process by test purposes. These traces are completed lately to become full Test Cases. This is called Test Case Completion, and its goal is to deal with the undeterminism contained in the specification. ATS are specified in a formal, abstract language (LOTOS in this case). ETS have been particularized for certain IUT, taking into account implementation details.

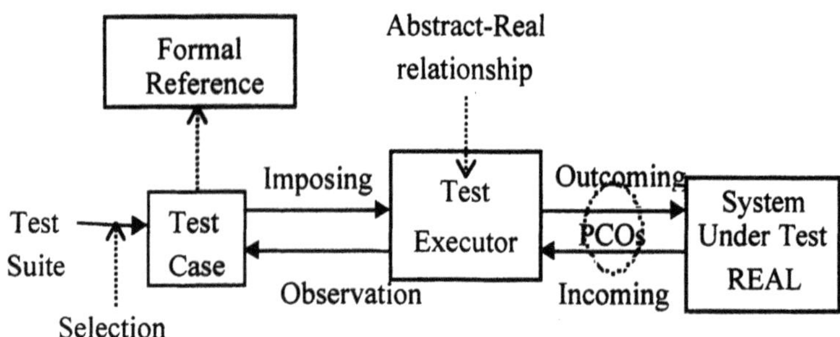

Figure 2: **Tester Functional Organization**

The advantages of this approach are:

1. Test Cases are generated automatically from the formal specification.
2. Test Cases are validated by construction.
3. Test Cases are formally defined.
4. Test Purposes are validated by composition with the Formal Specification.
5. Verdict Assignment may be automatically performed with the appropriate execution method.

Test Purposes are defined with respect to the formal specification, so means to identify elements within the specification should be provided. These elements can be actions and states. However, systems are usually conceived as state machines, extended or not, and this is the usual information used by humans. On the other hand, the same action may be performed in different states. Hence, a state identification mechanism should be provided, namely State Labeling. Relevant states are identified by assigning an unique label in the whole specification. These labels are special gates that the Test Executor will not consider as events to be sent to or to be received from the IUT. However, real events may be used to identify states if there is a biunique relation between a state and the event. This reduces the work of labeling a specification, making Test Purposes more expressive.

The execution of Test Case on a real implementation is shown schematically in figure 2. In this figure, the main concept is the interrelation between the Test Executor and the real implementation. These points separate the abstract, formal world, written in LOTOS with a specific semantic and the real interface with the IUT. This gap should be fulfilled with Adaptation Software, particular for each IUT. The requirements for this Adaptation Software are described in section 4.3.

Test Executor should decide which event should be imposed (controlled) to the IUT and which ones should be waited for (observed). The point in which this interaction takes place is called PCO (Point of Control and Observation). These PCOs may be physically separated from the IUT, and it is usually the case. In the ATAC project, as described below, the Implementation Access Points (IAPs) are inside a breadboard, connected with the Test Executor remotely. Hence, the Adaptation Software should take care also about delays in the communication with the IUT, so timeouts should be assigned to each event of the specification.

2.4 Selected methodology

The generation of Test Case from a LOTOS specification usually implies, for realistic systems, that the Test Case generator (CTG) should deal with indeterminism. This indeterminism is presented in several ways: different path (traces) to reach the same state, open (symbolic) values handled by the specification, unsolved predicates and/or guards, partial functions, etc. Using a specific methodology, or introducing more details of the system may solve some of them. The result is that a wider, longer tree than needed is obtained as seed for Test Cases. For large systems, performance problems will arise if no shortening is introduced.

However, the indeterminism due to open values still remains when the Test Case is generated. Usually, an intermediate phase of Test Parameterization is needed.

In the ATAC project, as a solution both to solve the indeterminism and to improve performances, the IUT runs in parallel with the CTG, thus providing the needed values for the Test Cases and narrowing the expanded tree obtained from the specification.

Among the advantages, this composition is able to deal with any kind of indeterminism. Also a great performance improvement is expected. The main drawbacks are that it tests a specific IUT, without previous test generation phase. Anyway, the results are logged, so they can be used as seed for test case completion.

2.5 Coverage

In this section, the coverage metric for LVARTS project is described. For a comparison among different metrics, please refer to [Hue95]. The metric should satisfy the need of tester specificators and developers in order to stop the test generation process or to decide if still more test are needed. The metric is also used to check the marginal coverage, i.e., the value increasing in a test battery when a new test is added.

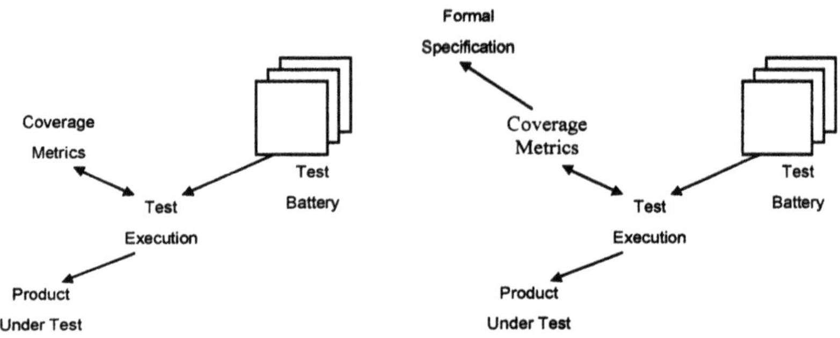

Figure 3: **Coverage Metric w.r.t. the Product (a) and w.r.t. the Specification (b)**

Commonly, coverage refers to which parts of a system have been tested. In Conformance Testing, the IUT is seen as a black box, so coverage metric should not (and could not) refer to any concept related with the implementation. However, a third element exists, the Reference Specification, written in some FDT, LOTOS for this project. The metrics defined within this project (and the chosen one, described below), take advantage of it, referring of which part of the specification has been tested in the IUT. Hence, when a test is executed, this metric quantifies how much of the specification has been exercised.

Besides, metric coverage can be calculated against the Reference Specification before executing the test, measuring the power of test cases. This value is called a priori measurement. On the other hand, when the test is executed not all of the possibilities are explored, resulting in some reduced values for coverage: it is the a posteriori value, which suits the common concept of execution coverage.

In an ideal world, every possible behaviour and every possible data value would be tested. However, both behaviour and data are often infinite objects, making unfeasible to exercise every trace of the system. Therefore, some reachable and desirable limit should be fixed, limit which allow us to get some decent confidentiality on the tests. Arbitrarily, some finite number is regarded as 100% coverage. Then, we start at 0% value, and as more and more test are executed on the implementation, the coverage grows until eventually it reaches the 100% coverage. If such is the case, no further test is needed, up to the criterion used. Obviously, a congruent coverage metric would accomplish that: $0 \leq C \leq 100$ for certain value of coverage for any test case.

2.5.1 Event components coverage

The semantic richness of LOTOS is the synchronization among processes, which compounds actions by means of involving one or more action denotation: the multi way rendez-vous.

Unfortunately, the number of components of a multi way rendez-vous cannot be statically determined: constraints may be generated dynamically. It is however true that potential components are finite, and may be determined statically. However, it is quite sensible to treat compound event as set of events (repetition is meaningless in coverage), which drive to a finite count. The sense of such affirmation is that repetition of a behaviour does not increase marginal coverage. In other words, confidence is not gained as the same event is repeated more and more: repetition of the same action denotation does not exercise new parts of the specification.

Even this finite measure may be difficult to count. If we want to know the different combinations of actions that may compose events, [symbolic] execution may be needed. As an alternative, we may extract the single action denotations in the text (as in action coverage), and then consider every possible combination of those, without repetition, and satisfying LOTOS rule for sort matching. In general, not all of these combinations are possible as not all action denotations are found in the same scope. Even with that, not all possible combinations are generated in a given compositional architecture. And more, there is an structural defect, since several important constructions become collapsed:

```
    g pattern ; g pattern
    g pattern || g pattern
g pattern [] g pattern
g pattern >> g pattern
```

That means that we are not only interested in the difference of every action denotation: if the specifier considered relevant to repeat some action denotation the coverage criteria should not hide such differences. That is easily achieved by means of considering the relative "position" of the events[1].

To summarize, this coverage metric is able:

- To check all possible synchronizations in a specification, and only those belonging to it, so the a priori and a posteriori values may coincide.
- To deal with different ways to compound an action.
- To deal with infinite behaviour due to the data types, as only the sort is considered relevant in the action.
- To manage infinite and recursive behaviours by means of avoiding repetitive components.

2.5.2 Formal Definition of the event component coverage

The event component coverage requires deriving all possible combination of events in order to set the 100% limit. Basically, it consists of the structural analysis of a LOTOS specification, which allows computing such set. This analysis will be used also to compute the coverage of a test execution, considering the offered actions belonging to the combination set.

First, let show a formal definition of the metric. The usual notation of Labeled Transition System is used [Bri86]. The following characteristics should be included in the definition:

- Unique identification of single event on the text of the specification.
- Characterization of the possible combination with the gate name and the associated sort list (from the experiments).

The first point is solved with the unique identification of the subexpressions in the specification. The second one is solved with an adequate representation of the combinations. As said above, this representation will be the gate name and the sort names of each action denotation.

Let see the following definitions:

2.5.2.1 Unique identification of subexpressions

A behaviour expression may be an action subexpression or the composition of behaviour with any LOTOS operator. For every action subexpression, an unique identifier may be assigned. In LOTOS, the following constructions are considered action denotations: internal actions (**i**), termination actions (**exit**) and other actions (gates with experiments, with or without predicates).

Some of these expressions may be identical. A possible way to assign the unique identifier is to denotate with numbers the position of a subexpression

1 An alternative way is to identify every LOTOS subexpression in the text.

in a global behaviour. The objective of this unique identifier is to differentiate actions whose syntactic compositions are not the same.

2.5.2.2 Composition Representation

A composition is defined as $g^{\{id_1,...,id_n\}}_{<s_1,...,s_n>}$, where their components are:

- g: the gate on which the synchronization occurs, with (g∟ $G \cup \{\tau,\delta\}$), and G is the set of gates defined in the specification.
- $<s_1, ... ,s_n>$: The list of sorts from the experiment list, if exists. It should be pointed out that in this way actual values are ignored, as the only interest is in the data types.
- $\{id_1, ... , id_m\}$: The set of unique identifier of each component in the external action. As it is a set, repetitions of the same event participating in an action are collapsed, representing the fact of it is not counted for the coverage the repetition of a event more than once.

2.5.2.3 Composition Set of a Behaviour Expression

The Composition Set of a behaviour expression are all the different compositions that the behaviour expression may offer. It is defined as:

$$PS(B) = \{o \mid \exists\sigma, B'; B = B'?\quad B'-o\ \blacklozenge\ \}$$

2.5.2.4 Stable state

Let be a specification S, and a state B1. B1 is a stable state if and only if any action after it, is observable.

2.5.2.5 Identifiable state

Let be a specification S, the initial state B_0, a state B_1, and σ a sequence of actions. B_1 is an identifiable state if and only if every application of A to the initial state B_0 ends in B_1.

$$\forall \mathrm{n} \mid \mathrm{B_n} = B_0 = \sigma\,?\qquad ?\ \ B_n\ ..B_1$$

2.5.2.6 Identifiable trace

A trace σ is identifiable in a behaviour expression *B*, and it is denoted as $\sigma\ id\ B$ if and only if:

$$\text{if } B = \sigma ? \quad ?$$
$$\forall Q_1, Q_2 \text{ such } B = \sigma ? \;\; Q_1, B = \sigma ? \;\; Q_2 \int Q_1 .. Q_2$$

Intuitively, a trace is identifiable if the specification ends in a unique state after its execution. In other words, the observations of such trace always reach the same state in the specification.

2.5.2.7 Terminating trace

A trace σ is terminating in a behaviour expression *B* and it is denoted as $\sigma \; ter \; B$ if and only if:

$$\text{if } B = \sigma ? \quad ?$$
$$\downarrow\exists \text{ a such } B = \sigma ? \;\; -a \;\blacklozenge$$

Intuitively, a trace is terminating when, if accepted in a behaviour expression, it may not show more actions. Now it is possible to establish when a test may be assigned a coverage value:

2.5.2.8 Measurable Tests

Test Cases should satisfy some conditions so coverage values can be assigned to them. There are two causes for it: 1) The test execution may end up without identification of a unique trace on the specification. This happens when the specification may reach different states from the same trace. 2) There may be sequences of actions that reach non-identifiable states in the specification, so coverage cannot be assigned accurately.

Therefore, every ending of a Test Case must reach an identifiable and stable state within the specification. On the other hand, coverage is assigned to the whole Test Case (a priori) or to a finished execution (a posteriori). Therefore, only those traces that reaches the end of a Test Case are taken into account to compute the coverage value. A measurable test on a specification (S) will be a behaviour expression *T* such:

$$\forall \sigma \llcorner \; Tr(T) \text{ if } \sigma \text{ ter } T \; ? \;\; \sigma \text{ id } S$$

where *Tr(T)* is the set of traces belonging to *T*.

2.5.2.9 Assignment of Coverage Value

Let be *T* a measurable test over a specification *S*. The coverage value of *T* would be[2]:

[2] *Card* stands for the set cardinality.

$$M(T,S) = \frac{Card(PS(T \parallel S))}{Card(PS(S))}$$

2.5.2.10 Suitability for LVARTS

Event component coverage is the closest concept to the offers that a specification may show. In other words, it tries to reflect how many offers have been executed among all possible ones in the system, without repeating behaviour.

The number of possible combinations is rather difficult to calculate: a specific semantic analysis of the behaviour is needed. This analysis should expand the behaviour tree and calculate the possible offers. That drives to the state explosion problem.

However, it is possible to establish the theoretical limit of combinations: all possible combinations of action denotations (see previous paragraph). This limit should be done taking into account the LOTOS rules for offer matching: the same gate name and the same sort list[3]. This limit is an upper bound for the number of real combinations that may exist in a specification. The conclusion is that this limit is suitable to be considered the coverage limit, knowing that it may not be reached as non-existence combinations may be generated. However, it is known that some of such non-existence combinations would be generated.

For test execution, it should be recorded the combinations of each performed action. This task is reduced to log the line number of all action denotations producing such action. As line numbers identify uniquely the action denotations, there is no need (maybe for debugging purposes) of registering the gate names and associated sort lists. LOLA provides the line numbers of the components of an offer.

Once the set of line numbers is generated (as a set, without duplicates), its cardinal divided by the number of combinations calculated as the limit, is the coverage figure for the test execution.

3. VALIDATION ENVIRONMENT DEVELOPMENT

As stated, the proposed approach provides a methodology with the objective of obtaining and executing a set of relevant ATCs. The described methodology is supported by different tools that cover the different phases of

[3] These lists are ordered. Besides, remember that specific values are not considered in behaviour coverage.

the testing process: 1) Formal Specification of the system, 2) Test Cases and Test Procedures generation and 3) Execution of the Test Procedures

The Validation Environment (VE) is composed of those tools and is conceived to be as general as possible for any IUT.

3.1 Functional Architecture

Following the discussion in section 2.1, the Validation Environment may be logically divided into four main components (see figure 4): 1) LOTOS transformational toolset 2) Conformance Tests Generator (CTG) 3) Test Executor (TEX) and 4) Adaptation Software (ASW).

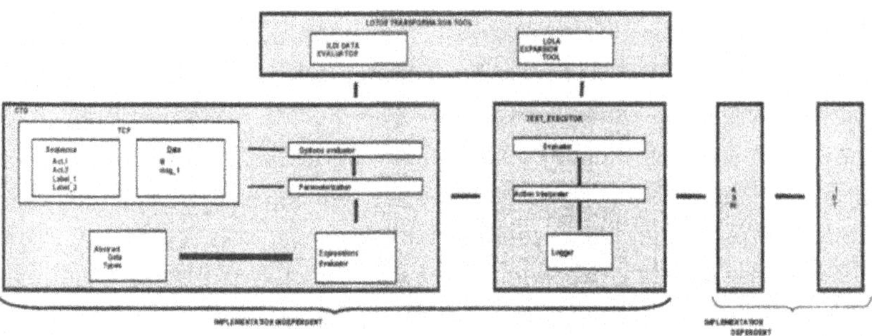

The transformational environment assists the user in the definition of the formal specification of the IUT and acts as the interface of the rest of the system with the formal specification. The CTG component produces Abstract Test Cases instantiated into Test Procedures with relevant data. The TEX interprets the test procedures and exercises the SUT with the actions in those test procedures.

All these components are implementation independent. They may be used to test any complex system. However these components are not enough to execute the test procedures. The specification is an abstract representation of the IUT. The interactions at specification gates must be translated into actions over the system interfaces. This is the role of the last component included in the Validation Environment: the ASW, which may be considered part of the SUT, and acts as the interface between the Test Executor and the IUT.

3.2 Transformational Environment

The validation of the LOTOS specification against its informal requirements requires a Transformational Environment (TE). Among other analysis, the following are performed: searching patter of actions, exploding the specification up to certain depth, proving that the specification shows a property, removing loops of internal actions, etc. LOLA was used for these objectives.

3.3 Conformance Test Generator

The purpose of *Conformance Test Generator* (CTG) is to derive from system specifications conformance test. In principle the CTG design was conceived for *any* LOTOS specification. Nevertheless, some assumptions must be followed when specifying the system.

The first approach was to take as much advantage as possible of the capability provided by the LOTOS tools to get into the specification. A specification is managed by the tools as a labeled transition system. Such a system is similar to an Extended Finite States Machine (EFSM). Obviously, the intention was to produce a subset of traces directly related with the objective of the Test Purpose. The use of symbolic values produced at generation time reduces the explosion in the number of states among other problems.

In the phase of ATCs parameterization, a value must be provided to the variables of the trace. In some points, neither the user nor the system itself is able to provide that value. A clear example is the "Switch" instruction. The specification of that instruction says that a task identifier of one of the candidates to run shall be returned. In that case the specification should offer a value, but in fact it *asks* for that value and keeps that value in a variable from that point. This value is managed from now as a symbolic value and the machine must explore all the open possibilities to follow depending on the content of that value that can only be supplied by the implementation, causing the state explosion problem.

To avoid this problem, the CTG executes the SUT in parallel with the specification expansion process. This process, when a value is needed, is driven by the SUT execution, which may provide specific values. It produces that all possible continuations of the specification expansion are reduced to those that match with the values provided by the SUT in a particular execution.

This approach involves merging three steps: ATCs generation, TProc generation and TProcs execution. Hence, no ATC is produced, that is, there is no intermediate product, though it may be logged for future uses.

With this approach the testing process may be summarized in two steps: 1) Test Purposes Definitions and 2) Step by step expansion of the specification and execution of the IUT.

3.3.1.1 Test Purposes Definition

Test purposes (TPs) are defined by a human expert taking into account his own experience and environment restrictions. The human tester uses the URD and the SRD as entries in test purposes definition. These test purposes are written in an informal way.

As the system goal is to automatically derive ATCs from TPs, the later must be formalized. The formalization process implies: 1) to identify relevant states into the specifications, 2) to mark relevant states with labels, and 3) to rewrite the TPs in terms of action and labels: Definition of Test Case Patterns (TCPs).

A formal TP establishes a logical relationship among several relevant states of the LOTOS specification. However, it is not possible to identify states in an explicit way inside a LOTOS specification. In LOTOS (as in any LTS), the elements on which we work are the transitions. In order to identify states, we may give labels (names) to relevant transitions and therefore, identify states by means of the transitions that may be performed from them.

Once the relevant states have been labeled, a TP may be formalized by means of a TCP. These TCPs are the representation of a test purpose as a chain of labels and/or actions. The setting of the labels to identify objectives and subobjectives must be as complete as possible. In this way, the state explosion can be avoided as far as the number of branches to examine is reduced.

The TCPs are the result of mapping TPs onto the specification. They must be oriented to the requirements captured in the specification.

A TCP shall contain the following information: the sequence of actions and labels that the ATC should include, the maximum number of steps between each action/label defined in that sequence, the labels and/or actions that the ATC shall avoid, the length of the ATC, the number of ATCs to produce, a pool of data for each gate, and an optional initial trace.

The sequence mentioned above is not a complete trace but just a pattern. That is, the user must specify which transitions (states) the ATC should go through. Therefore, this sequence is no more than an ordered list of actions and gates. The system takes care during the expansion process of traversing these transitions.

Obviously the more specific the sequence is, the easier for the system will be to find a test case that fulfils the pattern (or to decide that there is not such an ATC).

3.3.1.2 Step by step expansion and execution of the SUT

This phase includes the following actions: ATC generation, TProcs generation from the ATCs, TProc Expected Result Evaluation, and Execution of TProcs

These actions are performed in parallel in a stepwise fashion. At each step a new action of the ATC is produced. This action is chosen among all the actions that the specification can perform from the current state. If needed to, the free variables of this action are assigned a value, producing a step of the TProc. This step is composed of an action (an instruction) and some parameters. Both the specification and the SUT are fed with the same action and the same values in such a way that the expected result is always available from the specification.

When the specification responds with a data, the SUT is also asked for this data. The results are compared. If they are different, the TProc is said to have failed, otherwise, the process continues until a DEADLOCK is produced by the specification or the maximum depth specified in the TCP is reached.

Three kinds of actions may be offered by the specification during this process:

1. *Read* actions. The specification waits for a value from the environment.
2. *Write* actions. The specification offers a value to the environment.
3. *Indeterministic* actions. The specification waits for a value from the SUT.

When a *Read* action appears it means that both the specification and the SUT are waiting for a value from the environment. In our case, the environment would be the CPU that executes an instruction on ATAC with some parameters. For that reason, the CTG, which acts as the environment, must provide a value to both the specification and the SUT.

A *Write* action means that both the specification and the SUT offer a value to the environment. This is a checking point. Both returned values are compared. If they are equivalent, all the steps until the current one are supposed to be successfully executed and the process can go on.

If not, some step has failed. The VE can not know which step has failed. Remember that we are performing conformance testing and not module or unit testing. The data offered by the specification represent the expected result.

An *Indeterministic* action is slightly different to the ones. Both, the specification and the SUT must offer a value. But while the SUT knows exactly the value that must be offered, the specification merely knows that the value it must offer must be in a range or belong to a certain type. The only entity involved in the process that knows the value is the implementation. In consequence, this value is read from the SUT and offered to the specification. That is, for the specification, it is a *Read* action and for the SUT, it is a *Write* action. The specification checks the obtained value against the corresponding guards.

With this process no ATCs are produced but directly ETCs, the so-called TProcs. No state explosion is produced because the branches are reduced to a point where no symbolic values are managed. At every moment we have the expected result available.

If the generated TProcs do not contain points with indeterminism, they are totally general and may be applied to any implementation. Regression testing may be made just until the point where the TProc failed.

The TCPs drive the expansion process. That means that when the CTG has to choose among several options, the information extracted from the TCP is used in order to take a decision.

3.3.1.3 Data Values

At some steps the tool must provide a value both to the specification and to the SUT. Two different value sources were used: values provided by the user in the TCP and values extracted from the specification. This source is used when no values in the TCP are provided.

In general the chosen value must meet the guards in the option, that is, it is not enough with choosing a value but the value must agree with a given Boolean expression. For that reason an expression evaluator has been introduced within the CTG to facilitate the value assignment. The expression evaluator is based on SWI-PROLOG [Jan95] from the University of Amsterdam.

3.4 Test Executor

The *Test Executor* (TEX) covers the last phases of the testing process.

The CTG produces a set of TProcs according to the LOTOS specification of the system. These TProcs are a battery of instantiated ATCs for an abstract machine. The interface of this abstract machine with the outside world is based on the gates defined in the formal specification. The TProcs write/read to/from those gates. The values interchanged through the gates are those defined in the specification data definition. The TEX deals with those values and gates. It interprets the TProcs and commands the SUT. The TEX takes into consideration the Tproc expected result provided by the CTG assigning a verdict.

3.4.1.1 Test Execution

Once a TProc is obtained, the TEX must perform a set of operations aiming to exercise the SUT with the actions and values in the TProc. This TProc is expressed as a LOTOS process.

Summarizing, TProcs are obtained step by step by expansion of the specification. One step is produced at each step together with the expected result. The expected result may be a constant value offered by the specification or a range of values, expressed as a guard over a variable. And each step of the TProc together with the expected result is passed to the TEX. The TEX exercises the Implementation with the values in the TProc and checks that the Implementation is performing according to the TProc (or not) by observing the checking points

3.4.1.2 Verdicts Assignment

Verdicts are assigned to observations after according to the intention of the test (the expected test results). The Tprocs are completed with the required Verdict: *INCONCLUSIVE*, *PASS* or *FAIL*.

The objective to reach within the expansion process is to execute the last action in the pattern sequence having met all the checking points.

It is not possible to check at every action whether the TProc has been successfully executed by the IUT. Just at this points where the IUT and the specification return values it is possible to compare and assign a verdict. These points are the *Write* and *Indeterministic* actions. In the *Write* actions both the specification and IUT offer a value. This value may be compared. In the *Indeterministic* actions the IUT offers a value and the specification sets a restriction on that value: the value offered by the IUT must fulfil the restrictions imposed by the specification.

3.5 Adaptation Software

There exists a gap in the CTG between the Formal Specification (written in a formal language) and the SUT (real product). There are three aspects to be solved: 1) notation translation, 2) specification and SUT coordination (see previous section), and 3) communication management between the CTG and the SUT. The *Adaptation Software* (ASW) saves this gap.

The ASW isolates both the CTG and the TEX from the particular interfaces of a IUT, and implements particular features imposed by the IUT impossible to be covered by the specification.

In order to be executed the synchronization on specification gates must be translated into executable actions over the physical interface of the SUT. Depending on the abstraction level of the specification the translation process is more o less complicated.

The TEX interprets the TProcs and runs in the Host system. The TEX sends commands and data to the ASW that is running in the target system.

The ASW is configured for a twofold objective: 1) mapping the actions specified in the LOTOS system specification into actions the SUT understands and 2) controlling the communication between the Host and Target systems. The ASW receives commands from the TEX, translates them into atomic actions over the SUT and executes these actions. This architecture intends to isolate the Test Executor from the IUT so that just the ASW depends on the IUT.

4. APPLICATION TO ATAC

4.1 ATAC Overview

The Ada Tasking Coprocessor ATAC developed by R-Tech is a memory mapped coprocessor dedicated to significantly speed up Ada tasking operations, e.g. the rendezvous, and to provide a high degree of accuracy and predictability [Roos94a]. The purpose of ATAC is to avoid the drawbacks usually associated with the use of run-time Ada tasking, mainly: poor real-time response, lack of temporal predictability, and interference between interrupt handling and tasking.

ATAC implements the full Ada83 tasking semantics by hardware, running concurrently with the host CPU and unloading it from the whole management related to tasks. The coprocessor offers high-level instructions for operations that are normally in the run-time system and the tasking kernel of an Ada compiler system. These operations are seen from the CPU as memory loads and stores.

The ATAC handles data structures such as entry queues, delay queues, ready queues, tasks and hierarchies of tasks, in a private memory and incorporates special hardware for managing time and incoming interrupts, which replaces system components like the real-time clocks (RTC) and interrupt controller [Roos94b].

With all these elements, ATAC takes over all scheduling decisions and handles interrupts, delays and time of day in a completely integrated fashion [Roos94a, Roos94b]:

- ATAC controls entirely the Preemption, which forces the CPU to switch context by setting an interrupt to the host CPU, immediately and only when a higher priority task is activated.
- Incoming interrupts are handled like an intelligent high level Ada oriented interrupt controller: interrupts are managed as tasks

performing rendezvous sharing in this way in the global system of priority management.

- Exceptions are recognized in the situations that are handled by ATAC.
- The integrated RTC is used for Ada delay, the added delay_until and for time of day to support the Ada predefined calendar clock function. It can also be advanced from the CPU for debugging purposes.

The ATAC 2.0 supports a memory mapped hardware interface, and operations as seen from the microprocessor are normal memory read and write operations. In this sense, it can be said that ATAC 2.0 only acts as a slave element on the data bus, and so responding solely to explicit transactions. For more information, see [Roos94a].

4.2 Requirements Capture and Formalization

As said before, the starting point for Systems Validation is the Requirements Specification. Our concern was to capture in LOTOS the required observable external behaviour as the collection of all possible sequences of interactions with the environment.

As a consequence, the specification in LOTOS is *Indeterministic* in the sense that many different implementations may comply with a LOTOS specification. Development of LOTOS specifications describing complex system is very difficult. The final size was around 12.000 LOTOS lines.

In one hand, intermediate steps of design are needed in the way from "English" to LOTOS. In the other, a systematic architecture should be used with LOTOS to avoid inflexible designs difficult to modify and to follow by the persons not in charge of the activity.

The main problems found were the limited resources available in current computers and the validation of the LOTOS specification. With respect the first problem, the CTG process had to be harmonized with the state of the art in processor time. The problem of validating of the specification was minimized by using a specific architecture developed within the project.

4.3 System Under Test (SUT)

The system under test chosen in this experiment was an ATAC 2.0 based CPU (ATAC-CPU). The CPU used was the Instrument Control Processor (ICP) of the MIPAS-ICE instrument of ENVISAT-1 and it was adapted to embed ATAC 2.0 and associated logic within the current CPU architecture. The ATAC-CPU is able to work in three Operation Modes:

Test Mode: The ATAC-CPU is linked to the Host CPU via a serial RS-232 line. The ASW runs on the Breadboard CPU and communicates with the Host Computer using the serial link.

In this mode of operation, the ASW considers ATAC as a black box, which can be stimulated and observed. The functional external interfaces of ATAC are the bus interface (instructions to ATAC are bus read or write transactions on the bus), the ATAC outcoming ATTN line, the sixteen incoming interrupt lines and the RTC clock input.

Nominal Mode: Used to run a Real-Time Software Application compiled with the ATAC option. In this mode, ATAC is present in the CPU memory space, the ATAC ATTN line is connected to the CPU and the interrupt lines driven from the peripherals in the Breadboard are connected to ATAC in the nominal way. ATAC performs full interrupt management.

Transparent Mode: this mode is foreseen for developing and debugging purposes and for non-ATAC run of a Real-Time Software Application in Performance Testing. In this mode, ATAC is not present in the CPU memory space and neither the ATAC interrupt inputs nor the output ATTN line are connected.

The main components of this ATAC-CPU are: MA31750 Microprocessor [According to MIL-STD-1750A], AMA31751 Memory Management Unit, MC1031 Standard RBI Chip [OBDH interface], Real Time Clock, 8 KWords of Start-up PROM, 256 KWords of RAM, Serial Line Interface, PCC (Processor Companion Chip) [ICP Glue Logic], ATAC S/S [ATAC 2.0 Coprocessor + ATAC Glue Logic].

4.4 Tests Generation and Execution on ATAC

From the formal point of view, ATAC has several passive gates and one active gate (ATTN line). ATTN active Gate is the interface that can be executed autonomously by the IUT. The rest of interfaces or gates are called Passive Gates. To verify that an instruction has been executed correctly, it must be checked not only that the passive gates return the proper value but also that the active gate is not activated if not explicitly requested.

The process to generate and execute ATAC tests is as follows:

1) The CTG generates a test that is translated into ATAC instructions by the ASW.

2) The instructions are sent through a RS-232 serial line to the ATAC-CPU.

3) The ASW (ATAC-CPU) receives these instructions and commands them to the ATAC S/S.

4) Depending on the instruction the ATAC S/S returns or not some parameter values. It always returns some ATTN line information (e.g. if it has been produced or not).

5) These parameter values and ATTN information are sent by the ASW to RS-232 serial line.

6) The ASW transfer these data to the TEX that checks if these values are correct or not.

Only some ATAC instructions must activate the ATTN line. In the other instructions it must be checked that it is not produced. When the ATTN line has to be activated by an instruction, it must be done within a finite period of time (number of RTC pulses).

To be able to check the occurrence of ATTN in a specific period of time, and as explained in the previous paragraph, in Test Mode the ATAC has a software programmable clock input, so the time is fully controlled by the ASW. In this manner, it can be checked for each ATAC instruction if the ATTN line responds as it is expected.

5. CONCLUSIONS

5.1 Tools Generated

The Conformance Test Generator was conceived and developed to support the methodology exposed in this paper. It was complemented with a Test Executor as well as the needed Adaptation Software to apply the tools to the validation of ATAC. These tools have been built on top of existing tool to inherit part of the technology within the field of the Formal Language and in particular of the LOTOS language.

5.2 Results Obtained

The TCP battery used during the ATAC LOTOS specification validation was reused in the Conformance Testing activities. If the whole set of test covered all the numbered requirements in the URD, they were suitable to test the IUT. This fact had an added value: as the logs for stand-alone execution were available, results with FAIL verdict could be analyzed. The evolution of the model running stand-alone and on-line against the IUT could be compared in order to locate possible errors in the specification. The figures are:

Number of TCPs in the battery: 69
Number of PASS: 34
Number of FAIL: 25
Number of INCONCLUSIVE: 8
Others (execution errors): 2

Respect to the PASS results, it can be said that the model and the IUT evolve in the same way when they are stimulated with the same values by the TCP. The specification was previously validated against the URD corresponding to the test, therefore it can be affirmed that the IUT conforms to the testing requirement.

Further analysis is needed to evaluate the FAIL logs. It can be advanced that the misunderstanding cases between the specification and the ATAC chip are mainly due to bad interpretations of the ATAC documentation or errors in the own model conception. This last consideration is related to minor bugs during the edition phase of the specification or conceptual errors during the ATAC functionality capture. An URD is too abstract to describe a system as complex as ATAC and many functional details were recovered from the ATAC Compiler Adaptation Guide and even from the Ada code of the ATAC simulator. However, this documentation is not suitable enough to be sure the model correctness and completeness.

The evolution of the coverage value growing was registered and it was observed that the execution of TCPs with different purpose gave significant increments, but a similar TCP gave only a variation in the second or third decimal digit. When the number of participant TCPs were growing, the value of coverage started to not increase significantly and the value of less than 70% stabilized for a number of PASS tests next to 40% of the whole battery.

The interpretation of that 40% of the TCPs covers the 70% of the specification can surprise; however, it is a realistic figure: the used metric is a measurement if how many internal and external synchronization have been exercised. Each test needed a high number of internal actions. A big set of functionalities must be previously exercised and therefore there is a lot of common synchronization among requirements. For this reason it is supposed than an increment of test executions with PASS verdict was not associated to an increment of coverage value starting from a determinate number of PASS executions.

5.3 Possible Applications and Future

The methodology exposed can be applied to any complex system implemented in hardware, software or mixed. Candidates are systems with complicated behaviours at interface level, with strong relationship between their points of observation to the environment. Specially, those systems for which high reliability is required. Among others, we think in microprocessors, complicated ASICS or similar devices. Embedded Systems can be validated using this method at a level of coverage difficult to obtain in any other way.

References

[Bri86] Brinksma, Ed Scollo, G. *Formal Notions of Implementation and Conformance in LOTOS*. Univ. Twente, December 1986, Memorandum INF-86-13. Enschede, The Netherlands.

[Bri89] Brinksma, Ed, R. Alderden, R. Lngerak, J.v.d. Lagemaat and J. Tretmans. *A Formal Approach to Conformance Testing*. 2nd Workshop on Protocol Test Systems}. Berlin. October 1989.

[Hue95] Gabriel Huecas. *Contribución a la Formalización de la Fase de Ejecución de Pruebas*. PhD Thesis. Madrid, Spain, 1995.

[ISO:9646] ISO *Information Processing Systems -- Open Systems Interconnection -- OSI Conformance Testing Methodology and Framework* ISO, 1989.

[Jan95] Jan Wielemaker. *SWI-Prolog 2.0. Reference Manual.* University of Amsterdam.

[Rob90] Tomas Robles, Jose A. Mañas, Gabriel Huecas. *Specification and Derivation of OSI Conformance Test Suites*. IWPTS'92, Montreal, CANADA.

[Rob91] Tomas Robles *Contribución al Tratamiento Formal de la Fase de Pruebas del Ciclo Software en Ingenieria de Protocolos*. PhD Thesis. Madrid. Spain 1991.

[Roos94a]Roos J. *ATAC 2.0 Data Book*, Version 1.2, November 1994. R-Tech AB.

[Roos94b]Roos J. *Compiler Adaptation Guide*. Version 1.5, November 1994. R-Tech AB.

[Wall89] Dolores R. Wallace, Roger U. Fujii. *Software Verification and Validation: An Overview*. IEEE Software May 1989.

[Wez89] C.D. Wezeman. *The CO-OP Method for Compositional Derivation of Conformance Testers*. PSTV V. Protocol Specification, Testing and Verification V. June 1985. Editor: M. Diaz IFIP Elsevier Science B.V. North-Holland.

Full-Automatic Implementation of Protocol Programs for OSI Application Protocols over ROSE

Toru Hasegawa, Akira Idoue, Toshihiko Kato and Kenji Suzuki
KDD R&D Laboratories
2-1-15 Ohara, Kamifukuoka-shi, Saitama 356-8502, Japan
(Phone) +81-492-78-7368
(Fax.) +81-492-78-7510
(E-mail) hasegawa@hsc.lab.kdd.co.jp

Key words: ASN.1 (Abstract Syntax Notation One), ROSE (Remote Operation Service Element), OSI, Automatic Program Generation

Abstract: Recently, the automatic program generation from protocol specification comes to be used in order to increase the efficiency of protocol program implementation. In the field of OSI, it can be applied successfully to the application protocol programs over ROSE. However, the conventional RO program generators have problems that they cannot generate complete protocol programs. This paper proposes a full-automatic implementation method of OSI application protocols over ROSE. Our RO program generator supports the program generation for more than one application protocols over ROSE such as MHS P2/P7, and enables the presentation context handling. As a result, we have succeeded to generate a complete MHS P2 / P7 protocol program. This paper describes the detailed design of our RO program generator and the results of implementation of P2/P7 program and its performance evaluation. The work which we did for the implementation was just to specify 370 line P2 / P7 protocol specification. The automatically generated P2 / P7 program can provide about 100 operations per second. Therefore, the proposed implementation method is considered to achieve as high performance as applicable to the practical usage.

1. INTRODUCTION

Recently, the automatic program generation from protocol specification comes to be used in order to increase the efficiency of protocol program implementation. The stub generator of RPC (remote procedure call) is a typical example (Birrel and Nelson, 1984). In the field of OSI, an ASN.1 (Abstract Syntax Notation One) (ITU Rec. 208 and X.209, 1987) compiler, which automatically generates encoder and decoder programs from the ASN.1 specification of Presentation and Application PDUs (Protocol Data Units), is an example of program generation (Hasegawa, Nomura and Kato, 1992) (Neufeld and Yang, 1990). Since the encoding rule of ASN.1 is quite complicated, the ASN.1 compilers are indispensable in the OSI protocol program implementation. However, the program generation by ASN.1 compilers is limited to encoder and decoder programs because ASN.1 specifies only the data type of PDUs. Implementers need to write the rest parts of programs.

The program generation can be applied successfully to the OSI application protocol programs working over ROSE (Remote Operation Service Element) (ITU Rec. X.219 and X.229, 1988), which is considered as an extension of RPC in the OSI environment. In the case of those protocols, such as MHS (Message Handling System) P2/P7 protocols (ITU Rec.X.413, X.419 and X.420, 1988) and CMIP (Common Management Information Protocol) (ITU Rec. X.710 and X.711, 1988), the protocol behaviors are defined in ROSE, and the data types of PDUs and service primitives provided for their users are defined in the *RO-notation*. Therefore, it is possible to generate the protocol programs from the specification in RO-notation, including the programs for PDU encoding and decoding, and for protocol behavior handling, similarly with RPC stub generators. For example, the ISODE software (Rose, 1990), which is a public domain OSI software widely used for an experimental purpose, uses a Remote Operation (RO) program generator called *rosy* (Remote Operations Stub-generator (YACC-based)) for protocols over ROSE (Rose, Onions and Robbins, 1991).

However, the conventional RO program generators have some problems that they cannot generate complete protocol programs, in the following points.

(1) In the OSI application layer, there are some cases that more than one protocols are used over ROSE. In MHS P2/P7 protocols, P7 protocol, which specifies the requests and responses between UA (User Agent) and MS (Message Store), transfers interpersonal messages defined by P2 protocol as user data. In CMIP, data types of managed object information transferred by CMIP PDUs are defined in different specification from CMIP itself. In such cases, the program generation needs to be performed for more

than one protocol specifications. But, the conventional RO program generators cannot handle these cases.
(2) In the OSI communications, the data types of PDUs are managed by the abstract syntax, and the presentation context containing the object identifier value of abstract syntax needs to be defined for an individual presentation connection. The conventional RO program generators do not generate protocol programs handling the presentation context.

We have taken account of those problems and implemented our RO program generator which can generate protocol programs fully supporting application protocols over ROSE. The above problems come from the fact that the RO-notation and ASN.1 specification cannot describe the mapping between PDUs specified independently nor assignments of the object identifier of abstract syntax. Therefore, we have extended the RO-notation and developed a program generator for the specification. Our RO program generator is also designed so as to generate protocol programs which have more flexible user programming interface than conventional RO program generators, and which realize high performance by avoiding data copying.

This paper describes the detailed design of our RO program generator which realizes the full-automatic implementation of OSI application protocols over ROSE, and the results that we applied it to the implementation of MHS P2/P7 protocol programs. Section 2 briefly introduces ROSE protocols and section 3 describes the design principles and overview of our RO program generator. Section 4 describes the detailed design and section 5 shows the results of implementing MHS P2/P7 protocol programs.

2. BRIEF INTRODUCTION TO ROSE

ROSE (ITU Rec. X.218 and X.219, 1988) is an application service element of OSI, and uses a protocol stack illustrated in Fig.1.. Among the protocols in Fig. 1, an application protocol and ROSE which are shaded are targets of automatic implementation. ROSE provides a request-response style communication for application protocols such as MHS P2/P7 and CMIP, and this request-response is called a remote operation. An entity of application protocol (a requester) requests an operation which has some arguments using ROSE. Receiving the operation, the peer entity (responder) sends back a response which has a result of the operation to the requester using ROSE.

The data types of PDUs of application protocols over ROSE are defined using four macros of RO-notation : *operation*, *bind*, *unbind* and *error* macros. The operation macro is used to specify two kinds of application PDUs : an argument and a result of an operation. The bind and unbind macros are used to specify application PDUs transferred when an application association is established and released, respectively. The error macro is used to specify application PDUs which informs a requester of the error of operation. Among the application PDUs, those defined by the operation and error macros are transferred using ROSE protocol, and those defined by the bind and unbind macros are transferred using ACSE.

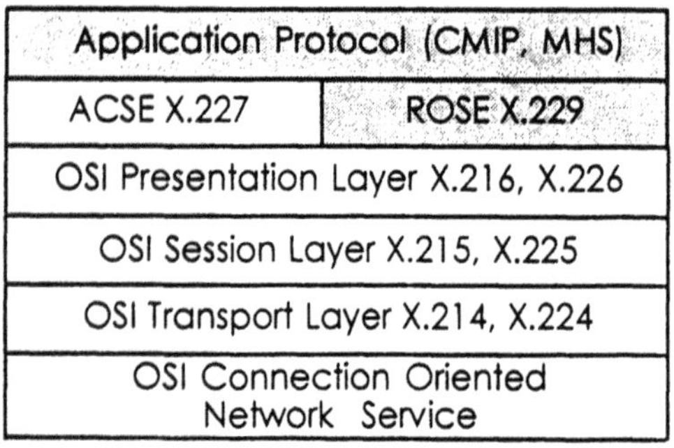

Application Protocol (CMIP, MHS)	
ACSE X.227	ROSE X.229
OSI Presentation Layer X.216, X.226	
OSI Session Layer X.215, X.225	
OSI Transport Layer X.214, X.224	
OSI Connection Oriented Network Service	

Figure 1. Protocol Stack

The ROSE protocol uses four ROSE PDUs : *ROIV (RO-INVOKE), RORS (RO-RESULT), ROER (RO-ERROR)* and *RORJ (RO-REJECT).* A ROSE PDU transfers not only an application PDU as user data, but also the following ROSE protocol control information: *operation value* which specifies a type of operation, and an *invoke identifier* which is used to map the request and response of an operation.

It should be noted that the service primitives of the application protocols over ROSE have the same parameters as their PDUs, and therefore, the data types of the service primitives are also defined by RO-notation.

3. DESIGN PRINCIPLES AND OVERVIEW OF RO PROGRAM GENERATOR

3.1 DESIGN PRINCIPLES

We have adopted the following principles for our RO program generator.

(1) As described above, RO-notation is not powerful enough for describing a complete application protocol specification. Therefore, our RO program generator allows the following specifications to be added into RO-notation.

- A data type contained in *ANY DEFINED BY* type and *OCTET STRING* type. This specification defines the mappings between PDUs of more than one protocols over ROSE.

- The object identifier value for the abstract syntax for the PDU definitions.
- A role of operation, i.e. an operation is requested by the initiator of the application association, the acceptor, or both.

(2) Our RO program generator implements protocol programs supporting application protocols and ROSE. The protocol program interfaces between user programs for the upper layer side, and ACSE (Application Control Service Element) and PL (Presentation Layer) programs for the lower layer side. Since the application protocols over ROSE often require the invocation of an operation during waiting for a response of another operation, the subroutine call based user programming interface which is used in RPC stub and ISODE is not sufficient for flexible programming. Therefore, the user programming interface of the protocol programs automatically implemented adopts the exchange of individual service primitives through queues between the user programs and the protocol programs.
(3) As described above, the service primitives and PDUs of the application protocols contain the corresponding parameters, and the user programming interface of the automatically implemented protocol programs uses the service primitives. In order to avoid data copying, the encoder and decoder programs generated by our RO program generator encode PDUs from service primitive data and decode PDUs into service primitive buffer without using any intermediate working buffers.

3.2 OVERVIEW

(1) Our RO program generator consists of the extended ASN.1 compiler and the protocol behavior routines as shown in Fig. 2. Since individual application protocols use the same protocol behavior defined by ROSE, the protocol behavior is realized by pre-implemented library routines, i.e. the protocol behavior routines. On the contrary, the extended ASN.1 compiler generates the following application protocol specific programs and data from a specification in the RO-notation: primitive data types, ASN.1 encoder / decoder routines, and an operation table containing control data.
(2) The structure of generated program is shown in Fig. 3. The program provides the programming interface based on exchanging service primitives through interface queues to user programs. The data types of service primitives are generated by the extended ASN.1 compiler. The behavior routines perform the application protocol behaviors using the encoder / decoder routines and the operation table. The encoder / decoder routines are used to encode and decode ROSE PDUs and application PDUs. The

operation table holds the static information for all operations, a bind and an unbind. For each operation, the data includes an operation value, pointers to encoder and decoder routines for the PDU and error list.

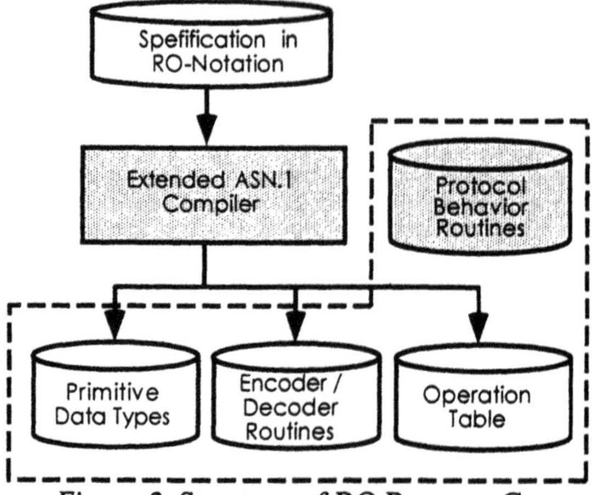

Figure 2. Structure of RO Program Generator

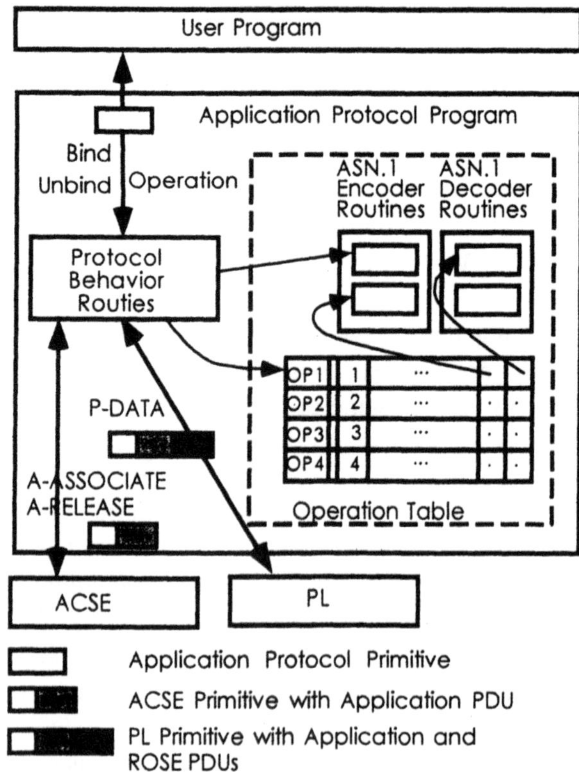

Figure 3. Structure of Generated Program

(3) The protocol behavior routines realize the application protocol processing in the following way.

- They use a control structure such that they receive a primitive from either a user program or a lower layer program (ACSE program or PL program), analyze the primitive, make a primitive and sends it to the lower layer program or the user program.
- When encoding a PDU from a service primitive and when decoding a PDU into a service primitive, they obtain a pointer to the encoder routine and the decoder routine by referring to the operation table.

They use two kinds of tables in order to manage the dynamic information. An association table is prepared for each application association. In each association table, a request-response table for mapping the request and response is provided.

4. DETAILED DESIGN

4.1 INPUT SPECIFICATION TO EXTENDED ASN.1 COMIPLER

An example of input specification to the extended ASN.1 compiler is shown in Fig. 4, which is a part of MHS P2 / P7 protocol specification. In this specification, a bind and an operation are defined by RO-notation extended as described above. *MSBind* is a bind, and *MessageSubmission* is an operation. The argument and result of MessageSubmission operation have *SubmissionArg* and *SubmissionRes* types, respectively, and the operation value is 3. These types are defined in ASN.1. For example, *SubmissionArg* type is defined using *SEQUENCE* type with two elements, *envelope* and *content*.

As described in section 3.2, the object identifier of the abstract syntax is specified at the line marked with *(a)* in Fig. 4. The line marked with *(b)* specifies the role of generated program for *MessageSubmission* operation. This operation is invoked by the initiator of an association. The line marked with *(c)* specifies that a P2 PDU whose data type is *InformationObject* is contained inside the *content* element whose data type is *OCTET STRING*. The line marked with *(d)* specifies the case that the *ANY DEFINED BY* type contains another data. In this case, the data types contained in the *values* element depends on the value of *atype* element whose data type is *OBJECT IDENTIFIER* type. The two lines followed by mark *(d)* specify pairs of the data type of contained data and the values of the *atype* element.

```
P2P7 DEFINITIONS
BEGIN
ABSSYN { 2 6 0 1 4 ) -- (a) presentation context
MSBind ::= BIND -- bind
  .....
ABSSYN { 2 6 0 2 1 } -- (a)
MessageSubmission OPERATION
!! INITIATOR  -- requester -- (b)
  ARGUMENT  SubmissionArg
  RESULT    SubmissionRes
  ERRORS     { ..... }
::= 3  -- operation value
-- P7 PDU
SubmissionArg ::= SEQUENCE {
  envelope   Envelope,
  content    Content }
Content ::= OCTET STRING {type = InformationObject } --(c)
Attribute ::= SEQUENCE {
  atype   AttributeType,
  values ANY DEFINED atype -- (d)
   ( heading  SeqOfHeading { 2 6 1 7 0 },
     body     SeqOfBody { 2 6 1 8 0 },       ...... )
AttribueType ::= OBJECT IDENTIFIER
SeqofHeading  ::= SEQUENCE OF Heading
Heading  ::= SEQUENCE { ... }
-- P2 PDU
InformationObeject ::= CHOICE {
  ipm   IPM,    ...... }
-- ROSE PDU
ROSEApdus ::= .........
```

Figure 4. Example Specification (P2/P7)

4.2 SERVICE PRIMITIVE TYPE GENERATION

The extended ASN.1 compiler generates two service primitive data types in C language from each operation, bind and unbind definitions. One data type is for a request primitive and an indication primitive, and the other is for a response primitive and a confirm primitive.

A service primitive data type consists of the following elements:

- header : common information for all primitives such as a layer identifier, an indicator of a request, indication, response or confirm, and an association identifier.
- parameters of lower layer protocols : information needed to encode and decode ROSE PDUs and parameters exchanged between lower layers and a user program.
- application PDU

The data type of an application PDU is specified in ASN.1, and the C data type corresponding to that is generated by the extended ASN.1 compiler. As described above, the application PDU field is used both as a working buffer in ASN.1 encoding / decoding and as a parameter of a primitive.

```
/* MessageSubmission Request / Indication Primitive */
struct MessageSubmission_req {
  prim_header_t   PrimHead;  /* header */
  long            InvokeId;  /* invoke identifier */
  SubmissionArg_t param;    /* application PDU :
                               argument of operation */ };
/* SubmissionArg Data Type */
struct SubmissionArg {
  Envelope_t   envelope;
  Contetnt_t   content; };
typedef struct SubmissionArg SubmissionArg_t;

/* MessageSubmission Response / Confirm Primitive */
struct MessageSubmission_rsp {
  prim_header_t   PrimHead;  /* header */
  long            InvokeId;  /* invoke identifier */
  long            resultFlag; /* Result : Success/Failure */
  union {
    SubmissionRes_t acc;    /* application PDU :
                               Response of operation */
    error_t         rej;    /* Error Response */;
                  } result; };
/* SubmissionRes Data Type */
struct SubmissionRes {
   ... };
typedef struct SubmissionRes SubmissionRes_t;
```

Figure 5. Service Primitive Data Types of Operation

Figure 5 shows the request and response primitive data types generated from an *MessageSubmission* operation specified in Fig. 4. The primitive data types are defined using C data type *struct* which holds the above mentioned information. For example, as for the request / indication primitive, *invokeID* element is an invoke identifier of ROSE PDU, and is used for encoding and decoding a ROSE PDU. *Param* element is a parameter of application PDU, i.e., an argument of operation, and the data type is specific to each operation. The response and confirm primitives include the following elements as well as invokeID element. *ResultFlag* element is a flag indicating whether the operation has been successfully performed by the responder or not. *Result* element is realized by C *union* structure and is either a result application PDU which is specific to each operation or an error application PDU. *Acc* element is the result application PDU, and *rej* element is the error application PDU.

It should be noted that the data types have the structure allocated in a contiguous memory area without using pointers as long as possible. The *SEQUENCE* and *SET* types which have nested structures do not use pointers to child components, but contain the values of child components themselves. In Fig. 5, C struct *SubmissioArg* corresponds to *SubmissionArg* of

SEQUENCE types, and have C struct themselves such as *Envelope_t* and *Content_t*.

The *SEQUENCE OF* and *SET OF* types which include ordered and unordered component values, respectively, have C data types which consist of an integer indicating a component value number, and a pointer to component values themselves. Figure 6 shows the C data types corresponding to *SeqOfHeading* type in Fig. 4. *Cnt* element and *data* element are the number of heading values and the pointer to the array of heading values. The memory area for storing heading values is allocated as a contiguous array, not as a list with individual heading values linked by pointers

```
struct Q_SeqOfHeading {
  u_long     cnt;    /* Number of Component Values */
  Heading_t  *data; /* Pointer to Array of Heading Values */ };
```

Figure 6. C Data Types of SEQUENCE OF Type

4.3 ENCODER / DECODER GENERATION

The extended ASN.1 compiler generates a C data type and encoder / decoder routines for a data type which is newly defined using structured types of ASN.1 such as *SEQUENCE* and *SET*. On the contrary, those for primitive types such as *INTEGER* and *OCTET STRING* types are pre-implemented as a library.

In order to deal with PDUs containing upper layer PDUs as user data, the mapping function is added to *ANY DEFINED BY* type and *OCTET STRING* type in the following way.

(1) A data type of upper layer PDU is defined using *DEFINED BY* phrase as shown in *(d)* of Fig. 4. The actual data type is dynamically chosen among the candidate data types. i.e, *SeqOfHeading* and *SeqOfBody* of Fig. 4. The data types are determined dynamically by an object identifier value at *atype* element. For example, when the object identifier value of type is {2 6 1 7 0}, the data type of *values* is *SeqOfHeading*.

(2) The data type of *ANY* with *DEFINED BY* phrase is a C data structure *union* which consists of all data types defined by *DEFINED BY* phrase, and it is generated by the extended ASN.1 compiler. Figure 7 shows the C data type generated from *Attribute* type which includes *ANY DEFINED BY* type. *Atype* element contains an object identifier value indicating data type of *values* element. *Values* element is a union of the C data types corresponding to the above candidate date types.

```
/* C data type for Attribute */
struct Q_Attribute
{   AttributeType_t atype;      /* Object Identifier */
    union                     /* Union of Candidate Types */
    { SeqOfHeading_t heading;  /* Candidate Type 1 */
      SeqOfBody_t body;        /* Candidate Type 2 */
    }       values;        };
```

Figure 7. C Data Type for Attribute

(3) When encoding the parameter of ANY, an encoder routine first checks an object identifier value of *atype* element of working buffer, and then calls the encode routine corresponding to the object identifier value. On the contrary, when decoding, a decoder routine first decodes the *atype* element of PDU, and then calls the decoder routine corresponding to the decoded object identifier value.

```
void decodeAttribute(var, edata, index,
     error, ida)
  char   *var;      STRING  *edata;
  u_long   *index; ERROR    *error;
  ID      *ida;
{ Attribute  *q_var;
  long        cnt = 0;
  u_long      next_obj = 0;
  /* decoding identifier and length octets */
  cnt = skp_idlenN(edata, index, &next_obj, ida, error);
  next_obj += *index;
  .....
  q_var = (Attribute *) var;
  /* decoding type */
  decodeobjectN(&(*q_var).type, edata, index,
       error, id_infoN + 31, (u_long) 0); }
  if ((*error).errlevel < 2) {
    if (obj_chk (&(*q_var).type, &headingId) == TRUE) {
          decodeSeqOfHeading(
           &(*q_var).value.heading,
           edata, index, error, id_info + 44); }
    else if (obj_chk (&(*q_var).type, &bodyId) == TRUE) {
          decodeSeqOfBody(&(*q_var).value.body,
          edata, index, error, id_info + 45); }
    else {
       ERRSETN(error, FATALN, 10, ida, *index); }
  .....
}
/* Object Identifier Values : Constants */
OBJID headingId { { 2, 6, 1, 7, 0}, 5};
OBJID bodyId    { { 2, 6, 1, 8, 0}, 5};
```

Figure 8. Decoder Routine of Attribute Type

Figure 8 shows a part of decoder routine of *Attribute* type in Fig.4 which includes the above *ANY DEFINED BY* type. Octets to be decoded are stored in *edata* variable, and the decode result is set in *var* variable. First, the decoder routine decodes identifier and length in the octets to be decoded using *skip_idlenN* library routine. Secondly, the decoder routine decodes object identifier of *atype* using *decodeobjectN* library routine. The decoded object identifier value is set in *atype* element of *var* variable. Thirdly, the routine compares the decoded object identifier value and candidate values using *obj_chk* library routine. The variables which hold candidate object identifier values are generated from an *ANY DEFINED BY* phrase by the extended ASN.1 compiler. *Headingld* and *bodyld* variables in Fig. 8 are the examples, and they are used as arguments to *obj_chk* routine. Eventually, the decoder routine finds the decoder routine corresponding to the object identifier value, and calls the decoder routine, i.e., either *decodeSeqOfHeading* routine or *decodeSeqOfBody* routine which are generated by the extended ASN.1 compiler.

5. MHS P2/P7 IMPLEMENTATION AND COMMUNICATION EXPERIMENT

5.1 MHS P2/P7 PROTOCOL PROGRAM GENERATION

MHS P2/P7 protocol, which is defined over ROSE, is used to transfer an IPM (Interpersonal Message) between UAs (User Agents) and MSs (Message Stores). P2 protocol defines the format of IPM. P7 protocol defines the protocol between UAs and MSs, such as the submission of a message to an MS and fetch of a message stored in an MS.

We have developed the P2 /P7 program using the proposed implementation method. We have written a P2 / P7 specification in the extended RO- notation and generated the program using the extended ASN.1 compiler. The specification supports three operations : *list*, *fetch* and *deletion* of a message, and its size is 370 lines. From the specification, about 11 K line C programs are generated. The details are shown in Table 1. Since protocol behavior routines are about 2.4 K lines, the total P2 / P7 program size is about 13.4 K lines.

Among the 370 lines of the P2 / P7 specification, the 350 lines have been just copied from the specification in the ITU Recommendations (ITU Rec. X.413, X.419 and X.420), and just the 20 lines which specify the mapping of PDUs, presentation context identifiers and so on have been additionally written. It has taken just two days to write the specification and to generate the programs from the specification.

Table 1. Sizes of Specification and Generated Program

specification	lines	program	lines
RO-notation	80	operation table	120
		primitive data types	240
ASN.1	290	encoder / decoder	10,640

5.2 COMMUNICATION EXPERIMENT

In order to run the P2 / P7 program, UA and MS programs for testing have been developed. The P2 / P7 program, UA or MS program are built into the previously developed OSI program which supports from ACSE to LLC protocols. The total OSI program, whose structure is shown in Figure 9, is run as a single process on UNIX operating system. The execution of the above programs is controlled by a pseudo-kernel program which provides the scheduling function and inter-program communication function. For evaluating the generated program, we performed the following experiments.

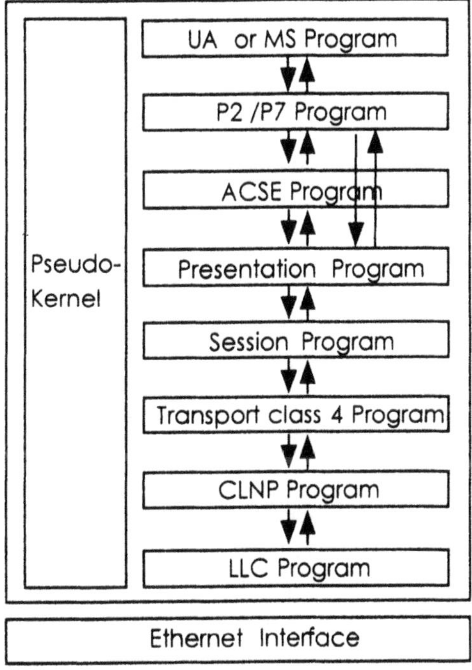

Figure 9. Program Structure

(1) Communication with Other MHS P2 / P7 Program

In order to validate the correctness of the automatically generated P2 / P7 program, we have performed the communication experiment between the generated P2 / P7 program with MS program and a UA program which we previously manually developed. The MS program is run on a Sun SS-20 workstation and the UA program is run on a personal computer. These are connected via a serial line. The message submit, list and fetch operations are executed between the both programs, and the generated P2/P7 program with the MS program provides the correct MS behavior.

(2) Overall Performance Measurement

In order to measure the performance, the communication experiments are performed between the generated P2 / P7 program with UA program and the P2 /P7 program with MS program. The network configuration is shown in Fig. 10. The two Sun SS-20 workstations (SuparSPARC II 60 MHz) which are connected via Ethernet LAN are used.

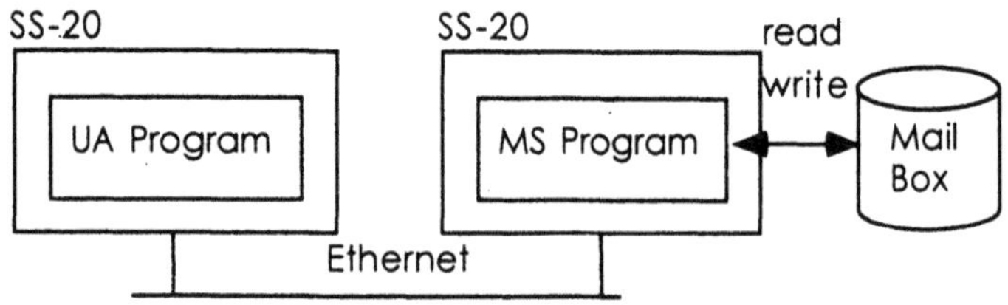

Figure 10. Network Configuration

We have measured the response times of list and fetch operations under the following conditions.

- list operation : The MS program searches the mail box, and returns a response which contains an originator name, a recipient name, a subject and a message identifier.
- fetch operation : The MS program reads a message whose body size is 1 K bytes, and returns a response which contains the message as a body.

The response times are measured by running 10,000 operations consecutively, and the measurement is performed ten times for each operation. A response time is a duration between the time when the UA program sends a request of operation and the time when it receives the response. The average response times of list and fetch operations are about 8.8 ms and 10.8 ms, respectively. As for a fetch operation, the processing tines of P2 / P7 programs of UA and MS sides are about 0.59 ms and 0.56 ms, respectively. As shown in the results, the automatically generated P2 / P7 program can support about 100 operations per second.

(3) Detailed Performance Measurement

We have evaluated the performance improvement caused by the proposed implementation method. The duration from the time of receiving a fetch request primitive from the MS program to the time of sending a P-DATA request primitive to the PL program is measured for both the proposed implementation method and the traditional method.

As for the proposed method, the duration is measured during the performance measurement of (2), and it is about 0.29 ms. On the contrary, as for the traditional method, we have developed a testing program for estimating the duration. The testing program uses a traditional method and a traditional ASN.1 compiler (Hasegawa, Nomura and Kato, 1992). The working buffer for an application PDU of fetch argument is not a parameter of a fetch request primitive. Therefore, the testing program converts the parameter of the fetch request primitive to the working buffer before it encodes using the ASN.1 encoder routine generated by the ASN.1 compiler. The duration is measured running the testing program on the same sun SS-20 workstation, and it is about 0.38 ms. The result shows that proposed direct encoding improves the P2/P7 program performance at about 30 %.

6. CONCLUSION

This paper has proposed a full-automatic implementation method of OSI application protocols over ROSE. We have realized a RO program generator consisting of the extended ASN.1 compiler and the protocol behavior routines. They support the program generation for more than one application protocols over ROSE such as MHS P2/P7, and enables the handling the presentation context. We have implemented MHS P2 / P7 protocol using our RO program generator. The results of implementation and communication experiment have made clear the following.

- The proposed method has achieved the full-automatic P2 / P7 protocol implementation. It took a few days to implement it since the required work was just to specify 370 line P2 /P7 protocol specification. Besides, since we did not need write any programming code at all, we did not encounter any program bug during the implementation.
- The automatically generated P2 / P7 program can provide about 100 operations per second. Besides, the processing time of P2 / P7 process is less than 1 ms. Therefore, the proposed implementation method is considered to achieve as high performance as applicable to the practical usage.

REFERENCES

Birrel, A. and Nelson, B. (1984) Implementing Remote Procedure Calls, ACM Trans. Comp. Syst. vol. 2, no. 1.

ITU Recommendations X.208 and X.209 (1987).

Hasegawa, T., Nomura, S. and Kato, T (1992) Implementation and Evaluation of ASN.1 Compiler, IPSJ Journal of INFORMATION PROCESSING, vol.15, no.2, 157-167.
Neufeld, G. and Yang, Y. (1990) An ASN.1 to Compiler, IEEE Trans. SE, vol.16, no.10, 1209-20..
ITU Recommendations X.219 and X.229 (1988).
ITU Recommendations X.413, X.419 and X.420 (1988).
ITU Recommendations X.710 and X.711 (1988).
Rose, M. T (1990) THE OPEN BOOK, Prentice-Hall.
Rose, M. T., Onions, J. P. and Robbins, C. J. (1991) The ISO Development Environment : User's Manual Volume 4 : The Application Cookbook.

BIOGRAPHY

Toru Hasegawa is a senior research engineer of High Speed Communications Lab. in KDD R&D Laboratories, Inc. Since joining KDD in 1984, he worked in the field of formal description technique (FDT) of communication protocols. From 1990 to 1991, he was a visiting researcher at Columbia University. His current interests include FDT, high-speed protocol and ATM. He received the B.S. and the M.S. Degree in information engineering from Kyoto University, Japan, in 1982 and 1984, respectively. He received IPSJ Convention Award in 1987.

Akira Idoue is a senior research engineer of High Speed Communications Lab. in KDD R&D Laboratories, Inc. Since joining KDD in 1986, he worked in the field of computer communication. His current research interests include implementation of high performance communication protocols and communication systems. He received the B.S. and M.E. Degrees of electrical engineering from Kobe University, Kobe, Japan, in 1984 and 1986 respectively.

Toshihiko Kato is the senior manager of High Speed Communications Lab. in KDD R&D Laboratories, Inc. Since joining KDD in1983, he has been working in the field of OSI, formal specification and conformance testing, distributed processing, ATM and high speed protocols. He received the B.S., M.E. and Dr. Eng. Degrees of electrical engineering from the University of Tokyo, in 1978, 1980 and 1983 respectively. From 1987 to 1988, he was a visiting scientist at Carnegie Melon University. Since 1993, he has been a Guest Associate Professor of Graduate School of Information Systems, in the University of Electro-Communications.

Kenji Suzuki is the executive vice president of KDD R&D Laboratories, Inc. Since joining KDD in 1976, he worked in the field of computer communication. He received the B.S., M.E. and Dr. Eng. Degrees of electrical engineering from Waseda University, Tokyo, Japan, in 1969, 1972 and 1976 respectively. He received Maejima Award from Communications Association of Japan in 1988, Achievement Award from the Institute of Electronics, Information and Communication Engineers in 1993, and Commendation by the Minister of State for Science and Technology (Persons of scientific and technological research merit) in 1995. Since 1993, he has been a Guest Professor of Graduate School of Information Systems, in the University of Electro-Communications.

Invited Talk III

A LOGICAL FRAMEWORK FOR DISTRIBUTED SYSTEMS AND COMMUNICATION PROTOCOLS

José Meseguer

Computer Science Laboratory
SRI International, Menlo Park
California 94025, USA
meseguer@csl.sri.com

WHY REWRITING LOGIC

I will introduce the main ideas of rewriting logic [11]—a logic for the specification, prototyping and programming of concurrent systems in which concurrent computation exactly corresponds to logical deduction—and will discuss some promising directions for its use as a logical and semantic framework for distributed computing and communication protocols. An important aspect is its *wide-spectrum* character. Thus, on the one hand it connects smoothly with—and provides a formal foundation for—notations suitable for the early phases of software design, such as architectural description languages and object-oriented modeling languages [14, 19]; and on the other hand it also provides a natural implementation path through subsets of the logic that are efficiently implementable as distributed or mobile languages [9]. Similarly, rewriting logic specifications, when supported by appropriate tools, can be used in a wide range of specification, prototyping, code generation, testing, formal analysis, and formal verification efforts to reach high levels of assurance about a system's correctness. All these capabilities seem potentially useful for specifying, prototyping, testing, validating, and implementing distributed systems in general and communication protocols in particular; and they offer the promise of substantially narrowing the gap between specification and code, and of reaching high assurance about the correctness of implementations that can themselves be realized in efficient subsets of rewriting logic.

For such applications, an attractive feature of rewriting logic is its simple and smooth integration of three levels of description that ultimately need to be addressed in a distributed system's formal specification, namely, the *data* level of data structures and functions, the *dynamic* level of processes and concurrent computation, and the *requirements* level of formal properties that must be satisfied by the system.

Indeed, a rewrite theory is a triple $\mathcal{R} = (\Sigma, E, R)$, with (Σ, E) an equational theory, and R a collection of labeled rewrite rules that are applied *modulo* the equations E; they axiomatize the basic *concurrent transitions* of the specified system. The two main kinds of axioms in a rewrite theory, namely, *equations* $t = t'$, and *rewrite rules* $r : t \longrightarrow t'$, specify, respectively, the data and dynamic levels. That is, the data types and the overall state space of the system are axiomatized as the initial algebra $T_{\Sigma,E}$ associated to the equational specification (Σ, E), and the rewrite rules[1] in R are understood as local rules of concurrent change in the intended system, so that rewriting logic deduction with such rules describes exactly those concurrent computations possible in the system. This one-to-one correspondence between concurrent computation and logical deduction is of course of great practical usefulness, because it means that, given an adequate implementation of rewriting logic, a rewrite theory $\mathcal{R}$ becomes an *executable specification* of the concurrent system that it formalizes. Such a specification can then be used for rapid prototyping, symbolic simulation, formal analysis, and formal verification, to uncover design errors very early in the design process; and can also be used to generate correct code by means of adequate semantics-preserving transformations.

Since a rewrite theory $\mathcal{R}$ has an *initial model* $\mathcal{T}_{\mathcal{R}}$ [11] that mathematically characterizes the system as a category whose objects are the system's states and whose morphisms are the system's concurrent computations, formal system properties are in essence *inductive properties* of such a model. They can be expressed either in a natural extension of rewriting logic allowing arbitrary quantification, or in a temporal or modal logic providing a convenient notation for such properties. In this way, the three levels of data, dynamics and properties are naturally and seamlessly addressed.

A REFLECTIVE LOGICAL AND SEMANTIC FRAMEWORK

Being a *semantic framework* means that rewriting logic, instead of building in a particular model of concurrency or distribution such as, for example, process algebras, allows a wide range of such models—including indeed process algebras and languages such as LOTOS, but including also many other models such as Petri nets, Actors, the Unity language, dataflow, concurrent objects, concurrent graph rewriting, the π-calculus, and real-time systems [12, 13]. Being a *logical framework* means that rewriting logic can be used as a *metalogic* in which many other logics can be naturally defined and executed [10]. Natural and simple faithful such representations have been defined for a wide range of logics including classical, intuitionistic and linear logic, temporal and modal logics, inductive equational logic, and any logics with a sequent calculus presentation.

In particular, logical representations of this kind have been used to define and build *theorem proving tools* in a rewriting logic language such as Maude [3]. The way in which a model of computation, a programming language, a design or architectural language, or a logic are represented in the rewriting logic framework is by means of *representation maps* of the form

$$\Phi : \mathcal{L} \longrightarrow RWLogic.$$

Such maps give a rewriting logic semantics to the language $\mathcal{L}$ in question.

A key property of rewriting logic is that it is *reflective* [6, 2], in the sense that rewriting logic can represent its own metalevel as well as those of other logics. In fact, there is a finitely presented rewrite theory $\mathcal{U}$ that is *universal* in the sense that for any finitely presented rewrite theory $\mathcal{R}$ (including $\mathcal{U}$ itself) we have the following equivalence

$$\mathcal{R} \vdash t \rightarrow t' \;\Leftrightarrow\; \mathcal{U} \vdash \langle \overline{\mathcal{R}}, \overline{t} \rangle \rightarrow \langle \overline{\mathcal{R}}, \overline{t'} \rangle,$$

where $\overline{\mathcal{R}}$ and $\overline{t}$ are terms representing $\mathcal{R}$ and t as data elements of $\mathcal{U}$. Specifically, $\overline{\mathcal{R}}$ is a term of sort `Module`, and $\overline{t}$ is a term of sort `Term` in $\mathcal{U}$. Since $\mathcal{U}$ is representable in itself as a term of sort `Module`, we can achieve a "reflective tower" with an arbitrary number of levels of reflection, since we have

$$\mathcal{R} \vdash t \rightarrow t' \Leftrightarrow \mathcal{U} \vdash \langle \overline{\mathcal{R}}, \overline{t} \rangle \rightarrow \langle \overline{\mathcal{R}}, \overline{t'} \rangle \Leftrightarrow \mathcal{U} \vdash \langle \overline{\mathcal{U}}, \overline{\langle \overline{\mathcal{R}}, \overline{t} \rangle} \rangle \rightarrow \langle \overline{\mathcal{U}}, \overline{\langle \overline{\mathcal{R}}, \overline{t'} \rangle} \rangle \ldots$$

For efficiency reasons, the Maude language provides key features of the universal theory $\mathcal{U}$ in a built-in module called `META-LEVEL` that supports an arbitrary number of levels of reflection [4].

An important use of reflection is defining and executing *within* rewriting logic itself representation maps $\Phi : \mathcal{L} \longrightarrow RWLogic$. Such maps associate to each module M in the language $\mathcal{L}$ a rewrite theory $\Phi(M)$ in *RWLogic*. In semantic framework applications the language $\mathcal{L}$ can be a model of computation, a programming language, or an architectural description language; in logical framework applications $\mathcal{L}$ is typically a logic. In all cases, both $\mathcal{L}$ and Φ are metalevel entities that are, in principle, outside rewriting logic; however, thanks to reflection, they can be *internalized*—or as it is sometimes said *reified*—within rewriting logic. The idea is that in the universal theory $\mathcal{U}$ rewrite theories are already reified in an algebraic data type `Module`; we can then define another such data type $\texttt{Module}_{\mathcal{L}}$ reifying the modules in $\mathcal{L}$, and can reify Φ as a function

$$\overline{\Phi} : \texttt{Module}_{\mathcal{L}} \longrightarrow \texttt{Module}.$$

Since typically Φ is a total computable function, by general results of Bergstra and Tucker [1] it can always be specified by Church-Rosser and terminating rewrite rules, and therefore can be defined and executed in a reflective rewriting logic language such as Maude.

Yet another important application of reflection is that rewriting logic computations can be controlled at the metalevel by means of *internal strategies* defined with rewrite rules [6, 2]. This is of great practical use for the formal analysis of highly concurrent systems such a communication protocols, because, once they have been specified in rewriting logic, we can define strategies that explore all the possible concurrent computations from some given initial state or states, and that check whether some key property is satisfied or violated in the new states. In this way, formal specifications can be analyzed and corrected very early in the design process by what amounts to symbolic model checking and symbolic testing.

RECENT AND FUTURE DEVELOPMENTS

Rewriting logic is at present a young international research program with over a hundred papers by authors in Europe, the US and Japan, and three language implementations (see the survey [13] and the upcoming Second Workshop Proceedings [8]). We are still in an early phase in the task of applying rewriting logic to distributed computing and communication protocols. However, in addition to the work on foundations, on models of concurrent computation, and on languages, some recent research focusing specifically on this area seems quite promising. For example, the paper [7] shows how cryptographic communication protocols and attackers can be specified, executed, tested, and analyzed in Maude, and how adequate execution strategies can search and find security violations. Similarly, Appendix B of [4] discusses a reliable broadcast protocol Maude specification currently being jointly developed by researchers at the University of California, Santa Cruz, Stanford University, and SRI International (Denker, García-Luna, Meseguer, Smith, Ölveczky, and Talcott); using executable specifications very early in the design process has quickly exposed a number of mistakes and deadlocks in the initial design. In the same vein, several fault-tolerant communication protocols have also been specified in Maude, and *active network algorithms* written in the PLAN language are currently being specified and formally analyzed in Maude as part of a joint collaboration between researchers at the University of Pennsylvania and at SRI International (Gunter, Meseguer, and Wang). Other very promising developments include the work of Najm and Stefani using rewriting logic to specify computational models for open distributed systems [15], Talcott's work on open distributed components [18], Pita's and Martí-Oliet's work specifying a network management system in rewriting logic [17], Nakajima's work on the semantics of the calculus of mobile ambients and on specifying a Java/ORB implementation of a network management system [16], and Wirsing's and Knapp's work on a systematic translation from object-oriented design notations to Maude specifications [19].

Much more work remains ahead. More experience should be gained through examples and case studies; compositional aspects—so that protocols and other distributed systems and services can be specified, reasoned about, and built in a much more modular way—should be systematically studied; specification and proof techniques for system properties at the requirements level—in adequate temporal or modal logics above rewriting logic—also need to be much further advanced; and efficient implementations of distributed and mobile rewriting logic languages need to be developed. All this will require a lively international cooperation between individual researchers and entire research teams that is already taking place and whose prospects seem very encouraging.

Acknowledgments

I wish to thank the organizers of FORTE/PSTV'98 for kindly suggesting the possibility of giving this talk. I also wish to thank all the researchers at SRI, Stanford University, UC Santa Cruz, the University of Pennsylvania and other research institutions with whom I am collaborating on these ideas for their many contributions. Some have been already mentioned by name above or in the references, and many other contributions are surveyed in [13]; I would like to thank also Jonathan Millen at SRI, who has given us many good suggestions about cryptographic protocols. This work has been supported by DARPA through Rome Laboratories Contract F30602-97-C-0312, by Office of Naval Research Contracts N00014-95-C-0225 and N00014-96-C-0114, and by National Science Foundation Grant CCR-9633363.

Notes

1. Both equations and rewrite rules can be *conditional*, but for simplicity I discuss the unconditional case.

References

[1] J. Bergstra and J. Tucker. Characterization of computable data types by means of a finite equational specification method. In J. W. de Bakker and J. van Leeuwen, editors, *Automata, Languages and Programming, Seventh Colloquium*, pages 76–90. Springer-Verlag, 1980. LNCS, Volume 81.

[2] M. Clavel. Reflection in general logics and in rewriting logic, with applications to the Maude language. Ph.D. Thesis, University of Navarre, 1998.

[3] M. Clavel, F. Durán, S. Eker, and J. Meseguer. Building equational logic tools by reflection in rewriting logic. In *Proc. of the CafeOBJ Symposium '98, Numazu, Japan.* CafeOBJ Project, April 1998.

[4] M. Clavel, F. Durán, S. Eker, J. Meseguer, and P. Lincoln. An introduction to Maude (beta version). Manuscript, SRI International, March 1998.

[5] M. Clavel, S. Eker, P. Lincoln, and J. Meseguer. Principles of Maude. In J. Meseguer, editor, *Proc. First Intl. Workshop on Rewriting Logic and its Applications*, volume 4 of *Electronic Notes in Theoretical Computer Science.* Elsevier, 1996. `http://www1.elsevier.nl/mcs/tcs/pc/volume4.htm`.

[6] M. Clavel and J. Meseguer. Reflection and strategies in rewriting logic. In J. Meseguer, editor, *Proc. First Intl. Workshop on Rewriting Logic and its Applications*, volume 4 of *Electronic Notes in Theoretical Computer Science.* Elsevier, 1996. `http://www1.elsevier.nl/mcs/tcs/pc/volume4.htm`.

[7] G. Denker, J. Meseguer, and C. Talcott. Protocol Specification and Analysis in Maude. In N. Heintze and J. Wing, editors, *Proc. of Workshop on Formal Methods and Security Protocols, 25 June 1998, Indianapolis, Indiana*, 1998.

[8] C. Kirchner and H. Kirchner (eds.). *Proc. 2nd Intl. Workshop on Rewriting Logic and its Applications*, ENTCS, North Holland, 1998.

[9] P. Lincoln, N. Martí-Oliet, and J. Meseguer. Specification, transformation, and programming of concurrent systems in rewriting logic. In G. Blelloch, K. Chandy, and S. Jagannathan, editors, *Specification of Parallel Algorithms*, pages 309–339. DIMACS Series, Vol. 18, American Mathematical Society, 1994.

[10] N. Martí-Oliet and J. Meseguer. Rewriting logic as a logical and semantic framework. In J. Meseguer, editor, *Proc. First Intl. Workshop on Rewriting Logic and its Applications*, volume 4 of *Electronic Notes in Theoretical Computer Science*. Elsevier, 1996. `http://www1.elsevier.nl/mcs/tcs/pc/volume4.htm`.

[11] J. Meseguer. Conditional rewriting logic as a unified model of concurrency. *Theoretical Computer Science*, 96(1):73–155, 1992.

[12] J. Meseguer. Rewriting logic as a semantic framework for concurrency: a progress report. In *Proc. CONCUR'96, Pisa, August 1996*, pages 331–372. Springer LNCS 1119, 1996.

[13] J. Meseguer. Research directions in rewriting logic. In U. Berger and H. Schwichtenberg, editors, *Computational Logic, NATO Advanced Study Institute, Marktoberdorf, Germany, July 29 – August 6, 1997*. Springer-Verlag, 1998.

[14] J. Meseguer and C. Talcott. Using rewriting logic to interoperate architectural description languages (I and II). Lectures at the Santa Fe and Seattle DARPA-EDCS Workshops, March and July 1997. `http://www-formal.stanford.edu/clt/ArpaNsf/adl-interop.html`.

[15] E. Najm and J.-B. Stefani. Computational models for open distributed systems. In H. Bowman and J. Derrick, editors, *Formal Methods for Open Object-based Distributed Systems, Vol. 2*, pages 157–176. Chapman & Hall, 1997.

[16] S. Nakajima. Encoding mobility in CafeOBJ: an exercise of describing mobile code-based software architecture. In *Proc. of the CafeOBJ Symposium '98, Numazu, Japan*. CafeOBJ Project, April 1998.

[17] I. Pita and N. Martí-Oliet. A Maude specification of an object oriented database model for telecommunication networks. In J. Meseguer, editor, *Proc. First Intl. Workshop on Rewriting Logic and its Applications*, volume 4 of *Electronic Notes in Theoretical Computer Science*. Elsevier, 1996. `http://www1.elsevier.nl/mcs/tcs/pc/volume4.htm`.

[18] C. L. Talcott. An actor rewrite theory. In J. Meseguer, editor, *Proc. First Intl. Workshop on Rewriting Logic and its Applications*, volume 4 of

Electronic Notes in Theoretical Computer Science. Elsevier, 1996. http://www1.elsevier.nl/mcs/tcs/pc/volume4.htm.

[19] M. Wirsing and A. Knapp. A formal approach to object-oriented software engineering. In J. Meseguer, editor, *Proc. First Intl. Workshop on Rewriting Logic and its Applications*, volume 4 of *Electronic Notes in Theoretical Computer Science*. Elsevier, 1996. http://www1.elsevier.nl/mcs/tcs/pc/volume4.htm.

PART EIGHT

Test 2

EXPLOITING SYMMETRY IN PROTOCOL TESTING

Judi Romijn[1]
Jan Springintveld[1,2]

[1] CWI
Amsterdam
The Netherlands
E-mail: {judi, spring}@cwi.nl

[2] Computing Science Institute
University of Nijmegen
The Netherlands

Abstract:
Test generation and execution are often hampered by the large state spaces of the systems involved. In automata (or transition system) based test algorithms, taking advantage of symmetry in the behavior of specification and implementation may substantially reduce the amount of tests. We present a framework for describing and exploiting symmetries in black box test derivation methods based on finite state machines (FSMs). An algorithm is presented that, for a given symmetry relation on the traces of as FSM, computes a subautomaton that characterizes the FSM up to symmetry. This machinery is applied to Chow's classical W-method for test derivation. Finally, we focus on symmetries defined in terms of repeating patterns.
Keywords: conformance testing, automated test generation, state space reduction, symmetry.
Note: The research of the first author was carried out as part of the project "Specification, Testing and Verification of Software for Technical Applications" at the Stichting Mathematisch Centrum for Philips Research Laboratories under Contract RWC-061-PS-950006-ps. The research of the second author was partially supported by the Netherlands Organization for Scientific Research (NWO) under contract SION 612-33-006. His current affiliation is: Philips Research Laboratories Eindhoven, Prof. Holstlaan 4, 5656 AA, Eindhoven, The Netherlands.

1 INTRODUCTION

It has long been recognized that for the proper functioning of components in open and distributed systems, these components have to be thoroughly tested for interoperability and conformance to internationally agreed standards. For thorough and efficient testing, a high degree of automation of the test process is crucial. Unfortunately, methods for automated test generation and execution are still seriously hampered by the often very large state spaces of the implementations under test. One of the ways to deal with this problem is to exploit structural properties of the implementation under test that can be safely assumed to hold. In this paper we focus on taking advantage of *symmetry* that is present in the structure of systems. The symmetry, as it is defined here, may be found in any type of parameterized system: such parameters may for example range over IDs of components, ports, or the contents of messages.

We will work in the setting of test theory based on finite state machines (FSMs). Thus, we assume that the specification of an implementation under test is given as an FSM and the implementation itself is given as a black box. From the explicitly given specification automaton a collection of tests is derived that can be applied to the black box. Exploiting symmetry will allow us to restrict the test process to subautomata of specification and implementation that characterize these systems up to symmetry and will often be much smaller. The symmetry is defined in terms of an equivalence relation over the trace sets of specification and implementation. Some requirements are imposed to ensure that such a symmetry indeed allows to find the desired subautomata. We instantiate this general framework by focusing on symmetries defined in terms of repeating *patterns*. Some experiments with pattern-based symmetries, supported by a prototype tool implemented using the OPEN/CÆSAR tool set [14], have shown that substantial savings may be obtained in the number of tests.

Since we assume that the black box system has some symmetrical structure (cf. the *uniformity hypothesis* in [15, 6]), it is perhaps more appropriate to speak of *gray* box testing. For the specification FSM it will generally be possible to *verify* that a particular relation is a symmetry on the system, but for the black box implementation one has to *assume* that this is the case. The reliability of this assumption is the tester's responsibility. In this respect, one may think of exploiting symmetry as a structured way of test case selection [13, 4] for systems too large to be tested exhaustively, where at least some subautomata are tested thoroughly.

This paper is not the first to deal with symmetry in protocol testing. In [20], similar techniques have been developed for a test generation methodology based on labeled transition systems, success trees and canonical testers [3, 24]. Like in our case, symmetry is an equivalence relation between traces, and representatives of the equivalence classes are used for test generation. Since our approach and the approach in [20] start from different testing methodologies, it is not easy to compare them. In [20], the symmetry relation is defined through bijective renamings of action labels; our pattern-based definition generalizes this approach. On the other hand, since in our case a symmetry relation has to result in subautomata of specification and implementation that characterize these systems up to the symmetry, we have to impose certain requirements, which are absent in [20].

In [18], symmetrical structures in the product automaton of interoperating systems are studied. It is assumed that the systems have already been tested in isolation and attention is focused on pruning the product automaton by exploiting symmetry arising from the presence of identical peers. In the present paper, we abstract away from the internal composition of the system and focus on defining a *general* framework for describing and using symmetries on FSMs.

This paper is organized as follows. Section 2 contains some basic definitions concerning FSMs and their behavior. In Section 3, we introduce and define a general notion of trace based symmetry. We show how, given such a symmetry on the behavior of a system, a subautomaton of the system can be computed, a so-called *kernel*, that characterizes the behavior of the system up to symmetry. In Section 4 we apply the machinery to Chow's classical W-method for test derivation. In Section 5 we will instantiate the general framework by focusing on symmetries defined in terms of repeating patterns. Section 6 contains an extensive example, inspired by [23]. Finally, we discuss future work in Section 7. Due to space limitations, proofs have been left out.

2 FINITE STATE MACHINES

In this section, we will briefly present some terminology concerning finite state machines and their behavior, that we will need in the rest of this paper.

We let N denote the set of natural numbers. (Finite) Sequences are denoted by greek letters. Concatenation of sequences is denoted by juxtaposition; ϵ denotes the empty sequence and the sequence containing a single element a is simply denoted a. If σ is non-empty then $first(\sigma)$ returns the first element of σ and $last(\sigma)$ returns the last element of σ.

If V and W are sets of sequences and σ is a sequence, then $\sigma\, W = \{\sigma\, \tau \mid \tau \in W\}$ and $V\, W = \bigcup_{\sigma \in V} \sigma\, W$. For X a set of symbols, we define $X^0 = \{\epsilon\}$ and, for $i > 0$, $X^i = X^{i-1} \cup X\, X^{i-1}$. As usual, $X^* = \bigcup_{i \in \mathbb{N}} X^i$.

Definition 2.1. A *finite state machine (FSM)* is a structure $\mathcal{A} = (S, \Sigma, E, s^0)$ where

- S is a finite set of *states*
- Σ a finite set of *actions*
- $E \subseteq S \times \Sigma \times S$ is a finite set of *edges*
- $s^0 \in S$ is the *initial state*

We require that $\mathcal{A}$ is *deterministic*, i.e., for every pair of edges (s, a, s'), (s, a, s'') in $E_{\mathcal{A}}$, $s' = s''$.

We write $S_{\mathcal{A}}$, $\Sigma_{\mathcal{A}}$, etc., for the components of an FSM $\mathcal{A}$, but often omit subscripts when they are clear from the context. We let s, s' range over states, $a, a', b, c, \ldots$ over actions, and e, e' over edges. If $e = (s, a, s')$ then $act(e) = a$. We write $s \xrightarrow{a} s'$ if $(s, a, s') \in E$ and with $s \xrightarrow{a}$ we denote that $s \xrightarrow{a} s'$ for some state s'. A *subautomaton* of an FSM $\mathcal{A}$ is an FSM $\mathcal{B}$ such that $s^0_{\mathcal{B}} = s^0_{\mathcal{A}}$, $S_{\mathcal{B}} \subseteq S_{\mathcal{A}}$, $\Sigma_{\mathcal{B}} \subseteq \Sigma_{\mathcal{A}}$, and $E_{\mathcal{B}} \subseteq E_{\mathcal{A}}$.

An *execution fragment* of an FSM $\mathcal{A}$ is an alternating sequence $\gamma = s_0\, a_1\, s_1 \cdots a_n\, s_n$ of states and actions of $\mathcal{A}$, beginning and ending with a state, such that for all $i, 0 \leq i < n$, we have $s_i \xrightarrow{a_{i+1}} s_{i+1}$. If $s_0 = s_n$ then γ is a *loop*, if $n \neq 0$ then γ is a *non-empty* loop. An *execution* of $\mathcal{A}$ is an execution fragment that begins with the initial state of $\mathcal{A}$.

For $\gamma = s_0 a_1 s_1 \cdots a_n s_n$ an execution fragment of $\mathcal{A}$, *trace*(γ) is defined as the sequence $a_1 a_2 \cdots a_n$. If σ is a sequence of actions, then we write $s \xrightarrow{\sigma} s'$ if $\mathcal{A}$ has an execution fragment γ with *first*(γ) $= s$, *last*(γ) $= s'$, and *trace*(γ) $= \sigma$. If γ is a loop, then σ is a *looping trace*. We write $s \xrightarrow{\sigma}$ if there exists an s' such that $s \xrightarrow{\sigma} s'$, and write *traces*($s$) for the set $\{\sigma \in (\Sigma_{\mathcal{A}})^* \mid s \xrightarrow{\sigma}\}$. We write *traces*($\mathcal{A}$) for *traces*($s^0_{\mathcal{A}}$). □

3 SYMMETRY

In this section we introduce the notion of symmetry employed in this paper.

We want to be able to restrict the test process to subautomata of specification and implementation that characterize these systems up to symmetry. In papers on exploiting symmetry in model checking [2, 8, 10, 11, 12, 17], such subautomata are constructed for explicitly given FSMs by identifying and collapsing symmetrical *states*. We are concerned with black box testing, and, by definition, it is impossible to refer directly to the states of a black box. In traditional FSM based test theory, FSMs are assumed to be deterministic and hence a state of a black box is identified as the unique state of the black box that is reached after a certain trace of the system. Thus it seems natural to define symmetry as a relation over *traces*.

Our basic notion of symmetry on an FSM $\mathcal{A}$, then, is an equivalence relation on $(\Sigma_{\mathcal{A}})^*$, such that $\mathcal{A}$ is *closed* under the symmetry, i.e., if a sequence of actions is symmetrical to a trace of $\mathcal{A}$ then the sequence is a trace of $\mathcal{A}$ too.

The idea is to construct from the specification automaton an automaton such that its trace set is included in the trace set of the specification and contains a representative trace for each equivalence class of the symmetry relation on the traces of the specification. In order to be able to do this, we need to impose some requirements on the symmetry. For the specification we demand (1) that each equivalence class of the symmetry is represented by a unique trace, (2) that prefixes of a trace are represented by prefixes of the representing trace, and (3) that representative traces respect loops. The third requirement means that if a representative trace is a looping trace, then the trace with the looping part removed is also a representative trace. This requirement introduces some state-based information in the definition of symmetry.

These requirements enable us to construct a subautomaton of the specification, a so-called *kernel*, such that every trace of the specification is represented by a trace from the kernel. Of course, it will often be the case that the symmetry itself is preserved under prefixes and respects loops, so the requirements will come almost for free.

For the black box implementation, we will, w.r.t. symmetry, only demand that it is closed under symmetry. So if tests have established that the implementation displays certain behavior, then by assumption it will also display the symmetrical behavior. In Section 4, where the theory is applied to Mealy machines, we will in addition need a way to identify a subautomaton of the implementation that is being covered by the tests derived from the kernel of the specification.

Definition 3.1. A *symmetry* S on an FSM $\mathcal{A}$ is pair $\langle \simeq, ()^r \rangle$ where $\simeq$ is a binary equivalence relation on $(\Sigma_{\mathcal{A}})^*$, and $()^r : (\Sigma_{\mathcal{A}})^* \to (\Sigma_{\mathcal{A}})^*$ is a *representative function* for $\simeq$ such that:

1. $\mathcal{A}$ is closed under $\simeq$: If $\sigma \in traces(\mathcal{A})$ and $\sigma \simeq \tau$, then $\tau \in traces(\mathcal{A})$.
2. Only traces of the same length are related: If $\sigma \simeq \tau$, then $|\sigma| = |\tau|$.
3. $()^r$ satisfies:
   (a) $\sigma^r \simeq \sigma$
   (b) $\tau \simeq \sigma \Rightarrow \tau^r = \sigma^r$
   (c) $()^r$ is prefix closed on $\mathcal{A}$: If $\sigma\, a \in traces(\mathcal{A})$ and $(\sigma\, a)^r = \tau\, b$, then $\sigma^r = \tau$
   (d) $()^r$ is loop respecting on representative traces: If $(\sigma_1\, \sigma_2\, \sigma_3)^r = \sigma_1\, \sigma_2\, \sigma_3 \in traces(\mathcal{A})$ and σ_2 is a looping trace, then $(\sigma_1\, \sigma_3)^r = \sigma_1\, \sigma_3$.

The class of traces τ such that $\tau \simeq \sigma$ is denoted with $[\sigma]_S$, or, if S is clear from the context, $[\sigma]$. □

As mentioned above, we will demand that there exists a symmetry on the specification, while the implementation under test is required only to be closed under the symmetry. Note that $(\sigma^r)^r = \sigma^r$.

Definition 3.2. Let $S = \langle \simeq, ()^r \rangle$ be a symmetry on FSM $\mathcal{A}$. A *kernel* of $\mathcal{A}$ w.r.t. S is a subautomaton $\mathcal{K}$ of $\mathcal{A}$, such that for every $\sigma \in traces(\mathcal{A})$, $\sigma^r \in traces(\mathcal{K})$. □

In the remainder of this section, we fix an FSM $\mathcal{A}$ and a symmetry $S = \langle \simeq, ()^r \rangle$ on $\mathcal{A}$. Figure 1 presents an algorithm that constructs a kernel of $\mathcal{A}$ w.r.t. S. The algorithm basically explores the state space of $\mathcal{A}$, while keeping in mind the trace that leads to the currently visited state. As soon as such a trace contains a loop, the algorithm will not explore it any further.

In Figure 1, *enabled*$(s, \mathcal{A})$ denotes the set of actions a such that $E_{\mathcal{A}}$ contains an edge (s, a, s'), and for such an a, *eff*$(s, a, \mathcal{A})$ denotes s'. Furthermore, *repr*(σ, E) denotes the set F of actions such that $a \in F$ iff there exists an action $b \in E$ such that $\sigma^r\, a = (\sigma\, b)^r$. We will only call this function for σ such that $\sigma^r = \sigma$. By definition of $()^r$, for some action c, $(\sigma\, b)^r = \sigma^r\, c = \sigma\, c$. So, since $\mathcal{A}$ is deterministic and closed under $\simeq$, $F \subseteq E$ and if E is non-empty, F is non-empty. This justifies the function *choose*(F) which nondeterministically chooses an element from F.

Theorem 3.3. Let $\mathcal{K}$ = Kernel($\mathcal{A}$,S). If $\sigma \in traces(\mathcal{A})$, then $\sigma^r \in traces(\mathcal{K})$.

4 TEST DERIVATION FROM SYMMETRIC MEALY MACHINES

In this section we will apply the machinery developed in the previous sections to Mealy machines. There exists a wealth of test generation algorithms based on the Mealy machine model [1, 5, 7]. We will show how Chow's classical W-method [7] can be adapted to a setting with symmetry. The main idea is that test derivation is not based on the entire specification automaton, but only on a kernel of it. A technical detail here is that we do not require Mealy machines to be minimal (as already observed by [19] for the setting without symmetry).

Definition 4.1. A *Mealy machine* is a (deterministic) FSM $\mathcal{A}$ such that

$$\Sigma_{\mathcal{A}} = \{(i/o) \mid i \in I_{\mathcal{A}} \wedge o \in O_{\mathcal{A}}\}$$

```
function Kernel(𝒜, S): FSM;
  var 𝒦;
    procedure Build_It(s, σ, Seen);
    var a, b, s, s', E, F;
    begin
          if s ∉ Seen
          then E := enabled(s, 𝒜);
               F := ∅;
               while E ≠ ∅
               do a := choose(repr(σ, E));
                  s' := eff(s, a, 𝒜);
                  S_𝒦 := S_𝒦 ∪ {s'};
                  Σ_𝒦 := Σ_𝒦 ∪ {a};
                  E_𝒦 := E_𝒦 ∪ {(s, a, s')};
                  Build_It(s', σ a, Seen ∪ {s});
                  F := F ∪ {a};
                  E := E \ {a};
                  for each b ∈ E . σ a ≃ σ b
                  do E := E \ {b};
                  od;
               od;
          fi;
    end;

  begin
        s⁰_𝒦 := s⁰_𝒜;
        S_𝒦 := {s⁰_𝒜};
        Σ_𝒦 := ∅;
        E_𝒦 := ∅;
        Build_It(s⁰_𝒜, ϵ, ∅);
        return 𝒦;
  end.
```

Figure 1 The algorithm Kernel

where $I_{\mathcal{A}}$ and $O_{\mathcal{A}}$ are two finite and disjoint sets of *inputs* and *outputs*, respectively. We require that $\mathcal{A}$ is *input enabled* and *input deterministic*, i.e., for every state $s \in S_{\mathcal{A}}$ and input $i \in I_{\mathcal{A}}$, there exists precisely one output $o \in O_{\mathcal{A}}$ such that $s \xrightarrow{(i/o)}$.

Input sequences of $\mathcal{A}$ are elements of $(I_{\mathcal{A}})^*$. For ξ an input sequence of $\mathcal{A}$ and $s, s' \in S_{\mathcal{A}}$, we write $s \overset{\xi}{\Longrightarrow}_{\mathcal{A}} s'$ if there exists a trace σ such that $s \xrightarrow{\sigma}_{\mathcal{A}} s'$ and ξ is the result of projecting σ onto $I_{\mathcal{A}}$. In this case we write $outcome_{\mathcal{A}}(\xi, s) = \sigma$; the execution fragment γ with $first(\gamma) = s$ and $trace(\gamma) = \sigma$ is denoted by $exec_{\mathcal{A}}(s, \xi)$. A *distinguishing sequence* for two states s, s' of $\mathcal{A}$ is an input sequence ξ such that $outcome_{\mathcal{A}}(\xi, s) \neq outcome_{\mathcal{A}}(\xi, s')$. We say that ξ distinguishes s from s'. □

In Chow's paper, conformance is defined as the existence of an isomorphism between specification and implementation. Since we do not assume automata to be minimal, we will show the existence of a *bisimulation* between specification and implementation. Bisimilarity is a well-known process equivalence from concurrency theory [21]. For minimal automata, bisimilarity is equivalent to isomorphism, while for deterministic automata, bisimilarity is equivalent to equality of trace sets.

Definition 4.2. Let $\mathcal{A}$ and $\mathcal{B}$ be FSMs. A relation $R \subseteq S_{\mathcal{A}} \times S_{\mathcal{B}}$ is a *bisimulation on $\mathcal{A}$ and $\mathcal{B}$* iff

- $R(s_1, s_2)$ and $s_1 \xrightarrow{a}_{\mathcal{A}} s_1'$ implies that there is a $s_2' \in S_{\mathcal{A}}$ such that $s_2 \xrightarrow{a}_{\mathcal{B}} s_2'$ and $R(s_1', s_2')$,
- $R(s_1, s_2)$ and $s_2 \xrightarrow{a}_{\mathcal{B}} s_2'$ implies that there is a $s_1' \in S_{\mathcal{A}}$ such that $s_1 \xrightarrow{a}_{\mathcal{A}} s_1'$ and $R(s_1', s_2')$.

$\mathcal{A}$ and $\mathcal{B}$ are *bisimilar*, notation $\mathcal{A} \leftrightarrows \mathcal{B}$, if there exists a bisimulation R on $\mathcal{A}$ and $\mathcal{B}$ such that $R(s^0_{\mathcal{A}}, s^0_{\mathcal{B}})$. We call two states $s_1, s_2 \in S_{\mathcal{A}}$ bisimilar, notation $s_1 \leftrightarrows_{\mathcal{A}} s_2$, if there exists a bisimulation R on $\mathcal{A}$ (and $\mathcal{A}$) such that $R(s_1, s_2)$. The relation $\leftrightarrows_{\mathcal{A}}$ is an equivalence relation on $S_{\mathcal{A}}$; a *bisimulation class* of $\mathcal{A}$ is an equivalence class of $S_{\mathcal{A}}$ under $\leftrightarrows_{\mathcal{A}}$. □

The main ingredient of Chow's test suite is a *characterizing set* for the specification, i.e., a set of input sequences that distinguish inequivalent states by inducing different output behavior from them. In our case, two states are inequivalent if they are non-bisimilar, i.e. have different trace sets. In the presence of symmetry we will need a characterizing set not for the entire specification automaton but only for a kernel of it. However, a kernel need not be input enabled, so two inequivalent states need not have a common input sequence that distinguishes between them. Instead we will use a characterizing set that contains for every two states of the kernel that are inequivalent in the original specification automaton, an input sequence that these states have in common in the specification and distinguishes between them.

Constructing distinguishing sequences in the specification automaton rather than in the smaller kernel is of course potentially as expensive as in the setting without symmetry, and may lead to large sequences. However, if the number of states of the kernel is small we will not need many of them, so test *execution* itself may still benefit considerably from the restriction to the kernel. Moreover, we expect that in most cases distinguishing sequences can be found in a well marked out subautomaton of the specification that envelopes the kernel.

Definition 4.3. A *test pair* for a Mealy machine $\mathcal{A}$ is a pair $\langle \mathcal{K}, W \rangle$ where $\mathcal{K}$ is a kernel of $\mathcal{A}$ and W is a set of input sequences of $\mathcal{A}$ such that the following holds. For every pair of states $s, s' \in S_{\mathcal{K}}$ such that $s \not\simeq_{\mathcal{A}} s'$, W contains an input sequence ξ such that $outcome_{\mathcal{A}}(\xi, s) \neq outcome_{\mathcal{A}}(\xi, s')$. □

The proof that Chow's test suite has complete fault coverage crucially relies on the assumption that (an upper bound to) the number of states of the black box implementation is correctly estimated. Since specification and implementation are also assumed to have the same input sets and to be input enabled, this is equivalent to a correct estimate of the number of states of the implementation that can be reached from the start state by an input sequence from the specification. Similarly, we will assume that we can give an upper bound to the number of states of the black box that are reachable from the start state by an input sequence from the kernel of the specification. We call the subautomaton of the implementation generated by these states the *image* of the kernel.

Technically, the assumption on the state space of the black box is used in [7] to bound the maximum length of distinguishing sequences needed for a characterizing set for the implementation. Since, like the kernel, the image of the kernel need not be input enabled, it may be that distinguishing sequences for states of the image cannot be constructed in the image itself. Thus, it is not sufficient to estimate the number of states of the image, but we must in addition estimate the number of steps distinguishing sequences may have to take outside the image of the kernel.

Definition 4.4. Let $\mathcal{A}$ and $\mathcal{B}$ be Mealy machines with the same input set and let $\mathcal{K}$ be a kernel of $\mathcal{A}$. A *$\mathcal{K}$-sequence* is an input sequence ξ such that $s^0_{\mathcal{K}} \stackrel{\xi}{\Longrightarrow}_{\mathcal{K}}$. A state s of $\mathcal{B}$ is called *$\mathcal{K}$-related* if there exists a $\mathcal{K}$-sequence ξ such that $s^0_{\mathcal{B}} \stackrel{\xi}{\Longrightarrow}_{\mathcal{B}} s$.

We define $im_{\mathcal{K}}(\mathcal{B})$ as the subautomaton (S, Σ, E, s^0) of $\mathcal{B}$ defined by:

- $S = \{s \in S_{\mathcal{B}} \mid s \text{ is } \mathcal{K}\text{-related}\}$
- $E = \{(s, a, s') \in E_{\mathcal{B}} \mid s, s' \in S\}$
- $\Sigma = \{a \in \Sigma_{\mathcal{B}} \mid \exists s, s'. \, (s, a, s') \in E\}$
- $s^0 = s^0_{\mathcal{B}}$

□

Definition 4.5. A subautomaton $\mathcal{B}$ of a Mealy machine $\mathcal{A}$ is *(m_1, m_2)-self-contained in $\mathcal{A}$* when the number of bisimulation classes Q of $\mathcal{A}$ such that $Q \cap S_{\mathcal{B}} \neq \emptyset$ is m_1, and for every pair of states s, s' of $\mathcal{B}$ such that $s \not\simeq_{\mathcal{A}} s'$, there exist input sequences ξ_1, ξ_2 of $\mathcal{A}$ of length at most m_1, m_2, respectively, such that $s \stackrel{\xi_1}{\Longrightarrow}_{\mathcal{B}}$, $s' \stackrel{\xi_1}{\Longrightarrow}_{\mathcal{B}}$, and $outcome_{\mathcal{A}}(\xi_1\xi_2, s) \neq outcome_{\mathcal{A}}(\xi_1\xi_2, s')$. □

The next lemma is a generalization of [7]'s Lemma 0.

Lemma 4.6. Let $\mathcal{A}$ and $\mathcal{B}$ be Mealy machines with the same input set I and let $\langle \mathcal{K}, W \rangle$ be a test pair for $\mathcal{A}$. Let $\mathcal{C} = im_{\mathcal{K}}(\mathcal{B})$. Suppose that:

1. $\mathcal{C}$ is (m_1, m_2)-self-contained in $\mathcal{B}$.
2. W distinguishes between n bisimulation classes Q of $\mathcal{B}$ such that $Q \cap S_{\mathcal{C}} \neq \emptyset$.

Then for every two states s and s' of $\mathcal{C}$ such that $s \not\simeq_{\mathcal{B}} s'$, $I^{m_1-n}\, I^{m_2}\, W$ distinguishes s from s'.

This result allows us to construct a characterizing set $Z = I^{m_1-n}\, I^{m_2}\, W$ for the image of the kernel in the implementation. The test suite resulting from the W-method consists of all concatenations of sequences from a *transition cover* P for the specification with sequences from Z.

Definition 4.7. A *transition cover* for the kernel of a Mealy machine $\mathcal{A}$ is a finite collection P of input sequences of $\mathcal{A}$, such that $\epsilon \in P$ and, for all transitions $s \xrightarrow{(i/o)} s'$ of $\mathcal{K}$, P contains input sequences ξ and $\xi\, i$ such that $s^0_{\mathcal{K}} \stackrel{\xi}{\Longrightarrow}_{\mathcal{K}} s$. □

Now follows the main theorem.

Theorem 4.8. Let *Spec* and *Impl* be Mealy machines with the same input set I, and assume $\langle \simeq, ()^r \rangle$ is a symmetry on *Spec* such that *Impl* is closed under $\simeq$. Let $\langle \mathcal{K}, W \rangle$ be a test pair for *Spec*. Write $\mathcal{C} = im_{\mathcal{K}}(Impl)$. Suppose

1. The number of bisimulation classes Q of *Spec* such that $Q \cap S_{\mathcal{K}} \neq \emptyset$ is n.
2. $\mathcal{C}$ is (m_1,m_2)-self-contained in *Impl*.
3. For all $\sigma \in P$ and $\tau \in I^{m_1-n}\, I^{m_2}\, W$

$$outcome_{Spec}(\sigma\, \tau, s^0_{Spec}) = outcome_{Impl}(\sigma\, \tau, s^0_{Impl})$$

Then $Spec \stackrel{\leftrightarrow}{-} Impl$.

5 PATTERNS

In this section we describe symmetries based on *patterns*. A pattern is an FSM, together with a set of permutations of its set of actions, so-called *transformations*. The FSM is a *template* for the behavior of a system, while the transformations indicate how this template may be filled out to obtain symmetric variants that cover the full behavior of the system.

In [18] an interesting example automaton is given for a symmetric protocol, representing the behavior of two peer hosts that may engage in the ATM call setup procedure. This behavior is completely symmetric in the identity of the peers. An FSM representation is given in Figure 2. Here, !<action>(i) means output of the ATM service to caller i, and ?<action>(i) means input from caller i to the ATM service. So, action ?set_up(1) denotes the request from caller 1 to the ATM service, to set up a call to caller 2. A set_up request is followed by an acknowledgement in the form of call_proc if the service can be performed. Then, action conn indicates that the called side is ready for the connection, which is acknowledged by conn_ack. A caller may skip sending call_proc, if it can already send conn instead (transition from state 3 to 5 and from 10 to 12 in Figure 2).

Here, a typical template is the subautomaton representing the call set up as initiated by a single initiator (e.g. caller 1), and the transformation will be the permutation of actions generated by swapping the roles of initiator and responder. Such a template is displayed in Figure 3.

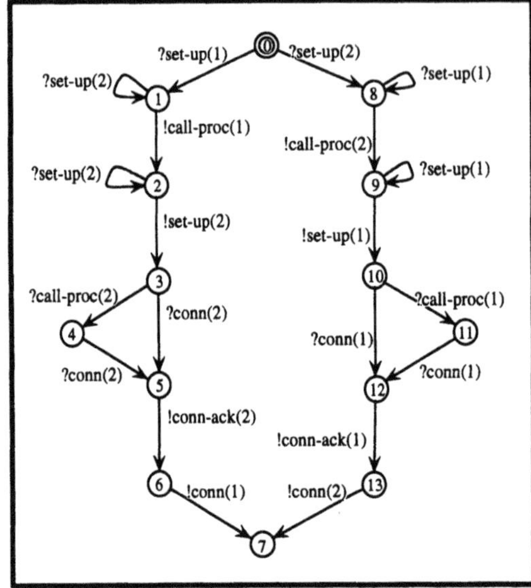

Figure 2 The ATM call setup protocol

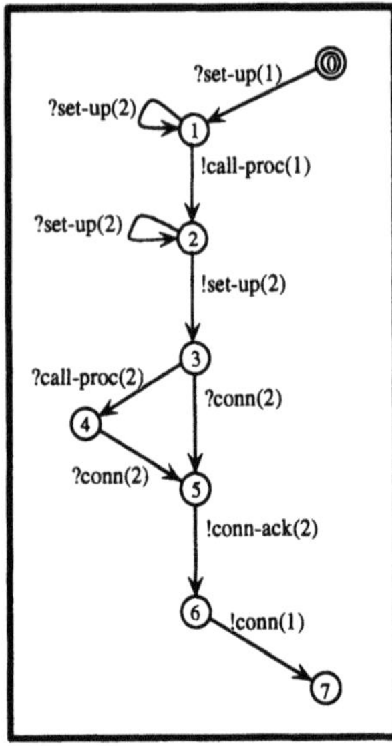

Figure 3 A template

In the example of Section 6, featuring a *chatbox* that supports multiple conversations between callers, the template will be the chatting between two callers, while the transformations will shuffle the identity of the callers.

The template FSM may be arbitrarily complex; intuitively, increasing complexity indicates a stronger symmetry assumption on the black box implementation.

To define pattern based symmetries, we need some terminology for partial functions and multisets. If $f : A \rightarrow B$ is a partial function and $a \in A$, then $f(a) \downarrow$ means that $f(a)$ is defined, while $f(a) \uparrow$ means that $f(a)$ is not defined. A *multiset* over A is a set of the form $\{(a_1, n_1), \ldots, (a_k, n_k)\}$ where, for $1 \leq i \leq k$, a_i is an element of A and $n_i \in \mathbb{N}$ denotes its *multiplicity*. We use $[f(x) | \operatorname{cond}(x)]$ as a shorthand for the multiset over A that is created by adding, for every single $x \in A$, a copy of $f(x)$ if the condition $\operatorname{cond}(x)$ holds.

Definition 5.1 *(Patterns).* A *pattern* $\mathcal{P}$ is a pair $\langle \mathcal{T}, \Pi \rangle$ where $\mathcal{T}$ is an FSM, called the *template* of $\mathcal{P}$, and Π is a finite set of permutations of $\Sigma_{\mathcal{T}}$, which we call *transformations*.

Given a sequence $\langle f_1, \ldots, f_n \rangle$ of (partial) functions $f_1, \ldots, f_n : \Pi \rightarrow E_{\mathcal{T}}$, we denote with $exec(\langle f_1, \ldots, f_n \rangle, \pi)$ the sequence of edges obtained by taking for each function f_i, $0 \leq i \leq n$, the edge e (if any) such that $f_i(\pi) = e$. □

In the remainder of this section, we fix an FSM $\mathcal{A}$ and a pattern $\mathcal{P} = \langle \mathcal{T}, \Pi \rangle$.

Below we will define how $\mathcal{P}$ defines a symmetry of the behavior of an FSM $\mathcal{A}$. Each transformation $\pi \in \Pi$ gives rise to a copy $\pi(\mathcal{T})$ of $\mathcal{T}$ obtained by renaming the actions according to π. Each such copy is a particular instantiation of the template. Intuitively, the trace set of $\mathcal{A}$ is included in the trace set of the parallel composition of the copies

$\pi(\mathcal{T})$, indexed by elements of Π, with enforced synchronization over all actions of $\mathcal{A}$. Using that traces of $\mathcal{A}$ are traces of the parallel composition, we will define the symmetry relation on traces in terms of the behavior of the copies and permutations of the index set Π.

The following definition rephrases the inclusion requirement above in such a way that the relation $\simeq$ and a representative function for it can be formulated succinctly. In particular, if $\mathcal{A}$ *is* the parallel composition of the copies of $\mathcal{T}$, the requirement in this definition apply.

Definition 5.2. Let $\sigma = a_1 \cdots a_n$ be an element of $(\Sigma_{\mathcal{A}})^*$. A *covering* of σ by $\mathcal{P}$ is a sequence $\langle f_1, \ldots, f_n \rangle$ of partial functions $f_i : \Pi \to E_{\mathcal{T}}$ with *non-empty* domain such that for every $\pi \in \Pi$ and $1 \leq i \leq n$:

1. If $f_i(\pi) = e$ then $a_i = \pi(act(e))$.
2. The sequence $exec(\langle f_1, \ldots, f_i \rangle, \pi)$ induces an execution γ_i of $\mathcal{T}$.
3. If the sequence $trace(\gamma_{i-1})\, a_i$ is a trace of $\pi(\mathcal{T})$ then $f_i(\pi) \downarrow$.

We say that $\mathcal{P}$ covers σ if there exists a covering of σ by $\mathcal{P}$.

We call $\mathcal{P}$ *loop preserving* when the following holds. Suppose $\sigma_1 \sigma_2 \in traces(\mathcal{A})$ is covered by $\langle f_1, \ldots, f_n, g_1, \ldots, g_m \rangle$ and σ_2 is a looping trace. Then for all $\pi \in \Pi$,

$$last(exec(\langle f_1, \ldots, f_n \rangle, \pi)) = last(exec(\langle f_1, \ldots, f_n, g_1, \ldots, g_m \rangle, \pi))$$

□

Intuitively, these requirements mean the following. The 'non-empty domain' requirement for the partial functions f_i ensures the inclusion of the trace set of $\mathcal{A}$ in the trace set of the parallel composition of copies of $\mathcal{T}$. Requirements 1 and 2 express that a covering should not contain 'junk'. Requirement 3 corresponds to the enforced synchronization of actions of the parallel composition.
Two traces σ and τ of the same length n that are covered by $\mathcal{P}$, are *variants* of each other if at each position i, $1 \leq i \leq n$, of σ and τ the following holds. The listings for σ and τ, respectively, of the copies $\pi(\mathcal{T})$ that participate in the action at position i, the states these copies are in before participating, and the edge they follow by participating, are equal up to a permutation of Π. Then, two traces of the same length are *symmetric* iff they are either both not covered by $\mathcal{P}$ or are covered by coverings that are variants of each other.

Definition 5.3. Let σ and τ be elements of $(\Sigma_{\mathcal{A}})^n$, which $\mathcal{P}$ covers by $cov_1 = \langle f_1, \ldots, f_n \rangle$ and $cov_2 = \langle g_1, \ldots, g_n \rangle$, respectively. Then cov_1 and cov_2 are said to be *variants* of each other if for every $1 \leq i \leq n$, $[f_i(\pi) \mid \pi \in \Pi] = [g_i(\pi) \mid \pi \in \Pi]$. We define the binary relation $\simeq_{\mathcal{P}}$ on $(\Sigma_{\mathcal{A}})^*$ by:

$$\begin{array}{lll} \sigma \simeq_{\mathcal{P}} \tau & \Leftrightarrow & \wedge \ |\sigma| = |\tau| \\ & & \wedge \ \vee \text{ both } \sigma \text{ and } \tau \text{ are not covered by } \mathcal{P} \\ & & \quad\ \vee\ \mathcal{P} \text{ covers } \sigma \text{ and } \tau \text{ by variant coverings} \end{array}$$

It is easy to check that $\simeq_{\mathcal{P}}$ is an equivalence relation. As in Section 3, we will write $[\sigma]$ for the equivalence class of σ and $\simeq$ instead of $\simeq_{\mathcal{P}}$. □

An important special case is the following. Suppose $\mathcal{A}$ consists of the parallel composition of components C_i, indexed by elements of a set I, that are identical up to their ID (which occur as parameters in the actions). Let σ and τ be traces of $\mathcal{A}$. If there exists a permutation ρ of the index set I such that for all indices $i \in I$, σ induces (up to renaming of IDs in actions) the same execution of C_i as τ induces in $C_{\rho(i)}$, then σ and τ are symmetric.

Now we define a representative function for $\simeq$. We assume given a total, irreflexive ordering $<$ on $\Sigma_{\mathcal{A}}$. Such an ordering of course always exists, but the choice for $<$ may greatly influence the size of the kernel constructed for a symmetry based on $\mathcal{P}$.

Definition 5.4. Let $<$ be a total, irreflexive ordering on $\Sigma_{\mathcal{A}}$. This ordering induces a reflexive, transitive ordering $\leq$ on traces of the same length in the following way:

$$a\,\sigma \leq b\,\tau \Leftrightarrow a < b \vee (a = b \wedge \sigma \leq \tau)$$

We define σ^r as the least element of $[\sigma]$ under $\leq$. □

Given the definition of $\simeq$, it is reasonable to demand that every trace of $\mathcal{A}$ is covered by $\mathcal{P}$. We will also need the following closure property. We call a binary relation R on $(\Sigma_{\mathcal{A}})^*$ *persistent on* $\mathcal{A}$ when $R(\sigma, \tau)$ and $\sigma\, a \in \mathit{traces}(\mathcal{A})$ implies that there exists an action b such that $R(\sigma\, a, \tau\, b)$.

The next result allows us to use the pattern-approach for computing a kernel. In our example of the ATM switch, we have computed the kernel from the FSM in Figure 2, using the symmetry induced by the template in Figure 3 and an ordering $<$ that obeys the relation ?set_up(1) $<$?set_up(2). Not surprisingly, the resulting kernel is identical to the template.

Theorem 5.5. Suppose $\mathcal{P}$ is a loop preserving pattern on $\mathcal{A}$ and let $<$ be a total, irreflexive ordering on $\Sigma_{\mathcal{A}}$. Let $()^r$ be as in Definition 5.4. Suppose every trace of $\mathcal{A}$ is covered by $\mathcal{P}$, $\mathcal{A}$ is closed under $\simeq$, and $\simeq$ is persistent on $\mathcal{A}$. Then $\langle \simeq, ()^r \rangle$ is a symmetry on $\mathcal{A}$.

6 EXAMPLE: TESTING A CHATBOX

In this section we report on some initial experiments in the application of symmetry to the testing of a *chatbox*. This example was inspired by the conference protocol presented in [23]. Part of the test generation trajectory was implemented: we used the tool environment OPEN/CÆSAR[14] for prototyping the algorithm Kernel from Section 3. We work with a pattern based symmetry (Section 5) and apply the test derivation method from Section 4.

A chatbox offers the possibility to talk with users connected to the chatbox. After one joins (connects to) the chatbox, one can talk with all other connected users, until one leaves (disconnects). One can only join if not already present, and one can leave at any time. For simplicity, we assume that every user can at each instance talk with at most one user. Moreover, we demand that a user waits for a reply before talking again (unless one of the partners leaves). Finally, we abstract from the contents of the messages, and consider only one message. The service primitives provided by the chatbox are thus the following; Join, Leave, DReq, and DInd, with the obvious meaning (see Figure 4).

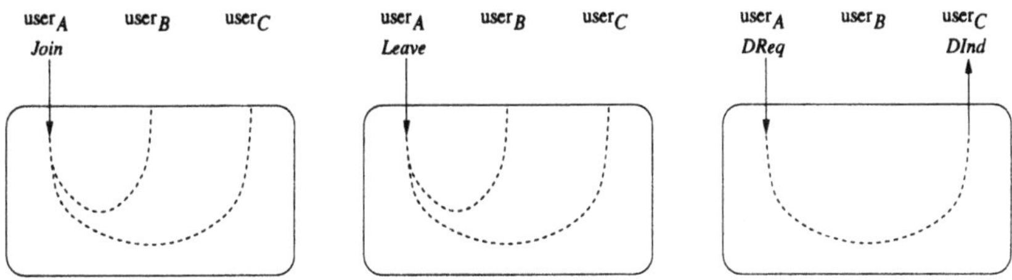

Figure 4 The chatbox protocol service

For lack of space, we do not give the full formal specification of the chatbox or its template.

What we test for is the service of the chatbox as a whole, such as it may be offered by a vendor, rather than components of its implementation, which we (the "customers") are not allowed to, or have no desire to, inspect.

The symmetry inherent in the protocol is immediate: pairs of talking users can be replaced by other pairs of talking users, as long as this is done systematically according to Definitions 5.2 and 5.3. As an example, the trace in which user 1 joins, leaves and joins again, is symmetric to the trace in which user 1 joins and leaves, after which user 2 joins. The essence is that after user 1 has joined and left, this user is at the same point as all the other users not present, so all new join actions are symmetric. Note that this symmetry is more general than a symmetry induced solely by a permutation of actions or IDs of users. Thus the template $\mathcal{T}$ used for the symmetry basically consists of the conversation between two users, including joining and leaving, while the transformations π in the set Π shuffle the identity of users. We feel that it is a reasonable assumption that the black implementation offering the service indeed is symmetric in this sense.

We have applied the machinery to chatboxes with up to 4 users. We also considered a (much simpler) version of the protocol without joining and leaving. Still, a chatbox with only 4 users and no joining or leaving already has 4096 reachable states.

We start the test generation by computing a kernel for these specifications. Our prototype is able to find a significantly smaller Mealy machine as a kernel for each of the models, provided that it is given a suitable ordering $<$ (see Definition 5.4) on the actions symbols for its representative function. The kernels constructed consist of interleavings of transformations of the pattern, constrained by the symmetry and the ordering $<$.

For instance, in a chatbox with 3 users and no joining and leaving, we take the ordering $<$ defined as follows. "Sending a message from i_1 to j_1" $<$ "sending a message from i_2 to j_2" if $(i_1 < i_2)$ or if $(i_1 = i_2$ and $j_1 < j_2)$, and "sending a reply from i_1 to j_1" $<$ "sending a reply from i_2 to j_2" if $(i_1 > i_2)$ or if $(i_1 = i_2$ and $j_1 > j_2)$.

Using this ordering, the kernel only contains those traces in which first messages from user 1 are sent, then messages from user 2 and finally messages from user 3, while the sending of replies is handled in the reverse order. Each trace with different order of sending messages can then be computed from a trace of this kernel, which is exactly what Theorem 3.3 states.

Our experiments so far have revealed that for chatboxes with joining and leaving, the kernel is approximately half the size. When considering chatboxes without joining and leaving, the size of the kernel is reduced much more. Of course, the algorithm should be run on more and larger examples to get definite answers about possible size reduction.

Given the computed kernels, we can construct test pairs by determining for each kernel a set of input sequences W that constitutes a *characterizing set* for the kernel (as defined in Definition 4.3). Although this part has not yet been automated, it is easily seen by a generic argument that for every pair of inequivalent (non-bisimilar) states very short distinguishing sequences exist. It is easy to devise a transition cover for a kernel, the size of which is proportional to the size of the kernel.

As shown in Theorem 4.8, the size of the test suite to be generated will depend on the magnitude of two numbers m_1 and m_2, indicating the search space for distinguishing sequences for the image of the kernel in the implementation. This boils down to the following questions: (1) What is the size of the image part of the implementation for this kernel? (2) What is the size of a minimal distinguishing experience for each two inequivalent (non-bisimilar) states in the image part of the implementation? (3) How many steps does a distinguishing sequence perform outside the image of the kernel? These questions are variations of the classical state space questions for black box testing. For practical reasons, these numbers are usually taken to be not much larger than the corresponding numbers for the specification.

The algorithm Kernel (see Figure 1) was implemented using the OPEN/CÆSAR [14] tool set. An interesting detail here is that the algorithm uses two finite state machines: one for the specification that is reduced to a kernel, and one for the template of the symmetry, which is used to determine (as an oracle) whether two traces are symmetric. To enable this, OPEN/CÆSAR interface had to be generalized somewhat so that it is now able to explore several labeled transition systems at the same time. We have the experience that OPEN/CÆSAR is suitable for prototyping exploration algorithms such as Kernel.

7 FUTURE WORK

We have introduced a general, FSM based, framework for exploiting symmetry in specifications and implementations in order to reduce the amount of tests needed to establish correctness. The feasibility of this approach has been shown in a few experiments.

However, a number of open issues remain. We see the following steps as possible, necessary and feasible. On the theoretical side we would like to (1) construct algorithms for computing and checking symmetries, and (2) determine conditions that are on the one hand sufficient to guarantee symmetry, and on the other hand enable significant optimizations of the algorithms. On the practical side we would like to (1) generate and execute tests for real-life implementations, and (2) continue prototyping for the whole test generation trajectory.

Acknowledgments

We thank Frits Vaandrager for suggesting the transfer of model checking techniques to test theory, and Radu Mateescu and Hubert Garavel for their invaluable assistance (including adding functionality!) with the OPEN/CÆSAR tool set. We also thank Jan Tretmans and the anonymous referees for their comments on this paper.

References

[1] A.V. Aho, A.T. Dahbura, D. Lee, and M.U. Uyar. An optimization technique for protocol conformance test generation based on UIO sequences and Rural Chinese Postman Tours. *IEEE Transactions on Communications*, 39(11):1604–1615, 1991.

[2] K. Ajami, S. Haddad and J-M. Ilie. Exploiting symmetry in linear time temporal logic model checking: One step beyond. In Steffen [22], pages 52–67.

[3] E. Brinksma. A theory for the derivation of tests. In S. Aggrawal and K. Sabani, editors, *Protocol Specification Testing and Verification*, Volume VIII, pages 63–74. North-Holland, 1988.

[4] E. Brinksma, J. Tretmans and L. Verhaard. A framework for test selection. In B. Jonsson, J. Parrow and B. Pehrson, editors, *Protocol Specification Testing and Verification*, Volume XI, pages 233-248. North-Holland, 1991.

[5] W.Y.L. Chan, S.T. Vuong, and M.R. Ito. An improved protocol test generation procedure based on UIOs. In *Proceedings of the ACM Symposium on Communication Architectures and Protocols*, pages 283–294, 1989.

[6] O. Charles, and R. Groz. Basing test coverage on a formalization of test hypotheses. In M. Kim, S. Kang, and K. Hong, editors, *Testing of Communicating Systems*, Volume 10, pages 109–124. Chapman & Hall, 1997.

[7] T.S. Chow. Testing software design modeled by finite-state machines. *IEEE Transactions on Software Engineering*, 4(3):178–187, 1978.

[8] E.M. Clarke, T. Filkorn, and S. Jha. Exploiting symmetry in temporal logic model checking. In Courcoubetis [9], pages 450–462.

[9] C. Courcoubetis, editor. *Proceedings 5th International Conference on Computer Aided Verification (CAV '93)*. Lecture Notes in Computer Science 697. Springer-Verlag, 1993.

[10] E.A. Emerson, S. Jha and D. Peled. Combining partial order and symmetry reductions. In E. Brinksma, editor, *Tools and Algorithms for the Construction and Analysis of Systems (TACAS '97)*, pages 19–34. Lecture Notes in Computer Science 1217. Springer-Verlag, 1997.

[11] E.A. Emerson and A.P. Sistla. Symmetry and model checking. In Courcoubetis [9], pages 463–478.

[12] E.A. Emerson and A.P. Sistla. Utilizing symmetry when model-checking under fairness assumptions: an automata-theoretic approach. *ACM Transactions on Programming Languages and Systems*, 19(4):617–638, 1997.

[13] S. Fujiwara, G. v. Bochmann, F. Khendek, M. Amalou and A. Ghedamsi. Test selection based on finite state models. *IEEE Transactions on Software Engineering*, 16(6):591–603, 1991.

[14] H. Garavel. OPEN/CÆSAR: An open software architecture for verification, simulation, and testing. In Steffen [22], pages 68–84. For more information on the tool set, see `http://www.inrialpes.fr/vasy/pub/cadp.html`.

[15] M.-C. Gaudel. Testing can be formal, too. In P.D. Mosses, M. Nielsen, and M.I. Schwartzbach, editors, *TAPSOFT'95: Theory and Practice of Software Development*, pages 82–96. Lecture Notes in Computer Science 915. Springer-Verlag, 1995.

[16] O. Grumberg, editor. *Proceedings* 9th *International Conference on Computer Aided Verification (CAV '97)*. Lecture Notes in Computer Science 1254. Springer-Verlag, 1997.

[17] V. Gyuris and A.P. Sistla. On-the-fly model checking under fairness that exploits symmetry. In Grumberg [16], pages 232–243.

[18] S. Kang and M. Kim. Interoperability test suite derivation for symmetric communication protocols. In T. Mizuno, N. Shiratori, T. Higashino, and A. Togashi, editors, *Formal Description Techniques and Protocol Specification, Testing and Verification (FORTE X/ PSTV XVII '97)*, pages 57–72. Chapman & Hall, 1997.

[19] A. Petrenko, T. Higashino, and T. Kaji. Handling redundant and additional states in protocol testing. In A. Cavalli and S. Budkowski, editors, *Protocol Test Systems*, Volume VIII, pages 307–322. Chapman & Hall, 1995.

[20] F. Michel, P. Azema, and K. Drira. Selective generation of symmetrical test cases. In B. Baumgarten, H.-J. Burkhardt and A. Giessler, editors, *Testing of Communicating Systems*, Volume 9, pages 191–206. Chapman & Hall, 1996.

[21] R. Milner. *Communication and Concurrency*. Prentice-Hall International, Englewood Cliffs, 1989.

[22] B. Steffen, editor. *Tools and Algorithms for the Construction and Analysis of Systems (TACAS '98)*. Lecture Notes in Computer Science 1384. Springer-Verlag, 1998.

[23] R. Terpstra, L. Fereira Pires, L. Heerink, and J. Tretmans. Testing theory in practice: A simple experiment. In *Proceedings of the COST 247 International Workshop on Applied Formal Methods in System Design*, 1996. Also published as Technical Report CTIT 96-21, University of Twente, The Netherlands, 1996.

[24] J. Tretmans. A theory for the derivation of tests. In *Formal Description Techniques (FORTE II '89)*. North-Holland, 1989.

Towards Automatic Distribution of Testers for Distributed Conformance Testing

Claude Jard, Thierry Jéron, Hakim Kahlouche & César Viho
IRISA - Campus de Beaulieu, F35042 Rennes, France
e-mail: {jard, jeron, kahlouch, viho}@irisa.fr

Abstract

This paper presents first steps towards automatic generation of distributed tests. We first define a characterization of the tests for which the property of unbias is preserved by the existence of an asynchronous environment. Then, starting from a centralized test case, we propose a method to derive automatically its corresponding distributed test case in an asynchronous environment. We prove that the generated distributed test case is not biased, it tests the same behaviors of an implementation and has the same testing power as the centralized test case.

Keywords

Conformance Test Generation, Unbiased Test Case, Distributed Testing

1 INTRODUCTION

In the context of black box conformance testing, an implementation under test (IUT) is tested in order to obtain the conviction that its behavior conforms with its specification. The testers stimulate the IUT by sending messages on points of control and observation (PCOs) and observe on these same PCOs the reactions of the IUT. Within sight of the reactions, a verdict (Fail, Pass or Inconclusive) is emitted. More recently, the conformance testing methodology has been formalized [1, 2, 3] and brought more confidence in the testing activity. Most of works in this field using formal methods concentrated on the production of a centralized single tester and make the implicite assumption of synchronicity between the tester and the IUT [4, 5]. In practice however, one cannot always avoid taking into account the test environment intercalated between the tester and the IUT. The most frequent example is that of remote testing architecture in which the tester reaches the IUT through a network connection. In this case, PCOs can be seen as composed of two FIFO queues, for each direction of the interaction: one speaks then about asynchronous interaction between the tester and the IUT. In the prolongation of this concept, we propose to aim at a test architecture in which several testers coexist and interact in an asynchronous way not only with the IUT but also between them to coordinate the testing activity. It is about a generalization of the multi-party testing technique which benefits from the various possibilities offered by the introduction of concurrency into TTCN [6]. By considering that the IUT itself could be distributed, we are in the general context of distributed testing of a distributed architecture, illustrated by figure 1. To face up to the problem of the design of such testers, we distinguish several subproblems.

The first main problem is that of the asynchronism on PCOs. The possibility of disorder on the observations collected on different PCOs as well as the possible collision on a PCO between a stimulus and an observation makes difficult the design of testers. During the examination of existing test suites, one realizes that these are the main reasons of non-validity of some tests: precisely a correct test not rejecting conformant implementations in a synchronous environment can be biased (i.e. it rejects a conformant implementation) when it is carried out in an asynchronous environment [7]. It is thus significant to know if a test is correct in an asynchronous environment. We will show that contrary to what is suggested in the ISO/9646 methodology [6, 8] with the concept of generic abstract test case, independent of the testing environment, this question depends on the specification of the system to be tested.

The second main problem is to design the distributed tester (a system of parallel testers) and their coordination procedure. We propose to start from a centralized test case and automatically to produce by compilation the communicating testers. Insofar as the initial centralized test case is sequential and describes a set of behaviors to be tested, we try to reproduce in a distributed way, exactly the same set of behaviors, by exploiting some ideas from protocol synthesis [18]. The possible concurrencies in the whole distributed test come from the desynchronization between the IUT and the testers and between local actions of the testers. To extract the potential concurrency of the initial test case is a prospect for our approach. This should impact the generation phase of test cases from formal specifications (as advocated in [15] using event structures), or demand some causal information at runtime (as suggested by [16, 17]).

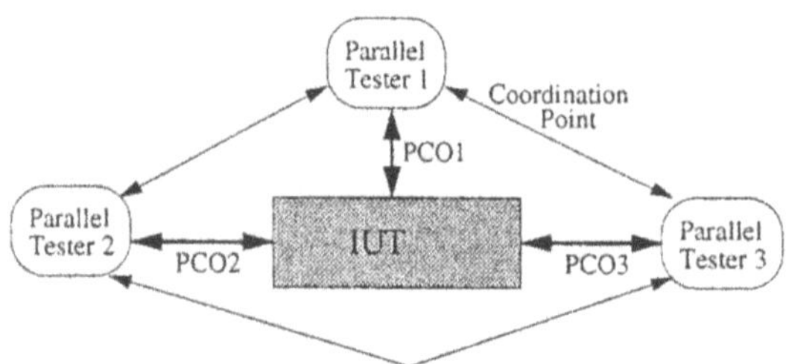

Figure 1 Distributed testing in an asynchronous environment

In order to found our approach mathematically, we use the standard model of input-output transition systems to represent all the objects (testers, specifications and IUTs) involved in the conformance testing. The notations and models used in this paper are described in the section 2. In section 3, we precisely state the condition under which the property of unbias of the test is preserved by the existence of an asynchronous environment. We show that it can be computed starting from the specifications of the IUT and the tester. Then we propose, in section 4, a compilation rule which produces the distributed tester starting from an initial centralized test case. The difficult question is the resolution of distributed choices. We show that under the assumption of an underlying service of election, one obtains a system whose traces reduced to the events on PCOs are the same ones as the traces defined by the centralized test case. A proof of the correctness of the distributed test case is given followed by an example of distribution. Concluding remarks and future works are presented in section 5.

2 MODELS FOR CONFORMANCE TESTING

In this section we describe the models used for the description of the different objects involved in the generation of test cases.

2.1 Input-Output labelled transition systems

The models used are all based on Input-Output Labelled Transition Systems (IOLTS) in which input and output actions on input and output ports are differentiated because of the asymmetrical nature of the testing activity.

Definition 21 *An IOLTS is a tuple* $M=(Q^M, P_I^M, P_O^M, A_I^M, A_O^M, T^M, q_{init}^M)$ *where* Q^M *is a set of states,* $q_{init}^M \in Q^M$ *is the initial state,* P_I^M *and* P_O^M *are finite sets of input and output ports,* A_I^M *and* A_O^M *respectively are finite input and output alphabets,* $A^M = P_I^M \times \{?\} \times A_I^M \cup P_O^M \times \{!\} \times A_O^M$ *is the alphabet of observable actions constructed from the sets of input-output ports and input-output alphabets,* $\tau \notin A^M$ *denotes an internal action,* $T^M \subseteq Q^M \times A^M \cup \{\tau\} \times Q^M$ *is the transition relation. We note* $q \xrightarrow{\alpha}_M q'$ *for* $(q, \alpha, q') \in T^M$.

An IOLTS is a particular Labelled Transition System (LTS). Consequently we adopt the following standard notations of LTS.

Let $\alpha_i \in A^M$, $\mu_i \in A^M \cup \{\tau\}$, $\sigma \in (A^M)^*$, $q, q', q_i \in Q^M$,

- $q \xrightarrow{\mu_1 \cdots \mu_n} q' =_\Delta \exists q_0 = q, q_1 \ldots, q_n = q', \forall i \in [1,n], q_{i-1} \xrightarrow{\mu_i} q_i$,
- $q \xrightarrow{\mu_1 \cdots \mu_n} =_\Delta \exists q', q \xrightarrow{\mu_1 \cdots \mu_n} q'$ and $q \not\xrightarrow{\mu_1 \cdots \mu_n} q' =_\Delta \mathbf{not}(q \xrightarrow{\mu_1 \cdots \mu_n})$,
- $q \stackrel{\epsilon}{\Rightarrow} q' =_\Delta q = q'$ or $q \xrightarrow{\tau \cdots \tau} q'$ and $q \stackrel{\alpha}{\Rightarrow} q' =_\Delta \exists q_1, q_2, q \stackrel{\epsilon}{\Rightarrow} q_1 \xrightarrow{\alpha} q_2 \stackrel{\epsilon}{\Rightarrow} q'$,
- $q \stackrel{\alpha_1 \cdots \alpha_n}{\Longrightarrow} q' =_\Delta \exists q_0 = q, q_1 \ldots, q_n = q', \forall i \in [1,n], q_{i-1} \stackrel{\alpha_i}{\Rightarrow} q_i$,

- $enable(q) =_\Delta \{\alpha \in A^M \mid \exists q' \text{ and } q \stackrel{\alpha}{\Rightarrow}_M q'\}$ is the set of observable actions possible in q,
- $In(q) =_\Delta \{a \in A_I^M \mid \exists p_I \in P_I^M, p_I?a \in enable(q)\}$ is the set of possible inputs in q and $Out(q) =_\Delta \{a \in A_O^M \mid \exists p_O \in P_O^M, p_O!a \in enable(q)\}$ is the set of possible outputs in q,
- q **after** $\sigma =_\Delta \{q' \in Q^M \mid q \stackrel{\sigma}{\Rightarrow}_M q'\}$ is the set of reachable states from q by the sequence of observable actions σ,
- $traces(q) =_\Delta \{\sigma \in (A^M)^* \mid q \textbf{ after } \sigma \neq \emptyset\}$
- if $\alpha \in A^M$ is an observable action, we note $\overline{\alpha}$ its mirror action defined by $\overline{p!a} = p?a$ and $\overline{p?a} = p!a$. This notation is extended to sequences of actions.

When needed, we will identify a transition system with its initial state.

2.2 Specifications, implementations, test cases and conformance

A **specification** is modelled by an IOLTS $S = (Q^S, P_I^S, P_O^S, A_I^S, A_O^S, T^S, q_{\text{init}}^S)$ If we want to model the absence of output in a specification, as in [9], we can model it by adding a loop with output $\delta \in A_O^S$ in each state where no output and no τ transition is possible. Here, for the sake of simplicity, we will not distinguish δ from other outputs.

Though an **implementation** of S is not necessarily a transition system (it may be a physical system), as in all testing theories, we have to reason formally about it and model its behaviour. As it is only considered by its interactions with the environment, it is also modelled by an IOLTS $I = (Q^I, P_I^I, P_O^I, A_I^I, A_O^I, T^I, q_{\text{init}}^I)$, with $P_I^I = P_I^S$ and $P_O^I = P_O^S$ and $A_I^S \subseteq A_I^I$, $A_O^S \subseteq A_O^I$. We will suppose that an implementation I can never refuse an input: $\forall \sigma \in (A^I)^*, In(I \textbf{ after } \sigma) = A_I^I$ *.

A **test case** is a set of sequences of actions describing all the interactions occurring between an IUT I and a tester which wants to verify that I conforms to its specification S. A test case is an IOLTS $T = (Q^T, P_I^T, P_O^T, A_I^T, A_O^T, T^T, q_{\text{init}}^T)$ such that $A_I^T = A_O^I$ i.e. every possible output of the implementation must be considered as an input of the test case, $A_O^T = A_I^S$ i.e. a test case should only send outputs that are waited by the system as described in the specification, $\{\textbf{pass}, \textbf{fail}\} \in Q^T$ with $enable(\textbf{pass}) = enable(\textbf{fail}) = \emptyset$, T is controllable i.e. in each state, either an output is enabled or all inputs are enabled which is formally described by $\forall \sigma \in (A^T)^*, In(T \textbf{ after } \sigma) = A_I^T$ or $\emptyset$, the state **fail** is directly accessible only by inputs i.e. $\forall q \in Q^T, \forall \alpha \in A^T, q \stackrel{\alpha}{\Rightarrow}_T \textbf{fail} \Longrightarrow \exists p \in P_I^T, a \in A_I^T, \alpha = p?a$ *. Notice that we do not suppose that testers are

*In a way identical to specifications, the absence of output is modeled by the output $\delta \in A_O^I$.
*In practice however $A_I^T = A_O^I$ is unknown. Thus only inputs not leading to **fail** can be mentioned, the other inputs are implicitly leading to fail or are mentioned by **? otherwise** with verdict **Fail**, like in TTCN.

deterministic as it would be to strong for distributed testers. Anyway, we only consider the traces of testers.

We consider a **conformance relation** quite similar to those in [10, 11, 9]. Informally, the conformance relation states that after a trace of the specification, outputs produced by the implementation must be foreseen by the specification. This authorizes the fact that outputs of the environment which are not accepted by the specification may be accepted by the implementation.

Let S and I be two IOLTS describing a specification and an implementation, I **ioconf** $S =_\Delta \forall \sigma \in traces(S),\ Out(I\ \textbf{after}\ \sigma) \subseteq Out(S\ \textbf{after}\ \sigma)$

Remark: $traces(I) \subseteq traces(I') \Longrightarrow (I'\ \textbf{ioconf}\ S \Longrightarrow I\ \textbf{ioconf}\ S)$.
But $traces(S) \subseteq traces(S') \not\Longleftrightarrow (I\ \textbf{ioconf}\ S \Longrightarrow I\ \textbf{ioconf}\ S')$
and $traces(S) \subseteq traces(S') \not\Longleftrightarrow (I\ \textbf{ioconf}\ S' \Longrightarrow I\ \textbf{ioconf}\ S)$.

3 CHARACTERIZATION OF UNBIASED TEST CASES

Our goal is to distribute test cases without loss of observation power. But before studying this problem we have to define what kind of test cases we want to distribute. So we start by giving a characterization of unbiased test cases with respect to a specification. This characterization supposes a synchronous communication between the tester and the implementation. As we want to distribute test cases for asynchronous architectures, we extend this characterization to testers which communicate asynchronously with implementations.

3.1 Synchronous testing

The synchronous application of a test case to an implementation is defined as a parallel composition $\|_s$ of the test case T and the implementation I.

$$\frac{I \xrightarrow{\tau} I'}{T\|_s I \xrightarrow{\tau} T\|_s I'} \qquad \frac{T \xrightarrow{\tau} T'}{T\|_s I \xrightarrow{\tau} T'\|_s I} \qquad \frac{T \xrightarrow{\alpha} T' \quad I \xrightarrow{\bar{\alpha}} I'}{T\|_s I \xrightarrow{\alpha} T'\|_s I'}$$

We can now define test failure and unbiased test cases with respect to **ioconf** and a given specification:

Definition 31 T **fails** $I =_\Delta \exists I', \exists \sigma, T\|_s I \overset{\sigma}{\Rightarrow} \textbf{fail}\|_s I'$
T is unbiased w.r.t. S $=_\Delta \forall I, T\ \textbf{fails}\ I \Rightarrow not(I\ \textbf{ioconf}\ S)$

Now the following proposition gives a characterization of unbiased test cases w.r.t. a specification in a synchronous environment. It says that after each trace of the test case leading to a state where an input is allowed, each possible output of the specification after the same trace should be taken into account and should not lead to a fail verdict.

Proposition 31 *T is unbiased w.r.t. S* $\iff$
$\forall \sigma \in traces(T)$ *s.t.* $In(T \textbf{ after } \sigma) \neq \emptyset$, $Out(S \textbf{ after } \sigma) \subseteq In_{\neg fail}(T \textbf{ after } \sigma)$
with $In_{\neg fail}(T') = \{a \in In(T') \mid \exists p, T' \overset{p?a}{\Rightarrow} T'' \wedge \textbf{fail} \notin T''\}$ *.

Proof. Let T be a test case such that $\forall \sigma \in traces(T)$ s.t. $In(T \textbf{ after } \sigma) \neq \emptyset$, $Out(S \textbf{ after } \sigma) \subseteq In_{\neg fail}(T \textbf{ after } \sigma)$. We have to prove that T is unbiased w.r.t. S. Suppose that for some I, T **fails** I . By definition, **fail** verdicts are only given on inputs of T (i.e. outputs of I), thus a trace leading to a **fail** verdicts ends with an input of T. T **fails** I can be written as $\exists I_1, I', \exists T_1, \exists \sigma', \exists p \in P_I^T, \exists a \in A_I^T$ s.t. $T \|_s I \overset{\sigma_1}{\Rightarrow} T_1 \|_s I_1 \overset{p?a}{\Rightarrow} \textbf{fail} \|_s I'$. Thus $a \notin In_{\neg fail}(T \textbf{ after } \sigma_1)$ and as $Out(S \textbf{ after } \sigma_1) \subseteq In_{\neg fail}(T \textbf{ after } \sigma_1)$, we also have $a \notin Out(S \textbf{ after } \sigma_1)$. But we have $I \overset{\sigma_1}{\Rightarrow} I_1 \overset{\overline{p?a}=p!a}{\Rightarrow} I'$ which implies $a \in Out(I \textbf{ after } \sigma_1)$. Consequently, $\exists \sigma_1 \in traces(S), \exists p, \exists a, a \in Out(I \textbf{ after } \sigma_1) \wedge a \notin Out(S \textbf{ after } \sigma_1)$ which proves that not (I **ioconf** S) and thus T is unbiased w.r.t. S.

Now suppose that $\exists \sigma \in traces(T), In(T \textbf{ after } \sigma) \neq \emptyset$ and $Out(S \textbf{ after } \sigma) \nsubseteq In_{\neg fail}(T \textbf{ after } \sigma)$. Then $\exists a \in A_I^T, \exists p$, $(S \textbf{ after } \sigma \overset{p!a}{\Rightarrow}) \wedge (T \textbf{ after } \sigma \overset{p?a}{\Rightarrow} \textbf{fail})$. Let $I = \sigma.p!a$. Clearly we have I **ioconf** S. But as $T \|_s I \overset{\sigma.p?a}{\Rightarrow} \textbf{fail} \|_s \emptyset$ we get T **fails** I and T is biased w.r.t. S. □

Remark: Proposition 31 exactly characterizes unbiased test cases with respect to a specification. This is defined for a particular conformance relation **ioconf** and with some hypothesis on the structure of test cases. But **ioconf** seems to be exactly what manual test developers have in mind and the hypothesis made on test cases are weaker than those made on manual test cases.

In the context of automatic generation of test cases it is important to ensure that test cases are unbiased and this kind of characterization should be verified by tools. Notice that the prototype tool TGV [3] ensures this by construction.

3.2 Asynchronous testing

Before extending the preceding characterization to asynchronous testing, we have to define what is asynchronous testing. It is defined from synchronous testing by considering a transformation of the specifications and the implementations by a queue context like in [12].

Let $M = (Q^M, P_I^M, P_O^M, A_I^M, A_O^M, T^M, q_{init}^M)$ be an IOLTS. We define an IOLTS $\mathcal{A}(M)$ which models the behavior of M in an asynchronous environment.
$\mathcal{A}(M) = (Q^{A(M)}, P_I^{A(M)}, P_O^{A(M)}, A_I^{A(M)}, A_O^{A(M)}, T^{A(M)}, q_{init}^{A(M)})$ with

- $Q^{A(M)} = Q^M \times \prod_{p \in P_I^M} A_I^{M*} \times \prod_{p \in P_O^M} A_O^{M*}$ and $q_{init}^{A(M)} =< M, (\epsilon \cdots \epsilon), (\epsilon \cdots \epsilon) >$

*Recall that $In(T') = A_I^T$ or $\emptyset$ due to the controllability property.

- $P_I^{A(M)} = P_I^A$ and $P_O^{A(M)} = P_O^A$, $A_I^{A(M)} = A_I^A$ and $A_O^{A(M)} = A_O^A$.
- $T^{A(M)}$ is described by the following operational rules defined for all $q, q' \in Q^M, a \in A_I^M, b \in A_O^M, p_{I_k} \in P_I^M, p_{O_l} \in P_O^M$:

R1 $< q, (p_{I_1} \cdots p_{I_k} = w \cdots), (p_{O_1} \cdots) > \overset{p_{I_k}?a}{\rightarrow}_{A(M)} < q, (p_{I_1} \cdots p'_{I_k} = w.a \cdots), (p_{O_1} \cdots) >$

R2 $< q, (p_{I_1} \cdots), (p_{O_1} \cdots p_{O_l} = b.w \cdots) > \overset{p_{O_l}!b}{\rightarrow}_{A(M)} < q, (p_{I_1} \cdots), (p_{O_1} \cdots p_{O_l} = w \cdots) >$

R3
$$\frac{q \overset{\tau}{\rightarrow}_M q'}{< q, (p_{I_1} \cdots), (p_{O_1} \cdots) > \overset{\tau}{\rightarrow}_{A(M)} < q', (p_{I_1} \cdots), (p_{O_1} \cdots) >}$$

R4
$$\frac{q \overset{p_{I_k}?a}{\rightarrow}_M q'}{< q, (p_{I_1} \cdots p_{I_k} = a.w \cdots), (p_{O_1} \cdots) > \overset{\tau}{\rightarrow}_{A(M)} < q', (p_{I_1} \cdots p_{I_k} = w \cdots), (p_{O_1} \cdots) >}$$

R5
$$\frac{q \overset{p_{O_l}!b}{\rightarrow}_M q'}{< q, (p_{I_1} \cdots), (p_{O_1} \cdots p_{O_l} = w \cdots) > \overset{\tau}{\rightarrow}_{A(M)} < q', (p_{I_1} \cdots), (p_{O_1} \cdots p_{O_l} = w.b \cdots) >}$$

Remark: The first rule allows the IOLTS $\mathcal{A}(M)$ to receive any input in any state. This leads to an infinite states IOLTS. When this transformation is used to model the behaviour of a specification in an asynchronous environment in order to derive test cases, this of course poses algorithmic problems. Thus it is reasonable to limit the behaviour of the specification in order to derive test cases. An example of limitation consists in allowing the environment to send only messages that are possibly waited and to limit the sum of lengths of input queues to 1. In this case **R1** could then be replaced by:

R1'
$$\frac{q \overset{p_{I_k}?a}{\rightarrow}_M}{< q, (\epsilon \cdots \epsilon \cdots), (p_{O_1} \cdots) > \overset{p_k?a}{\rightarrow}_{A(M)} < q, (\epsilon \cdots p_{I_k} = a \cdots), (p_{O_1} \cdots) >}$$

As in [13], we define the conformance of an implementation with respect to a specification in an asynchronous environment as the conformance of the implementation in its asynchronous environment with respect to the specification in the same asynchronous environment. This directly leads to the definition of test failure and unbias in an asynchronous environment and gives a corollary of **Proposition 31**:

Definition 32 *Let I be an implementation of a specification S, T a test case.*
I **ioconf**$_\mathcal{A}$ $S =_\Delta \mathcal{A}(I)$ **ioconf** $\mathcal{A}(S)$
T **fails**$_\mathcal{A}$ $I =_\Delta \exists \mathcal{A}', \exists \sigma, T \|_s \mathcal{A}(I) \overset{\sigma}{\Rightarrow}$ **fail** $\|_s \mathcal{A}'$. *Thus* T **fails**$_\mathcal{A}$ $I = T$ **fails** $\mathcal{A}(I)$.
T is **unbiased** *w.r.t S in an asynchronous environment* $=_\Delta \forall I, T$ **fails**$_\mathcal{A}$ $I \Rightarrow$ *not*(I **ioconf**$_\mathcal{A}$ S).
It is equivalent to say that test case T is unbiased with respect to $\mathcal{A}(S)$.

Corollary 31 *T is unbiased w.r.t. S in an asynchronous environment* $\Longleftrightarrow$ $\forall \sigma \in traces(T)$ *s.t.* $In(T$ **after** $\sigma) \neq \emptyset$, $Out(\mathcal{A}(S)$ **after** $\sigma) \subseteq In_{\neg fail}(T$ **after** $\sigma)$ *with* $In_{\neg fail}(T') = \{a \in In(T) | \exists p, T' \overset{p?a}{\rightarrow} T'' \wedge T'' \neq$ **fail**$\}$.

This corollary characterizes a class of test cases that are unbiased in an asynchronous environment. It is also easy to prove that an unbiased test case in an asynchronous environment is also unbiased in a synchronous environment. This uses the fact that $traces(M) \subseteq traces(\mathcal{A}(M))$ which relies on construction rules of $\mathcal{A}(M)$ and the property that if $traces(S) \subseteq traces(S')$ then, if T is unbiased w.r.t. S' then T is also unbiased w.r.t. S. The converse is of course false as was already noticed in [13].

4 AUTOMATIC DISTRIBUTION OF TEST CASES

In this section we develop the rule used to automatically distribute a centralized tester. Then, we give the correctness proof of that rule wich implies that the unbiased criterion is preserved by the distribution rule and that the distributed tester rejects the same IUTs as the original centralized tester. Some additional notations are needed. Let T be an IOLTS, $q \in Q^{\mathrm{T}}$ and $\alpha \in A^{\mathrm{T}}$:

- $P^{\mathrm{T}} = P_I^{\mathrm{T}} \cup P_O^{\mathrm{T}}$
- if $\alpha = p?a$ or $\alpha = p!a$, where $p \in P^{\mathrm{T}}$, then $port(\alpha) =_{\Delta} p$.
- $p\_enable(q) =_{\Delta} \{p \in P^{\mathrm{T}} \mid \exists \alpha \in enable(q) \quad port(\alpha) = p\}$
- if $\alpha = p?a$ then $present(\alpha)$ is a predicate which is true if the label a is present in the input queue associated to the port p, and it is false otherwise. If $\alpha = p!a$ then $present(\alpha)$ is true.
- The asynchronous parallel composition of $T_1, \cdots, T_n$ is denoted by $\|_{as} T_i = \|_s \mathcal{A}(T_i)$ where an output port of a tester may be unified with an input port of another tester to make the coordination possible. We also note $DT = \|_{as} T_i$ and Q^{DT} the set of global states of $\|_{as} T_i$. A global state $Q \in Q^{\mathrm{DT}}$ is a tuple $(q_1, \cdots, q_n, f_1, \cdots, f_n)$, where q_i is the state of T_i and f_i is the state of the input and output queues of T_i.

Starting from a test case T modellized by an IOLTS, the problem is to synthesize a set of asynchronously communicating IOLTSs $\{T_1, \cdots, T_n\}$ where n is the number of PCOs of T. The initial state of each tester T_i is q_{init}. For sake of clarity, we have considered that each tester is associated with each PCO. The extension to a general PCOs mapping is straightforward. In the following we consider that only the actions of A^{T} are observable.

4.1 Distribution rule

We present here bellow the rule used to automatically distribute a centralized tester. The result is a set of communicating testers. Starting from a pattern of the centralized tester defined by a state q with one or several outgoing transitions, the rule [R1] synthesizes a set of transitions of the distributed tester. These transitions can be decomposed into three parts. The part (A) defines the transitions of the testers T_i associated to the ports involved in the

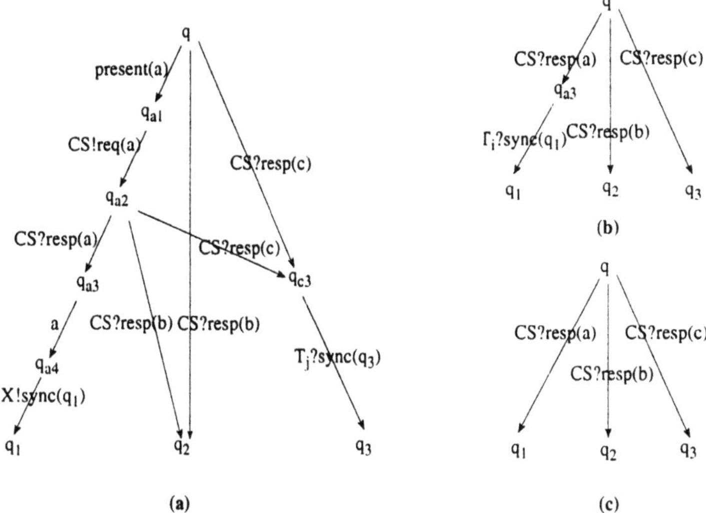

Figure 2 Distribution rule (a) Tester $T_i \mid i \in p\_enable(q)$ (b) Tester $T_k \mid k \notin p\_enable(q),\ k \in X$ (c) Tester $T_k \mid k \notin p\_enable(q) \cup X$

outgoing transitions of q, namely $p\_enable(q)$ (see Figure 2(a)). The part (B) defines the testers T_k which are not associated to the ports of $p\_enable(q)$ and must receive a synchronization message from T_i (see Figure 2(b)). The part (C) concerns the testers which are not associated to the ports of $p\_enable(q)$ and do not require a synchronization message from T_i (see Figure 2(c)).

$$[\mathrm{R1}]\ \frac{q \in Q^{\mathrm{T}}}{}$$

$$(A)\left\{\begin{array}{l}
\forall i \in p_enable(q) \quad \forall a \in enable(q) \mid port(a) = i,\ q \xrightarrow{a}_{\mathrm{T}} q_1 : \\
q \xrightarrow{present(a)}_{\mathrm{T}_i} q_{a1} \xrightarrow{CS!req(a)}_{\mathrm{T}_i} q_{a2} \xrightarrow{CS?resp(a)}_{\mathrm{T}_i} q_{a3} \xrightarrow{a}_{\mathrm{T}_i} q_{a4} \xrightarrow{X!sync(q_1)}_{\mathrm{T}_i} q_1 \\
\forall b \in enable(q) \mid port(b) \neq i,\ q \xrightarrow{b}_{\mathrm{T}} q_2,\ i \notin p_enable(q_2) : \\
q_{a2} \xrightarrow{CS?resp(b)}_{\mathrm{T}_i} q_2 \quad q \xrightarrow{CS?resp(b)}_{\mathrm{T}_i} q_2 \\
\forall c \in enable(q) \mid port(c) = j,\ j \neq i,\ q \xrightarrow{c}_{\mathrm{T}} q_3,\ i \in p_enable(q_3) : \\
q_{a2} \xrightarrow{CS?resp(c)}_{\mathrm{T}_i} q_{c3} \quad q \xrightarrow{CS?resp(b)}_{\mathrm{T}_i} q_{c3} \quad q_{c3} \xrightarrow{T_j?sync(q_3)}_{\mathrm{T}_i} q_3
\end{array}\right.$$

$$(B)\ \left\{ \forall k \notin p\_enable(q),\ k \in X : \quad q \xrightarrow{CS?resp(a)}_{\mathrm{T}_k} q_{a3} \xrightarrow{T_i?sync(q_1)}_{\mathrm{T}_k} q_1 \right.$$

$$(C)\ \left\{ \forall k \notin p\_enable(q),\ k \notin X : \quad q \xrightarrow{CS?resp(a)}_{\mathrm{T}_k} q_1 \right.$$

where $X = \{T_j \mid j \neq i,\ j \in p\_enable(q_1)\}$.

A **mapping function** F is introduced between the states of the distributed tester and the states of the centralized tester T. The states provided by F are called *observable states*: $F : \bigcup_{i=1}^{n} Q^{\mathrm{T}_i} \longrightarrow Q^{\mathrm{T}}$. According to the rule [R1]: $F(q_{a1}) = F(q_{a2}) = F(q_{a3}) = F(q_{c3}) = F(q) = q \quad and \quad F(q_{a4}) = F(q_1) = q_1$ This function ensures that the observable state of a tester should change

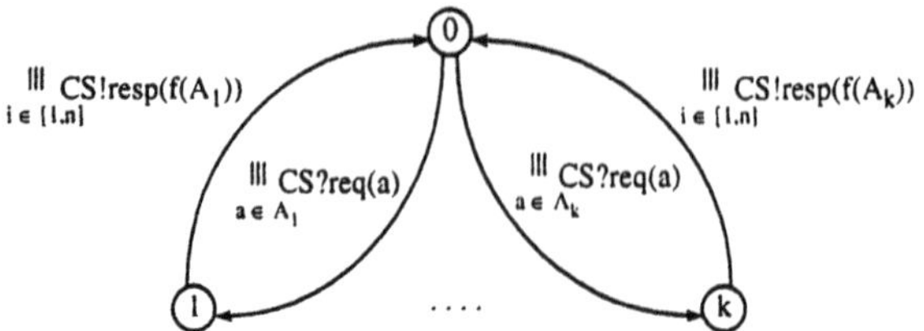

Figure 3 Election Service

only when the tester performs an observable action or when it receives a synchronization message. This function is introduced for correctness proof.

The testers use an **election service** when they attempt to perform an observable action (action of A^{T}). The service accepts at least one request from the testers and delivers to all the testers the observable action which has been selected. The interaction of each tester with the election service is synchronous and it is done through a coordination port named CS. An abstract specification is given in Figure 3. The transition labelled by $|||_{a \in A_i} CS?req(a)$ specifies all possible interleaving of the requests coming from the testers and concerning only with the actions of A_i (where $A_1, \cdots, A_k$ are all possible subsets of A^{T}). This transition specifies an hypercube. The transition labelled by $|||_{i \in [1,n]} CS!resp(a)$ is another hypercube which specifies the broadcasting of the selected* action among those of A_i.

4.2 Correctness proof

We present in this section the correctness proof of the distribution rule. Our purpose if to prove that the asynchronous parallel composition of the synthezised testers (i.e, the distributed tester) is trace-equivalent to the centralized tester. Under the assumption of an election service and reliable and FIFO communication channels between the testers, we start by proving two lemmas. The first one (Lemma 41) stipulates that when two successive observable actions occur in the distributed tester, the observable state in which the second action is done is the same as the one reached by the first action. The observable state of a tester is provided by the mapping function F. This lemma implies that every trace of the distributed tester is a trace of the centralized tester. The second lemma (Lemma 42) states that if the centralized tester can do an observable action in a given observable state, the distributed tester could necessarily do the same action (possibly surrounded by one or several τ-actions). This lemma implies that every trace of the centralized tester is a trace of the distributed tester. Using these two lemmas, we prove the trace-equivalence (Theorem 41).

*For sake of clarity of Figure 3, the selection is done using a function f.

Lemma 41 *$\forall Q, U, V, W \in Q^{DT}$, $\forall a, b \in A^T$ such that $Q \xrightarrow{a}_{DT} U \xrightarrow{\tau^*}_{DT} V \xrightarrow{b}_{DT} W$, $\exists q, x, w \in Q^T$ such that: $q = F(q_i) \xrightarrow{a}_T x = F(u_i) = F(v_i) \xrightarrow{b}_T w = F(w_j)$, where $i = port(a)$, $j = port(b)$, q_i and u_i are states of T_i resp. in Q and U, v_j and w_j are states of T_j resp. in V and W, and F is defined in Section 4.1.*

Proof. Let $Q, U, V, W \in Q^{DT}$, let $a, b \in A^T$ such that:

$$Q \xrightarrow{a}_{DT} U \xrightarrow{\tau^*}_{DT} V \xrightarrow{b}_{DT} W \tag{1}$$

$Q \xrightarrow{a}_{DT} U \Rightarrow \exists T_i \mid i = port(a)$ and $q_i \xrightarrow{a}_{T_i} u_i \Rightarrow F(q_i) \xrightarrow{a}_T F(u_i)$
$V \xrightarrow{b}_{DT} W \Rightarrow \exists T_j \mid j = port(b)$ and $v_j \xrightarrow{b}_{T_j} w_j \Rightarrow F(v_j) \xrightarrow{b}_T F(w_j)$
We should prove a first property (noted P1): $\exists q, x, w \in Q^T$ such that $q \xrightarrow{a}_T x \xrightarrow{b}_T w$, and then a second property (P2): $F(u_i) = F(v_j)$.
From the sequence (1), we know that $\exists q, x \in Q^T$ such that: $q \xrightarrow{a}_T x$. Suppose that: $\forall q, x \in Q^T$ such that $q \xrightarrow{a}_T x$, $\forall \alpha \in enable(x)$, $\alpha \neq b$. The rule [R1] guarantees that in the distributed tester after the observable action a we cannot have b, and this is in contradiction with (1). Therefore, (P1) is necessarily true. To prove (P2) we distinguish 2 cases:
Case 1 ($i \neq j$): From the sequence (1), we can extract the following sequences of the testers T_i and T_j:

$$q_i \xrightarrow{a}_{T_i} u_i \xrightarrow{\tau^*}_{T_i} v_i \tag{2}$$

$$u_j \xrightarrow{\tau^*}_{T_j} v_j \xrightarrow{b}_{T_j} w_j \tag{3}$$

According to rule [R1], and using (P1), among the τ-actions of (3), there exists necessarily a reception of a synchronisation message ($sync(x)$). So, (3) can be written as follow:

$$u_j \xrightarrow{\tau^*}_{T_j} u'_j \xrightarrow{\tau}_{T_j} u''_j \tag{4}$$

$$u''_j \xrightarrow{\tau^*}_{T_j} v_j \xrightarrow{b}_{T_j} w_j \tag{5}$$

The last transition of (4) is the reception of $sync(x)$. Using the mapping function F to that transition, we have: $F(u'_j) \xrightarrow{a}_T F(u''_j)$. By identification with $F(q_i) \xrightarrow{a}_T F(u_i)$, we obtain: $F(u'_j) = F(q_i)$ and $F(u''_j) = F(u_i)$. Since the medium is supposed FIFO, another reception among the τ-actions of (5) would imply another observation after a in (2) which is not true. So, there is no reception of synchronization message among the τ-actions of (5). Furthermore, because of the election service, there is no action $CS?resp(x)$ where $x \neq b$. So, (5) contains only the primitives which allow the election of b. From the definition of F, we deduce: $F(u''_j) = F(v_j)$, and consequently: $F(u_i) = F(v_j)$.

Case 2 ($i = j$): From (1), we can extract the following sequence of the tester T_i: $q_i \xrightarrow{a}_{T_i} u_i \xrightarrow{\tau^*}_{T_i} v_i \xrightarrow{b}_{T_i} w_i$. Since $port(a) = port(b) = i$ and according to rule [R1], there is no reception of synchronization message among the τ-actions of that sequence. Furthermore, as in Case 1, the sequence $u_i \xrightarrow{\tau^*}_{T_i} v_i$ contains only primitives which allow the election of b. Consequently, from the definition of F, we have: $F(u_i) = F(v_j)$ □

Lemma 42 *$\forall q, v \in Q^T$, $\forall a \in A^T$ such that $q \xrightarrow{a}_T v$, $\exists Q, V \in Q^{DT}$ such that: $\exists f_1, \cdots, f_n$ $Q = (q, \cdots, q, f_1, \cdots, f_n) \xrightarrow{\tau^* a \tau^*}_{DT} V = (v, \cdots, v, f_1, \cdots, f_n)$.*

Proof. By applying [R1] to the transition $q \xrightarrow{a}_T v$, we find the following sequences, respectively in T_i (where $i = port(a)$) and in T_j (where $j \neq i$):

$$\begin{cases} q \xrightarrow{\tau^3}_{T_i} q_{a3} \xrightarrow{a}_{T_i} q_{a4} \\ q \xrightarrow{\tau^m}_{T_j} v \quad where \quad m \in [1,2] \quad and \quad j \neq i \end{cases}$$

Using the above sequences we can find the following sequence in DT:
$Q = (q, \cdots, q, f_1, \cdots, f_n) \xrightarrow{\tau^3}_{DT} (q, \cdots, q_{a3}, \cdots, q, f_1, \cdots, f_n) \xrightarrow{a}_{DT} Q'$,
$Q' = (q, \cdots, q_{a4}, \cdots, q, f_1, \cdots, f_n) \xrightarrow{\tau^*}_{DT} (q, \cdots, v, \cdots, q, f'_1, \cdots, f'_n) = Q''$,
$Q'' \xrightarrow{\tau^*}_{DT} (v, \cdots, v, f_1, \cdots, f_n) = V$. The last τ^* sequence contains the receptions of the synchronization messages which have been sent after the action a (because the medium is supposed to be FIFO and reliable). □

Theorem 41 *If a distributed tester DT is synthesized from a centralized tester T using the rule [R1], then: $traces(DT) = traces(T)$.*

Proof. ($\subseteq$) Let $\sigma \in traces(DT)$, σ can be written as follow: $\sigma = \tau^* a_1 \tau^* a_2 \cdots \tau^* a_n \tau^*$ where $a_i \in A^T$ $\forall i \in [1, n]$. From $Q_{init} \xrightarrow{\sigma}_{DT} Q$ where $Q_{init}, Q \in Q^{DT}$, $\exists U_0, \cdots, U_{n-1}, Q_1, \cdots, Q_n \in Q^{DT}$ such that:

$$Q_{init} \xrightarrow{\tau^*} U_0 \xrightarrow{a_1} Q_1 \xrightarrow{\tau^*} U_1 \xrightarrow{a_2} Q_2 \cdots \xrightarrow{a_{n-1}} Q_{n-1} \xrightarrow{\tau^*} U_{n-1} \xrightarrow{a_n} Q_n = Q \quad (6)$$

From (6), $\exists T_{k_1}$ where $k_1 = port(a_1)$ such that: $q_{k_1} = q_{init} \xrightarrow{\tau^*}_{T_{k_1}} u^0_{k_1} \xrightarrow{a_1}_{T_{k_1}} q^1_{k_1}$, where $u^0_{k_1}$ and $q^1_{k_1}$ are the states of the tester T_{k_1} respectively in the global states U_0 and Q_1. Using the mapping function F, we get the first transition in T: $F(u^0_{k_1}) = q_{init} \xrightarrow{a_1}_T F(q^1_{k_1})$ (the τ-actions from q_{init} to $u^0_{k_1}$ contains only the primitives which allow the election of a_1 that is why $F(u^0_{k_1}) = q_{init}$). By applying the Lemma 41 to the sequence (6), we obtain:

$$\forall i \in [1, n] \quad F(u^{i-2}_{k_{i-1}}) \xrightarrow{a_{i-1}}_T F(q^{i-1}_{k_{i-1}}) = F(u^{i-1}_{k_i}) \xrightarrow{a_i}_T F(q^i_{k_i})$$

where $k_i = port(a_i)$. This implies that $a_1 a_2 \cdots a_n \in traces(T)$ and consequently, $traces(DT) \subseteq traces(T)$.

($\supseteq$) Let $\sigma \in traces(T)$, $\exists q_1, \cdots, q_n \in Q^T$ $\exists a_1, \cdots, a_n \in A^T$ such that: $\sigma = a_1 \cdots a_n$ *and* $q_{init} \xrightarrow{a_1}_T q_1 \xrightarrow{a_2}_T \cdots \xrightarrow{a_n}_T q_n$. By applying Lemma 42, $\exists Q_1, \cdots, Q_n$ such that:

$$Q_1 = (q_1, \cdots, q_1, f_1^1, \cdots, f_n^1) \xrightarrow{\tau^* a_1 \tau^*}_{DT} \cdots \xrightarrow{\tau^* a_n \tau^*}_{DT} Q_n = (q_n, \cdots, q_n, f_1^n, \cdots, f_n^n)$$

Consequently, $traces(DT) \supseteq traces(T)$ □

4.3 Possible optimizations

(a) Local choices

For sake of clarity, the compilation rule treats in a uniform way the choices in the initial tester by calling an election service. This election synchronizes the whole of the testers. However, there is no reason so that the testers non-concerned with the choice cannot evolve independently. This has two consequences: (1) it is necessary to add a compilation rule to treat the particular case of local choices, (2) to avoid blocking problems related to the existence of a choice between two τ-transitions from a local tester, it should be considered that the testers obtained are minimal (within the meaning of $\tau^* a$ minimization which eliminates all the τ). The additional rule R2 which is a simplification of R1 can be written:

$$[R2]\ \frac{q_1 \xrightarrow{a}_T q_2 \quad p\_enable(q_1) = \{i\}}{q_1 \xrightarrow{a}_{T_i} q_{12} \xrightarrow{X!sync(q_2)}_{T_i} q_2, \forall T_j \in X \quad q_1 \xrightarrow{T_i?sync(q_2)}_{T_j} q_2, \forall T_k \notin X \quad q_1 \xrightarrow{\tau}_{T_k} q_2}$$

where $i = port(a)$ and $X = \{T_j \mid j \neq i \wedge j \in p\_enable(q_2)\}$

(b) Synthesis of the election service

The distributed choices can be transformed into local choices by using the artifice illustrated in figure 4. The observable traces of the tester remain unchanged. The application of rule R2 produces distributed testers running without the general service of election. More precisely, the election is ensured by the circular transfer of control.

4.4 An example of distribution

Let us consider a test case whose model is presented in the left part of the figure 5. This tester decides to emit interaction a towards the IUT then awaits a reaction b or c. If it receives b, it then awaits c. One wishes to distribute this tester on two sites, one having the responsability of interactions a and b, the other of c.

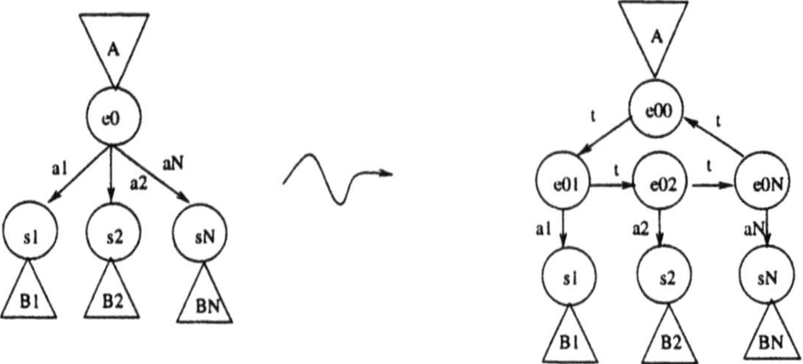

Figure 4 *Preliminary transformation of the test case leading to an integrated election service on a ring of testers where t denotes a τ action.*

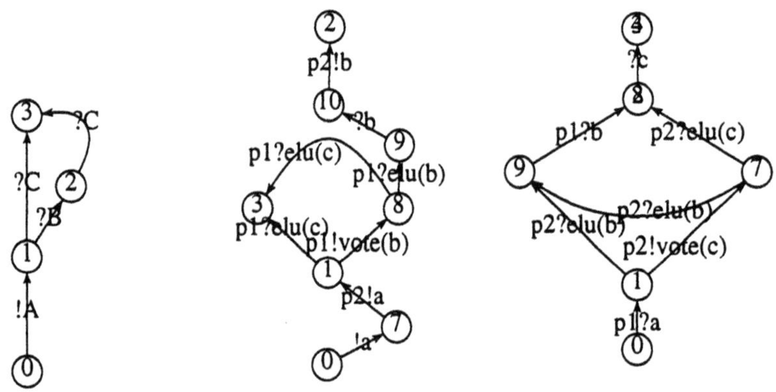

Figure 5 *The initial test case and the result of the synthesis rules.*

The application of the compilation rule R1 produces the two testers introduced in the figure 5: $!a$ and $?b$ are located on tester 1, $?c$ on tester 2. Notice that the synchronizations "P2!a → P1?a" and "P2!b → P1?b" (through the coordination points P1 and P2) ensure the sequencing of "!A?B?C" and "!A?C", as well as the use of the election service ensures that the testers make the same distributed choice. The figures in Figure 5 and 6 have been obtained using the Aldebaran Toolbox [14].

We consider now that these testers interact asynchronously and use synchronously (locally) the election service whose abstracted form is described in left side of figure 6 (the primitives elu() and vote() correspond respectively to req() and resp() in rule [R1]). The whole behaviors of the system obtained is given by the automaton in the right side of figure 6, which is trace-equivalent to the initial centralized tester.

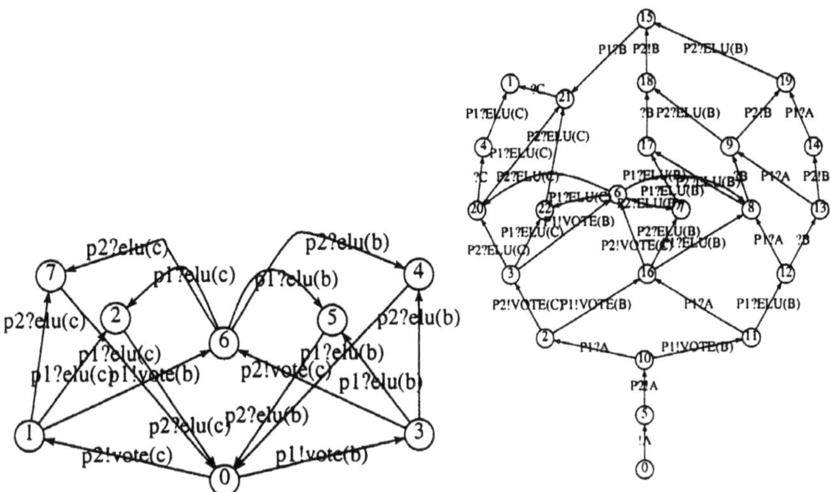

Figure 6 *The abstract spec. of the election service and the result of the composition, which is trace-equivalent to the initial automaton of figure 5*

5 CONCLUSION

In this paper, we gave the first results of our works concerning distributed testing of distributed architecture. We have proposed a method for an automatic distribution of a centralized test case. In this kind of test architecture, several testers coexist and interact in an asynchronous way to coordinate the testing activity. The testers communicate with the IUT, also in an asynchronous way via the PCOs. Under the assumption of an underlying service of election, we generate a distributed test whose traces reduced to the events on PCOs are the same ones as the traces defined by the centralized test case.

As a distributed testing is typically asynchronous, we have to be carefull on the validity of the original test case in an asynchronous environment. In fact, a correct test which does not reject conformant implementations in a synchronous environment can be biased (i.e. it rejects a conformant implementation) when it is carried out in an asynchronous environment. We have also defined a characterization of unbiased test cases preserved by the existence of an asynchronous environment. We show that it can be computed starting from the specifications of the IUT and the tester. Its calculation would be an interesting functionality of a tests generation environment.

As this work is just first steps towards automatic generation of distributed tests, several prospects remain open: (1) investigating a way to generate automatically the service of election as suggested in Section 4.3, (2) the possible concurrencies in distributed testing come from the desynchronization between the IUT and the testers. We will try to find a way to extract the potential concurrency in the test cases and to use it in the distribution, (3) reduction of the number of messages exchanged by the testers for their coordination.

REFERENCES

[1] J. Tretmans. *A formal approach to conformance testing.* PhD thesis, University of Twente, Enschede, The Netherlands, 1992.

[2] M. Phalippou. *Relations d'implantations et Hypothèses de test sur les automates à entrées et sorties.* PhD thesis, Université de Bordeaux, 1994.

[3] J.-C. Fernandez, C. Jard, T. Jéron, and C. Viho. Using on-the-fly verification techniques for the generation of test suites. *Conference on Computer-Aided Verification (CAV '96), USA*, LNCS 1102. Springer, July 1996.

[4] E. Brinksma. A theory for the derivation of tests. *Protocol Specification, Testing and Verification*, North Holland, 1988.

[5] R. Castanet and O. Koné. Deriving Coordinated Testers for interoperability. In *6th Int. Workshop on Protocols Test Systems*, North Holland, 1994.

[6] OSI Open Systems Interconnection. Information Technology - Open Systems Interconnection Conformance Testing Methodology and Framework. *International Standard ISO/IEC 9646-1/2/3*, 1992.

[7] J.-C. Fernandez, C. Jard, T. Jéron, and C. Viho. An experiment in automatic generation of test suites for protocols with verification technology. *Science of Computer Programming - Special Issue on Industrial Relevant Applications of Formal Analysis Techniques*, 1997.

[8] D. Hogrefe and J. Tretmans. Report on the standardization project : Formal methods in conformance testing. *9th International Workshop on Testing of Communicating Systems.* Chapman & Hall, September 1996.

[9] J. Tretmans. Test generation with inputs, outputs and repetitive quiescence. *Software - Concepts and Tools*, 17, 1996.

[10] L. Boullier, B. Kelly, M Phalippou, A. Rouger, and N. Webster. Evaluation of some test generation tools on a real protocol example. In *IWPTS'94 : Int. Workshop on Protocols Test Systems*, Tokyo, November 1994.

[11] J. Tretmans. Testing Labelled Transition Systems with Inputs and Outputs. In *8th Int. Workshop on Protocols Test Systems*, Evry, France, Sep 1995.

[12] L. Verhaard and J. Tretmans. A queue model relating synchronous and asynchronous communication. In *Protocol Specification, Testing and Verification, XII*, North Holland, 1992.

[13] L. Verhaard, J. Tretmans, P. Kars, and E. Brinksma. On Asychronous Testing. *5th International Workshop on Protocol Test Systems*, North Holland, 1993.

[14] J.Cl. Fernandez, H. Garavel, L. Mounier, A. Rasse, C. Rodriguez, and J. Sifakis. A Tool Box for the Verification of Lotos Programs. In *14th International Conference on Software Engineering*, Melbourne, Australia, May 1992.

[15] O. Henninger. On test case generation from asynchronously communicating state machines. *Testing of Communicating Systems X*. Chap. & Hall, 1997.

[16] C. Jard. How to observe interoperability at the service level of protocols. *Protocol Test Systems, VII*, pages 259-270, Tokyo, 1994. Chapman & Hall.

[17] M. Kim, S.T. Chanson, and S. Kang. An approach for testing asynchronous communicating systems. *Testing of Communicating Systems, IX*, pages 141–155, Darmstadt, Germany, 1996. Chapmann & Hall.

[18] H. Kahlouche, J.J. Girardot, A Stepwise Refinement Based Approach for Synthesizing Protocol Specifications in an Interpreted Petri Net Model. *Proc. of IEEE INFOCOM'1996*, San Francisco, 1996, pp. 1165-1173.

PART NINE

Hardware/Software Development

On the improvement of Estelle based automatic implementations

O. Catrina [a*], A. Nogai [b]
[a] *Institut National des Télécommunications, 9 Charles Fourier 91011 Evry France. e-mail: tavi@hugo.int-evry.fr*
[b] *Politehnica University of Bucharest, 1-3 Bv. I. Maniu, Bucharest 77202 Romania. e-mail: anogai@elcom.pub.ro*

Key words: Automatic protocol implementation, Estelle, transport protocol

Abstract: The main challenges for the tools which derive implementations from formal descriptions are to enhance the efficiency and facilitate the integration in various implementation contexts. We study in this paper the Estelle based implementations. Using the Estelle Development Toolset (EDT), we obtained a realistic implementation of a complex transport protocol, XTP 4.0, functionally comparable with the hand-coded reference implementation. This offered an experimental ground for a survey concerning the automated implementation. Based on runtime measurements, analysis of the Estelle model and of the tool's support, we studied the factors influencing the performance and the solutions to improve it.

1. INTRODUCTION

The use of the formal description techniques (FDT) in the development of the telecommunication systems is continuously extending. For protocol design, validation and test generation, the advantages of the FDT-based methods became obvious. The main challenge remains the complexity of the systems.

However, the ultimate goal of the protocol development process is to obtain the implementation. The automated derivation from the formal specification can ensure that the implementation complies with the specification and can greatly reduce the development and maintenance effort. Therefore, the

---

* On leave from Politehnica University of Bucharest, Romania. This work has been partially supported by the KIT-IDEMCOP action.

adequate support for implementation, as part of a consistent development methodology, can be the cutting edge of an FDT-based tool. Besides the conformance, the implementation must match the efficiency requirements of the application (runtime properties, code size). Although tool support has been available for a long time, the applications are still limited. The main cause is the relatively modest efficiency, compared with the current hand-coding methods. Furthermore, part of the implementation often cannot be obtained automatically: operations not supported by the specification language (e.g., checksum computation, packet encoding and decoding) or not detailed in the formal specification ("local matters" in the protocol definition, e.g., related to resource management). Also, a non-negligible effort is often necessary for integrating the implementation in a particular context (e.g., to interface it with other components, implemented by hand).

The aim of our study was to investigate these difficulties, by obtaining a realistic implementation from an Estelle specification, using EDT (Estelle Development Toolset) [11]. The Xpress Transport Protocol (XTP) 4.0 was an ideal example [15]. XTP 4.0 is a challenge for the implementer: it covers the functionality of existing protocols (TCP, UDP), includes new features needed by modern applications (e.g., multicast) and can accommodate various communication paradigms, media requirements and network properties. We already had a (virtually) complete Estelle specification, verified by simulation [3]. A hand-coded C++ implementation of the protocol, SandiaXTP, was available for comparison [14]. It runs on UNIX machines, in user space, on top of IP or UDP, and provides an application programming interface (API) adapted to the XTP extended functionality.

We used EDT to implement XTP 4.0 from the Estelle specification, for the same environment. The implementation, called E-XTP, provides full XTP functionality to real UNIX applications, can inter-operate with SandiaXTP and has a similar API (E-XTP is an enhanced version of the implementation presented in [4]). Section 2 introduces the EDT implementation model and section 3 contains an overview of the resulted XTP implementation.

The derivation of E-XTP offered a valuable experimental ground for our survey concerning the improvement of the Estelle based automated implementation. We made a joint analysis of the runtime measurements, of the Estelle model and of the tool's runtime environment, to identify the factors affecting the efficiency and to evaluate their influence. We studied pragmatic improvement solutions, including the integration of efficient hand-coding techniques, implementation aware specification styles, optimisation of the tool's libraries. The results are summarised in section 4, structured in two parts: communication issues and system management. Part of the solutions have been experimented during the derivation of the XTP implementation, others are currently being studied for future integration in EDT. Conclusions and further work are presented in section 5.

2. EDT AUTOMATED IMPLEMENTATION

2.1 The C code generation

EDT maps an Estelle system module instance (SYSTEMPROCESS or SYSTEMACTIVITY) to a process of the target operating system and implements the children modules as procedures. It generates a separate program for each body of an Estelle system module in the specification [12].

First, the Universal Generator tool (Ug) splits the original specification, e.g., *spec.stl*, in several separate specifications, one for each system module body, e.g., *mb_k.stl*. Each specification *mb_k.stl* contains a system module from *spec.stl*, embedded (as unique child) in a new Estelle system module, automatically generated by Ug. The role of the embedding module is to transfer messages between the original system module and the environment.

Next, for each specification *mb_k.stl*, the Estelle to C compiler (Ec) generates a C program and a *make* file. The C code generated by Ec is independent of the target implementation platform. Moreover, the same C code is used to obtain the simulator program, with the EDT simulation library, and the implementation, with the EDT implementation library. One can thus be confident that the implementation reproduces the behaviour verified by simulation.

We can continue either by immediately producing a prototype of the distributed system, or by further refinement, to obtain a final implementation. For the first case, EDT can automatically distribute the system's components on user selected hosts. The set of Estelle system module instances and communication links from *spec.stl* is mapped on a set of processes communicating via TCP/IP. Alternatively, we can adapt the implementation to a particular context and improve its efficiency, by customising the primitives in the tool's libraries. The XTP implementation was obtained using the second approach.

2.2 The interface with the environment

The embedding module added by Ug is the interface between the contained module and the environment. Each external interaction point of the original module is attached to an internal interaction point of the embedding module. The transitions of the added module perform the adaptation between the Estelle message based communication and any inter-process communication mechanisms provided by the target operating system (e.g., sockets, message queues, signals, etc.). For this purpose, the EDT implementation library offers four generic primitives, *mxinit*, *mxwhen*, *mxoutput* and *mxsend*, used by Ug as shown in this example:

```
INITIALIZE
  BEGIN
     mxinit;          { IPC mechanism(s) initialization }
  END;
```

```
TRANS {Outgoing interactions processing: for each interaction point and type}
  WHEN s_p_sap.Data_req
  BEGIN
      mxsend(1, 0); { send Data_req using an IPC mechanism }
  END;              { s_p_sap identifier =1, Data_req interaction identifier =0 }
{ ... and the transitions for the other outgoing interactions }

TRANS {Incoming interactions processing (2 interaction points in this example)}
  ANY x : 0 ..1              { for x = any of the interaction points }
  PROVIDED mxwhen(x)         { a message is received for interaction point x }
  BEGIN
      mxoutput(x);           { output the corresponding Estelle interaction
  END;                       { to the interaction point x }
```

This approach avoids any modification of the original Estelle specification and offers maximum flexibility for implementing the interface. The user can choose interface functions from the tool's library or develop customised versions. The restriction is that the IPC mechanisms used for input events must match those known by the kernel function which detects the events, described in the following.

2.3 The dispatcher function

The dispatcher function, provided by the EDT library, is the core of the process which implements an Estelle system module. It contains an infinite loop, in which the process waits for an event, selects the fireable transitions (system management phase) and then fires them (execution phase).

The process blocks waiting for the arrival of incoming messages from the environment or the expiration of running timers (corresponding to enabled delayed transitions). The detection of an event wakes up the process, which starts to execute the loop. A timer management function updates the list of running timers and marks the delayed transitions which become fireable. Then, a set of transitions is selected and fired. The set may contain several transitions, for the SYSTEM-PROCESS attribute, and a single transition, for the SYSTEMACTIVITY attribute

The process continues to run as long as fireable transitions exist, i.e., until it achieves the treatment of the event which woke it up and those which occurred in the mean time. Eventually, the process blocks waiting for another input or time-out event. During a run, the process fetches (at most) one incoming message from each input port and achieves their treatment before letting other messages in. Therefore, an implicit flow control on incoming messages is provided. This is a useful property for an implementation, which is not guaranteed by the Estelle semantics, due to the asynchronous communication model with unlimited queues. As a side effect, the parallelism between the Estelle sub-modules is reduced.

3. XTP AUTOMATIC IMPLEMENTATION

3.1 Overview of the specification and the implementation

The goal of the XTP design was to obtain a connection oriented transport protocol with extended functionality and flexible configuration, adaptable to the fast evolution of the networks and the applications (high speed networks, distributed applications with transactional, multipeer and multimedia communications). The protocol kernel is a toolkit of mechanisms with orthogonal functionality. They can be independently enabled, to obtain combinations permitting various communication paradigms. The protocol definition [15, 2] is not accompanied by a transport service definition. It only suggests how to use the toolkit to obtain a basic set of service types: unicast and multicast connection oriented or datagram service, unicast transaction service, etc.

We adopted an implementation context similar to that of SandiaXTP [14], the reference XTP 4.0 implementation, hand-coded in C++: UNIX environment, UDP/IP sockets at the data delivery service interface, UNIX domain sockets and XTP-specific application programming library at the transport service interface. Figure 1 shows E-XTP and SandiaXTP running on hosts connected to Internet. Figure 2 is a closer look at a host running E-XTP, with the XTP process providing connections for 2 transport service user processes.

The block XTP corresponds to the original XTP Estelle specification. The block XTPI is the embedding Estelle module produced by Ug. The Estelle module XTP describes an XTP entity. It only performs management tasks. For each communication, an XTP entity keeps a state record, called a *context*. A communication (any type of service) requires an *association* of contexts from the participating XTP entities. A CONTEXT module is the state machine which controls the activity at one association endpoint. The CONTEXT instances are dynamically created by XTP, for the duration of an association. The module PACKET_MNGM is the interface with the underlying connectionless network service. It submits outgoing packets to the network layer and decodes incoming packets and routes them to the destination CONTEXT instances.

The Estelle specification covers almost completely the XTP features presented in the informal specification. The mechanism for allocating CONTEXT instances to users, a "local matter", was not specified in Estelle (it was left to the C primitives of the interface). An interaction of the transport service interface just carries the index of the source or destination CONTEXT instance. The specification remained of moderate size (8000 lines), compared with the rich XTP functionality, by exploiting the XTP design principles (mainly for multicast) and by using a parametric transport service. The Estelle support for complex algorithms was essential for covering the XTP functionality, e.g., for managing the receiver group state in a multicast transmitter.

The core of the implementation consists of the C code generated by the Estelle compiler from XTPI (20000 lines) and the EDT library functions which implement the Estelle model (inter-module communication, transition selection, etc.). The size of the executable code is 210 kilobytes, similar to that of SandiaXTP 1.5.

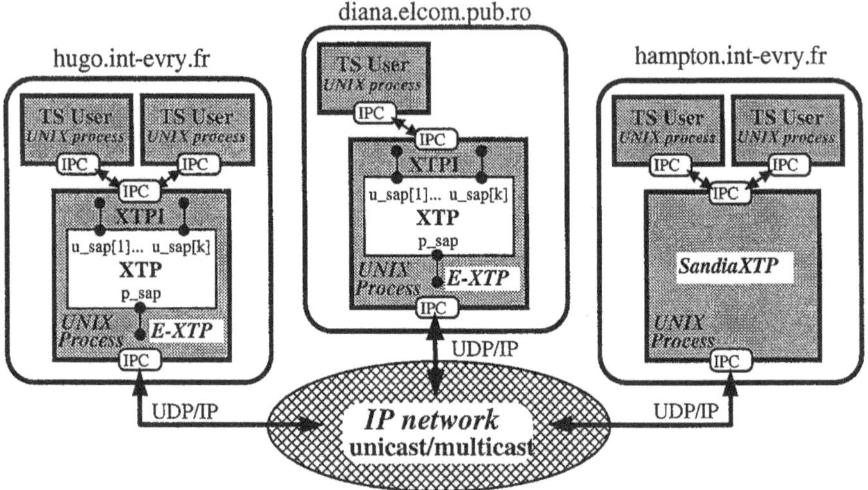

Figure 1. Hosts running E-XTP and SandiaXTP (TS = transport service).

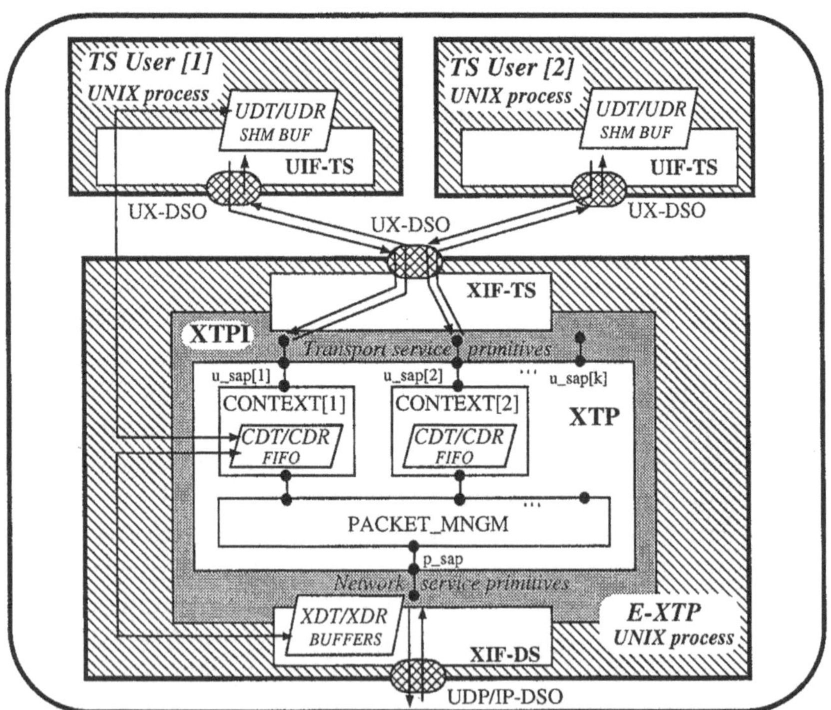

Figure 2. E-XTP structure and the interactions with the user processes.

UIF-TS is the interface of the user process for access to the transport service. It provides an API library [5] similar to that of SandiaXTP. *UIF-TS* is a state machine driven by messages received from E-XTP and by function calls made by the application program. Two communication mechanisms are involved: message transfer, via a UNIX domain datagram socket (*UX-DSO*), and shared memory, for sent and received data units. *UIF-TS* takes all the data types necessary for communicating with E-XTP from two C header files generated by the Estelle compiler from the specification.

XIF-TS is the interface of the XTP process with the transport service user processes. It consists of four C primitives, which are customised versions, for the XTP transport service interface, of the four EDT generic interface primitives (section 2.2). *XIF-TS* provides the adaptation between the Estelle inter-module communication mechanism and the UNIX domain datagram socket (*UX-DSO*), used for exchanging messages with the user processes. It is also responsible for the allocation of CONTEXT instances to users and for routing the messages between the user processes and the CONTEXT instances.

XIF-DS is the interface of the XTP process with the underlying data delivery service. It consists of four C primitives, customised versions for the UDP/IP interface, of the four EDT generic interface primitives. *XIF-DS* provides the adaptation between the Estelle inter-module communication mechanism and the UDP datagram socket (*UDP/IP-DSO*). It also performs the adaptation between the XTP packets representation in the Estelle specification and the real XTP packet formats, as well as the checksum computation.

UDT/UDR-SHM are shared memory buffers of the *UIF-TS* interface, used for data transmission and data reception, respectively. They are created by the user process and shared with the XTP process. These buffers permit the data transfer between the user process and an XTP context with a single data copy. *XDT/XDR-BUF* are collections of data buffers used by the *XIF-DS* interface. They store the data segments of the outgoing XTP packets and the received XTP packets, respectively. The transfer of the XTP data packets between the XTP contexts and the data delivery service requires a single data copy.

A CONTEXT instance contains the pair of buffers *CDT/CDR-FIFO*. *CDT* stores the transmitted data until acknowledgement. *CDR* stores the received data until the recovery of the lost segments and the delivery to the user. They existed in the Estelle specification, but the Pascal procedures for managing them are replaced by C primitives, aware of the *UIF-TS* and *XIF-DS* buffers.

3.2 E-XTP versus SandiaXTP

E-XTP and SandiaXTP provide similar XTP functionality, with similar API functions and can inter-operate. Both are user space implementations for UNIX environment and use sockets and shared memory for inter-process communication. However, there are important differences in the state machine design and some protocol functions are performed using different policies.

Figure 3 shows throughput measurement results for SandiaXTP and E-XTP. They are made for a one-way data transfer, with Sparc stations Ultra 1, running Solaris 2.5, on 10 Mbit/s Ethernet. The following protocol settings are used: flow control, error control and checksum enabled; rate control disabled; no segmentation, acknowledgement at end of window. For these typical settings, the two implementations offer quite similar throughput values. E-XTP seems better tuned for multicast communications. Running in user space is a major handicap for both of them. As shown in the next section, about 50% of the processing time per data unit goes to system calls.

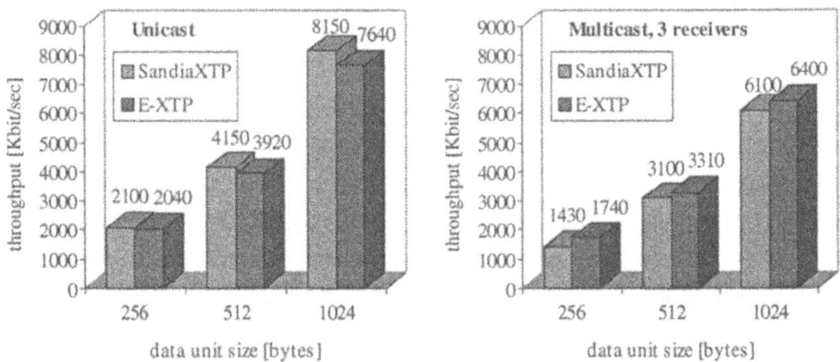

Figure 3. Throughput of E-XTP and SandiaXTP (user buffer to user buffer) as a function of the data unit size. The window size is 24 data units.

4. THE FACTORS AFFECTING THE EFFICIENCY

4.1 Communication mechanisms

Data and packets manipulation is the most obvious cause of overhead in a layered protocol architecture [6]. The various optimisation techniques used in hand-coded implementations [13] are based on the following principles:

- Avoid moving the data units by storing them in shared memory buffers and passing pointers to buffers between the communicating entities.
- Manage the variable size of the data units (insertion and removal of layer specific headers) without copying them, using either off-setting or scatter-gather. In the former case, a data unit is stored in contiguous memory, in a buffer accommodating the maximum data unit size, at a variable offset. In the latter case, the components of a data unit are kept in individual buffers and a descriptor indicates the number of buffers and their addresses.
- Integrate data processing (checksum computation, encryption, etc.) for multiple layers, to reduce the number of memory cycles [1].
- Anticipate the contents of the packet's or message's components, to minimise processing when sending them.

However, the integration of these optimisations in the automatic code generators is complicated by the FDT constraints and the difficulty to determine when and what some particular technique is useful and can be safely applied.

For a system modelled in Estelle, the communication consists (mainly) of asynchronous message exchange. The mapping of module instances to processes or procedures and the scheduling algorithm affect the implementation of the communication mechanisms. We refer in the following to the EDT model (a typical solution). Two kinds of communication can be identified:

- external communications, with the environment of the process;
- internal communications, with (sub)modules within the process.

For the first category, the implementation must rely on the inter-process communication (IPC) mechanisms offered by the target operating system. The tool must provide interface libraries for various operating systems and IPC mechanisms or/and enough flexibility to allow the user make the necessary adaptations. The latency of the IPC mechanisms is often the main cause of inefficiency. Shared memory and scatter-gather, if available and exploited by the tool's libraries, can avoid or minimise the data copying overhead.

The efficiency of the internal communications is mainly affected by the FDT restrictions. A general (always safe) solution cannot avoid copying the message parameters. For the implementations obtained with EDT, the transfer of an interaction implies a single copy of the parameters. First, an interaction buffer is allocated from a common pool and the parameters' values are copied to it. Next, the buffer's address is appended to the destination queue. The interaction's parameters are directly accessible to the receiving transition. The buffer is released when the execution of the receiving transition terminates.

However, the modules often just forward received information, possibly adding or removing some parts. This is due to the information transfer principles in layered protocol architectures and to message routing within the protocol entities. In such cases, the copy operation is redundant and very harmful to the implementation's performance (e.g., for a non-trivial protocol entity architecture, the packets are repeatedly copied). Queue management and message routing have a relatively more modest contribution.

Here is a simple example of packet transmission:

```
VAR p: pdu_type;
BEGIN
    build_pdu(p);
    OUTPUT ip_out.pdu(p);
END;
```

First, the packet is built in a local variable. Then, the OUTPUT statement is executed, allocates a buffer and copies the variable. The packet's fields (control, data segment) are copied twice, first to the variable, next to the interaction buffer. Obviously, the operations should be done in the reverse order: one should first allocate the buffer, then use it to build the packet. For instance, the local variable *p* could be mapped to an interaction buffer. This requires a code generator able to detect that the variable is used as a parameter in a single OUTPUT statement and it is not modified afterwards.

The remark also inspires a different specification style. It is often possible to have most of the contents of the packet already prepared, as part of the local state. One can define the components of the packet as parameters of the interaction and let the OUTPUT statement assemble them (which is also more favourable for optimisations using scatter-gather or off-setting). Here is another example, a data transmission using the suggested solutions:

```
WHEN ip_in.data_req(sdu)
   BEGIN
      put_buf(sdu);                { append the data to the transmission buffer }
      IF can_send THEN BEGIN       { flow control permits transmission }
         crt_header.sequence := (crt_header.sequence + 1) mod MaxSeq;
         OUTPUT ip_out.dt_pdu(crt_header, sdu);
      END; { else: send when a received CNTL packet advances the window }
   END;
```

With this specification style, the transfer of the interaction causes only one copy of the packet's fields. The example also introduces another element involved in data manipulation: the transmission buffer, responsible for storing the data until the transfer is acknowledged. The copy operations to/from this buffer can be avoided if the data are stored in shared memory buffers and both the message transfer management and the transmission buffer management are aware of this. The operations depend on the protocol rules, unknown to the code generator. However, there are solutions which do not require special code generator capabilities and satisfy the generality required from a specification. We can make the *sdu* type a buffer descriptor (which can be copied without harm) and use C primitives for protocol buffer management.

In conclusion, several strategies can be used for integrating optimised data manipulation techniques in the automatic implementations.

One approach relies on the code generator for detecting particular situations when it can use an optimisation, instead of the general (always safe) method. Such a solution is proposed in [9]. Alternatively, the code generator could be guided with annotations inserted in the specification, indicating explicitly the context of the desired application of some optimisation technique. We are currently studying this solution, which seems more flexible and realistic, as it relies on the user's knowledge of the protocol's functionality.

Another approach uses C primitives for buffer management provided by the tool's library. It is the only solution which can address particular features of the implemented system and of the implementation context. It can complement the previous approaches, limited to general message transfer issues.

The latter strategy was used for E-XTP. One goal was to optimise the data path between the user buffer and the UDP/IP interface, with external and internal communications and send/receive buffer management. Another goal was to treat packet manipulation issues: encoding, decoding, checksum computation. The FDT data types and operators do not allow the use of real packet formats in the specification. Hence, the protocol machine works with an internal packet format (also "readable" by the simulator, for validation). The adaptation between the internal format and the real XTP format and the checksum computation are made by the primitives of the UDP/IP interface.

However, these problems mainly concern the control information in the packets. For data segments, the size and the order are the only issues which interest the protocol state machine. Therefore, it only knows a descriptor indicating the size of the data unit and the buffer which holds it. A transmitted data segment joins the header only when moved to UDP/IP and a received packet is split as soon as it is retrieved from UDP/IP (using scatter-gather). All the user data manipulations are done by a set of C primitives, responsible for moving the required amount of data between the *CDT/CDR* ring buffers in the CONTEXT module and the *UDT/UDR* user buffers or the *XDT/XDR* buffers in the *XIF-DS* interface (figure 2). Two copy operations are implied by this scheme (solutions with less copy operations were difficult to apply to the existing specification, requiring the redesign of the data streams management). The buffer management primitives used in the CONTEXT module cooperate with the interface primitives in the XTPI module. Hence, efficient solutions offered by the implementation context could be used: shared memory buffers at the upper interface and scatter-gather at the UDP/IP interface.

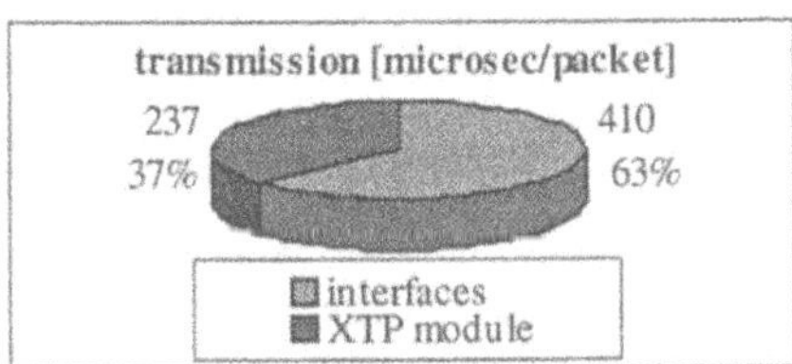

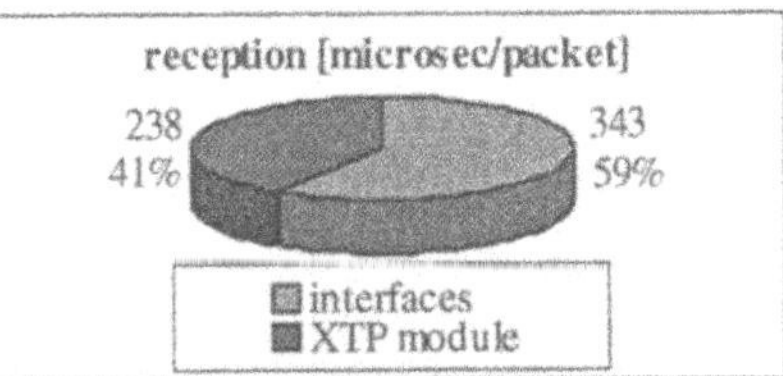

Figure 4. Contribution of the interfaces to the total processing time per data unit, for the XTP implementation (1024 bytes data units, Sparc station Ultra 1).

The experiments with E-XTP indicate a 300% increase of the user-to-user throughput, just by avoiding the copy operations for user data [4]. Figure 4 points out another major cause of throughput limitation: the latency of the socket interface. The values include three system calls: an exchange of messages with the user process, via the UNIX domain datagram socket, and the transmission or the reception of the XTP data packet, via the UDP socket.

E-XTP uses a connectionless underlying service. The lower interface just maps Estelle interactions to UDP/IP socket datagrams. This can be done by primitives from a general purpose library, which call user defined functions for application specific message processing. In general, the mapping can be much more complex, e.g., for a connection oriented service accessible through an API library like TCP sockets or TLI. The flexibility of the EDT approach to external communications can be an important advantage in such cases.

4.2 System management overhead

An Estelle specification describes an hierarchically structured collection of module instances. The system modules occupy the top of the hierarchy. They run independently and synchronise with each other only by message exchange. Within a system module, further synchronisation rules are imposed: priority of a parent module over the children modules and, depending on the module attributes, either parallel synchronous or asynchronous interleaving execution.

According to the operational semantics, a system module performs cyclically a management phase, to identify the fireable transitions, followed by a transition execution phase. This is the usual approach adopted by the Estelle-based implementation tools. The transition selection algorithm is complex and can cause an excessive overhead. Various solutions to reduce this overhead have been investigated [7, 8]: optimisation of the algorithm defined by the operational semantics, structural transformation of the specifications, identification of favourable specification styles, different scheduling policies, etc.

As pointed out in [13], two main classes of models are used in hand-coded implementations: the server model and the activity thread model. With the server model, each system component processes events in an endless loop. The set of components communicate asynchronously and a scheduler offers to each one a fair chance to run. With the activity thread model, the components are implemented by procedures and communicate synchronously. The output of a message from a component is implemented by a call to the procedure which treats that message in the destination component. The system treats one event at a time, as a sequence of procedure calls (i.e., the "activity thread").

The existing automatic implementation tools, including EDT, use the server model, which better matches the Estelle and SDL semantics. The activity thread model can provide a considerable performance improvement, when applicable. However, there are major contradictions between the Estelle model and the activity thread implementation model. Hence, excessive restrictions for the Estelle specifications are needed to permit its use [10].

The current EDT implementation kernel uses a (relatively) optimised transition selection algorithm. It performs a single-pass search in the tree of module instances. For each instance, it looks for fireable transitions using table-based decodification (input/state table) followed by programmed selection (checking of provided clauses, interaction point identity, priorities). The algorithm is fully compliant with the Estelle operational semantics and does not impose any restrictions concerning the Estelle language features.

One of our goals was to evaluate the influence of the system management overhead on the implementation's performance and its dependence on the Estelle model features and specification styles. The analysis was based on measurements made by probes inserted in the dispatcher loop (section 2.3). The time-stamps provided by these probes allow the calculation of pairs of values representing the execution duration for a transition and the duration of the preceding system management phase. The histograms in figure 5 show results of the measurements for E-XTP (average values). They were made on a Sparc station Ultra 1, for the following settings: SYSTEMACTIVITY attribute, one way transfer with 1024 bytes data units, error control, flow control and checksum enabled, no segmentation, one acknowledgement packet for 10 data packets. Figure 6 shows the overall contribution of the management phase to the processing time per data unit, done within the XTP module (computed from the collected samples, taking into account the relative firing rate).

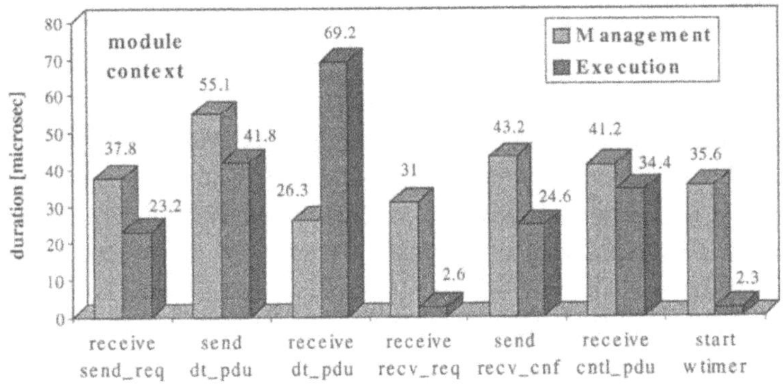

Figure 5. Management and transition execution duration for the data transfer, in a CONTEXT module instance.

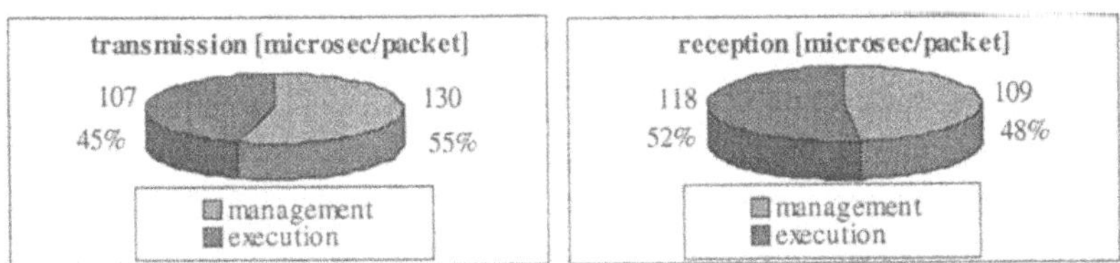

Figure 6. Contribution of the system management duration to the processing time per data unit, in the XTP module.

The joint analysis of various measurements, of the transition selection algorithm and of the specification, summarised in the following, points out the main causes of this relatively important overhead, as well as some solutions.

Due to the synchronisation constraints and the necessity to provide a certain fairness, the scheduler maintains a relatively complicated, tree-like, list of module instances, reflecting the hierarchy. A selection implies a traversal of this list, often partial, but always starting from the root. For each instance, the fast, table-driven decoding, based on main state and input (if any), offers in general two sets of transitions, not just a single transition. One set contains the input transitions for the received interaction(s). The other set contains spontaneous transitions (no input), including delayed transitions, used to model timers. A slower, programmed selection must be used to choose a transition from these sets, by checking the PROVIDED clause, the priority and (for input transitions only) the interaction point.

The influence of the synchronisation rules on the selection duration depends on the depth of the hierarchy and the module attributes. An adequate choice of the attributes can reduce the overhead per executed transition. The SYSTEMPROCESS or PROCESS attributes require the visiting of all the children instances (when the parent has no fireable transition), but all the offered transitions are executed before the next management phase. The SYSTEMACTIVITY or ACTIVITY attributes allow the termination of the search at the first module instance offering a transition (with some fairness rule). For a high degree of concurrency between children instances, the former attributes are the preferred choice, otherwise, the latter.

The spontaneous transitions, mainly the delayed ones, are a major cause of overhead. They should be avoided, but two Estelle features demand their use.

The first one is the constraint to indicate the next main state in the transition header, before processing the received interaction. If the next state cannot be determined by a simple condition, tested in a PROVIDED clause, a spontaneous transition must be added just to change the main state. This artificial solution can be avoided by relying on variables and the PROVIDED clause, instead of the STATE declaration and the clauses FROM and TO. However, in this case, the table-driven decodification works on input only and a larger set of transitions remains for programmed selection. The possibility to change the state in the transition's body and to use output interactions in procedures would reduce to a large extent the need to use spontaneous transitions.

The second source of spontaneous transitions is timer modelling, using delayed transitions. A timer is associated to each delayed transition and runs while the transition is enabled. The enabling of a delayed transition corresponds to a "start" timer command and the disabling to a "stop" timer command. The transition is fired when the timer expires. The main state and/or a provided clause can be used to control the timer's activity. The inefficiency is due to the fact that the start/stop commands are *implicit* and the scheduler must check systematically the conditions of the delayed transitions to detect if a timer must be started or, conversely, if a running timer must be stopped. It is often necessary to restart a timer before expiring. Such a "restart" timer command, can be obtained by combining a "stop" and a "start". However, two computation steps are necessary: the first one to disable the delayed transition, the second one to re-enable it. Another spontaneous transition must be added for the second step (e.g., the transition "start wtimer", in figure 5).

The presence of multiple interaction points, individual queues and transition priorities complicates the selection of the input transitions. First, there is a list of queues to be visited. Next, all the queues must be visited, even after finding an interaction, to check the priorities. Therefore, the preferable specification style is to indicate common queue as default and use individual queues only when really necessary. A common queue is especially important for multiplexer modules (e.g., PACKET_MNGM in the XTP specification) and usually there is no inconvenience. This kind of functionality is present very often in the systems and requires arrays of interaction points and transitions parameterised using the ANY clause. The selection algorithm and the code generator must treat these features with special care.

The excessive overhead of the outlined algorithm is due to the complex search of the transitions to fire. It is a general method, able to deal with the worst case. The solution we are studying is to replace the search by an event based management of a list of instances scheduled for (transition selection and) execution. If there are no spontaneous transitions, the list can be organised based on transferred messages: once a message is output, the receiving instance is inserted in the list. The solution can be extended for spontaneous transitions, if the PROVIDED clauses do not contain primitives or exported variables. In such a case, they can only be enabled by the execution of another transition of the same instance. The spontaneous transitions are checked only in this situation. If anyone is fireable, the instance qualifies for the scheduling list. It seems possible to obtain a considerable improvement for limited restrictions, acceptable by many applications.

5. CONCLUSIONS

The derivation of E-XTP from the XTP 4.0 Estelle specification, allowed us to asses the current EDT support for automatic implementation. E-XTP is functionally comparable with the hand-coded implementation SandiaXTP and inter-operates with it. They have similar, but relatively moderate performance.

The automatic implementation methods can match the requirements of the modern high speed network environment only if they integrate the highly efficient techniques already used in hand-coded implementations [9, 13]. Of main interest are the solutions used for data and packets manipulation and for mapping the system's components (e.g., entities in a layered protocol architecture) to processes or threads of the target implementation environment.

We experimented the optimisation of the data stream implementation using libraries of primitives which integrate the internal and external communications and the protocol buffer management. The approach is pragmatic and flexible. The libraries can be reused for applications having the same implementation context. Further benefits can be obtained by adding automatic code generator support for general message transfer optimisation.

Our analysis pointed out the Estelle features responsible for the excessive system management overhead. It also showed that the scheduling algorithm can be substantially improved, for widely acceptable constraints. The efficiency can approach that of the activity threads, without the excessive restrictions imposed by the strict application of this model [10].

The solutions used in the specification must take into account implementation issues and must be correlated with the support offered by the tool. A good basis for the implementation could be obtained without loss of the specification's generality, using just an adequate specification style. The adaptation in the implementation phase is expensive, harmful and should be avoided.

REFERENCES

[1] M. Abbot, L. Peterson. Increasing Network Throughput by Integrating Protocol Layers. *IEEE/ACM Transactions on Networking*, vol. 1, no. 5, 1993.

[2] J.W. Atwood, O. Catrina, J. Fenton, W.T. Strayer. Reliable Multicasting in the Xpress Transport Protocol. *Proceedings of the 21st Conference on Local Computer Networks (LCN'96)*, Minneapolis, Minnesota, USA, 1996.

[3] O. Catrina, E. Borcoci. Estelle specification and validation of XTP 4.0. Deliverable 2 for Task 2.1, Copernicus Project COP62 (COP#62/WP2/2), 1996.

[4] O. Catrina, E. Lallet, S. Budkowski. Implémentation automatique d'XTP à partir d'une spécification Estelle. *Ingénierie des Protocoles - CFIP'97.* Hermes, Paris, 1997.

[5] O. Catrina, E. Lallet, S. Budkowski. Automatic protocol implementation using Estelle Development Toolset. Research Report 971001, INT Evry, France, 1997.

[6] D. Clark, V. Jacobson, J. Romkey, H. Salwen. An Analysis of TCP Processing Overhead. *IEEE Communications*, 27(6), 1989, pp. 23-29.

[7] R. Gotzhein, et al. Improving the Efficiency of Automated Protocol Implementation Using Estelle. *Computer Communications* 19, 1996.

[8] T. Held, H. König. Increasing the efficiency of computer-added protocol implementations. *Protocol Specification Testing and Verification XIV.* Chapman & Hall, 1995.

[9] R. Henke, H. Konig, A. Mitschele-Thiel. Derivation of Efficient Implementations from SDL Specifications Employing Data Referencing, Integrated Packet Framing and Activity Threads. *SDL'97: Time for testing - SDL, MSC and trends.* Elsevier, 1997.

[10] R. Henke, A. Mitschele-Thiel, H. Konig. On the influence of semantic constraints on the code generation from Estelle specifications. *Formal Description Techniques X and Protocol Specification, Testing and Implementation XVII.* Chapmann & Hall, 1997.

[11] Estelle Development Toolset (version 4.1). Institut National des Télécommunications, Evry, France. http://alix.int-evry.fr/~stan/edt.html.

[12] E. Lallet, S. Fischer, J.-F. Verdier. A new Approach for Distributing Estelle Specifications. *Formal Description Techniques VIII.* Chapman & Hall, 1995.

[13] L. Svobodova. Implementing OSI Systems. *IEEE Journal on Selected Areas of Communication*, vol. 7, no. 7, 1989, pp. 1115-1130..

[14] T. Strayer. SandiaXTP User's Guide. SandiaXTP Reference Manual. Sandia National Laboratories, USA, 1996.

[15] Xpress Transport Protocol specification revision 4.0, XTP Forum, Santa Barbara, CA, USA, 1995. For the revised multicast procedures, see [2].

Hardware - Software co-design of embedded telecommunication systems

using multiple formalisms for application development

N. S. Voros[1], S. Tsasakou[1], C. Valderrama[2], S. Arab[2], A. Birbas[1], M. Birbas[1], V. Mariatos[1], A. Andritsou[3]

[1]*Department of Electrical and Computer Engineering – University Campus of Rion 26 500, Greece,Tel: +30 61 97312, Fax: +30 61 997333 , E-mail: voros@ee.upatras.gr*

[2]*TIMA/INPG 46 Ave Felix Viallet F-38031 Grenoble Cedex, France,Tel: +33 476 574 870, Fax : +33 476 473 814, E-mail: CA.Valderama@imag.fr*

[3]*INTRACOM S.A, P.O. Box 68 Paiania Attikis 190 02, Greece, Tel : +30 1 690082, Fax : +30 1 6860312, E-mail: anda@intranet.gr*

Key words: HW-SW co-design, multiple formalisms, embedded telecommunications systems

Abstract: In this paper a co-design methodology based on multiformalism modelling is presented. It defines a platform that integrates different notations and, the necessary mechanisms to handle different in nature models in a coherent way. The supported formalisms cover a wide area of application domains, allowing the designer to select the notations that are mostly appropriate for his/her application. The proposed co-design development cycle provides a full blown path from system specification to a virtual prototype of the system under construction. The methodology has been applied for the design and implementation of a MAC layer protocol, named MASCARA, for providing ATM QoS over wireless connections.

1. INTRODUCTION

During the last years, a new generation of design methods for embedded systems, referred to as co-design [8, 10, 12, 13, 14, 15, 17, 18], has emerged and matured. These methods are able to handle mixed hardware/software systems starting from behavioural-level down to implementation. They are based on co-design tools that are capable of providing concurrent design of the different parts of heterogeneous distributed systems, automation of H/W-S/W partitioning and automation during the integration steps at the lower levels of the development cycle.

In the next paragraphs we present a co-design development cycle that supports different specification formalisms. The goal is to propose a methodology that embraces a wide range of application domains [6] under a consistent co-design framework that leads to a virtual prototype of the final system. The proposed methodology framework is exhibited through the design of MASCARA protocol, a MAC layer protocol for wireless ATM networks.

2. CO-DESIGN USING MULTIPLE FORMALISMS FOR SYSTEM DEVELOPMENT

The increase of complexity in embedded systems and the demand for virtual prototypes as early as possible in the design cycle, are putting strict requirements on methodologies and the tools supporting them. The co-design approach proposed in this paper aims at building a coherent couple of methodology-toolset for the entire design cycle of embedded systems. Particular emphasis is given on the phases of system-level co-modelling, co-simulation , partitioning and co-targeting.

Co-modelling aims at formalising the specifications of an embedded system. Such specifications usually include several evolving models described by means of different notations. *Co-simulation* can be used at this stage, before partitioning, to conduct an early validation of the specifications. Co-simulation, as it says, involves simulation of at least two different models that interact among each other e.g. exchange data dynamically. Then comes *partitioning* to decide which parts of the system are going to be implemented in S/W and which parts in H/W. Partitioning takes into account the target platform (e.g. processors, distributed architecture etc.) to assign each system part. Co-simulation can also be used at this stage to estimate the validity of the choices. Finally, once the partitioning is agreed, *targeting* takes place. It involves S/W code generation, H/W synthesis, realisation of interfaces between H/W and S/W modules, and implementation of the whole system into the target platform.

The proposed methodology-toolset materialises a design environment for hybrid (continuous and discrete) H/W - S/W embedded systems. It supports a global design flow, based on multi-formalism modelling and executable prototypes to validate the system design. The toolset accompanying the proposed methodology has been built from existing products and prototypes, which have been adapted and extended for the needs of industrial end-users. Interfaces have also been developed for connecting external tools such as VHDL tools for H/W design. The co-design flow detailed in Figure 1 outlines the methodological stages involved in the design of embedded systems.

From the formalism point of view, Figure 2 depicts how the associated supporting tools are combined in the context of the proposed approach. The circles depict the co-modelling and co-simulation modules that have been developed for the need of our methodology. They implement MATRIXx-SCADE co-modelling , SDL-SDL co-execution, MATRIXx-Statechart co-simulation, SDL-MATRIXx-Statechart co-simulation and C-VHDL co-simulation [6]. Therefore multi-formalism co-simulation is possible both at the specification level and at the implementation level. Two more arrows have been drawn to indicate improvements dealing with software distribution using SEA and VHDL generation from SDL using COSMOS. The latter platforms are used in the partitioning phase: COSMOS produces C for S/W parts and VHDL for H/W parts, while SEA generates C from Statechart and MATRIXx models for distribution on a given H/W platform. Code generation and targeting is realised by taking into account the constraints attached to embedded systems such as e.g. communication efficiency and memory limitations.

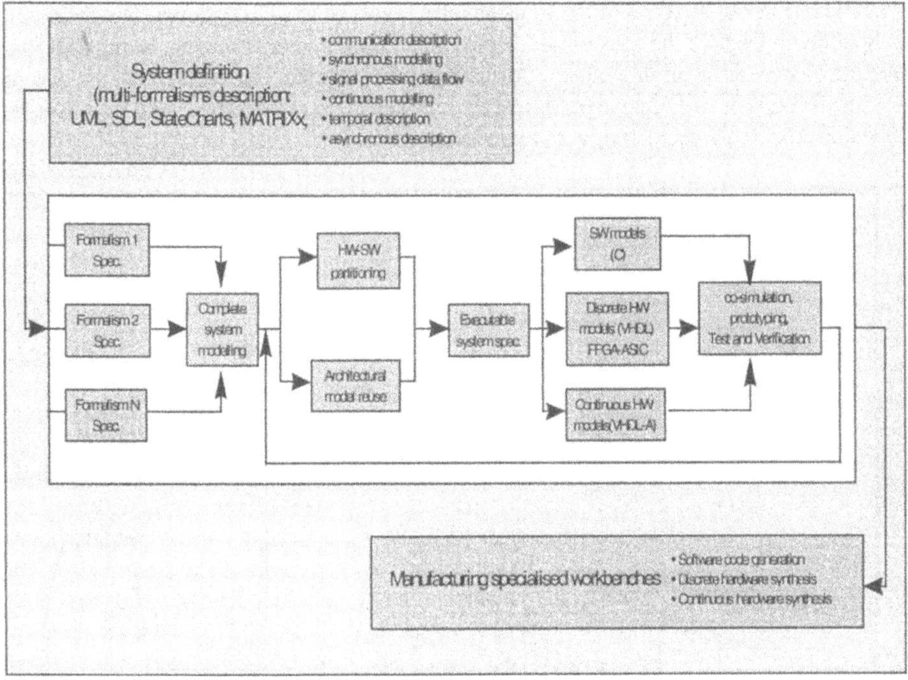

Figure 1. H/W-S/W co-design using multiple formalisms for embedded system design

Regarding the tools that match with the various phases of the proposed methodology, the following could be mentioned. For co-modelling, the user can employ four notations: Lustre [6], SDL [1, 3, 6], Statecharts [7, 6] and MATRIXx [6] diagrams. Each of these notations is supported by a corresponding tool and the proposed toolset provides for their efficient interconnection. With respect to partitioning, the user can rely either on the COSMOS [4, 11, 17] approach (covering SDL, Statechart, C and VHDL formalisms), or on the SEA [5] approach (involving MATRIXx, Statechart and C notations) for software distribution on hardware. As already mentioned, co-simulation is possible both at the specification and at the implementation levels, for the purpose of model validation and partitioning assessment respectively. Application targeting is generally covered by the partitioning tools and also by the CASE/CAD tools.

3. SDL-STATECHARTS CO-MODELLING PRIMITIVES

The effectiveness of the proposed approach has been tested over a number of different application domains including automotive, aerospace and telecommunications. Since the primary goal of this paper is to present the adaptation of the proposed methodology/toolset to embedded telecommunication systems, we will restrict our analysis to the tools usually involved in such systems. The following sections introduce the idea of developing telecommunication systems using SDL and Statecharts as complementary formalisms for system specifications. A set of transformation rules between Statechart and SDL is also presented, and proposes the required extensions to couple the two languages.

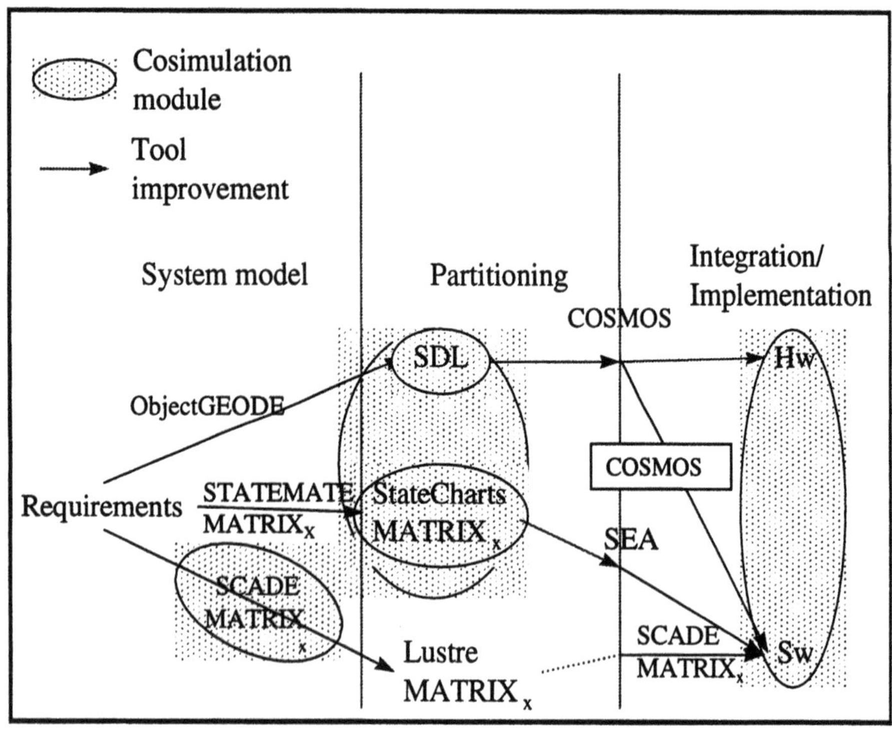

Figure 2. Tool coupling for multiformalism system specification

3.1.Statechart notation vs. SDL notation

The Statechart notation is an extension of the Finite State Machine notation. The primary objective of this modelling technique is to describe the system behaviour by means of a hierarchy of state diagrams. Concepts such as state refinement and concurrent activities are supported, allowing to represent a system behaviour at different levels of abstraction. The Statechart model is now popular and is present in well-know systems and software engineering techniques such as UML [2] or ROOM

[16]. It is used to model real-time systems that are centralised (i.e. without communication and synchronisation) and have a static architecture (i.e. no dynamic task creation).

The SDL notation has been defined to model and develop distributed real-time systems. The concept of dynamic objects that can be created/suppressed, is supported, so the system at run-time may behave as a dynamic network of communicating processes. For this purpose, SDL supports the necessary mechanisms for communication. The simulation and dynamic checking of SDL models are possible because of the underlying formal semantics of the language. The production of executable code is also possible, to be compared for example with the Statechart technique where C/C++ code has to be entered to describe the actions in the transitions. SDL is a graphical language as Statechart is, but does not include state abstraction facilities, and transition diagrams are purely flat diagrams, hierarchical states are not possible, neither entry and exit actions.

3.2.Complementary features between Statechart and SDL

From the conceptual point of view, the Statechart and SDL notations seem rather complementary than in competition. Statechart provides facilities for model abstraction, whereas SDL provides a complete graphical language to describe the dynamic behaviour at the lowest level. Statechart can be used early in the system development, to sketch a first description of the behaviour that can be given to non-expert users. Then this model can be refined with SDL at the detailed design stage, to provide a complete behavioural description. The latter can be transmitted to the development engineers for production, and to the validation teams for checking the conformance to the specification constraints.

Therefore a solution that integrates both notations to take advantage of their respective facilities is welcome. If most of the Statechart concepts are interesting for introduction in SDL, the concept of concurrent activity seems not complementary at all with SDL. The SDL designer usually describes the concurrent objects – that exist at run-time – by processes. Internal concurrency seems unnecessary for these objects; if parallel actions must be described, the right way would be to refine the object in two separate processes which can run in parallel. Besides it must be noted that internal concurrency is forbidden or not recommended in most Statechart techniques such as UML or ROOM. At last, it is worth to remind that internal concurrency is the unique way to structure the system in the original Statechart notation.

3.3.An approach to combine Statechart and SDL notations

The approach should take as basis the complementary nature of Statechart and SDL, as discussed above. The natural way of working is to use Statechart first for conceptual modelling. Basically, a Statechart model is a hierarchy of state diagrams. The next step is to transform them into SDL constructs and to derive an architectural model. Finally the latter will be completed in SDL in order to obtain a detailed design model that can be simulated. It must be noted that those considerations on the use of the concepts are independent of the support provided by the tools.

Therefore the main task is at language level to identify which Statechart constructs must be included in SDL, and how to transform them in standard SDL. For the ease of integration, assume a notation for Statechart that is a superset of UML, without preventing the use of the basic Statechart.

3.4.Mapping between Statechart and SDL

Table 1 makes the correspondence between Statechart constructs and SDL standard constructs. The basic primitives of mapping SDL constructs to Statecharts and vice versa are explained as follows:

Table 1. Correspondence between Statechart and SDL constructs

Statechart construct	Corresponding SDL construct
State name	State name
Entry action	No direct correspondence
Exit action	No direct correspondence
Activity	No direct correspondence
Initial state	Start
End state	Stop
Macro-state	No direct support
Internal concurrency	Not supported within the same process; basically supported between processes
Transition	Transition
Event with parameters	Signal with parameters
Provided (condition)	Enabling (condition)
Action	SDL transition body
Output	Output
Internal transition	Usual transition
Delay	Timer (set, reset and expiration)

- **Activities** : Activities i.e. sequential processing without the triggering of events, are not supported in SDL finite state machines. In SDL, actions are only performed on transitions. However it can be shown that activities can be easily modelled in the SDL notation extended by entry/exit actions as introduced below.
- **Internal Transitions with Entry/Exit Actions** : In order to support internal transitions i.e. execution of internal actions during all the time an object stays in the same state, it will be possible to inhibit automatic expansion of entry and exit actions by means of a specific syntactic rule for state and next state naming.
- **Internal Concurrency** : The usual way in SDL to express internal concurrency i.e. parallel activities defined within the same Statechart, is to refine the object in separate processes, each process implementing one parallel activity. Internal concurrency within the same SDL process has not been implemented.

- **Inheritance** : The Statechart notation does not support inheritance. Therefore inheritance will only be allowed in standard SDL constructs, but not used in the Statechart extensions that have been implemented.
- **Dynamic Semantics** :
  - Each concurrent class (process) has its own Statechart, i.e. a Statechart corresponds to a process. This differs from pure Harel's Statechart – one Statechart model for the whole model – but this conforms to the UML Statechart variant.
  - The dynamic semantics will conform to the SDL standard: asynchronous point-to-point communication, no broadcasting, each finite state machine has one queue that stores the events received but not yet consumed, dynamic process instance creation.

4. SYSTEM PARTITIONING

One of the key issues in the co-design of complex systems is the action of partitioning. The decision of which parts will be implemented in software and which will be transferred in hardware is crucial for the efficient implementation of the final system. Even though several prototype tools already exist for system partitioning, most of them are nothing but simple SDL-to-VHDL translators that produce a description that may feed existing RTL synthesis systems. In the context of the proposed methodology, the hardware-software partitioning is performed using COSMOS toolset, which is presented in the sequel.

4.1.User guided system transformational partitioning

The approach for SDL to C/VHDL proposed within COSMOS environment is built around an intermediate format called SOLAR [4, 11], designed to represent both the system concepts as well as those used by hardware description languages. The intermediate format SOLAR constitutes an intermediate representation permitting the unification of different specifications described in either hardware, software or system description languages. Thus, the different parts of software and hardware of a system can be unified within a single SOLAR format. SOLAR supports three levels of abstraction: the system level, the behavioural level and the Register Transfer Level. In addition, this model facilitates the reuse of existing sub-systems in the form of a library of components and communication models. SOLAR is used as an intermediate format during synthesis and is completely transparent to the designer.

The basic construction within SOLAR is a state table denoted by the keyword StateTable. It permits the specification of hierarchical and communicating finite state machines. In addition, other structures are added to give a modular specification and facilitate the communication between the processes. The design unit (DesignUnit) is introduced for structuring a system into a set of interconnected subsystems. The channel unit (ChannelUnit) permits the specification of communication between the communicating subsystems.

The translation approach adopted is based on the correspondence between the basic concepts in the two languages SDL and SOLAR. Concepts used in SOLAR and in SDL, permits us to observe certain similarities. Nevertheless, SDL describes the behaviour at a level sufficiently abstract, whereas, the SOLAR descriptions, even if they are at a high level, are implementation oriented. The translation process proposed essentially works with the behavioural, communication and structure aspects of the specified system. Since SDL design system is in front of COSMOS, basic concepts of SDL are supported in COSMOS. The following table lists the subset SDL supported by COSMOS. Synthesis is the final stage performed on the SOLAR representation and translation into VHDL, for the hardware components and into C, for the software components.

Table 2. SDL primitives to SOLAR primitives mapping

SDL		SOLAR
Structure	System, blocks, processes	Done
Communication		Done
Hierarchical specifications		Done
Declarations	Data type	Limited to: Dcl, synonym, newtype, literal, structure sort
	Exp/imp procedures	Done
	Shared variables	Done
	ADTs + operators	Done
Process/Procedure bodies	State tables	Done

The organisation of partitioning into several small steps reduces the complexity of the problem. The designer controls the partitioning history within an interactive environment, through fine grain control of the synthesis process. This methodology can be seen as a human guided compilation where the designer spends additional effort to produce an efficient implementation. To facilitate the user's interaction for incremental transformations, all these refinements make use of a set of primitives called *split*, *merge*, *move*, *flat* and *map* allowing to decompose and transform objects. Figure 3 summarises these partitioning primitives.

The user guided transformational partitioning assumes that the designer starts with an initial specification and an architectural solution in mind. System design from specification to implementation is performed through a set of primitives allowing the designer to transform the system, following an incremental refinement scheme, in a distributed model that matches the architectural solution. All the refinement transformations are performed automatically. Additionally, the decisions are made by the designer who uses his knowledge and experience to achieve the desired solution.

4.2.Architecture generation

The architecture generation translates the intermediate format used in COSMOS into executable code (C and VHDL) to allow the co-simulation and co-synthesis of the system. The virtual prototype generated is a heterogeneous architecture composed of a set of distributed modules, represented in VHDL for hardware elements and in C for software elements, communicating through communication modules from a library of components. The virtual prototype is a simulatable model of the system that implements or emulates the initial specification. The generated model allows to separate the behaviour of the modules (hardware and software) and the communication units. Inter-module interaction is abstracted using communication primitives that hide the implementation details of the communication units.

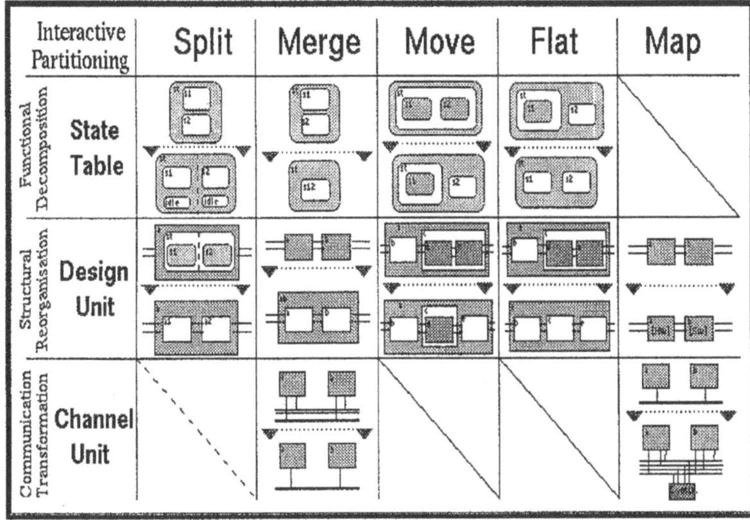

Figure 3. Partitioning primitives used in COSMOS

4.3.C/VHDL co-simulation interface

In the context of the proposed methodology, co-simulation is reckoned as the joint simulation of a system description based on heterogeneous notations, since different parts of the system may be described using different notations. The rationale of adopting multiformlism is that, (a) when dealing with overall system specification, the use of multiple formalisms is desirable since it allows describing different aspects of the system with the most appropriate notation, and (b) when dealing with the virtual prototype of the system we need at least two different notations for describing efficiently the hardware and the software parts respectively.

The COSMOS environment generates a distributed hardware/software co-simulation environment for heterogeneous systems prototyping. The environment provides as features: distributed hardware/software co-simulation and automatic hardware/software interface generation. Hardware components can be described at different levels of abstraction and generic/specific software debuggers can be used. Starting from a brief description of the interface of the interconnected modules, an operation named *communication synthesis* produces automatically the interface between hardware and software parts.

The distributed co-simulation environment is composed of a VHDL simulator (VSS: VHDL SYNOPSYS Simulator) for hardware components and C-debuggers (any tools in the market) for software modules communicating via a software bus. This bus relies on the UNIX IPC layer. The software bus and the elements used to interface VHDL and C modules during simulation are automatically created as result of communication synthesis.

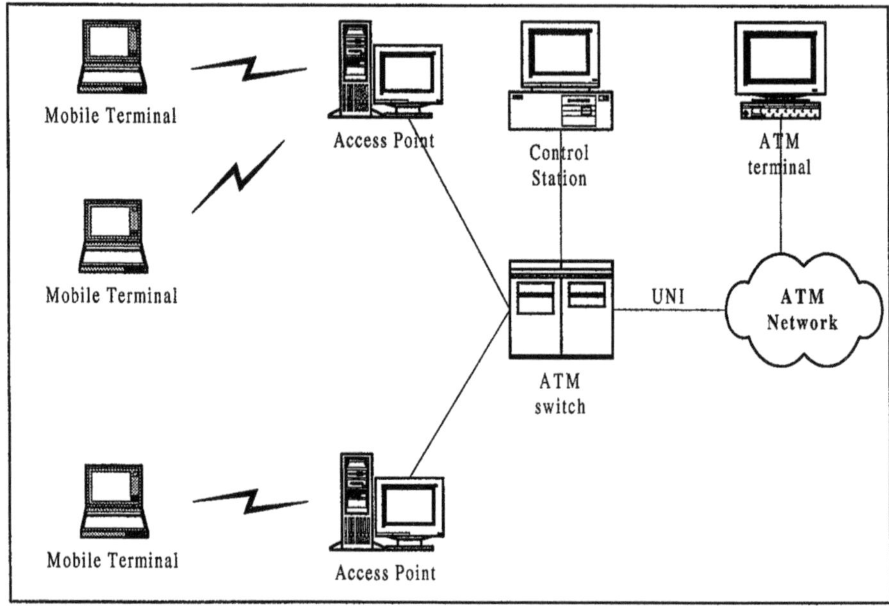

Figure 4. Overview of a wireless ATM system

5. DESIGN OF A MAC LAYER PROTOCOL USING MULTIPLE FORMALISMS FOR SYSTEM SPECIFICATION

The methodology proposed within this paper has been used to develop a MAC layer protocol named MASCARA [9] for offering ATM services over wireless connections. The wireless ATM architecture consists of *ATM switches*, *Access Points* and *Mobile Terminals* as shown in Figure 4. *ATM switch* is a standard customer premises access node, containing also mobility specific software and minimum hardware modifications; *Access Point* is the network element connected to the ATM switches with standard ATM connections; finally, *Mobile Terminal* is the end-user equipment that contains the *wireless ATM radio adapter* card, interfacing the 20 Mbits/sec air-interface.

For connection admission MASCARA provides information on link conditions including e.g. the available bandwidth and cell error ratio. The actual call admission is made taking into account existing connections, the radio link condition and spare capacity required for handovers. The scheduler of the protocol is designed to preserve the admitted connections as conformant as possible within the negotiated connection QoS parameters. The service policy implemented by the scheduler can be tuned to meet the needs of a particular installation or a particular set of users.

5.1.System specification

The first stage of the protocol design involves requirements capturing using UML [2]. The reason for using UML lies in the fact that significant parts of the protocol are already available as standards in UML. So having available standard parts of the protocol, like SAR (Segmenting and Reassembly) unit, we are able to reduce the amount of work needed for designing them (we use them either as they offered or with minor changes). Figure 5, presents a high level specification of MASCARA using UML, while Figure 6 depicts the internal behaviour of the "Control Segmenting and Reassembly" object using Statecharts (UML uses Statecharts for describing the dynamic behaviour of the objects composing a system specification). Due to limited space, we have restricted our specification only to the essential parts of the protocol, while for shake of simplicity we have intentionally eliminated the references among the constituent parts of the protocol. Among the objects composing the MAC layer, *ATM Layer* corresponds to the ATM layers above MASCARA that implement the classical ATM services; *Physical Layer* implements functions pertaining to the actual physical layer (in our case the transmission medium is air) of a traditional protocol stack; the remaining objects are part of the MASCARA specification and incorporate aspects concerning mobility.

The final stage of the specification involves description of the protocol's parts using SDL (Figure 7 reflects part of the UML specification presented in Figure 5). The objects have been transformed in blocks and their internal behaviour has been implemented using SDL structures (see Table 1). Having completed the system specification using objects, we transform them automatically to SDL adding the necessary implementation details for having a more detailed version of protocol's specification. This way, we combine of-the-shelf components described in UML with newly created components described in a powerful language like SDL. Having a detailed system specification in SDL, we are able to exploit the power of state-of-the-art tools like ObjectGeode which incorporates a set of powerful utilities for SDL-to-Statechart translation, exhaustive simulation, deadlock detection etc. So, we are able to verify/validate the system's specification early enough in the co-design flow, where the cost of repairing possible design flaws is relatively small.

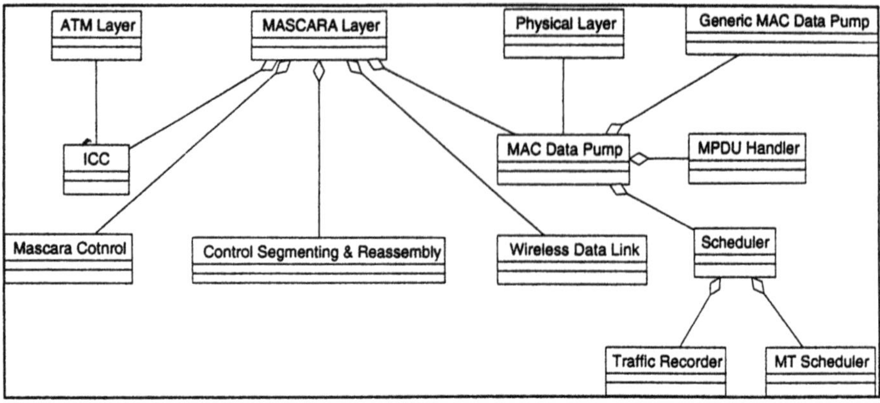

Figure 5. The constituent parts of the MASCARA protocol in UML

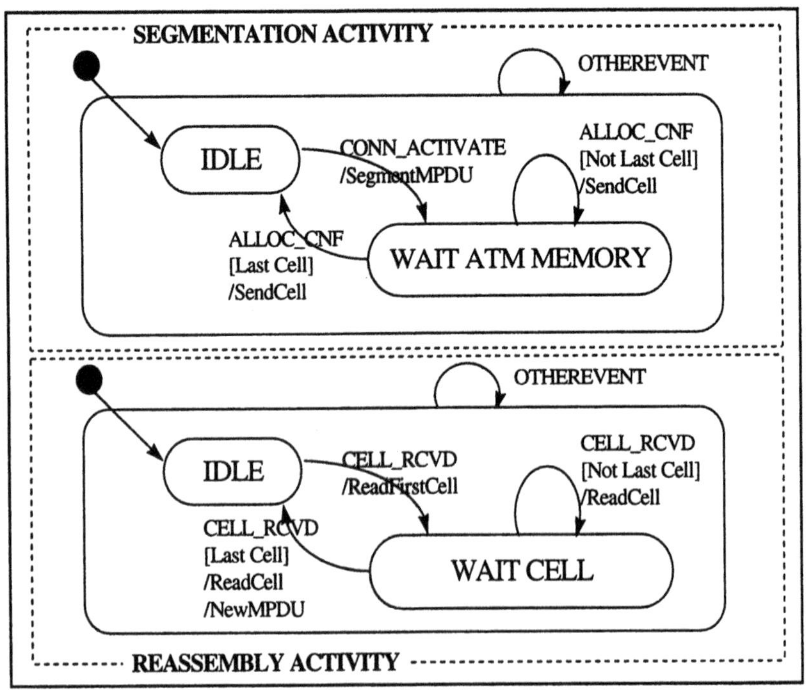

Figure 6. Statechart representing the dynamic behaviour of "Control Segmenting and Reassembly" object

5.2.System partitioning

This section refers to H/W-S/W partitioning of the MASCARA protocol. We will illustrate the overall design flow in COSMOS, from the system-level specification given in SDL to the distributive hardware/software architecture described in C/VHDL. The use of SDL allows for fivefold reduction of the size of system's specification when compared to distributed C/VHDL models. All the transformation applied in this section are fast enough to look instantaneous during an interactive session.

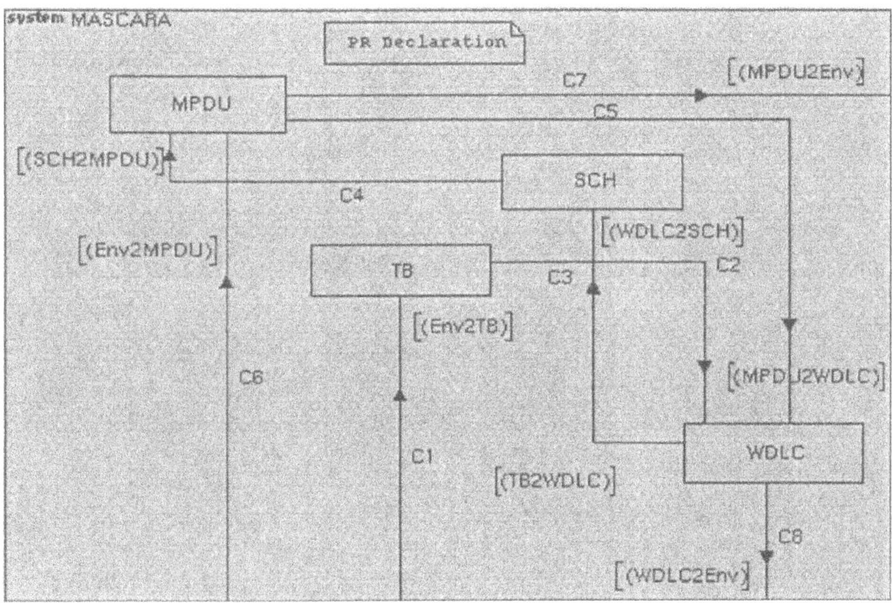

Figure 7. Specification of a small part of the MASCARA protocol in SDL

The SOLAR intermediate format is obtained by an automatic conversion from the initial SDL description. During the conversion, the following operations are performed: processes are converted to design units, processes behavior are converted to state stables and interconnections among SDL processes are converted to abstract channels. Figure 8 depicts the initial SOLAR model. The refinement steps performed for the production of the MASCARA protocol's virtual prototype, are listed in the sequel: (a) *structural flat* to keep out the hierarchy of all the design units, (b) *protocol selection*, where a communication protocol is chosen from a protocol library (the FIFO and rendez-vous protocols were chosen to implement all the abstract channels), (c) *channel map*, where the abstract channels were replaced by well defined signals interfaces, (c) *hardware/software assignment*, where (after a trial and error approach) the processes WDLC and SCH and the FIFO controller of WDLC are set to have a software implementation. The processes TB, MPDU and the others FIFO controllers are set to have an hardware implementation.

Summing up, the SDL specification has been transformed to SOLAR so as to enable efficient system partitioning. Each process has been represented as a design unit whose internal behaviour has been specified using a state table (which is the conceptual equivalence of the an FSM in SDL or a Statechart in UML). The communication scheme among the various design units have also been specified more accurately by selecting specific protocols for implementing their communication. For example, the concept of channels or signal routes that connect two system blocks in SDL are no longer abstract. In fact, the channel notion incorporates concepts such as communication protocols (e.g. FIFO, Rendez-Vous, etc) and specific bus (channel) arbitration schemes.

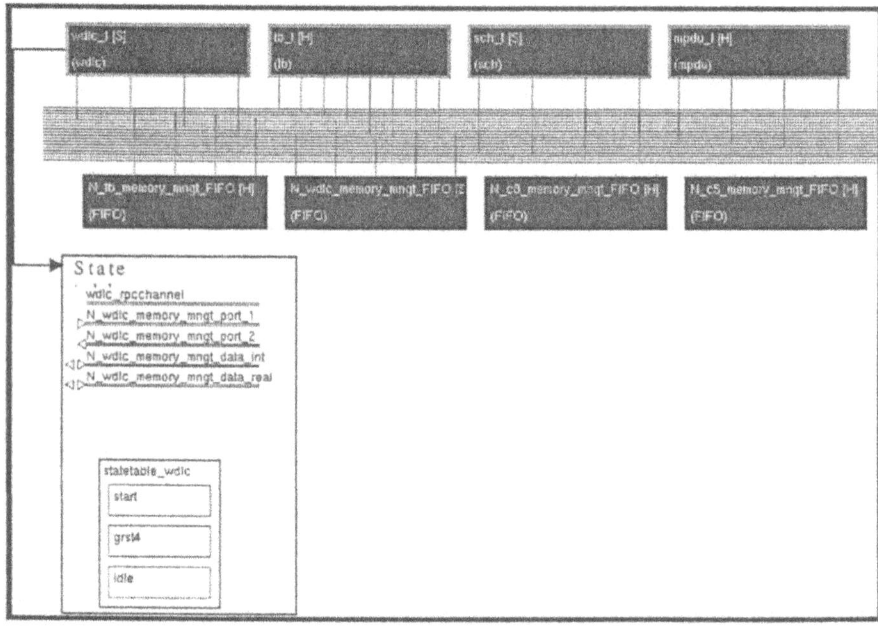

Figure 8. Intermediate model resulting from the partitioning

5.3. Virtual prototype

The distributed C/VHDL co-simulation environment is composed of a VHDL simulator for hardware modules and C-debuggers for software modules. Each C debugger is interconnected to the VHDL simulator through a software bus which relies on the UNIX IPC layer. The software bus and the elements used to interface VHDL and C modules during simulation are automatically created by the C/VHDL interface generation tool. The distributed C/VHDL co-simulation environment is detailed in Figure 9.

The VHDL structure is simulated through a VHDL simulator, while the programs, the design units and the controllers are executed by C debuggers (external boxes on Figure 9). Each program uses C procedures that implement the actual data exchange between the VHDL simulator and the C programs through a set of I/O primitives based on Unix-IPC channel.

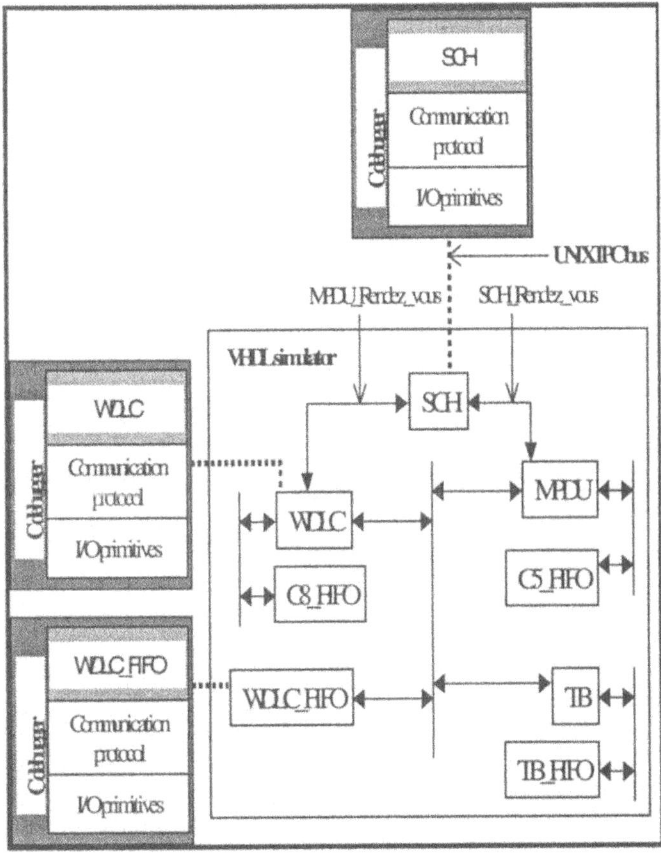

Figure 9. C/VHDL co-simulation scheme for the MASCARA protocol

6. CONCLUSIONS

In the previous sections we presented a co-design methodology for embedded systems design, where the concept of system specification using multiple formalisms was introduced. A platform for coupling different notations was introduced and, the necessary mechanisms to handle different in nature models in a coherent way were presented. The specification languages supported refer to a wide area of application domains, enabling selection of the mostly appropriate formalisms for a specific application. The co-design development cycle presented, provides a fully fledged methodology that leads from system specification to a virtual prototype of the system under construction, where the software parts are implemented in C and the hardware parts are implemented in VHDL. The efficiency of the proposed co-design approach is depicted through its use for the design and implementation of a MAC layer protocol, named MASCARA, for providing ATM QoS over wireless connections. The protocol has been designed using UML and SDL as complementary formalisms for requirement capturing. The system partitioning and the creation of a virtual prototype of the final system has been based on the COSMOS toolset and the system description in the SOLAR notation, the underlying formalism supported by COSMOS.

7. REFERENCES

[1] "ITU-T Specification and Description Language SDL", Recommendation Z.100 (SDL-92). ITU General Secretariat, Geneva, 1992. <http://www.itu.ch/ITU-T>.
[2] "UML Semantics & UML Notation Guide - Version 1.1", OMG Document, September 1997. <http://www.rational.com/uml>.
[3] "ObjectGeode: Method Guidelines", 1997
[4] A.A. Jerraya, K. O'Brien, "SOLAR: an intermediate format for system-level modelling and synthesis", in J. Rozenblit, K. Buchenrieder (Eds) IEEE Press, Piscataway, NJ., 1994.
[5] B. Kleinjohann, E. Kleinjohann, J. Tacken, "The SEA Language for System Engineering and Animation", Conference in Application and Theory of Petri Nets, LNCS 1091, Springer Verlag, 1996.
[6] COMITY project, Deliverable, ESPRIT/23015/WP2/JUN97/D20.1, "COMITY User Documentation, V1", June 1997
[7] D. Harel, "Statecharts: A visual formalism for complex systems", Science of Computer Programming 8(3), pp.231-274, July 1977.
[8] D.Gajski, F.Vahid, S.Narayan, J.Gong, Specification and Design of Embedded Systems, Prentice-Hall, Inc. Englewood Cliffs, New Jersey, 1994.
[9] Frederic Bauchot, Stephane Decrauzat, Gerad Marmigere, Lazaros Merakos, Nikos Passas, "MASCARA, a MAC Protocol for Wireless ATM", ACTS Mobile Summit, 1996, pp. 647-651
[10] J.Buck, S.Ha, E.Lee, Ptolemy: A framework for simulating and prototyping heterogeneous systems, International Journal of Computer Simulation, January 1994.

[11] Jean-Marc Daveau, Gilberto Fernandes Marchioro, Tarek Ben-Ismail, Ahmed Amine Jerraya, “Protocol selection and interface generation for H/W-S/W co-design”, IEEE transactions on VLSI systems, Vol. 5, No. 1, March 1997
[12] M. Srivastava, R.Brodersen, SIERA: A unified framework for rapid-prototyping of system-level hardware and software, IEEE Transactions on Computer-Aided Design of Integrated Circuits and Systems, pp. 676-693, June 1995.
[13] OMI/MCC, “Hardware Software Co-design Study Report”, October 1996.
[14] P.A.Subrahmanyam, W.Wolf, Hardware-Software Codesign for Embedded Systems, in ASP-DAC'95, Makuhari Messe, Chiba, Japan. pp. 1-72, 1995.
[15] R.Ernst, J.Henkel, Th.Benner, W.Ye, U.Holtmann, D.Herrmann, M.Trawny, The COSYMA Environment for Hardware/Software Cosynthesis, Journal of Microprocessors and Microsystems, Butterworth-Heinemann, 1995.
[16] Selic B., Gullekson G., Ward P., "Real-Time Object-Oriented Modelling" (ROOM), John Wiley, New York, NY, 1994.
[17] T.Ben-Ismail, K.O'Brien, A.A.Jerraya, PARTIF: Interactive System-level Partitioning, VLSI Design Vol. 3 no 3-4, pp. 333-345, 1995.
[18] W.Wolf, Object-Oriented Co-Synthesis of Distributed Embedded Systems, in Proceedings CHDL'95, IFIP, 1995.

HARDWARE SYNTHESIS FROM PROTOCOL SPECIFICATIONS IN LOTOS

Keiichi Yasumoto[1], Akira Kitajima[2],
Teruo Higashino[2], and Kenichi Taniguchi[2]

[1]Dept. of Information Processing and Management, Shiga Univ.,
Hikone, Shiga 522-0069, JAPAN
yasumoto@biwako.shiga-u.ac.jp

[2]Dept. of Information and Computer Sciences, Osaka Univ.,
Toyonaka, Osaka 560-8531, JAPAN
(kitajima,higashino,taniguchi)@ics.es.osaka-u.ac.jp

Abstract:

In this paper, we propose a technique for hardware implementation of protocol specifications in LOTOS. For the purpose, we define a new model called *synchronous EFSMs* consisting of concurrent EFSMs and a finite set of multi-rendezvous indications among their subsets, and propose a conversion algorithm from a subset of LOTOS. The derived synchronous EFSMs can be easily implemented as a synchronous sequential circuit where all the modules corresponding to the EFSMs work synchronously with the same clock. By applying our technique to the Abracadabra protocol, it is confirmed that the derived circuit handles multi-rendezvous efficiently.

1 INTRODUCTION

Due to the growth of computer networks, efficient implementation of communication protocols has been needed. Thus, the techniques for implementing protocols as hardware circuits have been stressed in recent years.

To specify hardware circuits formally, the description techniques LOTOS [1, 12], Estelle [15] and SDL [13] have been proposed. With these techniques, we can easily describe schemes for hardware circuits using predefined component libraries, and can verify/validate them. However for rapid prototyping, synthesis techniques from the specifications are desirable. Several ideas for hardware synthesis from formal specifications have been proposed [6]. For example, [15] has proposed a synthesis technique from Estelle. However, the technique does not deal with the highly structured specifications containing synchronization among concurrent modules like *multi-rendezvous*. In [7], although a technique to convert timed LOTOS specifications to VHDL specifications has been proposed, only two-way rendezvous between two processes is implemented. [11] has also proposed a technique to synthesize hardware circuits from LOTOS specifications, but focuses only on Basic LOTOS. Since LOTOS specifications are structurally composed of multiple sub-processes which are dynamically invoked, participants of each multi-rendezvous may be decided dynamically each time. In general, for efficient implementation it is desirable that we can transform complicated LOTOS specifications into ones on a flattened model like an EFSM which has no child-parent relationships between processes. It is also important to calculate in advance all the information about the combinations of synchronizing processes, the tuples of synchronizing events and their execution conditions. For the purpose, [10] has proposed a method to derive all possible multi-rendezvous instances from a LOTOS specification. However, the method requires the complete reachability analysis among all parallel processes, which needs time proportional to the product of the numbers of events in those processes.

In this paper, we propose a technique for hardware implementation of protocol specifications in a subclass of LOTOS where choice, synchronous/asynchronous parallel, interruption, sequential composition and dynamic process instantiation are specified with data. For the purpose, first we propose a new model called *synchronous EFSMs* [16] for representing LOTOS specifications. Synchronous EFSMs consist of concurrent EFSMs and a finite set of multi-rendezvous indications for their subsets (we call the set a *multi-rendezvous table*). In general, if we use all possible rendezvous instances (i.e., tuples of synchronizing transitions) as the multi-rendezvous table, the number of elements will be quite large. In our model, we reduce the number by composing each rendezvous indication of the tuple of possible event sets. Next, we give an algorithm to derive synchronous EFSMs from a protocol specification in our subclass of LOTOS. In the algorithm, we transform a given LOTOS specification to the parallel composition of EFSMs by introducing internal signals and by replacing each process instantiation with the corresponding behavior expression. We get a multi-rendezvous table statically from the information about transitions in each EFSM, and parallel operators specified among EFSMs by calculating all combinations of EFSMs synchronizing at each gate and by extracting all possible tuples of transitions.

To implement synchronous EFSMs as a synchronous sequential circuit, we compose a module to evaluate whether each rendezvous indication has an executable transition tuple or not. If several mutually exclusive multi-rendezvous become executable simultaneously for some combinations of EFSMs, we select one of them according to a priority order given in advance.

In Sect. 2, we introduce system design in LOTOS and give the definition of synchronous EFSMs. An algorithm to derive synchronous EFSMs is given in Sect. 3. Sect. 4 and 5 present a hardware implementation technique and its evaluation.

2 LOTOS AND SYNCHRONOUS EFSMS

2.1 System design in LOTOS

In a LOTOS specification, we specify a behavior expression of the protocol consisting of events and their temporal order. To specify the temporal order of events, we

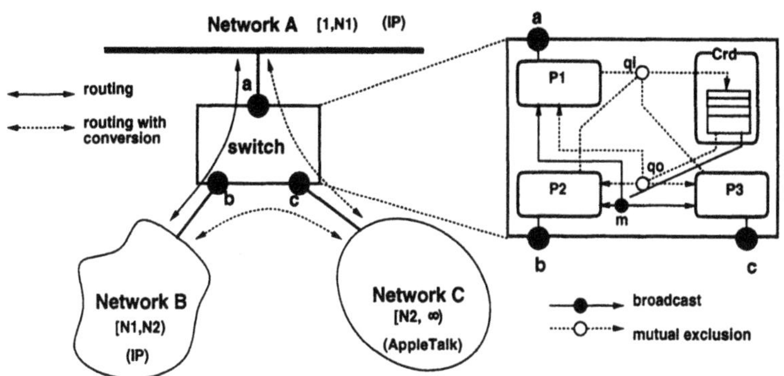

Figure 1 Network switch

use several operators in LOTOS such as action prefix (which combines events in sequential order) as well as choice, parallel, sequential and disabling between any two sub-behavior expressions. In a LOTOS specification, we can replace any sub-behavior expression by a process instantiation. It means that we can compose the whole behavior expression as the set of structured modules. (See [3] for details of LOTOS)

Especially, parallel operators and process instantiations make it easy to describe system specifications both structurally and simply. Multi-rendezvous enables group communication among any subset of concurrent processes, which drastically reduces communication actions in specifications. Multi-rendezvous also enables system specifications in constraint/resource oriented style [14] where we can design a system as a set of simple modules, and develop each module independently of the others.

For example, let us design a network switch in LOTOS which has the following requirements: the switch has three ports a, b and c which are connected to different networks A, B and C, respectively (see Fig. 1). Each data packet arriving at the switch should be forwarded to the appropriate network based on its destination. Here, we route the packets based on the intervals: i.e. if the packet's destination is in the interval $[1, N1)$, the packet should be directed towards network A (via port a); if in the interval $[N1, N2)$, towards network B; if in $[N2, \infty)$, towards network C. If the switch receives the packet with the destination 0 (i.e. broadcast), the packet should be broadcasted to all ports except for its reception port. In addition, we suppose networks A and B use the same protocol (e.g. IP), but network C uses a different protocol (e.g. AppleTalk). That means the switch should have the facility for protocol conversion of each packet from either A or B to C (and vice versa).

It is desirable to design the behavior of each port independently of the others. For a better response time, input and output behaviors for each port should be able to work in parallel. In addition, in order to allow asynchrony among ports we use a FIFO queue shared among them. Here, we introduce three internal ports qi, qo and m for the access to the queue. We suppose a new packet can be added to the queue via qi and the packet in the queue can be taken via qo in FIFO manner (see Fig.1). Although m is used in the same way as qo, it is dedicated to the purpose of broadcast.

According to the above discussion, LOTOS processes (`P1` and `P2`) for ports a and b are described as follows:

```
P1[a,qi,qo,m](l1,l2):=I1[a,qi]|||O1[a,qo,m](l1,l2)
where
  I1[a,qi]:= a?idt:IP;qi!idt;I1[a,qi]
  O1[a,qo,m](l1,l2):=qo?odt:IP[l1<=dst(odt)<l2];a!odt;O1[a,qo,m](l1,l2)
        [] m?odt:IP[l1<=src(odt)<l2];O1[a,qo,m](l1,l2)
        [] m?odt:IP[not(l1<=src(odt)<l2)];a!odt;O1[a,qo,m](l1,l2)
P2[b,qi,qo,m](l1,l2):=P1[b,qi,qo,m](l1,l2)
```

(here, $dst(data)$ and $src(data)$ are the ADT functions to get the destination and the source of the packet $data$, respectively. `l1 <= x < l2` represents that the value of x is in the range of $[l1, l2)$. `P2` is described as an instantiation of `P1`)

Inputs/outputs via port c needs protocol conversion. So, for port c, we slightly modify the above process as follows:

```
P3[c,qi,qo,m](l1,l2):=I3[c,qi]|||O3[c,qo,m](l1,l2)
where
  I3[c,qi]:=c?idt:Atk;qi!convIP(idt);I3[c,qi]
  O3[c,qo,m](l1,l2):=
    qo?odt:IP[l1<=dst(odt)<l2];c!convAtk(odt);O3[c,qo,m](l1,l2)
 [] m?odt:IP[l1<=src(odt)<l2];O3[c,qo,m](l1,l2)
 [] m?odt:IP[not(l1<=src(odt)<l2)];c!convAtk(odt);O3[c,qo,m](l1,l2)
```

(here, $convIP()$ means the conversion of the packet to IP, while $convAtk()$ to AppleTalk)

Next, we need a coordinator for the FIFO queue. We describe the queue and its operations by an ADT in LOTOS. The coordinator stores a new packet coming to qi to the queue, or outputs to port either qo or m the last entry in the queue based on its destination. The process Crd for the coordinator can be described as follows:

```
Crd[qi,qo,m](queue):=
     [size(queue)<MAX]->qi?data:IP;Crd[qi,qo,m](append(queue,data))
[] [dst(head(queue))==0]->m!head(queue);Crd[qi,qo,m](tail(queue))
[] [not(dst(head(queue))==0]->qo!head(queue);Crd[qi,qo,m](tail(queue))
```

(here, $size(), append(), head()$ and $tail()$ represent ADT functions for the queue operation. MAX is the maximum number of entries in the queue)

Finally, we specify the interaction among the above processes. In general, we have to design the mutual exclusion mechanism for the queue since some parallel processes may access it at the same time. However, in LOTOS, we can simply describe such a mechanism with multi-rendezvous as follows (here,`INF` denotes ∞):

```
specification Switch[a,b,c]:=
hide qi,qo,m in
  (P1[a,qi,qo,m](1,N1)|[m]|P2[b,qi,qo,m](N1,N2)|[m]|P3[c,qi,qo,m](N2,INF))
|[qi,qo,m]|
   Crd[qi,qo,m]
```

In the above specification, one of `P1`, `P2` or `P3` synchronizes with `Crd` to store/get a packet to/from the queue via internal ports qi/qo. When the packet's destination is 0, all of `P1`, `P2` and `P3` get the packet at the same time by multi-rendezvous on m.

2.2 Synchronous EFSMs

Synchronous EFSMs are the model where any subset of concurrent EFSMs can communicate with each other via gates by multi-rendezvous[16].

Synchronous EFSMs are given as a set of EFSMs $\{efsm_1, ..., efsm_n\}$ and a multi-rendezvous table $\mathcal{R}$. We suppose that each EFSM can have a finite number of registers,

Table 1 Synchronization condition.

P_i	P_j	condition	result
$g!E_i$	$g!E_j$	$val(E_i) = val(E_j)$	—
$g!E_i$	$g?x : t$	$val(E_i) \in domain(t)$	$x \leftarrow val(E_i)$
$g?x : t$	$g?y : u$	$t = u$	$x,\ y \leftarrow Ext,\ Ext \in domain(t)$

($val(E)$ is the normal form of E, Ext is the value input at gate g)

that a certain execution condition called a *guard* expression can be specified to each transition (i.e. edge), and that each transition can perform several substitutions for the registers in parallel.

In LOTOS, multi-rendezvous is specified by just giving abstract relationships among concurrent processes by parallel and other operators. For the efficient implementation of multi-rendezvous, we should calculate in advance the information about the combinations of synchronizing EFSMs, the tuples of synchronizing transitions (*synchronization tuples*) and their execution conditions. If we represent the multi-rendezvous simply by the set of all the combinations of transitions in synchronizing EFSMs, the number will be $O(k^n)$ where n and k are the number of EFSMs and the number of transitions in an EFSM, respectively.

Therefore, in our model, we represent all possible multi-rendezvous instances by a set of *rendezvous indications* where each indication is a tuple of transition sets on a gate for a combination of synchronizing EFSMs. Here, every combination of transitions (called *synchronous tuple*) in the sets has the possibility to be executed by multi-rendezvous (i.e. every synchronous tuple satisfies the condition in Table 1).

We denote each rendezvous indication by $\langle(E_1, \cdots, E_m), (A_1, \cdots, A_m)\rangle$ where $(E_1, \cdots, E_m)$ is a tuple of synchronizing EFSMs, and each A_i is the *synchronous transition set* which contains transitions executed in E_i for the rendezvous. We represent elements of A_i as the triples (a, p, I). Here, a is the transition name consisting of a gate name and input/output parameters, p is a guard expression, and I is the set of substitutions to undefined variables.

Criteria for each rendezvous indication

To implement multi-rendezvous efficiently, we adopt the following criteria for each rendezvous indication $\langle(E_1, ..., E_m), (A_1, ..., A_m)\rangle$: (1) all transitions in each A_i must be either all input transitions ($g?$) or all output transitions ($g!$); (2) each A_i must not contain transitions with different kind of output values which are executed at the same state (i.e. if two transitions $a!V_1$ and $a!V_2$ are executable at a state, we assign them to different rendezvous indications); and (3) at most one A_i of $(A_1, ..., A_m)$ can have a set of output transitions.

By the above criteria, we can easily know in each rendezvous indication what EFSM outputs some values and what other EFSMs expect the values as their inputs. This contributes to implement data paths to transfer the values among related EFSMs to evaluate the synchronization condition.

The above criteria are not restrictions since we can automatically get the rendezvous indications satisfying them as we will explain in Sect. 3.

By using the above technique, each rendezvous indication $\langle(E_1, \cdots, E_m), (A_1, \cdots, A_m)\rangle$ can represent a maximum of $\prod_{i=1}^{m} |A_i|$ rendezvous instances (here, $|A_i|$ means the number of elements in A_i). Consequently, the number of elements in the multi-rendezvous table and the time to calculate the table are bound to $O(p \cdot k \cdot n)$ where p is the maximum number of combinations of the synchronizing EFSMs on a gate (usually p can be considered as a constant) and $k \cdot n$ is the sum of the numbers of the output transitions with different values in EFSMs (in the worst case, $k \cdot n$ may be the number of all transitions in EFSMs). For example, for EFSM1$|[a, b]|$EFSM2 in Fig. 4, ten rendezvous instances could be produced. With the above technique, we can reduce the number of tuples to three as shown in Table 3.

Behavior of synchronous EFSMs

We call each transition e in an EFSM E_i an *asynchronous transition* if $e \notin A_i$ for every rendezvous indication $\langle(E_1, ..., E_m), (A_1, ..., A_m)\rangle \in \mathcal{R}$. We define an asynchronous transition to be *executable* when the current state has the corresponding outgoing edge.

For a rendezvous indication $r = \langle(E_1, \cdots, E_m), (A_1, \cdots, A_m)\rangle$, if each E_i is in a state to execute the transition $e_i \in A_i$ and the execution condition of e_i is *true*, then the synchronous tuple $(e_1, \cdots, e_m)$ can be executed. Whether a synchronous tuple $(e_1, \cdots, e_m)$ can be executed or not is decided by checking the existence of the rendezvous indication which contains such a synchronous tuple. For different two rendezvous indications $r_1 = \langle(E_1, \cdots, E_m), (A_1, \cdots, A_m)\rangle$ and $r_2 = \langle(E'_1, \cdots, E'_l), (A'_1, \cdots, A'_l)\rangle \in \mathcal{R}$, suppose that at least one EFSM is the common member of both r_1 and r_2. In that case, such an EFSM has to decide to execute the transition in either r_1 or r_2. We say that r_1 *conflicts* with r_2 if they can offer the different synchronous tuples at the same time.

Here, we explain how synchronous EFSMs work in cooperation, using an example in Fig. 2. In Fig. 2, the dotted line shows that one of EFSM1, EFSM3 and EFSM5 can synchronize with EFSM7 on gate *qi* at the same time (one of EFSM2, EFSM4, EFSM6 on gate *qo*); the solid line shows that EFSM2, EFSM4, EFSM6 and EFSM7 can synchronize with each other simultaneously on gate m. In the initial state (s_1, s_1, s_1) of EFSM1, EFSM5 and EFSM7, they have the outgoing transitions $a?idt, c?idt$ and $qi?data[size\ (queue) < MAX]$, respectively. The first two transitions are asynchronous, so they are executed independently of the other EFSMs when the input data come to the gates. When $a?idt$ is executed in EFSM1, the current state is changed to (s_2, s_1, s_1). In the state, EFSM1 and EFSM7 have the outgoing edges $qi!idt$ and $qi?data[size(queue) < MAX]$, respectively. Since $queue$ contains nothing initially, the execution condition $size(queue) < MAX$ holds. Therefore, the tuple $(qi!idt,\ qi?data[size(queue) < MAX])$ can be executed by the rendezvous indication (1) of Fig. 2. When the tuple is executed, the value of idt is assigned to the undefined variable *data*, and the current state is changed to (s_1, s_1, s_2) in EFSM1, EFSM5 and EFSM7.

In some state, there may be several synchronous tuples to be executable simultaneously. For example, in Fig. 2, when the current state is (s_1, s_1, s_1) for EFSM2, EFSM6 and EFSM7, a synchronous tuple $(qo?odt\ [dst(odt) < N1], qo!head(queue)$ $[dst(head(queue))! = 0])$ could be executed between EFSM2 and EFSM7 by the rendezvous indication (4) as well as $(qo?odt\ [N2 \leq dst(odt)],\ qo!head(queue)$ $[dst(head(queue))! = 0])$ between EFSM6 and EFSM7 by the rendezvous indication (6). In that case, one of them must be selected by consensus of the related EFSMs.

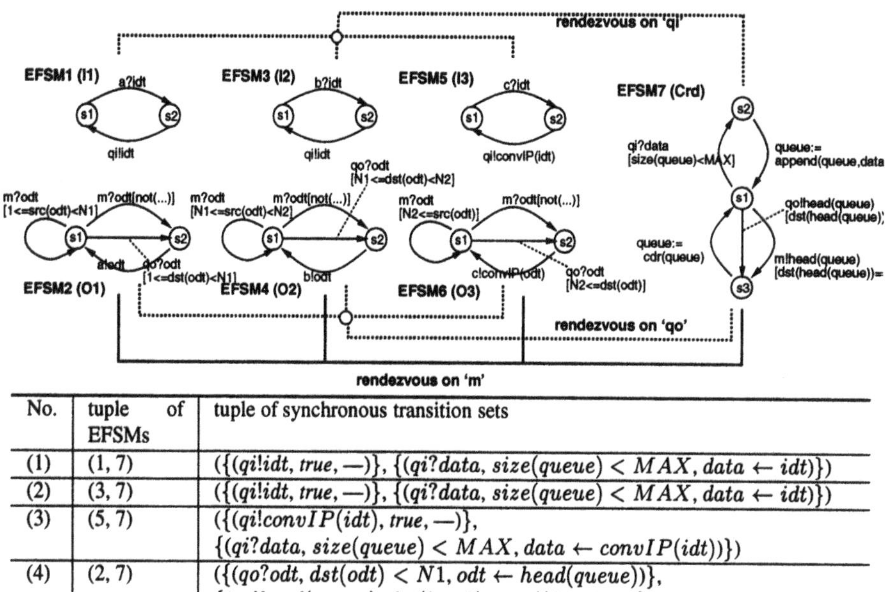

No.	tuple of EFSMs	tuple of synchronous transition sets
(1)	(1, 7)	$(\{(qi!idt, true, -)\}, \{(qi?data, size(queue) < MAX, data \leftarrow idt)\})$
(2)	(3, 7)	$(\{(qi!idt, true, -)\}, \{(qi?data, size(queue) < MAX, data \leftarrow idt)\})$
(3)	(5, 7)	$(\{(qi!convIP(idt), true, -)\},$ $\{(qi?data, size(queue) < MAX, data \leftarrow convIP(idt))\})$
(4)	(2, 7)	$(\{(qo?odt, dst(odt) < N1, odt \leftarrow head(queue))\},$ $\{(qo!head(queue), dst(head(queue))! = 0, -)\})$
(5)	(4, 7)	$(\{(qo?odt, N1 \le dst(odt) < N2, odt \leftarrow head(queue))\},$ $\{(qo!head(queue), dst(head(queue))! = 0, -)\})$
(6)	(6, 7)	$(\{(qo?odt, N2 \le dst(odt), odt \leftarrow head(queue))\},$ $\{(qo!head(queue), dst(head(queue))! = 0, -)\})$
(7)	(2, 4, 6, 7)	$(\{(m?odt, src(odt) < N1, odt \leftarrow head(queue)),$ $(m?odt, not(src(odt) < N1), odt \leftarrow head(queue))\},$ $\{(m?odt, N1 \le src(odt) \le N2, odt \leftarrow head(queue)),$ $(m?odt, not(N1 \le src(odt) \le N2), odt \leftarrow head(queue))\},$ $\{(m?odt, N2 \le src(odt), odt \leftarrow head(queue)),$ $(m?odt, not(N2 \le src(odt)), odt \leftarrow head(queue))\},$ $\{(m!head(queue), dst(head(queue)) == 0, -)\})$

Figure 2 Example of synchronous EFSMs

3 DERIVING SYNCHRONOUS EFSMS

3.1 Preliminaries

In this paper, we consider any LOTOS specifications represented in the class of Table 2 although we impose the following restrictions on process instantiations.

- Recursive processes are allowed when they are tail recursion (e.g. $P := B >> P$).
- Recursive processes which may produce infinite behavior such as $P := (B1 >> P >> B2)[]exit$ or $P := B\ op\ P$ ($op \in \{[], |[G]|, [>\}$), are not allowed. However, if the recursive process call is guarded and the guard expression can be evaluated statically (e.g. $P(x) := B|||([x < 100]-> P(x+1))$), we treat such a process.
- Mutually recursive processes are allowed as long as process calls are guarded and the guard expressions can be evaluated statically.

Let $\#Par(B)$ be the function that represents how many parallel processes can be activated at the same time in a behavior expression B. We say a behavior expression B is a *sequential behavior expression (SBE)* if $\#Par(B) = 1$ and B includes

Table 2 Target class of LOTOS

$B0$ ::=	hide G in $B1$ | let *value declarations* in $B1$ | $B1$
$B1$ ::=	$B1 >> B2 \mid B2$
$B2$ ::=	$B2[> B3 \mid B3$
$B3$ ::=	$B3\|\|\|B4 \mid B3\|\|B4 \mid B3\|[G]\|B4 \mid B4$
$B4$ ::=	$B4[]B5 \mid B5$
$B5$ ::=	$[boolean\ exp]-> B6 \mid B6$
$B6$ ::=	$event; B$ | stop | exit | $(B1)$ | $P[G](el)$
$event$ ::=	$gate \mid gate\ val \mid gate\ val\ [boolean\ exp]$
G ::=	$gate \mid gate, G$
val ::=	$?varname : sort\ val \mid !exp\ val$
exp ::=	(* every expression written in ACT ONE *)
el ::=	exp, el

only action prefixed sequences, their choices and iteration. We use $St(B)$ to refer the set of the initially executable events in B. We denote the behavior expression $B_1|[G_1]|...|[G_{m-1}]|B_m$ by $\prod_{i=1}^{m} B_i$, and $B_1[]...[]B_l$ by $\sum_{i=1}^{l} B_i$.

3.2 Transformation algorithm

We transform B_{main} to the parallel composition among SBEs by applying the following operations recursively to its sub-behavior expressions: (1) replace each process instantiation with its behavior expression unless the instantiation appears as a tail recursion (i.e. $P := B >> P$); (2) transform each action prefixed sequence B_{act} such that $\#Par(B_{act}) > 1$ into the parallel composition among sequential behavior expressions (SBEs); (3) transform each choice/disabling/sequential composition among sub behavior expressions into either an SBE or a parallel composition among SBEs.

For (1), we replace each process instantiation $P[G](V)$ appearing in its behavior expression B_P as a tail recursion, with the iteration of B_P by introducing **label(P[G]):** and **goto(P[G],X:=V)**. Although other process instantiations are replaced with their behavior expressions even if they are recursive processes, they are not infinitely instantiated by the restriction in Sect. 3.1.

We show the transformation algorithm $Trans$ below. Here, $pset$ is the set of process instantiations already replaced with their behavior expressions.

Algorithm $Trans(B, pset)$
begin
  if ($B = a_1; ...; a_l; B', B' \neq a'; B''$) then
    if ($\#Par(B') \geq 2$) then
      $Trans(B', pset)$
      $TransAct(B)$
    endif
  else if ($B = [guard]-> B'$) then
    if ($guard$ is $true$ or cannot be calculated statically) then
      $Trans(B')$
    endif
  else if ($B = \sum B_i$) then
    for each i, $Trans(B_i, pset)$
    $TransChoice(B)$
  else if ($B = B1|[G]|B2$) then
    $Trans(B1, \emptyset); Trans(B2, \emptyset)$

```
  else if (B = B1[> B2) then
    Trans(B1,pset);Trans(B2,pset);TransDis(B)
  else if (B = B_1 >> ... >> B_k) then
    for each i < k, Trans(B_i,∅)
    Trans(B_k,pset)
    TransSeq(B)
  else if (B = P[G](V)) then
    if (P[G] ∈ pset) then
      replace P[G](V) with goto(P[G],X:=V)
    else
      replace P[G](V) with label(P[G]): let X:=V in B_P[G]
      Trans(B_P[G],pset ∪ {P[G]})
    endif
  endif
end
```

The sub-procedures used above are given below.

$TransAct$: **transformation of action prefixed sequence.** For $TransAct$, $a_1; ...; a_l; (B_1 |[G_1]| ... |[G_{n-1}]| B_n) (\stackrel{def}{=} B_{act})$ is given. Here, each B_i is an SBE because $Trans$ transforms each sub-behavior expression which is not an SBE to the parallel composition of SBEs recursively. In B_{act}, each B_i should be activated after a_l is executed. Accordingly, we assign the event sequence $a_1; ...; a_l$ and the behavior expressions $B_1, ..., B_n$ to different SBEs, respectively. Then, we introduce an internal signal θ so that the SBE for $a_1; ...; a_l$ sends the signal θ to other SBEs after the execution of a_l and that the SBEs for $B_1, ..., B_n$ are activated when they receive θ. We specify the exchange of θ by multi-rendezvous among the SBEs. Consequently, B_{act} is represented by the parallel composition of SBEs as follows:

$$a_1; ...; a_l; \theta; exit \,|[\theta]|\, (\theta; B_1 \,|[\{\theta\} \cup G_1]|\, \theta; B_2 \,|[\{\theta\} \cup G_2]|\, ... \,|[\{\theta\} \cup G_{n-1}]|\, \theta; B_n)$$

$TransChoice$: **transformation of choice expression.** For $TransChoice(B_{cho})$, B_{cho} is given as $\sum_{j=1}^{n} B_j$. For each j, B_j can be represented by $\prod_{i=1}^{m_j} B'_{j,i}$ where each $B'_{j,i}$ is an SBE. If $m_j = 1$ for each i, B_{cho} is a SBE. Here, we consider the case $m_j > 1$ for some j (i.e. parallel operators are specified in B_j). We suppose each $B'_{j,i}$ to be an action prefixed sequence without losing generality. Let $B_j \stackrel{def}{=} \prod_{i=1}^{m_j} a_{j,i}; B_{j,i}$. We call each B_j the j-th group. Let mx be the maximum number among $m_1, ..., m_n$.

In choice, any pair from different groups cannot be executed at the same time. Therefore, we can assign B_{cho} to mx SBEs. Although there are various ways of assignment, here we extract an SBE from each group $j (1 \leq j \leq n)$, and compose a new SBE of the choice among the extracted n SBEs. For the sake of simplicity, we assign, to each new i-th SBE, i-th elements of all groups (if there is no i-th element, *exit* is used instead). In Fig. 3, the pairs from $B_1 [] B_2$: $(\mathrm{a}_{11}; \mathrm{B}_{11}, \mathrm{a}_{21}; \mathrm{B}_{21}), (\mathrm{a}_{12}; \mathrm{B}_{12}, \mathrm{a}_{22}; \mathrm{B}_{22})$ and (exit, $\mathrm{a}_{23}; \mathrm{B}_{23}$) are assigned to new SBEs, where $B1 := a_{11}; B_{11} \,|||\, a_{12}; B_{12}$ and $B2 := a_{21}; B_{21} \,|||\, a_{22}; B_{22} \,|||\, a_{23}; B_{23}$.

Next, we introduce a mechanism to keep the equivalence between the new behavior expression and the old one. Let us suppose that the j-th element has executed in one of the new mx SBEs. This means that the j-th group has selected in B_{cho}. So, we have

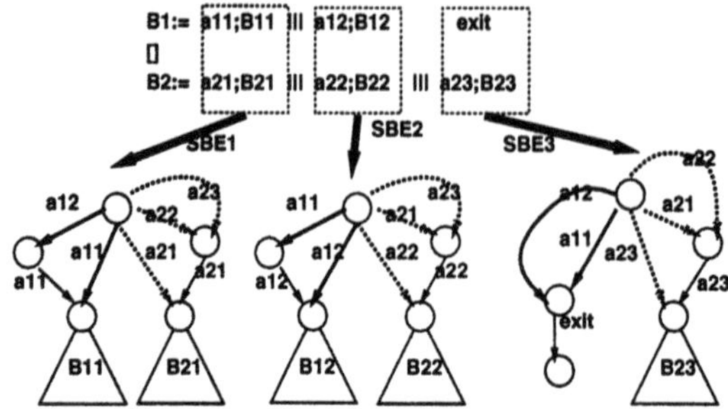

Figure 3 Assignment and resulting SBEs in a choice expression

to make only the j-th element execute in all the SBEs after that. To do so we add, to each new SBE, extra events for detecting what event has executed in other SBEs. We achieve those detections by the communication with multi-rendezvous among the new SBEs so that only first executable events in $St(B_{cho})$ synchronize among the SBEs. Accordingly, the following expression is produced for each SBE:

$$sbe_i \stackrel{def}{=} \sum_{j=1}^{n} a_{j,i}; B_{j,i} \, [] \sum_{a \in St(B_j), a \neq a_{j,i}} a; a_{j,i}; B_{j,i}$$

Fig. 3 shows the resulting SBEs from the above example. Here, a_{11} or a_{12} must be synchronized among the SBEs to detect that the group of $B1$ has been selected (thick solid transitions), while a_{21}, a_{22} or a_{23} detects the group of $B2$ (dotted transitions).

$TransDis, TransSeq$: **transformation for other operators.** In disabling expression $B_1[> B_2$ (here, $B_j \stackrel{def}{=} \prod_{j=1}^{m_j} sbe_{j,i}$), there is no possibility for SBEs in B_1 and those in B_2 to be executed simultaneously. So, we assign to a new SBE each pair of the i-th elements in B_1 and B_2. In each new SBE, we combine $sbe_{1,i}$ and $sbe_{2,i}$ with choice operators so that events in $sbe_{1,i}$ should be disabled if an event in $sbe_{2,i}$ is executed. Similar to the case of choice expressions, each SBE must be able to detect if such a disabling event is executed in other SBEs. To do so, we add to each SBE extra events for detecting the disabling events in $St(B_2)$, and specify multi-rendezvous so that those events should be synchronized among the new SBEs. By the above assignment each disabling expression is converted to the parallel composition of $max(\#Par(B_1), \#Par(B_2))$ SBEs.

For the sequential expression $B_1 >> B_2$, similarly we extract each pair of SBEs from B_1 and B_2 and compose a new SBE. Here, we introduce internal signal δ to indicate that all events in B_1 have been executed and to activate the behavior corresponding to B_2. With the multi-rendezvous of δ, we make the new SBEs finish the execution of B_1 at the same time as starting the execution of B_2. $B_1 >> \cdots >> B_l$ can be transformed in the same way.

If the sequential expression has tail recursion in the process instantiation such as $P[G](X) := B >> P[G](V)$ and if $\#Par(B) > 1$, all the SBEs extracted from B have to get to their initial states when **goto(P[G],X:=V)** is executed. In that case, for each SBE, we add the transition to the initial state from the state after executing δ with the corresponding subset of the substitutions of X:=V.

The details of the transformation algorithms and a simplified proof for its correctness are given in [16].

Calculation of Multi-Rendezvous Table

Only synchronization operators are specified among EFSMs after applying the transformation algorithm. In this section, we give a technique to get the multi-rendezvous table from transitions in each EFSM and operators specified among EFSMs. From the syntax tree of the operators among EFSMs and gate names used in each EFSM, we can get the combinations of synchronizing EFSMs on each gate. If a multi-rendezvous is specified among a subset of EFSMs $\mathcal{E}$ on gate g, we denote that by *Rend*$(\mathcal{E}, g)$. The finite set of all synchronous tuples as well as their execution conditions is statically determined if *Rend*$(\mathcal{E}, g)$ is given for all combinations of $\mathcal{E}$ and g. Let RI be the union of *Rend*$(\mathcal{E}, g)$ for all $\mathcal{E}$ and g. We calculate the set as follows.

For each *Rend*$(\mathcal{E}, g) \in RI, \mathcal{E} = \{E_1, \cdots, E_m\}$: (1) extract all transitions on gate g, for each $E_i \in \mathcal{E}$ (let *sync_ev*(E_i, g) be the extracted transitions for E_i); (2) calculate the set of tuples $\{(e_1, \cdots, e_m) | e_i \in$ *sync_ev*$(E_i, g)\}$ where each tuple can satisfy the synchronization condition in Table 1 (let $Tuples(\mathcal{E}, g)$ denote the set).

Next, we convert the synchronous tuples to rendezvous indications as follows, so that they satisfy the criteria in Sect. 2.2. For each $Tuples(\mathcal{E}, g)$:

(i) calculate the set of output values *OV* where each value is assigned to undefined variables by the synchronization.

(ii) for each $E_i \in \mathcal{E}$ and each $v \in$ *OV*, calculate the set of transitions A_i where each transition satisfies the synchronization conditions in Table 1 with v; and compose a rendezvous indication of the tuple $\langle \mathcal{E}, (A_1, \cdots, A_m) \rangle$.

(iii) If A_i includes both input (g?) and output (g!) transitions, we divide such a rendezvous indication $\langle \mathcal{E}, (A_1, \cdots, A_m) \rangle$ to $\langle \mathcal{E}, (A_1, \cdots, A_i^{in}, \cdots, A_m) \rangle$ and $\langle \mathcal{E}, (A_1, \cdots, A_i^{out}, \cdots, A_m) \rangle$ so that each A_i^{in} (A_i^{out}) includes only input (output) transitions. This operation is repeated until the elements of each $A_i (1 \leq i \leq m)$ are all input transitions or all output transitions.

(iv) If $E_i \in \mathcal{E}$ has two different output transitions (i.e. $g!V_1, g!V_2$) in A_i from the same state, the rendezvous indication $\langle \mathcal{E}, (A_1, \cdots, A_i, \cdots, A_m) \rangle$ is divided into $\langle \mathcal{E}, (A_1, \cdots, A_i - \{g!V_1\}, \cdots, A_m) \rangle$ and $\langle \mathcal{E}, (A_1, \cdots, A_i - \{g!V_2\}, \cdots, A_m) \rangle$. This operation is repeated until each A_i has only one output transition from each state in E_i.

(v) If a rendezvous indication $\langle \mathcal{E}, (A_1, \cdots, A_m) \rangle$ contains A_i and A_j such that the both sets have output transitions, we modify a transition in either A_i or A_j to an input transition. For example, $a!x \; |[a]| \; a!y \; |[a]| \; a?z$ is transformed to $a!x \; |[a]| \; a?w \, [w = y] \; |[a]| \; a?z$.

The above procedure could produce some rendezvous indications which contain impossible synchronous tuples since we do not use any reachability analysis technique. A synchronous tuple can be executed as long as the synchronization condition of the tuple holds for a rendezvous indication. Not all synchronous tuples in each rendezvous indication can be executed. That means the multi-rendezvous table itself is a sufficient condition for implying the possible multi-rendezvous instances. In our model, however, only the valid synchronous tuples become executable since the executability of each rendezvous indication is evaluated at each reachable state of EFSMs.

3.3 Example of conversion to synchronous EFSMs

After the algorithm $Trans$ is applied to the main behavior expression of $Switch$ in Sect. 2.1, seven SBEs are derived. We show part of them below:

SBE_{I1} := label(I1): a?idt:IP; qi!idt; goto(I1,—)
SBE_{O1} := label(O1): (qo?odt:IP[1 ≤ dst(odt) < N1]; a!odt; goto(O1,—)
[] m?odt:IP[1 ≤ src(odt) < N1]; goto(O1,—)
[] m?odt:IP[not(1 ≤ src(odt) < N1)]; a!odt; goto(O1,—))
SBE_{Crd} := label(Crd): (qi?data:IP[size(queue)<MAX]; goto(Crd,queue:=append(queue,data))
[] m!head(queue)[dst(head(queue))==0]; goto(Crd,queue:=cdr(queue))
[] qo!head(queue)[dst(head(queue))!=0]; goto(Crd,queue:=cdr(queue)))

In the above SBEs, process instantiations in the form of tail recursions are converted to the appropriate **goto** transitions. Each derived SBE can be converted to an EFSM easily. Fig. 2 depicts the resulting EFSMs and the multi-rendezvous table. Here, EFSM1 – EFSM7 are converted from SBE_{I1} – SBE_{Crd}, respectively.

4 HARDWARE SYNTHESIS FROM SYNCHRONOUS EFSMS

In this section, we give the technique to convert given synchronous EFSMs into a synchronous sequential circuit (our preliminary work can be found in [5]). Hereafter, we suppose the modules corresponding to EFSMs work synchronously with the same clock. In each clock cycle, each EFSM can execute a transition as long as its execution condition holds. We assume the components corresponding to ADT functions (e.g. guard expressions) are provided as combinational logic circuits and they can output the resulting values within a clock cycle. The circuit for each EFSM can be implemented easily by well-known techniques [9]. So, here we concentrate on the implementation of multi-rendezvous among EFSMs.

Given EFSMs and a multi-rendezvous table, we implement multi-rendezvous among EFSMs as the *multi-rendezvous circuit* consisting of the following three sub-parts: *(1) executability check part* checking whether there exist executable synchronous tuples for each rendezvous indication at each state; *(2) data transfer part* transferring the required data from a certain EFSM to the other EFSMs so that each EFSM can calculate the execution condition (guard) of its transition. *(3) conflict avoidance part* selecting a synchronous tuple among some mutually exclusive synchronous tuples;

Hereafter, we suppose that synchronous EFSMs are given as $\langle EFSM, \mathcal{R}\rangle$ where each rendezvous indication $r \in \mathcal{R}$ is represented as $\langle (E_1, ..., E_m), (A_1, ..., A_m)\rangle$ where each $E_i \in EFSM$.

Constructing executability check and data transfer parts

For the executability checking part, every EFSM in each rendezvous indication must check whether some transitions in its synchronous transition set are executable at the current state. So in each E_i, for every $r \in \mathcal{R}$, we provide a circuit generating an output signal r_i_ok which becomes *true* (i.e. 1) only when a transition in A_i becomes executable. Consequently, for the rendezvous indication r there exist some executable synchronous tuples if and only if r_1_ok, ..., r_m_ok (denoted by r_*_ok) are *true*.

For the data transfer part, EFSMs with input transitions and an EFSM with output transitions can be determined statically for each rendezvous indication by the criteria

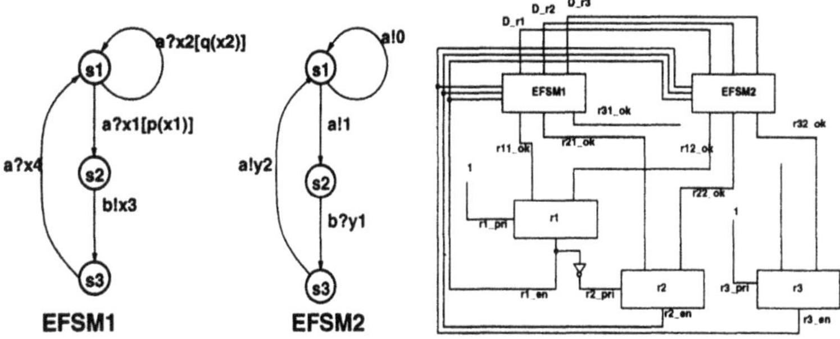

Figure 4 An example of EFSMs.

Figure 5 An example of derived circuit.

in Sect. 2.2. Hence, we provide a path D_r among EFSMs for each r so that an EFSM outputs an appropriate value to the path and the others obtain the value.

Constructing conflict avoidance part

The conflict avoidance part generates the signal r_en which becomes *true* when r has the right to execute its synchronous tuple, avoiding conflicts between r and other exclusive rendezvous indications. Although there can be some policies for avoiding conflicts, we adopt a policy that gives a priority (or total order) among rendezvous indications and selects a rendezvous indication by the priority.

Synchronous tuples of rendezvous indication r_1 and r_2 can conflict if and only if there exists the case that for a state set $(s_1, \cdots, s_p)$ of the common EFSMs in both rendezvous indications, the synchronous transition tuples of r_1 and r_2 are both executable at every state s_k $(1 \leq k \leq p)$. For all combinations of two rendezvous indications, we can statically determine whether they can conflict or not by checking the given EFSMs and the multi-rendezvous table.

Any synchronous tuples of r cannot be executed when another conflicting rendezvous indication with priority higher than r is ready to execute a synchronous tuple. Consequently, we construct r_en as follows:

$r\_en = r_1\_ok \wedge \cdots \wedge r_m\_ok \wedge pri_r$ where $pri_r = \neg r^1\_en \wedge \cdots \wedge \neg r^h\_en$

(Here, $\{r^1, \cdots, r^h\}$ are the rendezvous indications with higher priorities than r which conflict with r. pri_r means whether r has the right to execute its synchronous tuple or not).

Although we have introduced a priority based method, a module generating random numbers can also be used to select one of the conflicting rendezvous.

An example of derived circuit

In this section, we explain how we can derive the circuit in Fig. 5 from the synchronous EFSMs in Fig. 4 and the multi-rendezvous table in Table 3.

Hereafter, we denote the output signal from $EFSM_j$ for the rendezvous indication r_i as r_{ij}_ok. At the initial state (s_1, s_1), EFSM1 first calculates the output value r_{11}_ok for the rendezvous indication r_1 as follows. As EFSM1's current state is s_1, EFSM1 calculates the execution condition $p(x_1) \vee q(x_2)$ for the transitions $a?x_1\ [p(x_1)]$ and $a?x_2\ [q(x_2)]$ which are transitions in its transition set and that of r_1 respectively. Furthermore, since $p(x_1)$ and $q(x_2)$ need external values to calculate the conditions,

Table 3 An example of multi-rendezvous table.

Info. No.	EFSMs	synchronizing transition sets
r_1	(EFSM1, EFSM2)	$(\{(a?x_1, p(x_1), x_1 \leftarrow 1 org(y)),$ $(a?x_2, q(x_2), x_2 \leftarrow 1 \text{ or } g(y)), (a?x_3)\}, \{(a!1, true, -)\})$
r_2	(EFSM1, EFSM2)	$(\{(a?x_1, p(x_1), x_1 \leftarrow 0),$ $(a?x_2, q(x_2), x_2 \leftarrow 0), (a?x_3)\}, \{(a!0, true, -)\})$
r_3	(EFSM1, EFSM2)	$(\{(b!f(x_1), true, -)\}, \{(b?y, true, y \leftarrow f(x_1))\})$

EFSM1 uses the value from the data path D_{r1} (the data path for the rendezvous indication r_1) for the values of x_1 and x_2. So, it outputs the value of $p(D_{r1}) \vee q(D_{r1})$ to r_{11}_*ok*. On the other hand, as EFSM2's current state is s_1, EFSM2 can execute the transitions a!1 in its transition set for the rendezvous indication r_1. So it outputs *true* to r_{12}_*ok*. In addition, since $a!1$ is an output transition, it outputs the value 1 to the data path D_{r1}. For other rendezvous indications, EFSMs do the same operation. If r_{11}_*ok* and r_{21}_*ok* are both *true*, r_1 and r_2 are conflict with each other because both r_{12}_*ok* and r_{22}_*ok* are also *true*. In this case, the conflict avoidance part selects r_1 and outputs *true* only to r_1_*en* (since we assume that the priority $r_1 > r_2 > r_3$ holds).

Finally, if r_1_*en* is *true*, EFSM1 executes either $a?x_1$ or $a?x_2$ depending on which condition in $p(x_1)$ and $q(x_2)$ holds (when the both conditions are *true*, EFSM1 selects one of them by itself). If $a?x_2$ is executed, the value from D_{r1} is assigned to x_2.

Further optimization

Clock frequency is an important factor for efficient circuits. To what extent the frequency can go up depends on the *critical path* of the circuit, which is the most time consuming path of the logic gates used in a clock cycle.

In our technique, for each rendezvous indication r_i, (i) r_i_*ok* and (ii) r_i_*en* have to be evaluated in a clock cycle. The evaluation for (i) requires the transfer of data values among the related EFSMs and the evaluation of the guard expression in each EFSM. The evaluation for (ii) requires the logic gates of $log\, h$ depth where h is the number of conflicting rendezvous indications. To shorten the critical path, we can take the following approach: (1) divide each complicated ADT function into several sub-modules with registers to calculate the result in several clock cycles based on the technique in [2]. (2) solve each conflict among multiple rendezvous indications in several clock cycles, for example, by dividing them into several groups.

Another topic of the optimization is to reduce the number of data paths to simplify the resulting circuits. In general, several rendezvous indications can share a data path as long as they do not conflict with each other or cannot be executed at the same time. To share the data path, we allocate a bus available to the related EFSMs for some rendezvous indications. With this technique, all the required data path can be implemented by just N data buses where N is the maximum number of conflicting/simultaneous rendezvous indications.

5 EXPERIMENTAL RESULTS AND DISCUSSION

We have applied our technique to the LOTOS specification of Abracadabra protocol [4] and constructed the hardware circuit to show that the constructed circuit is reasonably small and fast. We have used a hardware synthesis system PARTHENON [8] which has been developed by NTT.

In general, the modules for ADT functions (e.g. execution conditions) depend on how to implement the data types in the target circuit (e.g. the size of each data). Therefore, in this experiment, we have mainly evaluated the derived multi-rendezvous circuit without the modules for calculating ADT functions in each EFSM.

The greatest effect in the performance will be the maximum depth of the logic gates in the circuit. Eight EFSMs were derived from the LOTOS specification of Abracadabra protocol using the algorithm in Sect. 3 (the derived synchronous EFSMs can be found in [5]). The number of rendezvous indications was 85, and the maximum number of the depth of the logic gates for the multi-rendezvous circuit was 7. The maximum depth became six after some optimization.

Another criterion should be the size of the resulting circuit. The size of the multi-rendezvous circuit grows in proportion to the number of rendezvous indications since each rendezvous indication has its own module. The time for selecting a synchronous tuple set grows in proportion to the maximum number of rendezvous indications which may conflict with each other. In general, the number of the logic gates in the circuit depends on the size of data.

We have synthesized the whole circuit with ADT data/functions using 8 bit data. We have used several hardware modules for implementing ADT functions such as integer comparison (e.g. $=, \leq$) and addition (e.g. $inc, +$).

In that case, the whole circuit obtained has about 5000 gates: about 500 gates for ADT functions; about 300 gates for the multi-rendezvous circuit; and the remainder for registers, selectors and control signals for EFSMs.

The maximum depth of logic gates for ADT functions was 15. In the experiment, our technique requires additional 6 logic gates in depth for the multi-rendezvous circuit. That means the whole circuit could work with the clock frequency at least 70 % as high as in hardware circuits without multi-rendezvous. In fact, as we explained in the previous section, the multi-rendezvous circuit could be optimized further for practical use as well as other modules. We can approximately estimate the optimized performance from the circuit automatically derived with our technique. According to the above discussion, we think our technique can be used for rapid prototyping.

6 CONCLUSION

In this paper, we have proposed a hardware implementation technique from LOTOS specifications. In the technique, by composing each rendezvous indication of the tuple of event sets, we can keep the information about all possible rendezvous instances in a reasonable space. It is important that our conversion algorithm does not require any reachability analysis among parallel processes. Such analysis needs plenty of time proportional to the product of the number of events in parallel processes. Through the experiment for the Abracadabra protocol, we have confirmed our technique can be used for the rapid prototyping. We are going to evaluate our technique by implementing various protocols/systems, and possibly develop the optimization techniques for synthesized circuits for their practical use. To apply the technique to a time extension of LOTOS is part of future work.

References

[1] Faci, M. and Logrippo, L.: "Specifying Hardware Systems in LOTOS", *Proc. IFIP Int. Conf. on Hardware Description Languages and Applications (CHDL'93)*: 305 – 312 (1993).

[2] Higashino, T., Yasumoto, K., Kitamichi, J. and Taniguchi, K. : "Hardware Synthesis from a Restricted Class of LOTOS Expressions", *Proc. 14th IFIP Int. Symp. on Protocol Specification, Testing, and Verification (PSTV-XIV)*: 355–362 (1994).

[3] ISO: "Information Processing System, Open Systems Interconnection, LOTOS - A Formal Description Technique Based on the Temporal Ordering of Observational Behaviour", *ISO 8807* (1989).

[4] ISO/IEC/TR 10167: "Information Technology – Open Systems Interconnection – Guidelines for the application of Estelle, LOTOS and SDL" (1991).

[5] Kitajima, A., Yasumoto, K., Higashino, T. and Taniguchi, K.: "A Method to Convert Concurrent EFSMs with Multi-Rendezvous into Synchronous Sequential Circuits", *IEICE Trans. on Fundamentals*, E81-A (4): 566 – 575 (1998).

[6] Kloos, C. D. and Damm, W. (Eds.): "Practical Formal Methods for Hardware Design", *Research Reports Esprit, Project 6128, FORMAT*, Vol. 1, Springer (1996).

[7] Lopez, A. M., Kloos, C. D., Valladares, T. R. and Moro, T. M.: "Generating VHDL code from LOTOS Descriptions", in [6]: 266 – 293 (1996).

[8] Nakamura, Y. : "An Integrated Logic Design Environment Based on Behavioral Description", *IEEE Trans. on CAD*, CAD–6 (3): 322–336 (1987).

[9] Ott, D. E. and Wilderotter, T. J. : "A Designer's Guide to VHDL Synthesis", *Kluwer Academic Publishers* (1994).

[10] Quemada, J., Larrabeiti, D. and Pavon, S. : "Compressing the State Space Representation of LOTOS Specifications", *Proc. 6th IFIP Int. Conf. on Formal Description Techniques (FORTE'93)*: 19–34 (1993).

[11] Sisto, R. : "A method to build symbolic representations of LOTOS specifications", *Proc. 15th IFIP Int. Symp. on Protocol Specification, Testing and Verification (PSTV-XV)*: 331–346 (1995).

[12] Turner, K. J. and Sinnott, R. O.: "DILL: Specifying Digital Logic in LOTOS", *Proc. 6th IFIP Formal Description Techniques (FORTE'93)* (1994).

[13] Csopaki, G. and Turner, K. J.: "Modelling Digital Logic in SDL": *Proc. Joint Int. Conf. on Formal Description Techniques and Protocol Specification, Testing and Verification (FORTE/PSTV'97)*: 367 – 382 (1997).

[14] Vissers, C. A., Scollo, G. and Sinderen, M. v.: "Architecture and Specification Style in Formal Descriptions of Distributed Systems", *Proc. 8th Int. Symp. on Protocol Specification, Testing, and Verification (PSTV-VIII)*: 189 – 204 (1988).

[15] Wytrebowicz, J. : "Hardware Specification Generated from Estelle", *Proc. 15th IFIP Int. Symp. on Protocol Specification, Testing and Verification (PSTV-XV)*: 435–450 (1996).

[16] Yasumoto, K.: "A method to derive concurrent synchronous EFSMs from protocol specifications in LOTOS", *Tech. Rep. # 1113 of Dept. IRO, University of Montreal* (1998).

PART TEN

Real-Time & Performance

INTEGRATING REAL TIME INTO SPIN: A PROTOTYPE IMPLEMENTATION

Dragan Bošnački[1,2] and Dennis Dams[1]

[1]Department of Mathematics and Computing Science,
Eindhoven University of Technology, PO Box 513, 5600 MB Eindhoven, The Netherlands
{dragan, wsindd}@win.tue.nl

[2]On leave from the Institute of Informatics, Faculty of Natural Sciences and Mathematics, University "Sts. Cyril and Methodius", Skopje, Macedonia

Abstract: We present a discrete-time extension of Promela, a high level modelling language for the specification of concurrent systems, and the associated Spin model checker. Our implementation is fully compatible with Spin's partial order reduction algorithm, which is indeed one of its main strengths. The real time package is for most part orthogonal to the other features of the tool, resulting in a modular extension. We have evaluated it by several experiments, with encouraging results.

1 INTRODUCTION

Promela is a high level modelling language for specification of concurrent systems. The models written in Promela serve as input to the *Spin* software package for their automated verification.

The time ordering of actions in a Promela program is implicit and depends on the (fixed) sequential composition of statements within each one of the component processes, as well as on the (unspecified) interleaving of statements from different processes. This time relation is only *qualitative*, meaning that we do not know the exact time interval that will elapse between two events. This can be a shortcoming when systems are to be verified whose correct functioning depends on timing parameters. Many such examples can be found among communication protocols that have to deal with unreliable transport media, where the duration of timeout intervals is important.

In this paper, we introduce an extension to Promela that allows to quantify the time elapse between events, by specifying the *time slice* in which they occur.

We describe a prototype implementation that integrates this extension into the Spin tool. Of particular concern is the compatibility of such an extension with the *partial order reduction* algorithm, which is an approach to alleviate the state-space explosion inherent in model checking, and indeed one of Spin's main strengths. We prove that Spin's partial order algorithm remains correct under the new timing constructs, and conduct a number of experiments that demonstrate its effectiveness.

The prototype implementation described in this paper grew out of an earlier implementation of discrete time in Promela and Spin [4] instigated by process algebra ACP [3] and clocked transition systems [15]. There, the timing constructs from the process algebra ACPdrt were modelled by macro definitions entirely on the level of Promela. An advantage of such an approach is that the real-time package thus obtained is orthogonal to Spin: when needed, it can be incorporated as a simple header-file inclusion, for the current and future versions of Spin. Furthermore, the resulting real-time extension is easily moderated, allowing to gain experience with several alternative syntactical constructs and semantic models of time, and also to study the interaction with Spin's other features, most notably its partial order reduction algorithm. An obvious drawback of such a "high-level" solution is its inefficiency, which is to be avoided particularly in the case of a model checker. The real-time extension of Promela described in the current paper still uses macro definitions for certain constructs. However, the operation that occurs most frequently, namely the advance of time ("tick"), has been implemented on a lower level, in the source code of the Spin tool.

Eventually, we envisage a more complete integration of the real-time package into Spin, not only regarding the level of implementation, but the integration of real time and partial order reduction as well. Also, the underlying mathematical model will be changed from discrete to dense time. While this complicates the implementation, it is commonly accepted (c.f. [1]) that the resulting model-checking algorithms for dense time have comparable computational complexity. Having said this, we do stress that also discrete-time models are sufficient for a broad range of practical applications [11].

A closely related development is the RT-Spin package of [21]. Although that extension of Spin is more general in that it can model Timed Automata ([1]), with real-valued clocks, it has a number of drawbacks. The most important one is that it is not compatible with the partial order reduction algorithm. Furthermore, Timed Automata lack a notion of urgency; instead, the fact that a transition with a deadline must be taken on time, has to be modelled explicitly in RT-Spin. Also, the package has not been kept up-to-date with Spin versions later than Version 2.0.

The extension of partial order reduction techniques to apply to real-time systems has recently been studied in [18] and [17]. Other model checkers that cover timed behaviour are, e.g., UPPAAL ([14]), COSPAN ([2]), and HyTech ([10]). The latter indeed is able to handle the more general class of *linear hybrid systems*.

2 THE SPIN MODEL CHECKER

Promela and Spin have been developed for the analysis and verification of communication protocols. The language syntax is derived from C, but also uses the denotations for communications from Hoare's CSP and control flow statements based on Dijkstra's guarded commands. The full presentation of Promela and Spin is beyond the scope of this paper. We suggest [12] as a reference to the interested reader; here, we only give a brief overview of Spin's verification capabilities.

In Promela, system components are modeled as *processes* that can communicate via *channels* either by buffered message exchanges or rendez-vous operations, and also through shared memory represented as *global variables*. The execution of statements is asynchronous and interleaved, which means that in every step only one *enabled* action is performed, without any assumptions of the relative speed of process executions.

Given as input a Promela model, Spin generates a C program that performs a verification of the system by scanning the state space using a depth-first search (DFS) algorithm. This way, both *safety* properties such as absence of deadlock, unspecified message receptions, invalid end states and assertions can be checked, as well as *liveness* properties such as non-progress cycles and eventual reception of messages. The so-called *never claims*, which are best seen as monitoring processes that run in lock step with the rest of the system, are the most general way to express properties in Spin. Being Büchi Automata, they can express arbitrary omega-regular properties. Spin provides an automatic translator from formulae in linear-time temporal logic (LTL) to never claims. When errors are reported, the trace of actions leading to an invalid state or cycle is saved, so that the erroneous sequence can be replayed as a guided simulation.

For large state spaces methods, Spin features state-vector compressing, partial-order reduction and bit-state hashing.

2.1 Partial order reduction

Partial order reduction (POR) is a technique to cope with the state explosion problem in model-checking. We give a brief introduction to the POR algorithm that is used in Spin, rephrasing some definitions from [13].

The basic idea of the reduction is to restrict the part of the state space that is explored by the DFS, in such a way that the properties of interest are preserved. To this purpose, the independence of the checked property from the possible interleavings of statements is exploited. More specifically, two statements a,b are allowed to be permuted precisely then, if for all sequences v,w of statements: if $vabw$ (where juxtaposition denotes concatenation) is an accepted behaviour, then $vbaw$ is an accepted behaviour as well. In practice, sufficient conditions for such permutability are used that can be checked locally, i.e., in a state. For this, a notion of "concurrency" of statements is used that captures the idea that transitions are contributed by different, concurrently executing processes of the system.

The semantics of a Promela program can be represented as a *labeled transition system* (LTS). An LTS is a triple (S, s_0, T), where S is a finite set of states, s_0 is a distinguished initial state, and $T \subseteq S \times S$ is a set of transitions. Every transition in T is the result of executing a statement in some process of the Promela program. We introduce a function *Label* that maps each transition to the corresponding statement. For a statement a, $Pid(a)$ denotes the process to which a belongs and $En(a)$ denotes the set of (global) states in which a is enabled. For $q \in En(a)$, $a(q)$ is state which is reached by executing a in state q.

Concurrent statements (i.e. statements with different *Pids*) may still influence each other's enabledness, whence it may not be correct to only consider one particular order of execution from some state. The following notion of *independence* defines the absence of such mutual influence. Intuitively, two statements are independent if in every state where they are both enabled, they cannot disable each other, and are commutative, i.e., the order of their execution makes no difference to the resulting state.

Definition 2.1.1 *The statements a and b are* independent *iff for all states q such that $q \in En(a)$ and $q \in En(b)$,*

- $a(q) \in En(b)$ *and* $b(q) \in En(a)$, *and*
- $a(b(q)) = b(a(q))$.

Statements that are not independent are called dependent.

Note that a and b are trivially independent if $En(a) \cap En(b) = \emptyset$. An example of independent statements are assignments to or readings from local variables, executed by two distinct processes.

Another reason why it may not be correct to only consider only one particular order of execution from state s of two concurrent statements a and b is that the difference between the intermediate states $a(s)$ and $b(s)$ may be observable in the sense that it influences the property to be checked. For a given proposition p that occurs in the property (an LTL formula), and a state s, let $p(s)$ denote the boolean value of the proposition p in the state s. Then, a is *invisible* iff for all propositions p in the property and all states $s \in En(a)$, we have $p(s) = p(a(s))$. a is said to be *safe* if it is invisible and independent from any other statement b for which $Pid(b) \neq Pid(a)$.

The reduction of the search space is now effected during the DFS, by limiting the search from a state s to a subset of the statements that are enabled in s, the so-called *ample set*. Such an ample set is formed in the following way: If there is a process which has only safe statements enabled and all those transitions lead to a state which is not on the DFS stack, then the ample set consists of all the statements from this process only. Otherwise, the ample set consists of all enabled statements in s. It can be proven [13, 19] that the reduced graph obtained in this way preserves the properties of the original LTS, stated as an

LTL formula. The condition that all transitions from the ample set must end out of the DFS stack, the so-called "cycle proviso", ensures that a statement that it is constantly enabled, cannot be "forgotten" by leaving it outside the ample set in a cycle of transitions.

While the cycle proviso is clearly locally checkable during a DFS, the condition that an enabled statement is safe is not, as the definition of safety requires independence from *any* concurrent statement. A sufficient condition for safety of a statement a that can be checked locally is that a does not touch any global variables or channels. Indeed, it is this condition that is implemented in Spin.

3 INTRODUCING DISCRETE TIME IN PROMELA AND SPIN

3.1 Real time Promela

In the discrete time model, time is divided into slices indexed by natural numbers. The actions are then framed into those slices, obtaining in that way a measure for the elapsed time between events belonging to different slices. Within a slice however, we only know the relative ordering between events, as in the time free case. The elapsed time between events is measured in ticks of a global digital clock that is increased by one with every such tick.

In state space enumeration methods like the one used in Spin, verification can be regarded as checking all possible simulations of the system. In that sense the simulation can be considered as a primitive for the validation, so we start with the description of the discrete time implementation for simulation purposes. The basic idea is to execute the system time slice by time slice. Recalling that the execution of statements in Spin is asynchronous and interleaved, the basic problem is how to avoid interleaving of actions belonging to different time slices. One way to solve this is by forcing each process to stop the execution after it has executed all the actions for the current time slice, waiting to receive a signal that the system has passed to a new time slice. This synchronization is done by a special "daemon" process that is not a part of the modelled system and waits in the background to become active only when all the other processes from the system are blocked. This daemon process is a pacemaker that transfers the system to the next time slice, by sending unblocking signals to the other processes.

We implement this synchronization scheme by extending Promela with a new variable type **timer** corresponding to discrete time countdown timers, three new statements **set**, **expire** and **tick** that operate on the new type, and a special timing process **Timers** which is the daemon process that uses **ticks** to decrease the timer values. The implementation can be done entirely on user level, without any additional changes in the Spin source code, for example, with the following Promela macro definitions and timer process:

```
#define timer int
#define set(tmr,val) (tmr=val)
#define expire(tmr) (tmr==0) /*timeout*/
#define tick(tmr) if :: tmr>=0 -> tmr=tmr-1 :: else  fi
```

```
proctype Timers()
{ do :: timeout -> atomic{ tick(tmr1); tick(tmr2) } od }
```

The first macro defines **timer** as a synonym for the integer type The **set** macro sets the value of the timer **tmr** to **val**. The **expire** macro is a test which becomes true when **tmr** becomes 0. The **tick** macro is used only in the **Timers** process and it decreases the value of **tmr** provided that it is active, i.e., its value is non-negative. Timers with negative values are considered as deactivated. **Timers** consists of an endless **do** iteration that realizes the passage of time. It is run concurrently with the other processes of the system. The key idea of the concept is the usage of **timeout** - a predefined Promela statement that becomes true when no other statement within the system is executable. By guarding **ticks** with **timeout**, we ensure that no process will proceed with an action from the next time slice until the other processes have executed all actions from the current time slice[1].

Within a process, statements are divided into time slices by putting **set** and **expire** at the beginning and end, respectively, of each time slice. For instance, assuming that we have declared **timer tmr**, and that **A**, **B** and **C** are nonblocking Promela statements, the sequence

```
set(tmr,1); A; B; expire(tmr); C
```

means that **A** and **B** will be executed in same time slice, while **C** belongs to the next time slice. The **expire** statement is a synchronization point where the process waits for **tmr** to become zero. This can be done only by **Timers**, i.e., only when all active timers are decreased. Thus, it is guaranteed that **C** will be executed in the next time slice. In fact, **set** is only a labelling of the time slice (some sort of reading of the global digital clock we assumed in our time model) and it can be permuted with the statements from the time slice that it labels (in our case **A** and **B**).

Also, in cases where we have empty time slices, we optimize sequences of the form

```
set(tmr,1);expire(tmr);set(tmr,1);expire(tmr);...;set(tmr,1);expire(tmr);
```

by `set(tmr,val); expire(tmr);` , where **val** is the number of **set-expire** pairs.

It is often convenient to use derived macros for modeling various delays. For instance, one tick delay and unbounded nondeterministic delay are implemented, by the following macros

```
#define delay(tmr,val) set(tmr,val); expire(tmr)
#define udelay(tmr) do :: delay(tmr,1) :: break od
```

In the unbounded nondeterministic delay, at each iteration a nondeterministic choice is made whether the loop will be broken and the process will proceed

---

[1] There is one **tick** for each timer in the system, In the example above we assumed that there are only two timers, **tmr1** and **tmr2**. The **ticks** are wrapped in **atomic** statement in order to avoid unwanted unblocking of some of the system processes before all timers are decreased.

with the execution of a new statement, or the decision will be delayed for the next time slice. In a similar way a nondeterministic bounded delay up to a certain number of ticks, or a nondeterministic delay within lower and upper bounds can be modeled.

The expressiveness of our extended Promela is the same as the one of timed automata interpreted in integral time (fictitious clocks, using the terminology of [1]). It is straightforward to show that using the aforementioned derived timing constructs one can model timed automata interpreted in integral time. Conversely, following the approach of [21], the semantics of the extended Promela can be given via timed automata.

It is noteworthy that, because of the independence of the timing implementation from the other parts of the Spin's source code, the expressivity of the timing framework is in fact augmented with the introduction of new features in Spin. For instance, a scheduler for several processes running on a single processor can be modeled in a natural way like an additional master process that synchronizes the other processes by giving them execution permission via Promela's `provided` declarator mechanism, which was recently introduced into Spin (from version 3.0).

3.2 Example

In this section we show how the discrete time can be used for the specification and verification of the Parallel Acknowledgment with Retransmission (PAR) protocol [20]. The choice of PAR was motivated by the fact that it is a relatively simple protocol, yet it is complex enough that its correct functioning depends on the duration of time intervals in a nontrivial way. PAR has been used in similar studies as this one, cf. [22, 16].

PAR is one-way (simplex) data-link level protocol intended to be used over unreliable transmission channels which may corrupt or lose data. There are four components in our implementation: a sender, a receiver, data channel K and acknowledgment channel L. The sender receives data from the upper level and sends them labeled with a sequence number that alternates between 0 and 1 over the channel K. After that it waits for an acknowledgment via the channel L. If this does not occur after some period of time, the sender times out and resends the old data. If the data are received undamaged and labelled with the expected sequence number, the receiver delivers them to the upper level and sends an acknowledgment.

Of crucial importance here is the duration of the time-out period which should be longer than the sum of the delays through the channels and the message processing time by the receiver. A premature timeout can cause the loss of a frame by the following scenario: The sender sends a frame and times out too early. This causes a duplicate of the just sent frame to be sent too. The receiver receives both frames and sends two acknowledgments. When the sender gets the acknowledgment for the first frame it thinks that it is for the second one (the duplicate). After receiving this first acknowledgment it sends next frame, which is lost by the data channel. In the meantime the second acknowledgment arrives (which was sent by the receiver as a result of successful receiving of the duplicate) and the sender mistakenly thinks that it is an acknowledgment for the last frame and never resends it.

The complete listing of the discrete time Promela model of PAR is given on the next page.

The Promela model starts with the already described preamble of macros for discrete time.

Then, after the definition of protocol specific parameters, the timers are declared. There is one timer per protocol entity, i.e., for sender (`sc`), receiver (`rc`), channel K (`xk`) and channel L (`xl`). Although used locally in this model, the timers are declared on global level in order to be accessible for `Timers`. The four defined Promela channels `A`, `B`, `C` and `D` have capacity 0 meaning that they are synchronous rendez-vous channels. `A`, `B`, `C` and `D` are used only as auxiliary channels, because the "real" unreliable channels are modeled as `proctypes`, i.e. as a separate processes. The channels `A` and `B` are the input and output, respectively, to the channel `K`.

They deal with messages consisting of two fields — one of type `byte` and one of type `bit`, corresponding to the message contents and the alternating bit. Similarly, `C` and `D` are the input and output for `L` and they can carry messages containing just one bit — the acknowledgment.

After the declarations of local variables, the `Sender` process begins with an unbounded start delay of the first operation which fetches a message from the upper level (user). In our model this is simplified by allowing `Sender` to generate the message itself by the statement `mt = (mt+1)%MAX`. The message followed by a sequence number is sent through the channel `K`. This event is "labeled" by an activation of the timer `sc` which is needed for the usual timeout mechanism (not to be confused with the `timeout` Promela statement) for unblocking the protocol in case of message or acknowledgment loss. After sending the message the `Sender` blocks waiting for the arrival of an acknowledgment or the expiration of the timeout timer. In the first case a nondeterministic choice [2] is made between signalling an error because of a corrupted acknowledgment and accepting the acknowledgment as correct.

`Receiver` starts by waiting for a message. This is achieved by the statement `B?mr,sn`, which denotes the reception of a message through the channel `B`. The receiving statements in our implementation are treated as executable as soon as possible, i.e. in the same slice when the corresponding sender is ready to send. After the label `S_h` an `assert` is used to ensure that the received message always matches the expected one. (Note that this statement is added to the model only because of validation reasons.) The message processing latency is modeled by a `delay` on the `Receiver`'s timer. Also the expected message and sequence number are updated. Successful reception is confirmed by sending

---

[2]The `if` statement is similar by its semantics to the do statement - a random choice is made between the alternatives and if none of them is enabled than the statement is blocked.

```
/*discrete time macros*/

#define timer int
#define set(x,y) x=y
#define expire(x) (x==0)
#define tick(x) if :: x>=0->x=x-1; :: else; fi
#define on(x) (x!=-1)


#define delay(x,y) set(x,y); expire(x);

#define udelay(x) do :: delay(x,1) :: break od

/*PAR time parameters*/
#define dK 3 /* delay along channel K */
#define dL 3 /* delay along channel L */
#define dR 1 /* receiver's message processing time */
#define To 9 /* sender's timeout value */
#define MAX 8 /*max number of different message contents*/

/*timers*/
timer sc, rc, xk, xl;

proctype Timers()
{
   do
   :: timeout ->
   atomic{tick(sc); tick(xk);
          tick(rc); tick(xl);}
   od;
}

/*channels*/
chan A = [0] of {byte, bit};
chan B = [0] of {byte, bit};
chan C = [0] of {bit};
chan D = [0] of {bit};

/*protocol entities*/

proctype Sender()
{
  byte mt; /* message data */
  bit  sn=0; /* sequence number*/
  R_h:
    /*message can be sent in any time slice*/
     udelay(sc);
    mt = (mt+1)%MAX;
  S_f:
    A!mt,sn; /*sand through channel K*/
    set(sc,To);
  W_s:
    do
    :: D?_ ->
       if
       :: atomic{skip; delay(sc, 1); sn=1-sn; goto R_h;};
       :: atomic{printf("MSC: ACKerr\n"); goto S_f};
       fi;
    :: expire(sc) -> goto S_f;  /*timeout*/
    od;

}

proctype Receiver()
{

 byte mr, me=1; /* received and expexted message*/
 bit rsn, esn=0; /*received and expected sequence number*/
 W_f:
   B?mr,rsn;
   if
   :: rsn == esn -> goto S_h; /*correct message and seq. num */
   :: rsn == 1-esn -> goto S_a; /*correct message, wrong  seq. num */
   :: atomic{printf("MSC: MSGerr\n");
                          goto W_f;};
   fi;
 S_h:
   /*Out!mr*/
   assert(mr == me);
   atomic{delay(rc,dR);
   /*message processing delay*/
   esn = 1-esn; me = (me+1)%MAX};
 S_a:
   C!1; /*send ack through channel L*/
   atomic{delay(rc, 1); goto W_f;};
}
```

```
/*channel processes*/

proctype K()
{
  byte rd; /*received chunk*/
  bit rab; /*received and expected bits*/

  do
  :: A?rd,rab;
     delay(xk, dK);
     if
     :: skip; B!rd,rab;
     :: skip;
     fi;
  od;
}

proctype L()
{
  do
  :: C?_;
     delay(xl,dL);
     if
     :: skip; D!1;
     :: skip;
     fi;
  od;
}

init
{   atomic{set(sc,-1); set(rc,-1);
           set(xk,-1); set(xl,-1);
           run Timers();
           run Sender();
           run K();
           run L();
           run Receiver();}
 }
```

a message through L (via Promela channel C) and after a latency delay the Receiver starts waiting again to a new message.

Channel processes model the lossy channels K and L. Both processes have the same structure — after a delay they nondeterministically chose between delivering or losing the message. The skip is always executable. It is used here to give equal precedence for both choices.

All processes are started from a special process called init via run statements. By enclosing the run statements within an atomic it is ensured that they start simultaneously.

Verification of the model can be done like in the standard Spin. Among the other properties, the validator is able to find out and display the scenario of losing a message, in case of an erroneous combination of the timing parameters.

3.3 Low level implementation

In order to achieve more efficiency we get down from the high-level implementation to a lower level, by realizing the tick construct in the Spin's source code. Also, the new semantics of tick is adopted so as to decrease all the timers in the system together. The implementation of the new statement in Spin is completely modular and does not interfere with other features.

The timing procedure now becomes much simpler:

```
proctype Timers
{ do :: timeout -> tick od }
```

As it is universal for all models, it can be made part of the standard header file that contains the basic macros. Another advantage of the new tick is that the timers can be declared locally. Like ordinary local variables, local timers can be dynamically created and deleted together with the processes they belong to. To the best of our knowledge this possibility is not present in any tool for validation of timed systems. Variable number of timers automatically means saving of the state space, due to Spin's variable size state representation.

The most important benefit of the local usage of timers is the partial order reduction which is, as already emphasized, one of the main strengths of the enumerative methods. The integration of partial order reduction with the discrete time model is one of the main results of this paper and it is the topic of the next subsection.

3.4 Compatibility with partial order reduction

In this section we prove that the partial order reduction algorithm of (standard, untimed) Spin is correct for the timed extension as well. This algorithm is based on selecting, during the DFS exploration of the state space, only statements from the ample set. The construction of an ample set is based on the notion of *safety* of statements: given a criterion to determine which statements are safe, the underlying theory, that was presented in Section 2.1, guarantees that the reduction strategy is correct, i.e. that (LTL) properties are preserved. For untimed Spin, this criterion is that a statement does not read or write any global variables or channels. We will now show that this criterion is still sufficient in the presence of timed statements.

Lemma 3.4.1 *Any statement that is safe in standard, untimed Promela, remains safe in the presence of the new timed statements* **`set`**, **`expire`**, *and* **`tick`**. *Furthermore, when applied to local*[3] *timers, the statements* **`set`** *and* **`expire`** *are safe, and* **`tick`** *is independent from any other statement.*

PROOF In order to show that a statement is safe, we have to show that it is *invisible*, and *independent* from any concurrent statement (i.e. any statement with a different *Pid*).

Let a be a statement in untimed Promela that is safe. Then it is clearly still invisible in timed Promela. Also, a is obviously still independent from any concurrent untimed statement. a's independence from any ("new") timed statement that is concurrent follows from the fact that the set of timer variables, that are the only variables which can be read or written by timed statements, is disjoint from the set of ordinary variables which can be affected by a.

Next, we let a be one of the statements `set` and `expire`, applied to a local timer. a is invisible because the local timer variable may not occur in the property to be checked. Let b be a concurrent statement different from `tick`. Then a is obviously independent from b because of the disjoint sets of variables they operate on. Independence between `tick` and a, and indeed independence between `tick` and any timed or untimed statement, follows from the fact that `tick` can never be enabled simultaneously with another statement. This is because of the way `tick` is implemented: it is only enabled in the case of a `timeout` condition.

As `tick` is implemented in the Spin source code, is is not subject to reduction by the partial order algorithm. However, this does not have any repercussions: any `tick` statement, regardless whether it is acting on a local or on a global timer, is the only statement that is enabled, when it is enabled at all. So, no reduction would have been possible anyway.

4 EXPERIMENTAL RESULTS

We have tested the implementation on various models known in the literature (e.g. Train Gate Controller, Seitz Circuit, Leader Election Protocol). We focus our attention on three of them that illustrate the effectiveness of the implementation and particularly the partial order reduction: Fischer's mutual exclusion protocol, the already presented PAR protocol and the Bounded Retransmission Protocol (BRP). All tests were performed on a Sun SPARC Station 5 with 64 MBytes of memory. Besides partial order reduction we used as an additional option *minimized automata*, a technique for reduction of the state space recently included in the standard Spin distribution. In the *options* column of the

[3]Recall that a local timer is a timer variable that is declared local to a Promela process. In timed Spin, it can be manipulated by the declaring process or by the `tick` operation.

tables below, "n", "r" and "ma" denote verifications without POR, with POR, and with POR together with minimized automata, respectively.

The version of the Fischer's protocol that was verified is a translation of the same model written in Promela with real (dense) time of [21], with virtually the same timing parameters. The obtained results for the verification of the mutual exclusion property are shown in Table 1 (N is the number of processes). As

Table 1 Results for Fischer's and PAR protocol.

N	*option*	*states*	*transitions*	*memory* [MB]	*time* [s]
2	n	528	876	1.453	0.1
	r	378	490	1.453	0.1
	ma	378	490	0.342	0.2
3	n	8425	10536	1.761	0.8
	r	3813	4951	1.596	0.4
	ma	3813	4951	0.445	0.3
4	n	128286	373968	7.085	15.5
	r	34157	44406	3.132	2.8
	ma	34157	44406	0.650	33.7
5	r	288313	377032	17.570	36.2
	ma	288313	377032	1.059	332.5
6	ma	2.35e6	3.09e6	2.118	6246.1

time parameters								
dK	dL	dR	To	*option*	*states*	*transit.*	*mem.* [MB]	*time* [s]
3	3	1	9	n	1318	1822	1.453	0.2
				r	1116	1533	1.493	0.3
30	30	10	90	n	7447	9994	1.761	0.5
				r	7295	9705	1.801	0.6
300	300	100	900	n	68737	91714	5.135	4.9
				r	68585	91425	5.420	4.9
				ma	68585	91425	2.221	143.7

expected, the state space growth is exponential and we were able to validate the model without using POR up to 4 processes. For 6 processes even POR was not enough and it had to be strengthened with minimized automata. Nevertheless, the profit from POR is obvious and even becomes more evident as the number of processes grows. While for $N = 2$ the number of states is reduced to 72% compared to the case without POR, and the transitions are reduced to 56%, for $N = 4$ the reduction increases to 27% and 12% for states and transitions, respectively.

It is difficult to compare our implementation with the one from [21], because they are based on different time models. Nevertheless, having in mind that the property that was checked was a qualitative one, for which discrete time suffices [11], one can safely say that after $N = 4$ our implementation has better performance. In fact, for $N = 5$ the validator from [21] runs out

Table 2 Results for timed and untimed BRP.

option	*states*	*transitions*	*memory* [MB]	*time* [s]
n	32967	71444	4.116	6.7
r	12742	14105	2.517	3.2
ma	12742	14105	1.571	26.4

option	*states*	*transitions*	*memory* [MB]	*time* [s]
n	199454	658852	7.515	37.6
r	113566	294546	5.547	19.0

of memory. Obviously, POR is the decisive factor, because without POR our implementation is also incapable to handle cases for $N > 4$.

In the case of the PAR protocol the reduction is very small and even diminishes with the increase of parameters (Table 1). The insensitivity of PAR model to reduction can be explained by observing that PAR is in fact a sequential algorithm. The protocol entities take turns during the execution of the protocol, most of the time only one of them being active while the others are waiting passively for some trigger-event. Very little concurrent activity results in a bad behaviour of the POR algorithm which deals with concurrent actions from different processes. It is an opposite situation compared to Fischer's protocol where the higher degree of concurrency led to an improvement of the effects of POR.

Note the sensitivity of the state space to the increase of parameter values. This sensitivity to parameters is analogous to the same effect observed in the region automaton implementation of the dense time system based on timed automata [1, 2]. At least for the type of protocols like PAR, when the sequential component of the system is predominant, this problem can be solved by grouping together consecutive ticks into one tick to avoid generation of spurious intermediate states.

Our implementation of BRP is based on [6], where the protocol was treated using a combination of Uppaal (for timing aspects) and Spin (for consistency of the specification). We give the results for safety properties (absence of deadlock and assertions). Using the same parameters as in [6], all the properties verified there can be checked with a memory consumption of at most 5 Mbytes and less than a minute of CPU time. This is a much better result than the one in [6]. In addition, our model has the advantage that the validation is done completely within Spin.

The benefit of POR reduction is obvious — the number of states is reduced to 39%, while the transitions are reduced to 20%. In order to get an indication how well the partial order reduction combines with our time extension in Spin, we have compared the above reductions for the BRP to the partial order reduction that is achieved on a slightly different version of the protocol that does not involve timers (see e.g. [8] or [7]). The result is given in Table 2. The effect for the timed version turns out to be even better: 39% and 20% versus 57% and

45%. It has to be said though that in order to model the effect of timeouts in the untimed case, we need an additional global variable, which may influence the effects of the reduction. Anyway, this is a promising result having in mind that, unlike previous two examples which are more academic, BRP is just slightly simplified version of an industry protocol used by Philips, which gives us hope that POR could work for real-world applications.

5 CONCLUSIONS AND FUTURE WORK

We presented an extension of Promela and Spin with discrete time, using integer variables as timers to stamp the time slices. The core idea was to use a special background daemon process to implement the passage of time. For this purpose we used the Promela `timeout` predefined statement as a mechanism for process (i.e. timer) synchronization. We first showed how the concepts can be implemented using only the existing features of Promela and Spin, by extending the language with new time statements as Promela macro definitions. The usage of timing features was demonstrated on the specification and validation of the PAR protocol.

As one step further toward a more efficient integration of the timing framework into Spin, we implemented timing on the level of the validator's source code. With this low level implementation, we gain the possibility to use the timers locally. This allows for dynamic change of the number of timers in the system — a feature which to the best of the authors' knowledge does not exist in any implementation of verification of real-time systems. This is in accord with the possibility of dynamic creation of processes already present in Spin. But, the most important benefit of the local usage of timers, and one of the main results of this paper, was the adaptation of the existing partial order reduction algorithm to the discrete time setting. The urgency that is a necessary feature in the modeling of a broad class of interesting timing dependent systems is achieved in a natural way, without usage of invariants. The low level implementation is a seamless and completely compatible extension of the standard Spin validator which allows one to use all the existing features of the untimed validator in a standard way. We showed the efficiency of the partial order reduction and the implementation in general on several examples, of which we emphasize the verification of the Bounded Retransmission Protocol, as an industrial protocol used by Philips. The main future tasks will certainly be to test the approach and accumulate experience by doing new verifications on systems known in the literature or some new real-world (industry) cases which are time dependent.

Besides the improvement of the existing prototype, the main direction for the future work will be the extension of Promela and Spin with dense time, which will provide the possibility to express and verify a more general class of properties. The full description of the implementation will be presented in a forthcoming paper. Here we give an outline of the basic idea, assuming that the reader is familiar with the theory of timed automata (see for instance [1]). In the dense-time extension we consider each Promela process in the system as a timed automaton, thus, the system as a whole is then represented as a parallel composition of timed automata. For this purpose we have to extend Promela with a new variable type `clock` whose value can be tested (compared) and/or set to zero. We are going to implement the discretizations of the timed automata, called region automata. Once we have these discretizations we can reuse the same basic concepts from the discrete-time implementation presented in this paper. The key idea is that to represent the time passage for the dense time one can again use a *discrete* global clock which ranges over *clock regions* (instead over integer valued vectors, as it was for the discrete time). The number of regions is always finite [1] which guarantees termination of the validation procedure. Thus, the whole concept of a special procedure that implements the time passage can be applied, like for the discrete-time case. Although there are many technical details that we omit because of the space limitations, it is obvious that the same concept of `timeout` as a clock tick can be also reused to obtain an extension which can be easily integrated with the other features of Spin. With regard to the efficiency it is promising that the partial order reduction algorithm can be adapted in the same way as for discrete time.

It would be very interesting to incorporate the ideas about data and time abstraction of [5] both in the existing discrete time prototype, as well as in the future dense time implementation, in order to obtain verifications that are independent of the concrete parameter values, and to see how general is their applicability.

Another promising idea is to extend the concept of time managing process to a class of hybrid systems called discrete time rectangular automata [9]. This can be done by introducing, besides timers (clocks), new time dependent variables and allowing the timing process to change their value also with steps greater than one.

Acknowledgments

The authors would like to thank Stavros Tripakis, Bart Knaack and the anonymous referees for their helpful comments.

References

[1] Alur, R., Dill, D.L., *A Theory of Timed Automata*, Theoretical Computer Science, 126, pp.183-235, 1994.

[2] Alur, R., Kurshan, R.P., *Timing Analysis in Cospan*, Hybrid Systems III, LNCS 1066, pp.220-231, Springer, 1996.

[3] Baeten, J. C. M., Bergstra, J. A., *Discrete Time Process Algebra*, Formal Aspects of Computing 8 (2), pp. 142-148, 1991.

[4] Bosnacki, D., *Implementing Discrete Time in Promela and Spin*, International Conference on Logic in Computer Science, LIRA '97, University of Novi Sad, Yugoslavia, 1997. (Copy also available from the author)

[5] Dams, D., Gerth, R., *Bounded Retransmission Protocol Revisited*, Verification of Infinite Systems Infinity, Bologna, Italy, 1997.

[6] D'Argenio, P. R., Katoen, J.-P., Ruys, T., Tretmans, J., *The Bounded Retransmission Protocol Must Be on Time!*, TACAS'97, 1997.

[7] Graf, S., Saidi, H., *Construction of Abstract State Graphs with PVS*, Computer Aided Verification '97, LCNS 1254, pp. 72-83, Springer, 1997.

[8] Groote, J. F., van de Pol, J., *A Bounded Retransmission Protocol for Large Data Packets*, in Wirsing, M., Nivat, M., ed., Algebraic Methodology and Software Technology, LCNS 1101, pp. 536-550, Springer-Verlag, 1996.

[9] Henzinger, T. A., Kopke, P. W., *Discrete-Time Control for Rectangular Automata*, International Colloquium on Automata, Languages and Programming (ICALP 1997), LNCS 1256, pp. 582-593, Springer-Verlag, 1997.

[10] Henzinger, T. A., Ho P.-H., Wong-Toi, H., *A Model Checker for Hybrid Systems*, Computer Aided Verification 97, LNCS 1254, pp.460-463, Springer-Verlag,1992.

[11] Henzinger, T. A., Manna, Z., Pnueli, A., *What good are digital clocks?*, Proceedings of the ICALP'92, LNCS 623, pp.545-558, Springer-Verlag, 1992.

[12] Holzmann, G. J., *Design and Validation of Communication Protocols*, Prentice Hall, 1991. Also: `http://netlib.bell-labs.com/netlib/spin/whatispin.html`

[13] Holzmann, G., Peled, D., *An Improvement in Formal Verification*, FORTE 1994, Bern, Switzerland, 1994.

[14] Larsen, K. G., Pettersson, P., Yi, W., *UPPAAL: Status & Developments*, Computer Aided Verification CAV 97, LNCS 1254, pp.456-459, Springer-Verlag, 1992.

[15] Kesten, Y., Manna, Z., Pnueli, A., *Verifying Clocked Transition Systems*, Handouts of School on Embedded Systems, Veldhoven, The Netherlands, 1996.

[16] Klusener, A. S., *Models and Axioms for a Fragment of Real Time Process Algebra*, Ph.D. Thesis, Eindhoven University of Technology, 1993.

[17] Knaack, B., *Real-time extension of SPIN - Ongoing Research*, Abstracts from SPIN'97, Third SPIN Workshop, Twente University, Enschede, The Netherlands, Also: `http://netlib.bell-labs.com/netlib/spin/ws97/papers.html`

[18] Pagani, F., *Partial Orders and Verification of Real Time Systems*, Formal Techniques in Real Time and Fault Tolerant Systems FTRTFT 96, LNCS, Springer, 1996.

[19] Peled, D., *Combining Partial Order Reductions with On-the-Fly Model Checking*, Computer Aided Verification 1994, LCNS 818, pp. 377-390, 1994.

[20] Tanenbaum, A., *Computer Networks*, Prentice Hall, 1989.

[21] Tripakis, S., Courcoubetis, C., *Extending Promela and Spin for Real Time*, TACAS '96, LCNS 1055, Springer Verlag, 1996.

[22] Vaandrager, F., *Two Simple Communication Protocols*, in Baeten, J.C.M., ed., Applications of Process Algebra, pp.23-44,Cambridge University Press, 1990.

CMC: A TOOL FOR COMPOSITIONAL MODEL-CHECKING OF REAL-TIME SYSTEMS

François Laroussinie[1]
and Kim G. Larsen[2]

[1]Laboratoire Spécification et Vérification,
ENS de Cachan, CNRS URA 2236
61, av. Pdt Wilson, 94235 Cachan Cedex, France
fl@lsv.ens-cachan.fr

[2]Basic Research in Computer Science
Centre of the Danish National Research Foundation.
Aalborg University, Fredrik Bajers Vej 7-E, DK-9220 Aalborg, Denmark
kgl@cs.auc.dk

Abstract: In this paper we present a tool (CMC) for compositional model-checking of real-time systems. CMC is based on a completely different method compared to existing real-time verification tools (HYTECH, KRONOS, UPPAAL). After a description of the method, we illustrate its efficiency by considering two examples : the Fischer's mutual exclusion protocol and a railroad crossing system.

INTRODUCTION

Within the last decade, model-checking has turned out to be a useful and successful technique for the verification of temporal properties in finite state systems. More recently, serious attempts have been made to extend the success of model-checking to the setting of real-time systems, with timed automata [2] as models. The major obstacle for the model-checking approach is the well-known state explosion problem due to parallel composition (as in the untimed case) and also to time encoding. Several heuristics have been proposed to overcome this problem : symbolic model-checking [10], on-the-fly technique [6, 19], efficient data-structures for time constraints [17, 4]. These and other techniques have been implemented into tools (KRONOS [22], UPPAAL [19] and HYTECH [11]) and a number of industrial case studies have been analyzed and verified using these tools [12, 13, 5, 20, 9].

In [16] we proposed a first *compositional model-checking* technique, which in a number of practical examples seems to avoid the state explosion problem. In this method, we never construct nor examine the global (symbolic) state space of a real-time system modeled as a network $(A_1|\ldots|A_n)$ of timed automata. Rather, when checking the system for a property φ, the components $A_1, \ldots, A_n$ are gradually removed from the network and incorporated into the formula; thus, the component A_n could be initially transfered from the network into the property φ yielding a new *quotient* property φ/A_n which expresses the precise property which has to be satisfied by the remaining network $(A_1|\ldots|A_{n-1})$ in order that the original property φ holds for the complete system. Now by repeatedly quotienting components from the network into the formula, we will finally be faced with the problem of establishing that the process *nil* (a process capable of performing no action at all) satisfies the last formula. A key point of this approach is to develop efficient simplification strategies in order to keep a small size of the quotiented formula.

In this paper we extend the compositional method in several ways. Firstly, we develop the method to a larger class of timed automata (with invariants and atomic propositions). Secondly we introduce a completely new (and considerably more efficient) quotient construction and specific simplification techniques adapted to our formalisms. Finally we present CMC a new real-time verification tool implementing the compositional method.

Plan of the paper: we first introduce the formalisms used in CMC : networks of timed automata used for the modeling of real-time systems, and the timed modal logic L_ν for specifying properties. We then present the quotient construction together with the simplification strategies emphasizing those dealing with clock constraints. Finally we illustrate the method with two examples. The first example is Fischer's mutual exclusion protocol [1, 21], which is a classical benchmark in this domain. Here our results both demonstrate the efficiency as well as the limits of the compositional approach. The second example is the railroad crossing system. In this example, L_ν proves to be an extremely convenient specification language, and our results proves that compositional method is useful for arbitrary systems and not only for networks with a symmetrical structure as in Fischer's protocol.

NETWORKS OF TIMED AUTOMATA

Let $\mathcal{A}$ be a finite set of actions and AP a finite set of atomic propositions. We denote by $\mathbf{N}$ the set of natural numbers and by $\mathbf{R}$ the set of non–negative real numbers. Θ denotes the set of delay actions $\{\epsilon(d) \mid d \in \mathbf{R}\}$. If C is a set of clocks, $\mathcal{B}(C)$ denotes *clocks constraints* over C, that is the set of formulas built using boolean connectives over atomic formulas of the form $x \bowtie m$ or $x - y \bowtie m$ with $x, y \in C$, $m \in \mathbf{N}$ and $\bowtie \in \{=, <, >, \leq, \geq\}$. Moreover $\mathcal{B}^-(C)$ denotes the

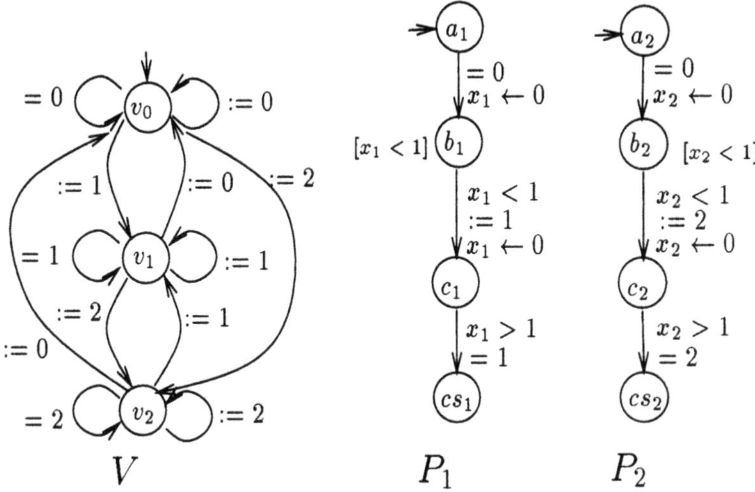

Figure 1.1 The Fischer's mutual exclusion protocol modeled with timed automata

restriction of $\mathcal{B}(C)$ to the boolean value *true* and to the conjunctions of atomic formulas of the form $x \bowtie m$ with $x \in C$ and $\bowtie \in \{<, \leq\}$. A time assignment v for C is a function from C to $\mathbf{R}$. We denote by $\mathbf{R}^C$ the set of time assignments for C. For $v \in \mathbf{R}^C$ and $d \in \mathbf{R}$, $v + d$ denotes the time assignment which maps each clock x in C to the value $v(x) + d$. For $C' \subseteq C$, $[C' \mapsto 0]v$ denotes the assignment for C which maps each clock in C' to the value 0 and agrees with v over $C \backslash C'$. Whenever $v \in \mathbf{R}^C$, $u \in \mathbf{R}^K$ and C and K are disjoint, vu denotes the time assignment over $C \cup K$. Given a condition $g \in \mathcal{B}(C)$ and a time assignment $v \in \mathbf{R}^C$, we note $v \models g$ when g holds for v. We define the timed automata (TA):

Definition 1 *A timed automaton A over $\mathcal{A}$ is a tuple $\langle N, \eta_0, C, E, l, Inv\rangle$ where N is a finite set of nodes, η_0 is the initial node, C is a finite set of clocks, $E \subseteq N \times \mathcal{B}(C) \times \mathcal{A} \times 2^C \times N$ corresponds to the set of edges: $e = \langle \eta, g, a, r, \eta' \rangle \in E$ represents an edge from the node η to the node η' with action a, r denotes the set of clocks to be reset and g is the enabling condition (the guard) over the clocks of A. We use the notation $\eta \xrightarrow{g,a,r} \eta'$. Moreover $l : N \mapsto 2^{\mathsf{AP}}$ is a labeling function of nodes with atomic propositions. Finally $Inv : N \mapsto \mathcal{B}^-(C)$ is a function assigning an invariant to each node.*

Example 1 *Consider the automata V, P_1 and P_2 in Figure 1.1. The automaton P_1 has four nodes, a_1, b_1, c_1 and cs_1, one clock x_1 and three edges. The edge between b_1 and c_1 has ":= 1" as action, $\{x_1\}$ as reset set and the enabling condition for the edge is $x_1 < 1$. Node b_1 has the invariant $x_1 < 1$ to enforce progress from the node before one time-unit elapses. The default invariant (for nodes a_1, c_1 and cs_1) is true.*

Note that the automaton V is a "classical" automaton (without clock). □

A *state* or a *configuration* of an automaton A is a pair (η, v) where η is a node of A and v a time assignment for C. Informally, the system starts at node η_0 with all its clocks initialized to 0 (i.e. with the time assignment v_0). The values of the clocks may increase synchronously with time as long as the invariant requirement $Inv(\eta_0)$ holds for the current values of the clocks . At any time, the automaton whose current node is η can change node by following an edge $\langle \eta, g, a, r, \eta' \rangle \in E$ provided the current values of the clocks satisfy g. With this transition the clocks in r get reset to 0. Formally the semantics of A is defined as a *labeled timed transition system*:

Definition 2 *A labeled timed transition system over $\mathcal{A}$ is a tuple $\mathcal{S} = \langle S, s_0, \rightarrow, l \rangle$, where S is a set of states, s_0 is the initial state, $\rightarrow \subseteq S \times \mathcal{A} \cup \Theta \times S$ is a transition relation and $l : S \mapsto 2^{\mathsf{AP}}$ a labeling function. We require that for any $s \in S$ and $d \in \mathbf{R}$, there exists at most one state s^d such that $s \xrightarrow{\epsilon(d)} s^d$. Moreover if s^d exists, then $s^{d'}$ exists for any $0 \leq d' \leq d$. Finally $(s^d)^e = s^{d+e}$.*

The labeled timed transition system associated with A is $\mathcal{S}_A = \langle S_A, \sigma_0, \longrightarrow_A \rangle$, where S_A is the set of states of A, σ_0 is the initial state (η_0, v_0), and $\longrightarrow_A$ is the transition relation defined as follows:

$$(\eta, v) \xrightarrow{a} (\eta', v') \quad \text{iff} \quad \exists \langle \eta, g, a, r, \eta' \rangle \in E \text{ s.t. } v \models g \text{ , } v' \models Inv(\eta') \text{ and } v' = [r \mapsto 0]v$$

$$(\eta, v) \xrightarrow{\epsilon(d)} (\eta', v') \quad \text{iff} \quad \eta = \eta' \text{ and } v' = v + d \text{ and } (v + d) \models Inv(\eta)$$

The labeling with atomic propositions is defined by: $l(\langle \eta, v \rangle) = l(\eta)$. Note that the condition $(v + d) \models Inv(\eta)$ in the second rule implies that for any d' s.t. $0 \leq d' \leq d$ we have $(v + d') \models Inv(\eta)$ because invariants are conjunctions of $x < k$ or $x \leq k$ formula.

Real-time systems may be modeled as parallel composition of timed automata. Following [14] we suggest a composition parameterized with a synchronization function generalizing a large range of existing notions of parallel compositions. Let $A_1, \ldots, A_n$ be n timed automata. A *synchronization function* f is a partial function $(\mathcal{A} \cup \{\bullet\}) \times \ldots \times (\mathcal{A} \cup \{\bullet\}) \hookrightarrow \mathcal{A}$, where $\bullet$ denotes a distinguished no–action symbol (for any i and $\eta \in N_i$, we assume $(\eta, \mathtt{tt}, \bullet, \emptyset, \eta) \in E_i$). In fact, f is an n-ary synchronization function with renaming. We denote by $(A_1 | \ldots | A_n)_f$ the parallel composition of $A_1, \ldots, A_n$ w.r.t. the synchronization function f. A network configuration is a pair $\langle \bar{\eta}, v \rangle$ where $\bar{\eta} = (\eta_1, \ldots, \eta_n)$ is vector of nodes and v is a valuation for $C = \cup_i C_i$, i.e. the clocks of the network (we denote by v_i the restriction $v_{|C_i}$).
The semantics of $(A_1 | \ldots | A_n)_f$ can be defined as a labeled timed transition system whose states are the configurations of the network and the transitions are given by the two following rules:

$$\langle \bar{\eta}, v \rangle \xrightarrow{\epsilon(d)} \langle \bar{\eta}, v + d \rangle \quad \text{iff} \quad \forall i \in \{1, \ldots, n\}, \langle \eta_i, v_i \rangle \xrightarrow{\epsilon(d)} \langle \eta_i, v_i + d \rangle$$

$$\langle \bar{\eta}, v \rangle \xrightarrow{b} \langle \bar{\eta}', v' \rangle \quad \text{iff} \quad \forall i \in \{1, \ldots, n\}, \langle \eta_i, v_i \rangle \xrightarrow{a_i} \langle \eta_i', v_i' \rangle \text{ s.t. } a_i \in \mathcal{A} \cup \{\bullet\} \text{ and } v' = v_1' \ldots v_n' \text{ and } f(a_1, \ldots, a_n) = b$$

Note that the first rule stipulates that all clocks increase synchronously. The atomic proposition labeling is done in a natural way: $l(\langle \bar{\eta}, v \rangle) = \cup_i l(\eta_i)$.

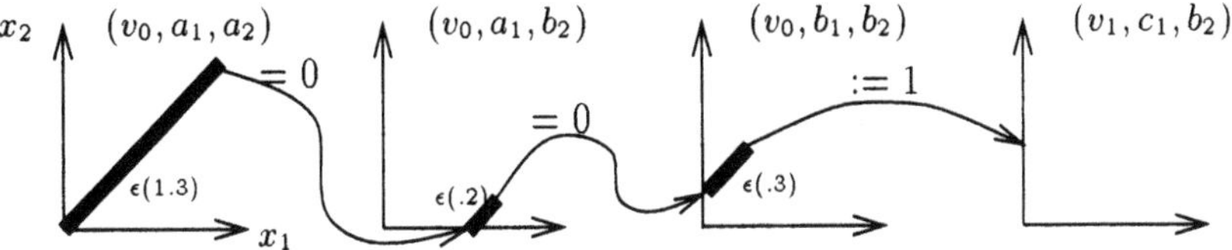

Figure 1.2 A prefix of an execution for FP.

Example 2 *Consider again the automata V, P_1 and P_2 of figure 1.1. This system describes Fischer's mutual exclusion protocol: two processes (modeled by P_1 and P_2) both try to reach a critical section (cs_1 and cs_2 resp.). To guarantee mutual exclusion of the critical sections, P_1 and P_2 share a variable (modeled by the process V) which can be assigned to 0,1 or 2 by the processes. Moreover the current value of V can be read by the two processes.*
The protocol can now be modeled by the parallel composition $FP = (V|P_1|P_2)_f$ with f the ternary synchronization function defined as follow: $f(=0,=0,\bullet) \stackrel{def}{=} f(=0,\bullet,=0) \stackrel{def}{=}= 0$, $f(a,a,\bullet) \stackrel{def}{=} a$ for any a in $\{=1,:=1\}$ and $f(a,\bullet,a) \stackrel{def}{=} a$ for any a in $\{=2,:=2\}$. The values of f correspond to the name of the visible action.
The coordinate systems in Figure 1.2 indicates (some of) the states of the system FP. Each point of the coordinate systems represents a unique time assignment for x_1 and x_2, with the coordinate systems representing states involving the nodes configurations (v_0, a_1, a_2), (v_0, a_1, b_2), (v_0, b_1, b_2) and $(v1, c1, b2)$ respectively. We indicated the following transition sequence : $((v_0, a_1, a_2), (0,0)) \stackrel{\epsilon(1.3)}{\longrightarrow} ((v_0, a_1, a_2), (1.3, 1.3)) \stackrel{=0}{\longrightarrow} ((v_0, a_1, b_2), (1.3, 0)) \stackrel{\epsilon(0.2)}{\longrightarrow} ((v_0, a_1, b_2), (1.5, 0.2)) \stackrel{=0}{\longrightarrow} ((v_0, b_1, b_2), (0, 0.2)) \stackrel{\epsilon(0.3)}{\longrightarrow} \ldots$ □

SPECIFYING TIMED SYSTEMS IN CMC

In our tool CMC, specifications of real-time systems are given as properties of the timed modal logic L_ν [16]. This logic may be seen as a fragment of the logic T_μ of [10] allowing only maximal fixed-points. Even though this restriction only enables safety properties to be specified and rules out liveness properties in general, it is still expressive enough to specify time-bounded liveness properties. In many real-time applications this is precisely what we wanted. Avoiding alternations of fixed-points also simplifies our model-checking algorithm.

Definition 3 *Let K a finite set of clocks, Id a set of identifiers, AP a set of atomic propositions, and $\mathcal{A}$ a set of actions. The set L_ν of formulae over K, Id, AP and $\mathcal{A}$ is generated by the following grammar:*

$$L_\nu \ni \psi, \varphi \quad ::= \quad \varphi \wedge \psi \mid \varphi \vee \psi \mid \langle a \rangle \varphi \mid [a] \varphi \mid \langle \delta \rangle \varphi \mid [\delta] \varphi \mid Z$$
$$\mid x \text{ in } \varphi \mid x + m \bowtie y + m' \mid x \bowtie m \mid ap \mid \neg ap$$

where $a \in \mathcal{A}$; $x, y \in K$; $m, m' \in \mathbf{N}$; $\bowtie \in \{=, <, \leq, >, \geq\}$, $Z \in \mathsf{Id}$ and $ap \in \mathsf{AP}$.

$\langle \bar{\eta}, v, u \rangle \models_{\mathcal{D}} \varphi \wedge \psi$	iff	$\langle \bar{\eta}, v, u \rangle \models_{\mathcal{D}} \varphi$ and $\langle \bar{\eta}, v, u \rangle \models_{\mathcal{D}} \psi$
$\langle \bar{\eta}, v, u \rangle \models_{\mathcal{D}} \langle a \rangle\ \varphi$	iff	$\exists \langle \bar{\eta'}, v', u \rangle.\ \langle \bar{\eta}, v \rangle \xrightarrow{a} \langle \bar{\eta'}, v' \rangle$ and $\langle \bar{\eta'}, v', u \rangle \models_{\mathcal{D}} \varphi$
$\langle \bar{\eta}, v, u \rangle \models_{\mathcal{D}} \langle \delta \rangle\ \varphi$	iff	$\exists d \in \mathbf{R}$ s.t. $\langle \bar{\eta}, v \rangle \xrightarrow{\epsilon(d)} \langle \bar{\eta}, v + d \rangle$ and $\langle \bar{\eta}, v + d, u + d \rangle \models_{\mathcal{D}} \varphi$
$\langle \bar{\eta}, v, u \rangle \models_{\mathcal{D}} [\delta]\ \varphi$	iff	$\forall d \in \mathbf{R}$, $\langle \bar{\eta}, v \rangle \xrightarrow{\epsilon(d)} \langle \bar{\eta}, v + d \rangle$ implies $\langle \bar{\eta}, v + d, u + d \rangle \models_{\mathcal{D}} \varphi$
$\langle \bar{\eta}, v, u \rangle \models_{\mathcal{D}} x \bowtie m$	iff	$u(x) \bowtie m$
$\langle \bar{\eta}, v, u \rangle \models_{\mathcal{D}} x \text{ in } \varphi$	iff	$\langle \bar{\eta}, v, [\{x\} \mapsto 0]u \rangle \models_{\mathcal{D}} \varphi$
$\langle \bar{\eta}, v, u \rangle \models_{\mathcal{D}} Z$	iff	$\langle \bar{\eta}, v, u \rangle$ belongs to the maximal solution of $Z = \mathcal{D}(Z)$
$\langle \bar{\eta}, v, u \rangle \models_{\mathcal{D}} p$	iff	$\exists i$ s.t. $p \in l(\eta_i)$
$\langle \bar{\eta}, v, u \rangle \models_{\mathcal{D}} A_j\!:\!p$	iff	$p \in l(\eta_j)$

Table 1.1 Semantics of L_ν.

Note that we can easily define new operators as tt, ff, $\varphi \Rightarrow \psi$ (if no identifier occur in φ). The meaning of the identifiers is specified by a declaration $\mathcal{D}$ assigning a formula of L_ν to each identifier. Given $S = (A_1|\ldots|A_n)_f$ a network of TA, we interpret the formulas of L_ν with respect to *extended configurations* $\langle \bar{\eta}, v, u \rangle$, where $\langle \bar{\eta}, v \rangle$ is a configuration of S and u is a time assignment for K. Whereas the classical modal operators $\langle a \rangle$ and $[a]$ deal with *action* transitions, the operator $\langle \delta \rangle$ (resp. $[\delta]$) denotes existential (resp. universal) quantification over *delay* transitions.
The K clocks are so-called formula clocks, used as stopwatches for measuring the time elapsing between states (specified as properties) of the system. The formula clocks increase synchronously with the automata clocks. The formula (x in φ) initializes the formula clock x to 0 and the formula ($x \bowtie m$) are used to compare the value of x in the current extended configuration with an integer value.
Finally, an extended configuration satisfies an identifier Z if it belongs to the maximal solution of the equation $Z = \mathcal{D}(Z)$. The atomic propositions can be of the form p or $A_j\!:\!p$ (to specify that p holds for component A_j) and they can be in the scope of a negation. The formal definition of L_ν semantics is given in table 1.1 ($\vee$ and $[a]$ are dual cases of $\wedge$ and $\langle a \rangle$).
Within L_ν we may express a number of classical temporal (and timed) properties. Let $\mathcal{B}$ be a subset of $\mathcal{A}$. We have the following derived operators :

- ALWAYS$_\mathcal{B}(\varphi)$ expresses that φ holds from any state reachable using actions of $\mathcal{B}$ and delay transitions. The AG operator of CTL corresponds to ALWAYS$_\mathcal{A}$. ALWAYS$_\mathcal{B}(\varphi)$ can be defined by the following equation: $X \stackrel{\text{def}}{=} \varphi \wedge \bigwedge_{a \in \mathcal{B}} [a]X \wedge [\delta]\ X$.

- φ_1Until$_\mathcal{B}\varphi_2$ corresponds to the weak until operator of temporal logic ("φ_1 holds unless φ_2 holds, when following delay transitions and actions within

$\mathcal{B}$"). To define $\mathsf{Until}_{\mathcal{B}}$, we can use the equation:

$$X \stackrel{\mathrm{def}}{=} \varphi_2 \vee \Big(\varphi_1 \wedge [\delta]\ X \wedge \bigwedge_{a\in\mathcal{B}} [a]X\Big).$$

- $\varphi\mathsf{UpTo}\ d$ ("φ will hold during at least d time units, whatever the performed transitions") can be expressed with: $x\ \mathsf{in}\ (\varphi\mathsf{Until}_{\mathcal{A}}(x > d))$.

Example 3 *Let* $\mathsf{Mut}_{(1,2)}$ *be the specification stating that* P_1 *and* P_2 *can never reach the critical section at the same time (see example 2). To write such a specification we need to label the* cs_i *nodes by the atomic proposition* CS. *We have:* $\mathsf{Mut}_{(1,2)} = \mathsf{ALWAYS}_{\mathcal{A}}(\neg P_1 : CS \vee \neg P_2 : CS)$ *with* $\mathcal{A} = \{=0, =1, =2, :=1, :=2\}$. □

COMPOSITIONAL MODEL-CHECKING

Quotient Principles

Recall from the Introduction that we are interested in constructing φ/A_n such that $(A_1|\ldots|A_n) \models \varphi$ if and only if $(A_1|\ldots|A_{n-1}) \models \varphi/A_n$. A first quotient construction for L_ν (dealing with timed automata without invariant nor atomic proposition) has been defined in [16] and a more efficient quotienting for a sublogic L_s has been given in [18, 15]. Here we present this new quotienting extended to L_ν.

A network of timed automata is described by using a n-ary synchronization function but it is easier to consider binary synchronization between A_n and the remaining part to define the quotient construction. Since our synchronization function allows renaming, we can easily decompose a parallel composition $(A_1|\ldots|A_n)_f$ as: $((A_1|\ldots|A_{n-1})_{f''}|A_n)_{f'}$ where f' (resp. f'') is a binary (resp. $(n-1)$-ary) synchronization function.

In table 1.2, we define the quotient construction (the cases $[a]\varphi$ and $\varphi\vee\psi$ are dual cases of those presented), with the following convention: $\langle\bullet\rangle\,\varphi = [\bullet]\varphi = \varphi$. Moreover the formula $r\ \mathsf{in}\ \varphi$ with $r = \{c_1, \ldots, c_l\} \subseteq C$ stands for $c_1\ \mathsf{in}\ \ldots c_l\ \mathsf{in}\ \varphi$. We have the following theorem [1]:

Theorem 1 *Let* $A_1,\ldots,A_n$ *be n timed automata* $\langle N_i, \eta_{i,0}, C_i, E_i, l_i, Inv_i\rangle$. *Let* φ *be a* L_ν *formula over the set of clocks* K, *all over an action set* $\mathcal{A}$. *Let* $\langle(\bar{\rho},\eta), vw, u\rangle$ *be an extended configuration of* $((A_1|\ldots|A_{n-1})_{f'}|A_n)_f$ *with* $v \in \mathbf{R}^{\bigcup_{i<n} C_i}$, $w \in \mathbf{R}^{C_n}$ *and* $u \in \mathbf{R}^K$, *then we have:*

$$\Big\langle (\bar{\rho},\eta),\, v\,w,\, u\Big\rangle \models_{\mathcal{D}} \varphi \quad \textit{if and only if} \quad \Big\langle \bar{\rho},\, v,\, w\,u\Big\rangle \models_{\mathcal{D}'} \varphi/_f\,\eta$$

Note that the clocks in C_n become formula clocks after the quotienting. Moreover the quotienting rule for variables may increase the number of equations (in the declaration $\mathcal{D}'$ compared with the declaration $\mathcal{D}$) since it creates a new variable Z^η for any variable Z in φ and any node η of the quotiented automaton. In order to avoid trading the acknowledged explosion problem of the global state space with a similarly harmfull explosion problem in the size of quotient formulas, it is essential that we find strategies for simplifying formulas. Note

$(\varphi_1 \wedge \varphi_2)/_f\, \eta = (\varphi_1/_f\, \eta) \wedge (\varphi_2/_f\, \eta)$	$Z/_f\, \eta = Z^\eta$
$\big(\langle a \rangle \varphi\big)/_f\, \eta = \bigvee_{\eta \xrightarrow{g,c,r} \eta' \in E_n \text{ s.t. } f(b,c)=a} g \wedge (r \text{ in } Inv(\eta')) \wedge \langle b \rangle \big(r \text{ in } \varphi/_f\, \eta'\big)$	
$\langle \delta \rangle\, \varphi/_f\, \eta = \langle \delta \rangle \big(Inv(\eta) \wedge \varphi/_f\, \eta\big)$	$(x \bowtie c)/_f\, \eta = (x \bowtie c)$
$[\delta]\, \varphi/_f\, \eta = [\delta] \big(Inv(\eta) \Rightarrow \varphi/_f\, \eta\big)$	$(x \text{ in } \varphi)/_f\, \eta = x \text{ in } \big(\varphi/_f\, \eta\big)$
$(P:\alpha)/_f\, \eta = \begin{cases} \mathtt{tt} & \text{iff } P = A_n \wedge \alpha \in l_{A_n}(\eta) \\ \mathtt{ff} & \text{iff } P = A_n \wedge \alpha \notin l_{A_n}(\eta) \\ (P:\alpha) & \text{iff } P \neq A_n \end{cases}$	
$(\alpha)/_f\, \eta = \begin{cases} \mathtt{tt} & \text{iff } \alpha \in l_{A_n}(\eta) \\ \alpha & \text{otherwise} \end{cases}$	

Table 1.2 Structural definition of quotient $\varphi/_f\, \eta$ with η a node of A_n.

that contrary to the quotient defined in [16] we do not quotient the formula for any node and any clock region since we insert explicitly guards into the formula (and not only their truth value w.r.t. the regions). *This explain why such a quotient is much more compact and more efficient than the first one.*

Example 4 *The quotient of* $\mathsf{Mut}_{(1,2)}$ *by the automaton* V *is* $(X_0, \mathcal{D}')$ *with* $\mathcal{D}'$*:*

$$\begin{array}{lcl} X_0 & \stackrel{def}{=} & (\neg P_1 : CS \vee \neg P_2 : CS) \wedge [=0]X_0 \wedge [:=1]X_1 \wedge [:=2]X_2 \wedge [\delta]\ X_0 \\ X_1 & \stackrel{def}{=} & (\neg P_1 : CS \vee \neg P_2 : CS) \wedge [=1]X_1 \wedge [:=1]X_1 \wedge [:=2]X_2 \wedge [\delta]\ X_1 \\ X_2 & \stackrel{def}{=} & (\neg P_1 : CS \vee \neg P_2 : CS) \wedge [=2]X_2 \wedge [:=1]X_1 \wedge [:=2]X_2 \wedge [\delta]\ X_2 \end{array}$$

□

Now, let π be the binary synchronization function completely specified by $\pi(\bullet, a) = a$ for all $a \in \mathcal{A}$ (i.e. the left argument to π is completely ignored). Then it is easy to see that any timed automaton A is strongly bisimilar [2] to the network $(nil|A)_\pi$ and hence can be proved to satisfy the same formula of L_ν. Using this observation, the quotient construction can be used to obtain alternative model–checking algorithms for L_ν as follows: $A \models \varphi \Leftrightarrow (nil|A)_\pi \models \varphi \Leftrightarrow nil \models \varphi/_\pi A$. Due to the projective nature of π it is clear that $\varphi/_\pi A$ contains no action modalities, and it is easy to build a special purpose model-checker for the simple automaton nil.

Note that the quotient definition would support the introduction of negation in the logic (by a simple propagation) and this would allow to use minimal fixpoint in the specification and then increase its expressive power. Nevertheless such an extension would require to adapt the simplification rules.

Simplification Strategies

We have seen in the previous section that repeated quotienting leads to an explosion in the formula. This phenomena was already observed in [16, 18, 15] and in the untimed case in [3]. An efficient way to avoid this explosion consists in applying *simplification rules after each quotient.*

First we need simplifications to tackle the explosion due to the parallel composition (occurring with the quotienting of the variables). This can be done by using the classical rules already used in the untimed case:

• *Boolean Simplification.* Formulas may be simplified using the following simple boolean equations and their duals: $\mathsf{ff} \wedge \varphi \equiv \mathsf{ff}$, $\mathsf{tt} \wedge \varphi \equiv \varphi$, $\langle a \rangle \mathsf{ff} \equiv \mathsf{ff}$, $\langle \delta \rangle \mathsf{ff} \equiv \mathsf{ff}$, $x \text{ in } \mathsf{ff} \equiv \mathsf{ff}$, $\langle a \rangle \varphi \wedge [a]\mathsf{ff} \equiv \mathsf{ff}$.

• *Constant Propagation.* Identifiers which are declared to either tt or ff may be removed while substituting their definitions in other equations.

• *Trivial Equation Elimination.* Equations of the form $X \stackrel{\text{def}}{=} [a]X$ are easily seen to have $X = \mathsf{tt}$ as solution. More generally, X can be replaced by tt when $\varphi[\mathsf{tt}/X]$ [3] (with $X \stackrel{\text{def}}{=} \varphi$) can be simplified to tt.

• *Equivalence Reduction.* If two identifiers X and Y are equivalent (i.e. are satisfied by the same timed transition systems) we may collapse them into a single identifier.

Secondly we have to add special simplifications to deal with the clocks constraints. Indeed the major problem in the real-time systems analysis consists in the treatment of time encoding: simplifying test or reducing the number of clocks used in the specification can be crucial to succeed in the verification of a system. We developed several strategies which have been evaluated with CMC. Here we present two simplifications which can be combined together and which lead to interesting results. The main idea is that we are interested in the truth value of a specification φ for extended configurations (s, u_0) (with s a network configuration) with all formula clocks equal to zero. Using this information, it is sometimes possible to simplify the tests inside φ:

• *Hitzone Reduction.* Consider the formula $\varphi = [a]\ (x > 1)$. Clearly φ is equivalent to $[a]\mathsf{ff}$ for any extended configuration (s, u_0) because no a-transition can modify the value of the formula clock x (which is still equal to 0 after the transition). A more complex example is: $\psi = [\delta]\ (x > 1 \Rightarrow [a]\, x < 1)$; for the same reason, ψ is equivalent to $[\delta]\ (x > 1 \Rightarrow [a]\,\mathsf{ff})$.
Given a specification φ, the principles of *Hitzone reduction* is to compute for any test t occurring in φ the clock constraints S_t (i.e. the set of time assignments for the formula clocks) which are effectively needed to evaluate the truth value of φ for any (s, u_0). Given this information (obtained by a fixpoint computation) we can simplify by tt (resp. ff) the tests t for which the truth value is true (resp. false) for every time assignment in S_t.

A more interesting example is the simplification of $\varphi = (X_0, \mathcal{D})$ with the following declaration $\mathcal{D}$:

$$\begin{aligned} X_0 &\stackrel{\text{def}}{=} [a]\,(x_1 \text{ in } X_1) \wedge [\delta]\ X_0 \\ X_1 &\stackrel{\text{def}}{=} [b]\,(x_2 \text{ in } X_2) \wedge [\delta]\ X_1 \\ X_2 &\stackrel{\text{def}}{=} x_1 \geq x_2 \wedge [\delta]\ X_2 \end{aligned}$$

An observation of the declaration shows that the truth value of X_2 is always considered from X_1 after a reset of x_2 or from its own definition. Then the test $x_1 \geq x_2$ in X_2 will always be equal to $\mathtt{tt}$, therefore X_2 becomes a trivial equation and therefore X_1 too. Finally X_0 can be reduced to $\mathtt{tt}$.

• *Constraint Reduction.* Another way to simplify the tests inside the formula is to compute the *constraint* associated to each variable X, that is the set of timed valuations which can satisfy the tests in its definition. For example, the constraint of $X \stackrel{\text{def}}{=} x < 1 \wedge \langle a \rangle (y < 2)$ is $[x < 1 \wedge y < 2]$: for any extended configuration (η, v, u) satisfying X, we have $u \models x < 1 \wedge y < 2$. Each L_ν operator has its own rule to propagate the constraint and a fixpoint computation is necessary to obtain the final constraint of any variable. For example, the constraint of a variable $X \stackrel{\text{def}}{=} x < 1 \wedge \ldots \wedge [\delta]\ X$ will be empty and X is then equal to $\mathtt{ff}$.

Finally it is possible to combine the Hitzone and the Constraint redusctions to simplify the tests inside the formula. Note that these two simplifications assume that the formula clocks and automata clocks are disjoint: we assume that the action transitions do not modify the value of formula clocks. Without these simplifications, we could use automata clocks in formula.

Example 5 *Recall Example 2. Without simplification (except boolean reductions) the quotient specification $mut_{12} /\!_f\, V /\!_f\, P_1 /\!_f\, P_2$ contains 18 identifiers. Applying the minimization strategies of the* CMC *tool we find that $mut_{12} /\!_f\, V /\!_f\, P_1 /\!_f\, P_2$ is equivalent to* $\mathtt{tt}$. *Then the property holds for the system.* □

∘ **Why it is efficient ?** Quotienting immediately followed by the various simplification strategies often allows to dismiss parts of the quotiented component which are irrelevant to the specific property in question. Also, the iterative strategy allows to deal with clock constraints after each quotienting i.e. before considering all clocks of the system, thus making simplification easier and more efficient (this will be clearly illustrated by Fischer's mutual exclusion protocol).

EXPERIMENTAL RESULTS WITH CMC

The compositional method has been implemented in the CMC [4] tool (written in C++) : it contains a command language and an environment to define networks of TA and L_ν specifications, and several functions to build quotient, to simplify formula (every reductions described in the previous sections has been implemented) ...Moreover it contains a **Check** procedure to decide if a formula holds for the *nil* process. It is also possible to verify if two timed automata without invariant are bisimilar.

In this section we present two examples verified with CMC. The first one is Fischer's mutual exclusion protocol, which is a benchmark in the timed verification area. This example is interesting because it shows both the power of compositional approach and its limits. Here we address the two classical versions of the protocol and we present a third one.
The second example is the train crossing example. Its purpose is to show that L_ν is an expressive specification language: we write a short specification including safety properties, bounded liveness property etc. Moreover, whereas Fischer's protocol has a symmetrical structure potentially making the compositional approach more efficient, the train crossing example has no special structure thus showing that the method is not limited to special systems but can be applied to arbitrary networks of timed automata.

Fischer's mutual exclusion protocol

Fischer's mutual exclusion protocol described in the example 2 is a benchmark of the verification of timed systems. Consider n processes P_i, the problem is to verify that at any time at most one process can access its critical section: we want to verify that for any $i, j \in \{1, \ldots, n\}$ with $i \neq j$, $\mathsf{Mut}_{(i,j)}$ holds for the initial configuration of the system. Remember that $\mathsf{Mut}_{(i,j)} \stackrel{\text{def}}{=} \mathsf{ALWAYS}_{\mathcal{A}}(\neg P_i : cs_i \vee \neg P_j : cs_j)$.

In order to verify $\mathsf{Mut}_{(1,2)}$ efficiently, we need to carefully choose the order in which the processes are quotiented. In fact, since $\mathsf{Mut}_{(1,2)}$ only directly deals with the processes V, P_1 and P_2, we may hope that quotienting these three components will suffice for determining the truth value of the formula. Using CMC, we start by quotienting V, then P_1 and finally P_2. The simplification rules allow to deduce $\mathtt{tt}$! To evaluate the efficiency of the method we give the size of the specification (i.e. the number of variables). For example, we have the following computation [5] for $n = 100$:

Quotient by ...	V_{100}	P_1	P_2
Size of specification:	$202 \rightsquigarrow_s 102$	$404 \rightsquigarrow_s 202$	$800 \rightsquigarrow_s 0(\mathtt{tt})$

These results can be read as follows: The quotient of $\mathsf{Mut}_{(1,2)}$ by V_{100} gives a specification with 202 variables which are reduced to 102 after the simplifications, then the quotient by P_1 gives 404 variables which are reduced to 202, and finally the quotient of P_2 gives 800 variables which are reduced to the single formula $\mathtt{tt}$.
Of course a complete verification needs to do such a computation for any $\mathsf{Mut}_{(i,j)}$ but by choosing the right order for removing the components the previous result shows that the method can be successfully applied. Note that the

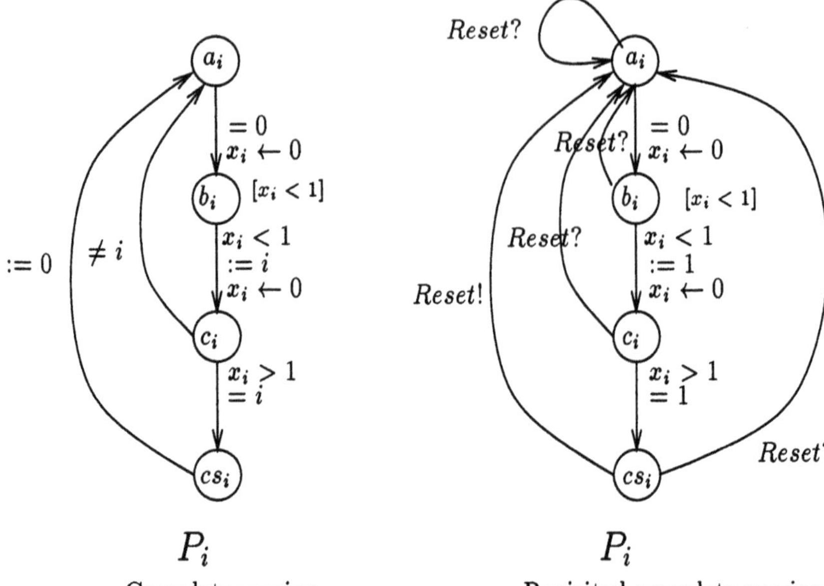

Figure 1.3 The complete version of Fischer's mutual exclusion protocol and a variant.

compositional approach is very efficient in this case: other existing real-time verification tools cannot verify this protocol with more than 15 processes.

In fact the previous version of Fischer's mutual exclusion protocol is not the complete protocol. The complete version is cyclic and is described on the left part of figure 1.3. The main point is that a process using the critical section can leave it and then reset the variable V. This version is much more difficult to verify. Indeed if we quotient $\mathsf{Mut}_{(1,2)}$ (in which we add a term $[:= 0]X$ in the conjunction) by V, P_1 and P_2, we **cannot** deduce $\mathtt{tt}$ (contrary to what we claim in [15]) because at this step we know, thanks to the synchronization function, that other processes can perform the $:= 0$ action and we have no information about **when** they can perform it. Typically if P_3 was just a process able to reset V at any time, then P_2 could access the critical section at the same time as P_1. Then for this version of the protocol, we need to quotient all processes to finally deduce $\mathtt{tt}$. And in this case, we cannot avoid the explosion of the size of the specification and we are limited to small systems. We obtain the following computation for 6 processes:

V_6	P_1	P_2	P_3	P_4	P_5	P_6
$14 \leadsto_s 8$	$30 \leadsto_s 20$	$74 \leadsto_s 29$	$106 \leadsto_s 45$	$162 \leadsto_s 69$	$242 \leadsto_s 101$	$338 \leadsto_s 0$

Note that for this version, the newest BDD approach proposed in [7] for KRONOS can deal with 14 components. Nevertheless we can consider a third version of the Fischer protocol. It is described in the right part of the figure. The main idea is to see the reset of the variable as a reset of the protocol: when a process leaves the critical section, then it resets the variable and all processes receive

Train_1	Train_2	Controller	Counter	Gate	visible action
$App!$	•	$App?Incr!$	$Incr?$	•	App
–	$App!$	$App?Incr!$	$Incr?$	•	App
$Exit!$	•	$Exit?Reset!$	$Reset?$	•	$Exit$
$Exit!$	•	$Exit?Decr!$	$Decr?$	•	$Exit$
•	$Exit!$	$Exit?Reset!$	$Reset?$	•	$Exit$
•	$Exit!$	$Exit?Decr!$	$Decr?$	•	$Exit$
•	•	$GoUp!$	•	$GoUp?$	$GoUp$
•	•	$GoDown!$	•	$GoDown?$	$GoDown$
i	•	•	•	•	i
•	i	•	•	•	i
•	•	•	•	i	i

Table 1.3 Synchronization table for (Train_1 | Train_2 | Controller | Counter | Gate).

this signal and go back to their initial state. To obtain such protocol it is sufficient to synchronize the *Reset* action over all processes. Note that the protocol is still cyclic and if such a change simplifies the protocol, it remains quite reasonable. In this case, we succeed [6] in applying the compositional approach:

Quotient by ...	V_{100}	P_1	P_2
Size of specification:	$202 \rightsquigarrow_s 102$	$404 \rightsquigarrow_s 301$	$1194 \rightsquigarrow_s 0$

The train crossing example

We now present a version of the classical railroad crossing example to illustrate the expressive power of L_ν. The network of timed automata (Train_1 | Train_2 | Controller | Counter | Gate$)_f$ in figures 1.4, 1.5 and 1.6 models a train crossing: The trains can send an *App*! (resp. *Exit*!) message to a controller to inform it of their arrival in (resp. departure from) the section before (resp. after) the gate. The gate can receive orders from the controller and can change its position (open or close) in at most 10 seconds.
The controller uses the same strategy as the one presented in [8]: the idea is to count the number of trains which are staying between the *App* point and the *Exit* point and to use this information to decide if an *Exit*! message has to be followed by a *GoUp*! order. To model this strategy we use the automaton Counter whose nodes correspond to the possible values (bounded by 2). The synchronization function is given in table 1.3 and has to be read as follows: An *App* action of the system corresponds to a *App*! action performed by one of the trains, a *App*?*Incr*! action of the controller and a *Incr*? action of the counter. Note that the distinction between *Reset*? and *Decr*? allows us to deduce whether there remains only a single train in the section.
In this example, we use name of nodes as atomic propositions.

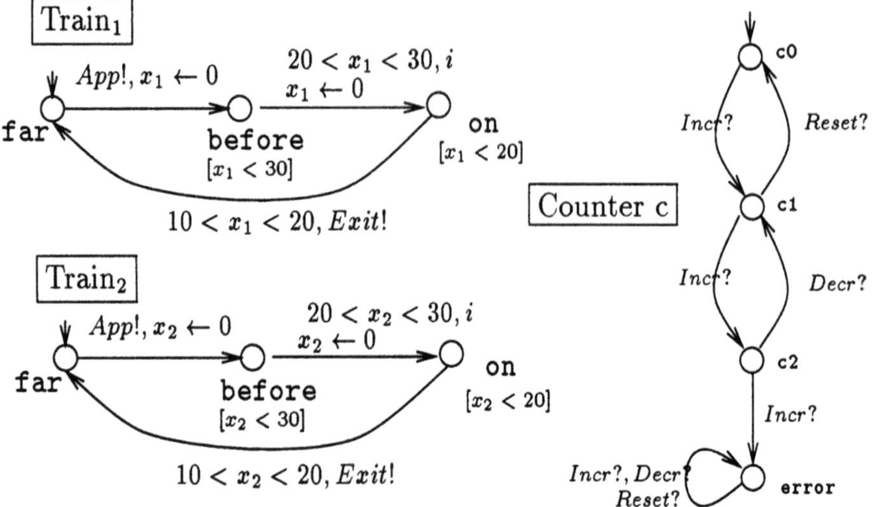

Figure 1.4 The trains and the counter.

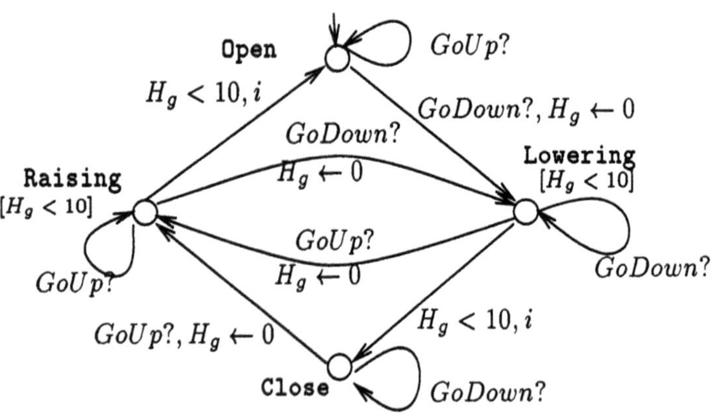

Figure 1.5 The gate.

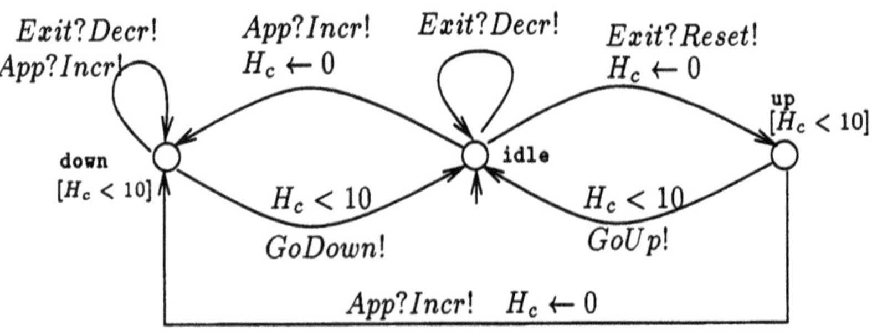

Figure 1.6 The controller.

Specifying with L_ν. An important point of CMC is the expressivity of its specification language, L_ν, which allows application specific properties to be expressed concisely. The aim of this part is to illustrate this point by specifying the train crossing example. Clearly we expect the following two properties to hold:

- "When a train is **on** the rail crossing, the gate is **close**". This property can be expressed by the formula:
  $\varphi_{\texttt{safe}} = \mathsf{ALWAYS}_{\mathcal{A}}\Big((\mathrm{Train}_1\!:\!\mathbf{on} \vee \mathrm{Train}_2\!:\!\mathbf{on}) \;\Rightarrow\; \mathrm{Gate}\!:\!\texttt{close}\Big)$

- "The gate is open in at most δ_l seconds after an **Exit** signal if no train approaches the railroad crossing inbetween". We can use the formula:

$$\varphi_{\texttt{live}}(\delta_l) = \mathsf{ALWAYS}_{\mathcal{A}}\Big[[Exit]\,\Big(z \text{ in } [(z<\delta_l)\mathsf{Until}_{\mathcal{A}\setminus App}\mathrm{Gate}:\mathbf{open}]\Big)\Big]$$

  Note that the restriction on the set of allowed actions in Until is useful in capturing the "if" part of the informal statement of the property.

However, in addition to the application specific properties above, we also want to ensure that the model is *realizable* or *physically valid* and possesses no unreasonable behavior:

- "There is no deadlock in the model: it is always possible to wait for some delay such that an action (*App*, *Exit*, *i*, *GoUp*, *GoDown*) is enabled". We can use $\varphi_{\texttt{no\_dl}}$ with:

$$\varphi_{\texttt{no\_dl}} \stackrel{\text{def}}{=} \mathsf{ALWAYS}_{\mathcal{A}}\Big[\langle\delta\rangle\Big(\langle i\rangle\mathtt{tt}\vee\langle App\rangle\mathtt{tt}\vee\langle Exit\rangle\mathtt{tt}\vee\langle GoDown\rangle\mathtt{tt}\vee\langle GoUp\rangle\mathtt{tt}\Big)\Big]$$

- Bounding the counter with 2 is sufficient: "the **error** state is not reachable". This can be expressed by $\varphi_{\texttt{err}} \stackrel{\text{def}}{=} \mathsf{ALWAYS}_{\mathcal{A}}(\neg\mathrm{Counter}:\texttt{error})$.

- Actions *App?*, *Exit?*, *GoDown?* and *GoUp?* correspond to the reception of signals: they have to be enabled when a corresponding signal can be sent (unless this could prevent possible behaviors). The property clearly holds for *App* and *GoDown* but it is unclear for *Exit?* because there is no transition *Exit?* from **up** node in the controller. We can use : $\varphi_{\texttt{signal}} \stackrel{\text{def}}{=} \mathsf{ALWAYS}_{\mathcal{A}}(\langle Exit!\rangle\,\mathtt{tt} \;\Rightarrow\; \langle Exit?\rangle\,\mathtt{tt})$.
  Note that to verify this property we have to add in the synchronization table that an *Exit!* (resp. *Exit?*) signal can be performed by one of the train without synchronization (resp. by the controller with actions *Exit?Reset!* or *Exit?Decr!* with a synchronization over Counter actions). Being able to express such a property with L_ν is important because it gives a good help for the model construction.

Results. For our train example, we have the following results [7]:

	/Train$_1$	/Train$_2$	/Controller	/Counter	/Gate	Check	Res.
$\varphi_{\mathtt{safe}}$	12$\leadsto$,7	30$\leadsto$,27	65$\leadsto$,59	48$\leadsto$,40	116$\leadsto$,0(tt)	–	Yes
$\varphi_{\mathtt{live}}(70)$	17$\leadsto$,17	55$\leadsto$,55	141$\leadsto$,127	96$\leadsto$,76	216$\leadsto$,0(tt)	–	Yes
$\varphi_{\mathtt{live}}(65)$	17$\leadsto$,17	55$\leadsto$,55	141$\leadsto$,127	96$\leadsto$,76	216$\leadsto$,77	17 it.	No
$\varphi_{\mathtt{no\_dl}}$	14$\leadsto$,14	66$\leadsto$,50	154$\leadsto$,112	98$\leadsto$,65	237$\leadsto$,83	7 it.	Yes
$\varphi_{\mathtt{err}}$	6$\leadsto$,6	27$\leadsto$,27	65$\leadsto$,59	48$\leadsto$,0(tt)	–	–	Yes
$\varphi_{\mathtt{signal}}$	11$\leadsto$,10	40$\leadsto$,33	79$\leadsto$,68	61$\leadsto$,42	124$\leadsto$,51	5 it.	Yes

The column "Check" contains the number of iterations (of the fixpoint computation) used by the **Check** procedure to decide if φ/Train$_1$/Train$_2$/Controller /Counter/Gate holds for the *nil* process when the formula has not already been reduced to tt or ff by the simplifications. The last column gives the final answer (Yes or No) for $S \models \varphi$.
Note that for $\varphi_{\mathtt{err}}$ it is not necessary to quotient all automata: the analysis of a subpart of the system allows to verify that this formula holds for the system.

CONCLUSION

Since the introduction of the compositional method in [16], substantial improvements have been achieved. In particular, the tool CMC presented in this paper offers a collection of new simplification heuristics adding to the efficiency of the overall method. Partially based on the experimental results of this paper, our conclusion is that (1) the compositional method is efficient and not limited to special types of networks of timed automata and (2) the timed modal logic L_ν constitutes a convenient specification language.

The future development will concern the elaboration of new simplifications (based on partial evaluation of formula) and an optimization of the data structure to encode the clocks constraints.

Acknowledgments

We want to thank Béatrice Bérard and Philippe Schnoebelen for their helpful comments about this work.

Notes

1. The proof is omitted due to space limits but it is very similar to the proofs of [16, 18] but a long version of this paper including the proof is available by email.
2. We extend the notion of strong bisimulation to require that delay transitions are matched by delay transitions of equal time-quantity.
3. $\varphi[\mathtt{tt}/X]$ is the formula obtained by substituting all occurrences of X in φ with the formula tt.
4. CMC can be obtained by sending an email to `fl@lsv.ens-cachan.fr`.

5. This has been done in few minutes on a SUN Ultra Sparc 1 with 64 Mb memory

6. This computation takes approximately one hour on a SUN Ultra Sparc 2 with 512 Mb.

7. All these results have been computed on a Sparc Ultra 1 with 64 Mb memory in few seconds except for $\varphi_{\texttt{live}}(65)$ which needs approx 6 minutes.

References

[1] M. Abadi and L. Lamport. An Old-Fashioned Recipe for Real-Time. In *Proc. of REX Workshop "Real-Time: Theory and Practice",LNCS 600*, 1993.

[2] R. Alur and D. Dill. A theory of timed automata. *Theoretical Computer Science*, 126(2):183–235, April 1994.

[3] H. R. Andersen. Partial Model Checking. In *Proc. of LICS'95*, 1995.

[4] E. Asarin, M. Bozga, A. Kerbrat, O. Maler, A. Pnueli, and A. Rasse. Data-Structures for the Verification of Timed Automata. In *Proc. of HART'97, LNCS 1201*, 1997.

[5] J. Bengtsson, W. O. David Griffioen, K. J. Kristoffersen, K. G. Larsen, F. Larsson, P. Pettersson, and W. Yi. Verification of an audio protocol with bus collision using uppaal. In *Proc. of the 8th International Conference on Computer-Aided Verification, LNCS 1102*, pages 244–256, 1996.

[6] A. Bouajjani, S. Stavros, and S. Yovine. On-the-Fly Symbolic Model Checking for Real-Time Systems. In *Proc. of the 18th IEEE Real-Time Systems Symposium, RTSS'97.* IEEE Computer Society Press, December 1997.

[7] M. Bozga, O. Maler, A. Pnueli, and S. Yovine. Some progress in the symbolic verification of timed automata. In *Proc. of the 9th International Conference on Computer-Aided Verification, LNCS 1254.* Springer Verlag, 1997.

[8] N. Halbwachs, F. Maraninchi, and Y.E. Proy. The railroad crossing problem Modeling with Argos - Analysis with Polka. In *Proc. of 2nd European Workshops on Real-Time and Hybrid Systems*, 1995.

[9] K. Havelund, A. Skou, K. G. Larsen, and K. Lund. Formal modelling and analysis of an audio/video protocol: An industrial case study using uppaal. In *Proc. of the 18th IEEE Real-Time Systems Symposium*, 1997.

[10] T. A. Henzinger, X. Nicollin, J. Sifakis, and S. Yovine. Symbolic model checking for real-time systems. *Information and Computation*, 111:193–244, 1994.

[11] T.A. Henzinger, P.H. Ho, and H. Wong-Toi. A user guide to hytech. In E. Brinksma, W.R. Cleaveland, K.G. Larsen, T. Margaria, and B. Steffen, editors, *Proceedings of TACAS'95*, volume 1019 of *LNCS*, pages 41–71. SPRINGER, 1995.

[12] T.A. Henzinger and H. Wong-Toi. Automated analysis of an audio control protocol. In P. Antsaklis, A. Nerode, W. Kohn, and S. Sastry, editors, *Hybrid systems II*, volume 999 of *LNCS*, pages 265–293. SPRINGER, 1995.

[13] T.A. Henzinger and H. Wong-Toi. Using HyTech to synthesize control parameters for a steam boiler. In J.R. Abrial, E. Börger, and H. Langmaack, editors, *Formal methods for industrial applications. Specifying and programming the steam boiler control*, volume 1165 of *LNCS*, pages 265–282. SPRINGER, 1995.

[14] H. Hüttel and K. G. Larsen. The use of static constructs in a modal process logic. *Lecture Notes in Computer Science, Springer Verlag*, 363, 1989.

[15] K.J. Kristoffersen, F. Laroussinie, K.G. Larsen, P. Pettersson, and W. Yi. A Compositional Proof of a Real-Time Mutual Exclusion Protocol. In M. Bidoit and M. Dauchet, editors, *Proc. TAPSOFT'97, Lille, France, LNCS 1214*, pages 565–579. Springer-Verlag, April 1997.

[16] F. Laroussinie and K.G. Larsen. Compositional Model-Checking of Real Time Systems. In *Proc. CONCUR'95, Philadelphia, USA, LNCS 962*, pages 27–41. Springer-Verlag, August 1995.

[17] K. G. Larsen, F. Larsson, P. Pettersson, and W. Yi. Efficient verification of real-time systems: Compact data structure and state-space reduction. In *Proc. of the 18th IEEE Real-Time Systems Symposium, RTSS'97*. IEEE Computer Society Press, December 1997.

[18] K. G. Larsen, P. Pettersson, and W. Yi. Compositional and symbolic model-checking of real-time systems. In *Proc. of the 16th IEEE Real-Time Systems Symposium, RTSS'95*, 1995.

[19] K. G. Larsen, P. Pettersson, and W. Yi. UPPAAL in a Nutshell. *Journal of Software Tools for Technology Transfer*, 1(1/2), October 1997.

[20] O. Maler and S. Yovine. Hardware timing verification using KRONOS. In *Proc. 7th Israeli Conference on Computer Systems and Software Engineering*, Herzliya, Israel, June 1996.

[21] N. Shankar. Verification of Real-Time Systems Using PVS. In *Proc. of the 5th Int. Conf. on Computer Aided Verification, LNCS 697*, pages 210–224, 1993.

[22] S. Yovine. Kronos: A Verification Tool for real-Time Systems. *Journal of Software Tools for Technology Transfer*, 1(1/2), October 1997.

TWOTOWERS: A TOOL INTEGRATING FUNCTIONAL AND PERFORMANCE ANALYSIS OF CONCURRENT SYSTEMS

M. Bernardo[1], W.R. Cleaveland[2], S.T. Sims[3], W.J. Stewart[2]

[1]Università di Bologna, Dipartimento di Scienze dell'Informazione
Mura A. Zamboni 7, 40127 Bologna, Italy
bernardo@cs.unibo.it

[2]North Carolina State University, Department of Computer Science
Raleigh, NC 27695-7534, USA
rance@eos.ncsu.edu, billy@eos.ncsu.edu

[3]Naval Research Lab, Center for High Assurance Computer Systems
Washington, DC 20375, USA
sims@itd.nrl.navy.mil

Abstract: We present TwoTowers, a tool for analyzing functional and performance properties of concurrent systems expressed as terms in the stochastically timed reward process algebra EMPA_r. TwoTowers builds on two existing tools, CWB-NC and MarCA, that have been retargeted to carry out functional and performance analysis (respectively) of EMPA_r system specifications. As an example, we describe the application of TwoTowers to the Lehmann-Rabin randomized distributed algorithm for the dining philosopher problem.

INTRODUCTION

The desirability of taking account of the performance aspects of a system in the early stages of its design has been widely recognized [23, 10, 13, 7]. Nevertheless, it often happens that a concurrent system is tested for efficiency only after it has been fully designed and tested for functionality. This results in two problems. On the one hand, the detection of poor performance causes the system to be designed again, so the cost of the project increases. On the other hand, since the performance-related analysis is done on a model of a system extracted from the implementation while the functionality-related analysis is usually performed on a model derived from a system design, great care must be taken to ensure that these models are consistent with one another, i.e. do reflect (different aspects of) the same system.

In order to solve the two problems above, in [3, 4] a two-phase approach for modeling and analyzing both functional and performance characteristics of concurrent systems has been proposed, which is based on stochastically timed process algebras and stochastically timed Petri nets in order to exploit their complementary advantages:

- The first phase requires the designer to specify the concurrent system as a term in the stochastically timed process algebra. Because of compositionality, the designer is allowed to develop the algebraic representation of the system in a modular way: every subsystem can be modeled separately, then these models can be put together through the operators of the algebra. The analysis of the algebraic term is then carried out on its underlying integrated interleaving semantic model, which is a transition system labeled with both action types and durations. The integrated interleaving semantic model can be analyzed as a whole by a notion of integrated equivalence or projected on a functional semantic model and a performance semantic model that can be analyzed by means of existing tools.

- The second phase consists of deriving from the algebraic representation of the system an equivalent representation in the form of a stochastically timed Petri net. The net representation turns out to be useful whenever a less abstract representation is required highlighting dependencies, conflicts, and synchronizations among system activities, and helpful in detecting some properties that can be easily checked only in a distributed setting. Additionally, the net representation is usually more compact than the integrated interleaving semantic model resulting from the algebraic representation, since concurrency is kept explicit instead of being simulated by alternative computations obtained by interleaving actions of concurrent components. The functional and performance analysis of the net representation can then be undertaken using existing tools.

Since the two phases above are complementary, the choice between them is made according to the adequacy of the related representation with respect to the analysis of the system under consideration and the availability of the corresponding tools. In any case, the designer is suggested to start with an algebraic representation of the system in order to take advantage of compositionality of algebras and avoid graphical complexity of nets.

This paper describes our efforts to provide tool support for the first phase of this approach through the implementation of a tool called TwoTowers. To use TwoTowers, the designer first specifies the concurrent system as an algebraic term. The algebra adopted in the tool is EMPA_r [1], an extension of the stochastically timed process algebra EMPA [5] allowing for the specification of performance measures based on rewards, which augments the expressive power of a classical process algebra with additional features for expressing probabilities, priorities, and timing characteristics of systems. Because EMPA_r supports compositional modeling, the designer may develop the algebraic representation of the system in a modular way; every subsystem can be modeled separately, then combined using the appropriate operators. Another useful feature of $EMPA_r$ is that many functional properties of a system may be examined independently of performance properties and vice versa. Our implementation of TwoTowers exploits this fact by using two existing tools (that we have retargeted to $EMPA_r$) for the analysis of these types of properties. CWB-NC [9] provides a variety of different types of analysis (e.g. model checking, equivalence checking, preorder checking, reachability analysis) of the functional behavior of systems, while MarCA [21] conducts stationary and transient performance analysis, as depicted in Fig. 1.1.

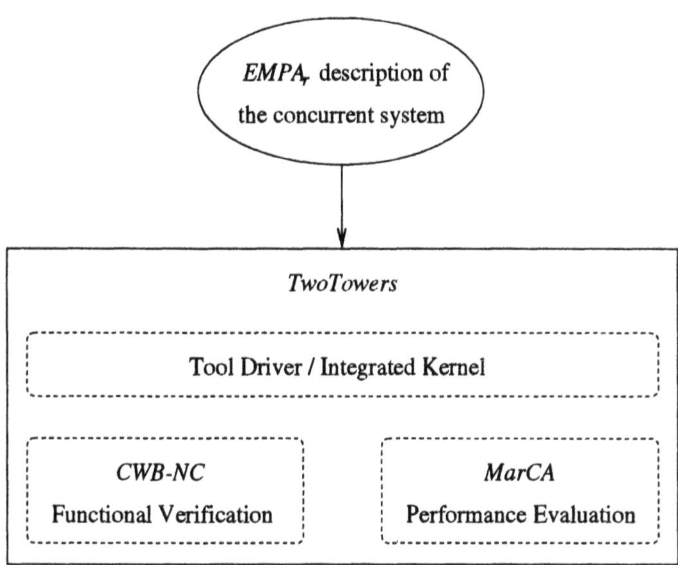

Figure 1.1 TwoTowers builds on CWB-NC and MarCA.

The purpose of TwoTowers is similar to that of other stochastically timed process algebra based tools, such as the PEPA Workbench [11] and the TIPP-tool [14]. However, besides the difference in the expressive power of the algebras used by these tools [5], the underlying philosophy is quite different. TwoTowers profits from two well-known software tools to carry out the functional and performance analysis. This is advantageous both from the point of view of the implementors and from the point of view of the users. We needed not to write code for implementing the analysis routines, thereby making the development of TwoTowers itself easier and faster, and users are provided with a wide range of automated techniques helping them during the study of systems.

The remainder of the paper is organized as follows. In the second section we present the language of TwoTowers, i.e. $EMPA_r$. In the third section we present the architecture of TwoTowers. In the fourth section we describe our use of the tool to study the Lehmann-Rabin randomized distributed algorithm for the dining philosopher problem. Finally, in the fifth section we report some concluding remarks about related work and future extensions of TwoTowers.

THE LANGUAGE OF TWOTOWERS: EMPA$_R$

The building blocks of EMPA$_r$ [1] are *actions*. An action is a quadruple $<a, \tilde{\lambda}, \tilde{r}_1, \tilde{r}_2>$ where:

- a is the *type* of the action. Types divide actions into *external* and *internal* (denoted by action type τ) depending on whether they can be seen by an external observer or not.

- $\tilde{\lambda}$ is the *rate* of the action, i.e. a concise way to represent the random variable specifying its duration. Based on rates, we have the following classification:

  - *Exponentially timed actions* are actions whose rate is a positive real number. Such a number is interpreted as the parameter of the exponentially distributed random variable specifying the duration of the action.
  - *Immediate actions* are actions whose rate, denoted by $\infty_{l,w}$, is infinite. Such actions have duration zero and each of them is given a *priority level* $l \in \mathbb{N}_+$ and a *weight* $w \in \mathbb{R}_+$, useful for expressing prioritized and probabilistic choices respectively.
  - *Passive actions* are actions whose rate, denoted by $*$, is undefined. The duration of a passive action is fixed only by synchronizing it with a nonpassive action of the same type.

  The restriction to exponentially distributed durations implies that the semantics can be defined in the interleaving style, thanks to the memoryless property of exponential distributions, and that the underlying performance models are Markov chains.

- $\tilde{r}_1$ and $\tilde{r}_2$ are, respectively, the *yield reward* and the *bonus reward* of the action [15]. Both of them are included within actions to allow for a high level specification of the performance measures of interest. The former expresses the speed at which gain is accumulated at the state executing the action, whereas the latter expresses the gain awarded upon executing the action. Performance measures are computed as weighted sums of state probabilities and transition frequencies, where rewards are used as weights.

We denote by

$$Act_r = \{<a, \tilde{\lambda}, \tilde{r}_1, \tilde{r}_2> \in AType \times ARate \times AYReward \times ABReward \mid \tilde{\lambda} = * \Longleftrightarrow \tilde{r}_1 = \tilde{r}_2 = *\}$$

the set of actions, where $AType$ is the set of types, $ARate = \mathbb{R}_+ \cup Inf \cup \{*\}$, with $Inf = \{\infty_{l,w} \mid l \in \mathbb{N}_+ \wedge w \in \mathbb{R}_+\}$ is the set of rates, $AYReward = \mathbb{R} \cup \{*\}$ is the set of yield rewards, and $ABReward = \mathbb{R} \cup \{*\}$ is the set of bonus rewards.

Let $Const$ be a set of *constants* and $ARFun = \{\varphi : AType \longrightarrow AType \mid \varphi(\tau) = \tau \wedge \varphi(AType - \{\tau\}) \subseteq AType - \{\tau\}\}$ be a set of *action relabeling functions*. The set $\mathcal{L}_r$ of *process terms* of EMPA$_r$ is generated by the following syntax

$$E ::= \underline{0} \mid <a, \tilde{\lambda}, \tilde{r}_1, \tilde{r}_2>.E \mid E/L \mid E[\varphi] \mid E + E \mid E \|_S E \mid A$$

where $L, S \subseteq AType - \{\tau\}$ and $A \in Const$. We denote by $\mathcal{G}_r$ the set of guarded and closed terms of $\mathcal{L}_r$.

The *null term* "$\underline{0}$" represents a termination or deadlocked state. The *prefix operator* "$<a, \tilde{\lambda}, \tilde{r}_1, \tilde{r}_2>.\_$" denotes the sequential composition of an action and a term. The *functional abstraction operator* "$\_/L$" hides the actions whose type belongs to L by changing their type to τ. The *functional relabeling operator* "$\_[\varphi]$" changes the types of actions according to φ. The *alternative composition operator* "$\_ + \_$" expresses a choice between two terms: the choice is resolved according to the *race policy* whenever exponentially timed actions are involved (the fastest one succeeds) or according to the *preselection policy* whenever immediate actions are involved (only the actions having the highest priority level are executable and each of these is given an execution probability proportional to its own weight), while the choice is purely *nondeterministic* whenever passive actions are involved. The *parallel composition operator* "$\_ \|_S \_$" expresses the concurrent execution of two terms according to the following synchronization discipline: two actions can synchronize if and only if they have the same type belonging to S and at most one of them is not passive. Finally, constants together with their corresponding defining equations of the form $A \stackrel{\Delta}{=} E$ allow for recursion.

Given a term $E \in \mathcal{G}_r$, its *integrated interleaving semantics* $\mathcal{I}_r[\![E]\!]$ is represented as a transition system with action-labeled edges. Because of the presence of actions with different priorities, as well as the fact that immediate actions take precedence over exponentially timed ones due to the race policy, the integrated interleaving model is generated by a two-layer semantics: first all the potential moves are derived, then those with lower priority are discarded. The states of $\mathcal{I}_r[\![E]\!]$ are divided into *tangible*, *vanishing*, *open*, and *absorbing* depending on whether they have outgoing exponentially timed transitions, outgoing immediate transitions, outgoing passive transitions, or no outgoing transitions. $E \in \mathcal{G}_r$ is said to be *performance closed* if $\mathcal{I}_r[\![E]\!]$ does not contain passive transitions; we denote by $\mathcal{E}_r$ the set of performance closed terms of $\mathcal{G}_r$.

From the integrated interleaving semantic model, two projected semantic models can be obtained. The *functional semantics* of $E \in \mathcal{G}_r$ is still represented as a transition system $\mathcal{F}_r[\![E]\!]$ now labeled by action types only, derived from $\mathcal{I}_r[\![E]\!]$ by dropping action rates and rewards. The *performance semantics* of $E \in \mathcal{E}_r$ is represented as a homogeneous Markov reward chain basically derived from $\mathcal{I}_r[\![E]\!]$ by dropping action types and by lifting yield rewards from transitions to states: the yield reward earned by a state is the sum of the yield rewards earned by its outgoing transitions. To be more precise, the performance model is given by a homogeneous continuous time Markov reward chain or a homogeneous discrete time Markov reward chain depending on whether $\mathcal{I}_r[\![E]\!]$ contains only exponentially timed or immediate transitions. If both kinds of transition coexist, the performance model is a homogeneous continuous time Markov reward chain built by applying to $\mathcal{I}_r[\![E]\!]$ a suitable graph reduction rule, which eliminates vanishing states together with their outgoing immediate transitions and modifies rates of transitions entering such states accordingly.

For a more formal presentation of EMPA$_r$ semantics, the interested reader is referred to [1, 2].

ARCHITECTURE OF TWOTOWERS

The architecture of TwoTowers is depicted in Fig. 1.2. TwoTowers is composed of a graphical user interface, a tool driver, an integrated kernel, a functional kernel, and a performance kernel.

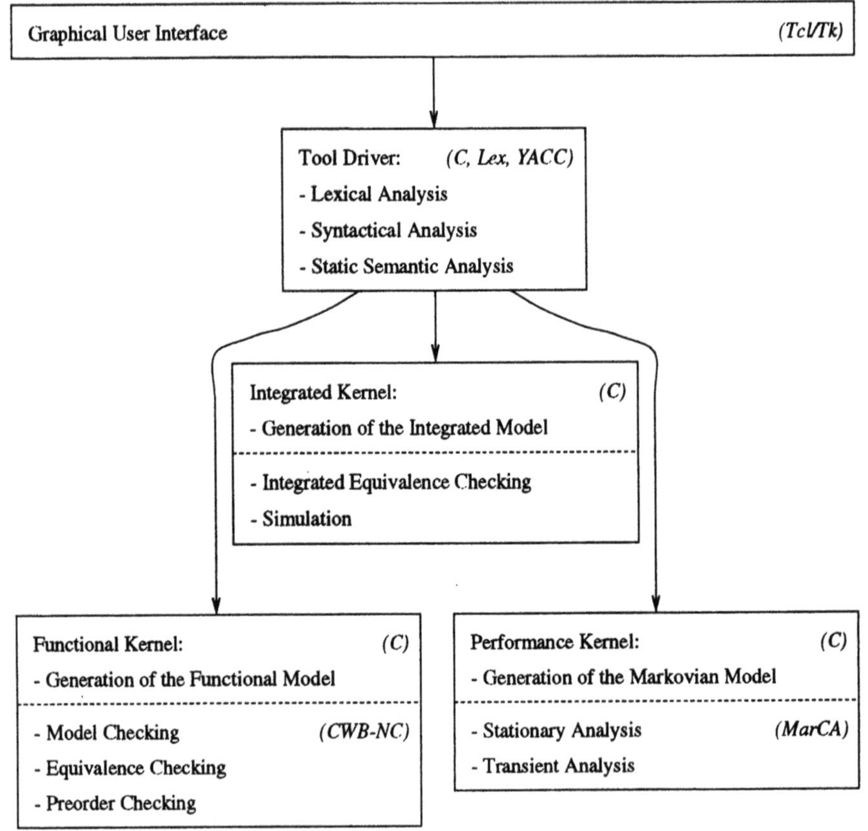

Figure 1.2 Architecture of TwoTowers.

The graphical user interface is written in Tcl/Tk [20] and allows the user to edit $EMPA_r$ specifications, compile them, and run the various analysis routines.

The tool driver, which is written in C [17] and uses Lex [19] and YACC [16], includes routines for parsing $EMPA_r$ specifications and performing lexical, syntactic and static semantic (closure, guardedness, finiteness) checks on the specifications. If there are errors, a message indicates the number of errors and a file is generated pinpointing the errors in the specification source file. When there are no errors, the syntax tree of the specification is produced for analysis by the three kernels.

The purpose of the integrated kernel is to perform those analyses that require both functional and performance information and therefore cannot be performed by factoring out information to be passed to either the functional or performance kernels. The integrated kernel, which is implemented in C, contains the routines to generate the integrated interleaving semantic model of correct EMPA_r specifications. The model generation employs the transition caching strategy proposed in [8] to store certain semantic information during model construction thereby avoiding recomputation of this information and saving time and memory. After generating a model the tool reports the number of states (classified as tangible, vanishing, open, or absorbing) as well as the number of transitions (classified as observable or invisible according to their types, and as exponentially timed, immediate, or passive according to their rates). Moreover, the integrated kernel contains a routine to check two EMPA_r specifications for strong extended Markovian reward bisimulation equivalence [1], as well as a facility for simulating an EMPA_r specification which supports the derivation of performance measures according to the method of independent replications [22].

The functional kernel, which is written in C, generates the functional semantic model of correct EMPA_r specifications. The analysis of the labeled transition systems as well as the terms themselves is then performed by a version of the Concurrency Workbench of North Carolina (CWB-NC) [9] that was retargeted for EMPA_r using the Process Algebra Compiler of North Carolina (PAC-NC) [8]. The functional semantics of EMPA_r along with a syntactic description of EMPA_r were encoded as input to the PAC-NC which then produced the code to retarget the CWB-NC to EMPA_r. The TwoTowers functional kernel therefore comprises all the analysis capabilities of the CWB-NC including interactive simulation of systems, reachability analysis to check safety properties, model checking in the μ-calculus or CTL, equivalence checking, and preorder checking.

Finally, the performance kernel, which is written in C, generates the Markovian semantic model of correct performance closed EMPA_r specifications. The analysis of the Markov reward chains is then carried out by means of the Markov Chain Analyzer (MarCA) [21]: transient and stationary probability distributions are calculated, then performance indexes are derived using the rewards expressed in the specifications themselves.

ANALYSIS OF A DISTRIBUTED ALGORITHM

In this section we describe our use of TwoTowers to analyze the Lehmann-Rabin randomized distributed algorithm for the dining philosopher problem, which is characterized as follows. Suppose there are n philosophers P_i, $0 \leq i \leq n-1$, sitting at a round table each with a plate in front of him/her. Furthermore, let there also be n chopsticks C_i, $0 \leq i \leq n-1$, each shared by two neighbor philosophers and used to get the rice at the center of the table. Given that philosophers can be thought of as concurrent processes sharing resources represented by the chopsticks, the mutual exclusive use of the chopsticks must be guaranteed in such a way that no adjacent philosophers are simultaneously eating and deadlock does not occur.

A fair and elegant solution to this problem has been proposed in [18]. The idea is that P_i flips a fair coin to choose between C_i and C_{i+1}, gets the chosen chopstick as soon as it becomes free, and gets the other chopstick if it is free. If the second chopstick is not available, the philosopher replaces the first one and repeats the procedure. If we denote by "$\_ +_n \_$" ("$\_ -_n \_$") the addition (subtraction) modulo n, and let $think_i$ be the action type "P_i is thinking", eat_i be "P_i is eating", pu_i be "C_i is being picked up", and pd_i be "C_i is being put down", this algorithm can be modeled as follows with EMPA$_r$:

$$\begin{array}{rcl} DP_n & \equiv & (P_0 \,\|_\emptyset \ldots \|_\emptyset\, P_{n-1}) \,\|_{\{pu_i, pd_i \mid 0 \le i \le n-1\}} (C_0 \,\|_\emptyset \ldots \|_\emptyset\, C_{n-1}) \\ P_i & \stackrel{\Delta}{=} & <think_i, \lambda_i, 0, 0>.P'_i \\ P'_i & \stackrel{\Delta}{=} & <\tau, \infty_{1,1/2}, 0, 0>.<pu_i, \infty_{1,1}, 0, 0>.(<pu_{i+_n 1}, \infty_{2,1}, 0, 0>.P''_i + \\ & & \qquad <pd_i, \infty_{1,1}, 0, 0>.P'_i) + \\ & & <\tau, \infty_{1,1/2}, 0, 0>.<pu_{i+_n 1}, \infty_{1,1}, 0, 0>.(<pu_i, \infty_{2,1}, 0, 0>.P''_i + \\ & & \qquad <pd_{i+_n 1}, \infty_{1,1}, 0, 0>.P'_i) \\ P''_i & \stackrel{\Delta}{=} & <eat_i, \mu_i, 1, 0>.<pd_i, \infty_{1,1}, 0, 0>.<pd_{i+_n 1}, \infty_{1,1}, 0, 0>.P_i \\ C_i & \stackrel{\Delta}{=} & <pu_i, *, *, *>.<pd_i, *, *, *>.C_i \end{array}$$

where the τ actions correspond to coin flips with the weights reflecting the respective probabilistic outcomes. It is worth noting that only actions with type $think_i$ or eat_i are relevant from the performance viewpoint and so they have been given an exponential timing, whereas the time it takes P_i to flip the coin or to pick up or put down a chopstick can be neglected so the corresponding actions are immediate. We also point out that priority levels of immediate actions have proven to be quite helpful in the description of the choice between getting the second chopstick and releasing the first one in order to model the fact that, whenever the second chopstick is available, the philosopher does pick it up. This is achieved in the subterms of term P'_i enclosed in parentheses by assigning priority level 2 to the action with type pu and priority level 1 to the action with type pd.

Now, we apply TwoTowers to the algebraic description of the algorithm in order to automatically verify that no adjacent philosophers be simultaneously eating and deadlock be avoided (functional properties ensuring the correctness of the algorithm), and to automatically assess the mean number of philosophers who are simultaneously eating (a performance index reflecting the degree of parallelism of the algorithm). To verify the two functional properties, we can resort to model checking by proving that every state satisfies respectively

$$\phi_1 = \bigwedge_{i=0}^{n-1} \neg(\langle eat_i \rangle tt \wedge (\langle eat_{i+_n 1} \rangle tt \vee \langle eat_{i-_n 1} \rangle tt))$$

and

$$\phi_2 = \langle - \rangle tt$$

The mean number ν of philosophers who are simultaneously eating is instead obtained by summing up the steady state probabilities of any state of the underlying Markov reward chain multiplied by the number of philosophers eating in that state, which is derived by yield reward 1 assigned to action having type eat_i in term P''_i.

In Table 1.1 we summarize the results we have obtained by varying the number n of philosophers from 3 to 6, where $\lambda_i = 4$ and $\mu_i = 0.75$ for any $0 \le i \le n-1$. The columns show, respectively, the number of philosophers, the number of states of the integrated interleaving semantic model, the number of tangible states (which coincides with the number of states of the Markov reward chain), the number of vanishing states, the number of transitions, the number of exponentially timed transitions, the number of immediate transitions, the absence of adjacent philosophers simultaneously eating

Table 1.1 Results of the analysis of the Lehmann-Rabin algorithm.

n	*states*	*ts*	*vs*	*transitions*	*ett*	*it*	ϕ_1	ϕ_2	ν
3	109	13	96	147	27	120	yes	yes	0.999084
4	387	35	352	620	100	520	yes	yes	1.758340
5	1101	81	1020	1655	285	1370	yes	yes	1.975830
6	3199	199	3000	5070	846	4224	yes	yes	2.530878

(verified by means of CWB-NC), the absence of deadlock (verified by means of CWB-NC), and the mean number of philosophers who are simultaneously eating (computed by means of MarCA).

CONCLUSION

In this paper we have presented TwoTowers, a tool based on the stochastically timed reward process algebra $EMPA_r$ for the analysis of functional and performance properties of concurrent systems. TwoTowers generates the semantic models of algebraic specifications; then the task of studying such models is transferred to two already existing tools: CWB-NC for functional verification, MarCA for performance evaluation. Relying on existing tools is advantageous from the point of view of both implementors and users. We needed not to write code for implementing the analysis routines, thereby making the development of TwoTowers easier and faster, and users are provided with a wider range of automated techniques.

In this paper we have applied TwoTowers to the Lehmann-Rabin randomized distributed algorithm. More complex systems have been analyzed by means of TwoTowers, such as the alternating bit protocol [4], CSMA/CD [2], ATM switches supporting different services [2], and an adaptive mechanism for trasmitting packetized audio over the Internet [6]. When dealing with larger state spaces, suitable techniques should be employed to reduce the cost of the analysis:

- As far as functional verification is concerned, TwoTowers provides the user with the techniques implemented in the CWB-NC, such as on-the-fly model checking and compositional minimization.

- From the performance evaluation standpoint, all the techniques implemented in MarCA are available in TwoTowers, together with the support for simulative analysis which allows infinite state systems to be handled. Moreover, in the future capabilities for the detection at the syntax level of product form solutions (see e.g. [12]) shall be added.

- The planned implementation of the second phase of the approach proposed in [4] should help in the analysis, because this would be conducted on truly concurrent models given by stochastically timed Petri nets, instead of interleaving models.

Acknowledgments

This research has been partially funded by Italian MURST and CNR Progetto Strategico "Modelli e Metodi per la Matematica e l'Ingegneria", NSF Young Investigator Award CCR-9257963, NSF Award CCR-9403807, NSF Award INT-9603441, AFOSR Grant F49620-95-1-0508, and NSF Grant CCR-9413309.

References

[1] Bernardo, M. (1997) An Algebra-Based Method to Associate Rewards with EMPA Terms, in *Proc. of the 24th Int. Coll. on Automata, Languages and Programming (ICALP '97). Lecture Notes in Computer Science*, **1256**, 358–368.

[2] Bernardo, M. (1998) Formal Specification of Performance Measures for Process Algebra Models of Concurrent Systems. Technical Report UBLCS-98-08, University of Bologna (Italy).

[3] Bernardo, M., Donatiello, L. and Gorrieri, R. (1994) Integrated Analysis of Concurrent Distributed Systems using Markovian Process Algebra, in *Proc. of the 7th Int. Conf. on Formal Description Techniques for Distributed Systems and Communication Protocols (FORTE '94)*. Chapman & Hall, 455-457.

[4] Bernardo, M., Donatiello, L. and Gorrieri, R. (1998) A Formal Approach to the Integration of Performance Aspects in the Modeling and Analysis of Concurrent Systems, to appear in *Information and Computation.*

[5] Bernardo, M. and Gorrieri, R. (1998) A Tutorial on EMPA: A Theory of Concurrent Processes with Nondeterminism, Priorities, Probabilities and Time, *Theoretical Computer Science*, **202**, 1–54.

[6] Bernardo, M., Gorrieri, R. and Roccetti, M. (1998) Formal Performance Modeling and Evaluation of an Adaptive Mechanism for Packetized Audio over the Internet. Technical Report UBLCS-98-09, University of Bologna (Italy).

[7] Bochmann, G.V. and Vaucher, J. (1988) Adding Performance Aspects to Specification Languages, in *Proc. of the 8th Int. Conf. on Protocol Specification, Testing and Verification (PSTV VIII)*. North Holland, 19–31.

[8] Cleaveland, W.R., Madelaine, E. and Sims, S. (1995) A Front-End Generator for Verification Tools, in *Proc. of the 1st Int. Conf. on Tools and Algorithms for the Construction and Analysis of Systems (TACAS '95). Lecture Notes in Computer Science*, **1019**, 153–173.

[9] Cleaveland, W.R. and Sims, S. (1996) The NCSU Concurrency Workbench, in *Proc. of the 8th Int. Conf. on Computer Aided Verification (CAV '96). Lecture Notes in Computer Science*, **1102**, 394–397.

[10] Ferrari, D. (1986) Considerations on the Insularity of Performance Evaluation. *IEEE Trans. on Software Engineering*, **12**, 678–683.

[11] Gilmore, S. and Hillston, J. (1994) The PEPA Workbench: A Tool to Support a Process Algebra-based Approach to Performance Modelling, in *Proc. of the*

7th Int. Conf. on Modelling Techniques and Tools for Computer Performance Evaluation. Lecture Notes in Computer Science, **794**, 353–368.

[12] Harrison, P.G. and Hillston, J. (1995) Exploiting Quasi-Reversible Structures in Markovian Process Algebra Models. *Computer Journal*, **38**, 510–520.

[13] Harvey, C. (1986) Performance Engineering as an Integral Part of System Design. *BT Technology Journal*, **4**, 143–147.

[14] Hermanns, H., Mertsiotakis, V. and Rettelbach, M. (1996) A Construction and Analysis Tool Based on the Stochastic Process Algebra TIPP, in *Proc. of the 2nd Int. Conf. on Tools and Algorithms for the Construction and Analysis of Systems (TACAS '96). Lecture Notes in Computer Science*, **1055**, 427–430.

[15] Howard, R.A. (1971) *Dynamic Probabilistic Systems*. John Wiley & Sons.

[16] Johnson, S.C. (1975) *YACC - Yet Another Compiler Compiler*. Computing Science Technical Report 32, AT&T Bell Labs.

[17] Kernighan, B.W. and Ritchie, D.M. (1988) *The C Programming Language*. Prentice Hall.

[18] Lehmann, D. and Rabin, M. (1981) On the Advantage of Free Choice: A Symmetric and Fully Distributed Solution to the Dining Philosophers Problem, in *Proc. of the 8th Symp. on Principles of Programming Languages (POPL '81)*. ACM Press, 133–138.

[19] Lesk, M.E. (1975) *Lex - A Lexical Analyzer Generator*. Computing Science Technical Report 39, AT&T Bell Labs.

[20] Ousterhout, J.K. (1994) *Tcl and the Tk Toolkit*. Addison-Wesley.

[21] Stewart, W.J. (1994) *Introduction to the Numerical Solution of Markov Chains*. Princeton University Press.

[22] Welch, P.D. (1983) The Statistical Analysis of Simulation Results, in *Computer Performance Modeling Handbook* (ed. S.S. Lavenberg), Academic Press, 267–329

[23] Yemini, Y. and Kurose, J. (1982) Towards the Unification of the Functional and Performance Analysis of Protocols, or Is the Alternating-Bit Protocol Really Correct?, in *Proc. of the 2nd Int. Conf. on Protocol Specification, Testing and Verification (PSTV II)*. North Holland.

INDEX OF CONTRIBUTORS

GPSR Compliance
The European Union's (EU) General Product Safety Regulation (GPSR) is a set of rules that requires consumer products to be safe and our obligations to ensure this.

If you have any concerns about our products, you can contact us on

ProductSafety@springernature.com

In case Publisher is established outside the EU, the EU authorized representative is:

Springer Nature Customer Service Center GmbH
Europaplatz 3
69115 Heidelberg, Germany

www.ingramcontent.com/pod-product-compliance
Ingram Content Group UK Ltd.
Pitfield, Milton Keynes, MK11 3LW, UK
UKHW022322190726
13856UKWH00001B/147

* 9 7 8 1 4 7 5 7 5 2 6 1 8 *